RADIATING NONUNIFORM TRANSMISSION-LINE SYSTEMS AND THE PARTIAL ELEMENT EQUIVALENT CIRCUIT METHOD

RADIATING NONUNIFORM TRANSMISSION-LINE SYSTEMS AND THE PARTIAL ELEMENT EQUIVALENT CIRCUIT METHOD

Jürgen Nitsch, Frank Gronwald and Günter Wollenberg
Otto-von-Guericke-University Magdeburg

WILEY
A John Wiley and Sons, Ltd., Publication

This edition first published 2009

Registered office
John Wiley & Sons Ltd, The Atrium, Southern Gate, Chichester, West Sussex, PO19 8SQ, United Kingdom

For details of our global editorial offices, for customer services and for information about how to apply for permission to reuse the copyright material in this book please see our website at www.wiley.com.

Library of Congress Cataloging-in-Publication Data

Nitsch, Jürgen.
Radiating nonuniform transmission-line systems and the partial element equivalent circuit method / Jürgen Nitsch, Frank Gronwald, Günter Wollenberg.
p. cm.
Includes bibliographical references and index.
ISBN 978-0-470-84536-3 (cloth)
1. Electromagnetic compatibility–Mathematical models. 2. Electric lines–Mathematical models. 3. Electronic circuit design–Data processing. 4. Electronic apparatus and appliances–Design and construction–Data processing. I. Gronwald, Frank. II. Wollenberg, Günter. III. Title.
TK7867.2.N58 2009
621.382′24–dc22

2009031448

A catalogue record for this book is available from the British Library.

ISBN 978-0-470-84536-3 (H/B)

Typeset in 10/12pt Times by Aptara Inc., New Delhi, India
Printed in Great Britain by CPI Antony Rowe, Chippenham, Wiltshire

To

Ulla, Daniel, Sebastian
and Felicitas

Monika and Wolfgang

Evdokiya, Alexander and Karin

"As far as the laws of mathematics refer to reality, they are not certain, and so far as they are certain, they do not refer to reality."

Albert Einstein (1879–1955)

Contents

Preface

The content of the book is organized as follows. Chapter 1 provides an account of the foundations of classical electrodynamics. It yields the basic equations of electromagnetics that are needed to solve problems together with their physical interpretation. The presentation is original, unique and believed to be of appreciable pedagogical value. It combines an *axiomatic approach* with the *classical gauge field approach*. The axiomatic approach has been outlined in a monograph [1] which is based on a careful study of relevant literature (see also [2] and [3]). The classical gauge field approach was established by Weyl [4] and nowadays constitutes a cornerstone of elementary particle physics [5]. A formulation of the gauge field approach is used which stresses the physical importance of the gauge potentials and has been found useful both in gravity and electromagnetics [6–8]. After a comparison of the axiomatic and the classical gauge field approach, special attention is paid to the *dynamical properties of the electromagnetic field*. In the course of this it will be essential to introduce the split of the electromagnetic field into an irrotational and a rotational part. This split will also lead to a distinction between *Coulomb fields* and *radiation fields*. As a further consequence it will be seen that Coulomb fields and radiation fields are inseparably intertwined. It is this circumstance that leads to many conceptual and practical problems in electrical engineering applications such as antenna theory or transmission-line theory. Therefore, the presentation of the foundations of classical electrodynamics is beneficial in order to recognize the link between basic electromagnetic field properties and fundamental difficulties that are encountered during the solution of actual engineering problems. On the basis of functional analysis an elegant approach is derived for the *Green's function method* which is of the utmost importance for a proper formulation of transmission-line and antenna theory. Green's functions are essential to represent the electromagnetic interaction either in free space or in the presence of boundaries. For the formulation of equations that are appropriate to determine transmission-line and antenna currents the focus is on the class of linear wire antennas and how to obtain relevant *electric field integral equations*.

Chapter 2 deals with the coupling of electromagnetic fields to transmission lines, modeled by integral equations. It will outline how to put the Maxwell equations in the form of *generalized telegrapher equations* and how to arrive at the *classical transmission-line theory*. In the course of this it will turn out to be useful to introduce *antenna-mode currents* and *transmission-line currents*. For uniform transmission lines both types of current decouple in free space and may also approximately decouple within a cavity. The benefit of such a decoupling is that the transmission-line mode can, approximately, be calculated from the comparatively simple classical transmission-line theory. However, the classical transmission-line theory is not suitable to characterize general tranmission-line structures. In this case it is necessary to solve the integral equations of antenna theory without restricting approximations. The mixed potential integral equation (MPIE) is chosen to derive the *transmission-line supertheory* (Carl E. Baum, private communication, 2002). Besides new telegrapher equations, one also obtains equations for the determination of the per-unit-length parameters and the source terms. The theory is directly based on Maxwell's theory. There are no restrictions (like an exclusive TEM mode) or simplifications, although a transition from the current and charge densities to the conductor currents and charges is performed. The new theory is applicable to nonuniform transmission lines consisting of one or more wire-like conductors. It describes the propagation of electromagnetic waves along those lines as well as the coupling of external electromagnetic fields to the lines. It automatically includes all field modes and thus covers all physical effects, including radiation losses, that can occur on a nonuniform transmission line. The principal structure of the telegrapher equations is preserved; the generalized telegrapher equations are still a system of first-order differential equations. However, because, in the general case, the voltage is not the difference between two values of the scalar potential anymore (it is the path-dependent integral of the electric field strength) this quantity is replaced by the charge per unit length. The telegrapher equations arc formulated in terms of the current and the charge per unit length. The parameters not only depend on the cross-sectional shape of the line but also depend on the whole geometry of the conductors and are determined by the solution of an integral equation. Due to frequency-dependent field modes and radiation losses the parameters become complex valued and frequency dependent, even for a simple superconducting line. The theory is represented by a system of first-order ordinary differential equations with nonconstant coefficients. The solution method for this kind of equation is well known and given by the product-integral or matrizant. However, the calculation of the parameters and of the solution for current and potential is quite involved and executed by iteration. Subsequently several numerical and semi-analytical evaluation methods of the matrizant are discussed and compared. Some interesting applications of the new theory are added.

In Chapter 3 it is shown how transmission lines are integrated into topological networks. The starting point is the topological representation of complex systems introduced in [9]. These ansaetze are extended to allow the inclusion of networks of nonuniform transmission lines. A braided cable shield and the shielding of an anisotropic spherical shell are presented as examples for an elementary and proper surface, respectively, in a topological concept.

Finally, Chapter 4 represents a homogeneous derivation of the PEEC theory. It is a numerical method which is suitable for establishing full wave models of three-dimensional, electrical interconnection structures of arbitrary geometry. It is based on an electric field integral equation that is written in the form of a mixed potential integral equation. It also takes advantage of the Galerkin method and, thus, exhibits features similar to the method of moments. However, electric currents and electric charges, that depend on each other via the continuity equation, are kept as unknowns. This involves the discretization of the current and charge carrying domains. After discretization, a linear system of equations is obatined which can be interpreted as a circuit model where the circuit equations are given in modified nodal analysis (MNA)-form or modified loop analysis (MLA)-form. This allows electrical interconnection structures to be modeled as electrical circuits which involve circuit elements R, L and C as well as potential coefficients P and controlled current and voltage sources. These circuits can be analyzed by means of standard network theory and its corresponding tools such as numerical circuit solvers. In contrast to other numerical methods of electromagnetic field computation, the PEEC method, due to its circuit interpretation which involves currents and potentials/voltages, is closer to the terminology of practical engineers, and therefore leads to a better understanding. Even though the properties of the PEEC method will be described in detail, a number of interesting features should be pointed out: the interconnection structures can be given by perfect or imperfect conductors; generally, the cross sections of wire-like structures can be of arbitrary form; of practical interest are circular and rectangular cross sections as well as the thin-wire approximation. It is also possible to treat plane conducting structures and to take into account the skin and proximity effects. Moreover, the PEEC method offers the possibility of modeling perfect or lossy dielectrics that are part of interconnection structures or their environment. The above mentioned term 'full wave model' implies that radiation effects that are due to the interconnection structures and the effect of external fields on the connecting structures are properly taken into account during the modeling.

JÜRGEN NITSCH, FRANK GRONWALD
AND GÜNTER WOLLENBERG

Magdeburg and Ingolstadt, Germany
March, 2009

References

[1] Hehl, F.W. and Obukhov, Y. *Foundations of Classical Electrodynamics: Charge, Flux and Metric*, Birkhäuser, Boston, USA 2003.

[2] Truesdell, C. and Toupin, R.A. 'The classical field theories', in *Handbuch der Physik, vol. III/1*, (ed. S. Flügge), Springer, Berlin, Germany, 1960, 226–793.

[3] Post, E.J. *Formal Structure of Electromagnetics: General Covariance and Electromagnetics*, North Holland, Amsterdam, Holland, 1962, and Denver, New York, USA, 1997.

[4] Weyl, H. 'Electron and graviton', *Zeit. f. Phys.*, **56**, 1929, 330–352.

[5] Chen, T.-P. and Li, L.-F. *Gauge Theory of Elementary Particle Physics*, Clarendon Press, Oxford, UK, 1984.

[6] Gronwald, F. 'Metric-affine gauge theory of gravity I. fundamental structure and field equations', *Int. J. Mod. Phys. D*, **6**, 1997, 263–303.

[7] Gronwald, F. and Nitsch, J. 'The structure of the electromagnetic field as derived from first principles', *IEEE Antennas and Propagation Mag.*, **43**, 2001, 64–79.

[8] Gronwald, F. and Nitsch, J. 'Universeller Zusammenhang: Wie das elektromagnetische Feld die Welt verbindet', (in German), *Magdeburger Wissenschaftsjournal*, **1/2**, 2001, 19–28.

[9] Baum, C.E., Liu, T.K. and Tesche, F.M. 'On the analysis of general multiconductor transmission line networks', *Interaction Note 350*, 1978.

Acknowledgments

This book contains the results of our research work in the field of electrical engineering over many years, and it has been developed from the cooperation of several people. They contributed with the results of their research work, which has been conducted at the departments of J.N. and G.W., to the content of this book. We cite the collaboration over a period of more than five years with Drs Görisch, Haase, Kochetov, Steinmetz and Tkachenko on coupling processes into linear structures. The results of their doctorial theses and habilitations are partly considered. We thank them for fruitful and valuable discussions, hints and their permanent interest in this work. We also wish to acknowledge helpful advice from and discussions with Prof. Dr. Dr.-Ing. E.h. Baum, Dr. Giri, Prof. Dr. Hehl and Prof. Dr. Tesche.

J.N., F.G. and G.W.

List of Symbols

$\mathbf{A}$, $\underline{\mathbf{A}}$, A, A_i	vector gauge potential, magnetic vector potential
${A_{ij}}^k$	interaction field
$\underline{\mathbf{A}}$, $\mathbf{A}^{\mathrm{T}}$	matrix, transpose of **A**
$\overline{\mathbf{A}}$	supermatrix of 1st order
$\overline{\overline{\mathbf{A}}}$	supermatrix of 2nd order
$\boldsymbol{a}$	two-dimensional surface
a	cross-sectional area, wire radius, distance
$\mathbf{a}$	vector of incoming waves
$d\boldsymbol{a}$, da	two-dimensional surface element
B^i, $\mathbf{B}$, $\boldsymbol{B}$, B, $\underline{\mathbf{B}}$	magnetic field strength, magnetic flux density
$\mathbf{B}^{\mathrm{u}}$	binormal unit vector
$\boldsymbol{b}$	three-component vector function
b	normal mode variable of electromagnetic field
$\mathbf{b}$	vector of outgoing waves
b	spatial distance
$\mathbf{b}^{\mathrm{c}}$	vectorial basis function for current expansion
b^{P}	basis function for potential (charge) expansion
$\mathbf{C}$	spatial vector to conductor 'center'
$\mathcal{C}$	integration path
C, $\mathbf{C}$	capacitance, matrix of capacitance coefficients
$\mathbf{C}_{\mathrm{p}}$	matrix of partial capacitances
C', $\mathbf{C}'$	per-unit-length capacitance, per-unit-length capacitance matrix
C^+	excess capacitance
$\mathbb{C}$	set of complex numbers
c	velocity of light
$\boldsymbol{c}$	pilot vector, curve in three-dimensional space, three-component vector function
c	normal mode variable of electromagnetic field
dc^i	one-dimensional line element
cond.	condition number

D^i, $\mathbf{D}$, $\boldsymbol{D}$, D, $\underline{\mathbf{D}}$	electric excitation, electric flux density
$\mathbf{D}$	damping coefficient, diagonal matrix of eigenvalues
D	dimensional
$D_{\mathcal{L}}$	domain of operator $\mathcal{L}$
d	eigenvalues of the parameter matrix, distance, metric
$\boldsymbol{d}$	three-component vector function
$d_{\mathbf{J}}$, d_ρ	current density distribution, charge density distribution
$\mathcal{E}$, E_i, $\mathbf{E}$, $\boldsymbol{E}$, E, $\underline{\mathbf{E}}$	electric field strength
$\mathbf{E}^{\mathrm{inc}}$, $\mathbf{E}^{\mathrm{i}}$	incident electric field
$\mathbf{E}$, $\overline{\overline{\mathbf{E}}}$	unit matrix
E	energy, Ewald parameter
$Ei_1(t)$	exponential integral
$\mathbf{e}$	unit vector
$\boldsymbol{e}_i$	vector frame
e_n	orthonormal basis of a function space
e	elementary charge, measure of error in function space
erf	error function
exp	exponential function
$e^{\mathbf{A}}$	matrix exponential function
F^i, $\mathbf{F}$, F	general vector field
F_n^i	component in function space
F_i	Lorentz force
$\boldsymbol{F}_n$	transverse eigenfunctions
$\mathfrak{F}$, $\mathfrak{F}^{-1}$	Fourier transform, inverse Fourier transform
$\mathfrak{F}_{\mathrm{s}}$, $\mathfrak{F}_{\mathrm{s}}^{-1}$	spatial Fourier transform, inverse spatial Fourier transform
f	general scalar function, element of function space, frequency
$\boldsymbol{f}$	general function with values in $\mathbb{C}^m$
f_{e}	upper end of the extended frequency range
f_{m}	maximum frequency of interest
f_n	basis of a function space
$\boldsymbol{f}_n$	eigenfunction with values in $\mathbb{C}^m$
G, G'	conductance, per-unit-length conductance
$\mathbf{G}$, $\mathbf{G}'$	conductance matrix, per-unit-length conductance matrix
G, g	Green's function (general)
$\underline{G_0}$	free space Green's function of Helmholtz equation
$\overline{\boldsymbol{G}}_0$	dyadic free space Green's function of Helmholtz equation
G_1, G_2, G_3	integrals of the Green's function
$\overline{\overline{\mathbf{G}}}^A$, $\overline{\boldsymbol{G}}^A$	dyadic Green's function of the magnetic vector potential $\mathbf{A}$
$\overline{\overline{\mathbf{G}}}^E$, $\overline{\boldsymbol{G}}^E$	dyadic Green's function of the electric field strength $\mathbf{E}$
$\overline{\overline{\mathbf{G}}}^B$, $\overline{\boldsymbol{G}}^B$	dyadic Green's function of the magnetic field strength $\mathbf{B}$
$\underline{G^\Phi}$, G^φ	Green's function of the scalar electric potential Φ, φ
$\overline{\overline{\mathbf{G}}}_{\mathrm{H}}$, G_{H}, g_{H}	Green's function of the half space

g_{s}	static free space Green's function
g	determinant of metric field, element of function space, general scalar function
$\boldsymbol{g}$	general function with values in $\mathbb{C}^m$
g_{ij}, g^{ij}	metric field
$g(t)$	impulse response function
H	Hilbert space, linear transfer function
H_n	nth order nonlinear transfer function
H_i, $\mathbf{H}$, $\boldsymbol{H}$, H, $\underline{\mathbf{H}}$	magnetic excitation, magnetic field strength
$\mathcal{H}$	amplitude of magnetic excitation
$H_0^{(2)}$	Hankel function of 0th order and 2nd kind
$H(j\omega)$	approximation function
h	Planck constant, element of function space, geometric parameter
$\hbar$	Planck constant divided by 2π
$h(t)$	Heaviside step function
I, $\underline{I}$	electric current
$\mathbf{I}$	matrix, vector, result of an integration
$I'_{i,u}$, $I''_{i,u}$	transmission line scalar Green's functions for currents
$\overline{\mathbf{I}}$	supermatrix, result of an integration
$\overline{\boldsymbol{I}}$, $\overline{\overline{\mathbf{I}}}$	unit/identity dyad
$\overline{\overline{\mathbf{I}}}_{\mathrm{t}}$	transverse unit/identity dyad
$\mathcal{I}$	mathematical functional
$i(t)$	electric current in time domain
$\Im$	imaginary part
J^i, $\mathbf{J}$, $\boldsymbol{J}$, J, $\underline{\mathbf{J}}$	electric current density
J_0	Bessel function of 1st kind
$\mathbf{J}_{\mathrm{s}}$	electric surface current density
$\mathbf{J}_{\mathrm{C}}$	conduction current density
$\mathbf{J}_{\mathrm{P}}$	polarization current density
$\mathbf{J}_{\mathrm{D}}$	displacement current density
J_i^{Φ}	magnetic flux current
j	imaginary unit
K_0	modified Bessel function of 2nd kind
$\overline{\mathbf{K}}$	supermatrix of integral kernel matrices
$\mathbf{K}_{\mathrm{p}}$	matrix of partial reluctances
$K_{\mathrm{p}\alpha,n}$	partial reluctance
k_c, k_l, $\mathbf{k}_c$, $\mathbf{k}_l$	integral kernels and corresponding matrices
k	wave number
$\mathbf{k}$, $\hat{\mathbf{k}}$	wave vector, normalized wave vector
k', k''	real, imaginary part of wave number
d^3k	volume element of reciprocal space

L	Lagrangian, inductance, antenna length
L', $\mathbf{L}'$	per-unit-length inductance, per-unit-length inductance matrix
$L_{\mathrm{p}\alpha,n}$, $\mathbf{L}_{\mathrm{p}}$	partial inductance, matrix of partial inductances
$\mathbf{L}_{\mathrm{p}}^{-1}$	inverse of $\mathbf{L}_{\mathrm{p}}$
$L^{+}_{\mathrm{p}\alpha,n}$	partial inductance with increased accuracy
$\boldsymbol{L}_n$, $\boldsymbol{L}_{mnp}$	longitudinal eigenfunction
$\boldsymbol{L}^p(\Omega)^m$	space of Lebesgue integrable functions
$\mathcal{L}$	Lagrangian density, linear operator
$\mathcal{L}_{\mathrm{D}}$	linear differential operator
$\mathcal{L}_{\lambda}$	the operator $\mathcal{L} - \lambda I$
$\mathcal{L}^{\star}$	adjoint operator
l	length
Δl	discretization length
l_u	Lie derivative with respect to velocity field u^i
dl, $d\mathbf{l}$	line element
LM	layered media
M	subspace of a Hilbert space
$M^{\perp}$	orthogonal complement of M
$\boldsymbol{M}_n$, $\boldsymbol{M}_{mnp}$	transverse eigenfunction
$\mathcal{M}_{\xi_0}^{\xi}\{\overline{\boldsymbol{P}}\}$	product integral/matrizant of $\overline{\boldsymbol{P}}(\xi)$
$\overline{\mathbf{M}}^{\mathrm{w}}$, $\overline{\mathbf{M}}^{vi}$	matrizant for the transmission line in wave and voltage–current representation, respectively
$\mathbf{M}_{12}^{\mathrm{w}}$, $\mathbf{M}_{12}^{vi}$	blockmatrix (partial matrix) of order (n_L, n_L) of $\overline{\mathbf{M}}^{\mathrm{w}}$ and $\overline{\mathbf{M}}^{vi}$, respectively
m	mass
N	number of conductors in transmission lines
N	order of a matrix
$\mathbf{N}^{\mathrm{u}}$	normal unit vector
N	normalization function
$\boldsymbol{N}_n$, $\boldsymbol{N}_{mnp}$	transverse eigenfunction
P	probability
P, p	active power
P_k	pulse function
$\overline{\boldsymbol{P}}$	per-unit-length parameter matrix
$\mathbf{P}$, $\mathbf{P}^{\star}$	per-unit-length parameter matrix, blocks of a supermatrix
$\overline{\mathbf{P}}$, $\overline{\mathbf{P}}^{\star}$	per-unit-length parameter supermatrix
$\mathbf{P}$, $\underline{\mathbf{P}}$	electric polarization
$\mathbf{P}$, $\mathbf{P}^{-1}$	matrix, inverse matrix of potential coefficients
$P_{i,m}$	potential coefficients
$P^{+}_{i,m}$	potential coefficients with increased accuracy
p_i, p^i	momentum
Q	electric charge, reactive power, quality factor
$\mathbb{Q}$	set of rational numbers

q, $\underline{q}$	electric charge, per-unit-length electric charge
q'	per-unit-length electric charge
$\mathbf{q}$	electric charge vector, per-unit-length electric charge vector
qd	superscript 'quasi-dynamic'
qs	superscript 'quasi-static'
R	radius, resistance, distance between two points
R', $\mathbf{R}'$	per-unit-length resistance, per-unit-length resistance matrix
$\boldsymbol{R}$	resistance matrix, position vector
$R_{\mathcal{L}}$	range of operator $\mathcal{L}$
$\mathbb{R}$	set of real numbers
$\Re$	real part
Res	residuum
r	radius, distance
$\mathbf{r}$, $\mathbf{r}'$, $\mathbf{r}''$	position vectors
dr, $d\boldsymbol{r}$	line element
d^2r	area element
d^3r	volume element
S	two-dimensional surface, action, topological space
$\underline{S}$	complex power
S_n	Sommerfeld integral of nth order
S_k	sinusoidal basis function
s	surface area, complex frequency
$s(t)$	step response function
∂S	closed boundary of two-dimensional surface
ds	surface element
$d\mathbf{s}$	line element
$\mathbf{T}$	tangential vector
$\mathbf{T}^{\mathrm{u}}$	tangential unit vector
$T^{\pm}$	adjacent triangles
t	time, line parameter
$\underline{U}$	voltage
$\mathbf{U}$	voltage vector
$\underline{U}^{\mathrm{inc}}$	voltage due to $\underline{\mathbf{E}}^{\mathrm{inc}}$
u^i, $\boldsymbol{u}$	velocity field
$u(t)$	voltage in the time domain
u	perimeter, length of tangential vector
V	three-dimensional volume, voltage in frequency domain
v	volume, voltage in time domain
$\mathbf{v}$	voltage vector
v^i, $\boldsymbol{v}$	velocity field
v_i	covariant vector

$\boldsymbol{v}$	three-dimensional volume
∂V	closed surface of three-dimensional volume
$V'_{i,u}$, $V''_{i,u}$	transmission line scalar Green's functions for voltages
dV, dv	three-dimensional volume element
W	electromagnetic energy
$W(j\omega)$	approximation function
$\mathbf{W}$	dyadic product of eigenvectors
w	electromagnetic energy density
$\mathbf{w}$	eigenvector of parameter matrix, vectorial weighting function
w^i	contravariant vector density
$\mathbf{w}_+$, $\mathbf{w}_-$	positive, negative traveling waves
X	reactance
$\boldsymbol{X}$	voltage–current vector
x^i	spatial coordinate
$\mathbf{x}$	vector
$\overline{\mathbf{x}}$	supervector of 1st order
$\overline{\overline{\mathbf{x}}}$	supervector of 2nd order
$\overline{\mathbf{x}}^{(s)}$	supervector of distributed sources
$\hat{\mathbf{x}}$	spatial vector to the surface of a conductor
$\mathbf{x}$, $\mathbf{x}'$	spatial vectors to observation and source points
Y_0	Bessel function of 2nd kind
Y, $\underline{Y}$	admittance
$\underline{Y}_s$	surface admittance
$\overline{\mathbf{Y}}$	admittance supermatrix
$y(t)$	impulse function response
y^i	spatial coordinate
Z, $\underline{Z}$	impedance
$\underline{Z}^s$	local mean surface impedance
$\underline{Z}_{GSI}$	global surface impedance
Z_{mn}	self ($m = n$), mutual ($m \neq n$) impedance
$\mathcal{Z}_{int}$	intrinsic impedance
$\mathcal{Z}_s$, $\underline{Z}_s$	surface impedance
$\mathbf{Z}$	impedance matrix
$\mathbf{Z}'$	per-unit-length impedance matrix
$\mathbf{Z}_c$	normalization impedance matrix, characteristic impedance matrix
z, $\mathbf{z}$	surface impedance, surface impedance matrix
z^i	spatial coordinate
α	angle, real or complex scalar, index variable, parameter
α_i	general covariant vector
α_n, $[\alpha]$	set of real or complex scalars

β	angle, velocity divided by velocity of light, real or complex scalar
β^j	general contravariant vector density
$\boldsymbol{\beta}$	velocity field divided by velocity of light
β_n, $[\beta]$	set of real or complex scalars
β_{R}	reference phase
γ	angle, scalar density, Euler's constant
γ^0, γ^i	Pauli matrix
$\gamma^i{}_j$	magnetoelectric material tensor
Γ	boundary of volume Ω
$\boldsymbol{\Gamma}$	propagation matrix
δ	delta function, skin depth, infinitesimal parameter
δ_i^n	unit tensor
δ_{mn}	Kronecker symbol
δ_{t}, δ_{x^i}	translation in time and space
δ_ϵ	gauge transformation
$\delta_{\omega_i{}^j}$	rotation in space
$\Delta^\pm$	area of the triangle $T^\pm$
ϵ, $\underline{\epsilon}$, ϵ_{r}, $\underline{\epsilon}_{\mathrm{r}}$, ε, ε_r	absolute, relative permittivity
ε_0, ϵ_0	vacuum permittivity
ϵ	gauge parameter
ϵ_{0N}	Neumann's factor, equal to 1 for $N = 0$, equal to 2 for $N = 1, 2, \ldots$
ϵ_{ijk}, ϵ^{ijk}	Levi-Civita symbol
ε^{ij}	permittivity tensor
ε	infinitesimal parameter
ζ	coordinate, curve parameter
ϑ, θ	angle
θ, θ^{R}	variable, phase of wave function and its component
κ	curvature
λ	wavelength, complex number
$\lambda_{\min}$	minimum wavelength
λ_n	set of eigenvalues
Λ_{mn}	abbreviation, denotes $\int_\Gamma \big(\boldsymbol{e}_n \times \boldsymbol{H}_m(\boldsymbol{r})\big) \cdot \big(\boldsymbol{e}_n \times \boldsymbol{H}_n(\boldsymbol{r})\big)\, d^2r$
$\boldsymbol{\Lambda}$	matrix of generalized partial inductances
$\underline{\Lambda}_{\alpha,n}$	generalized partial inductance
μ, μ_0, μ_{r}	permeability (absolute, vacuum, relative)
μ_{ij}^{-1}	impermeability tensor
ν	parameter
ξ	parameter, complex function

$\mathbf{\Pi}$	matrix of generalized potential coefficients
$\underline{\Pi}_{i,m}$	generalized potential coefficient
ρ, $\underline{\rho}$	electric charge density
ρ_b	density of bound electric charges
ρ_{f}	density of free electric charges
ρ_{mag}	magnetic charge density
ρ_{s}	electric surface charge density
$\rho(\mathcal{L})$	resolvent set
$\boldsymbol{\rho}^{\pm}$	local position vector for triangle $T^{\pm}$
σ, σ_{e}	electric conductivity, equivalent electric conductivity
$\sigma(\mathcal{L})$	spectrum
τ, $\tilde{\tau}$	propagation time, parameter
τ_1, τ_2	position variable along wire
φ, $\boldsymbol{\varphi}$	scalar electric potential, quasivoltage, phase
${\varphi_j}^k$	interaction field
φ_n	function basis
φ_n, φ_{mnp}	eigenfunction corresponding to Poisson equation
Φ	magnetic flux, parameter
χ_{e}	electric susceptibility
χ_n, χ_{mnp}	eigenfunction of scalar Helmholtz equation
ψ	general scalar function
ψ_n, ψ_{mnp}	eigenfunction of scalar Helmhotz equation
Ψ, Ψ_0, Ψ^{R}	wave function, its amplitude and its component
Ψ	digamma function, parameter
ω	angular frequency
ω_{m}	maximum angular frequency of interest
ω', ω''	real and imaginary part of angular frequency
Ω	subset of $\mathbb{R}^n$
Ω, $\hat{\Omega}$	coefficient
∂_i, ∂_t	partial derivative
D_u/Dt	material derivative with respect to velocity field u^i
D_i^A, D_t^{φ}	gauge covariant derivative
$\boldsymbol{\nabla}$	nabla operator
$\boldsymbol{\Delta}$	delta operator
$\Vert\ \Vert$	norm
$[\ ]$	matrix
$\langle\ ,\ \rangle$	inner product
$\langle\ ,\ \rangle_{\mathrm{p}}$	pseudo inner product
$\langle a, b\rangle_{\mathrm{p}}$	reaction between field strength $\boldsymbol{E}^a$ and current density $\boldsymbol{J}^b$

$\parallel$	longitudinal component
$\perp$	transverse component
$*$	complex conjugate
$()^{(i)}$	incident
$()^{(k)}$, $()^{(k+1)}$	quantity after the kth or $(k+1)$th iteration
$()^{\mathrm{u}}$	unit vector
$()_{\mathrm{s}}$	source term
$()_{\mathrm{t}}$	coefficient of trial function
$()_{\mathrm{rad}}$	radiated
(x, y, z)	Cartesian coordinates
(r, φ, ϑ)	spherical coordinates
(u, v, w)	normalized nonorthogonal coordinates
ADC	associated discrete circuit
CTLT	classical transmission-line theory
DGF	dyadic Green's function
DGFLM	dyadic Green's function for layered media
EFIE	electric field integral equation
EMC	electromagnetic compatability
EMT	electromagnetic topology
FD	frequency domain
FDTD	finite differences in time domain
FIT	finite integrals
FSCM	full spectrum convolution macromodeling
GSI	global surface impedance
IC	integrated circuit
MLA	modified loop analysis
MNA	modified nodal analysis
MoM	method of moments
MOR	model order reduction
MPIE	mixed potential integral equation
MSI	mean surface impedance
OVF	orthonormal vector fitting
PCB	printed circuit board
PEC	perfect electric conductor
PEEC	partial element equivalent circuit
PEI	perfect electric insulator
PO	physical optics
RWG	Rao–Wilton–Glisson
TD	time domain
TEM	tranverse electromagnetic mode
TLM	transmission-line matrix
TLST	transmission-line super theory
TLT	transmission-line theory

Introduction

Over the last few decades the number of electric and electronic components in industrial products, devices and installations has increased continuously. Since more and more of these components become concentrated in technical objects for information exchange, electromagnetic compatability (EMC) becomes increasingly important as a product quality. Rising working frequencies, increasing communication – connected by wire or wireless – and the progressional miniaturization of electronic structures contribute to the promotion of this trend. Due to the increasing number of built-in components the EMC analysis of an entire system becomes very complicated. On the one hand there is an increasing mutual interference of subsystems and on the other hand one needs increasing interconnection structures for the information and energy transfer.

A further challenge is the virtualization of the classical development process. Nowadays the time period for the development of a new product is very short. Therefore, one cannot wait for the solution of EMC problems until the completion of prototypes. Rather EMC has to be taken into consideration in the design process. For that purpose numerical analysis or simulation is necessary.

For the assessment of the EMC of systems and technical installations it is of prime importance to know how interconnecting structures influence the electromagnetic behavior due to interfering radiation as well as to its perturbation sensitivity.

The impressive technical progress over recent years in the development of computer science is reflected in the efficiency of modern computers and in the development of various powerful software packages for the simulation of electromagnetic fields. These program packages have been established in the field of the design of integrated electronic circuits, in particular in high-frequency technology. They enable the behavior of products to be predicted in the design stadium and to make optimizations. For example one may mention well-known methods like the finite differences in time domain (FDTD), the method of moments (MoM), physical optics (PO), the method of finite integrals (FIT) or the method of partial element equivalent circuits (PEEC). For any of these methods there exist quite powerful implementations. Any of these methods has its own advantages and disadvantages for specific problems [1–8].

In spite of this massive progress predictions on the EMC-behavior of complex systems are very complicated on the basis of numerical simulations. On the one hand this is due to large electrical extensions of the objects to be investigated. The order of magnitude of the systems of interest range to the 100-fold of a wavelength. On the other hand, the characteristic geometries of systems differ considerably so that the ratio of the total geometry to the smallest dimensions that have to be considered may be 10^{-5}. For example, typical cross sections of leads in a cable are of the order of magnitude of 1 mm, in printed circuits even one order of magnitude smaller. The typical dimension of the total system varies from several centimetres up to many metres.

These different discretization requirements for the entire system make a field-numerical calculation of complex systems with cables and conductors very inefficient. Hybrid methods that, combine numerical field methods with transmission-line theory, for example, facilitate the calculations [1,9].

Thus transmission-line theory becomes an important instrument for the investigation of electromagnetic phenomena, even after more than 130 years of history. Since at least the 18th century electrical transmission lines have been a fascinating as well as a challenging object for scientists and engineers. In that period Sir William Watson tried to determine the speed of electricity. For this he conducted experiments involving a 6.4 km (12 276 ft) long wire line. Within the accuracy of the measuring equipment of that time, he found that the 'velocity of electricity was instantaneous'. However, he noticed the electrical resistance. His results were published in [10,11]. Driven by the development of telegraphs and the desire for overseas communication, a mathematical description of the signal propagation along transmission lines was indispensable. A cornerstone was the contribution from Sir William Thomson (Lord Kelvin). He established the 'Law of retardation' in 1854. However, he only took into account the capacity and the resistance of the cable [12].

In 1857 Gustav Kirchhoff published two papers [13,14] (see also [15,16] for an English translation) where he set up equations describing the motion of electricity in wires and conductors. In the first paper he formulated equations for the current and the charge based on an electric and a magnetic potential, which are comparable to today's telegrapher equations. He also found that in a wire with negligible resistance, electricity propagates at the speed of light.

Later on, in 1874, Oliver Heaviside developed the well-known transmission-line theory and the telegrapher equations [17]. These equations are valid for a broad range of transmission lines. After these pioneering achievements numerous scientists contributed to the transmission-line theory and this was published in numerous excellent textbooks and reviews, for example, [18–29]. The theory was successfully applied to several fields in electrical engineering.

Although the transmission-line theory has been around for more than 130 years, it did not loose its relevance to current situations. High operating frequencies of modern electronic equipment turn even the smallest piece of wire into a transmission line with signal retardation, dispersion, attenuation and distortion. Moreover, in today's electromagnetic environment transmission lines can pick up external electromagnetic fields which generate noise currents which are superimposed on the useful signals. The lines act not only as receiving antennas but can also radiate parts of the signal energy into the environment. Some of these effects are covered by the 'classical transmission-line theory' (cTLT), some are not.

Heaviside showed that the transmission-line theory is a special case of Maxwell's theory. The key is the transverse electromagnetic mode (TEM mode), where the electric and magnetic fields are perpendicular to each other and perpendicular to the direction of propagation. With the assumption of an exclusive TEM mode propagating along a transmission line one can, without any further approximations, find the telegrapher equations.

The exclusive TEM mode is only possible on an ideal transmission line. This line would be infinitely long, lossless, with parallel conductors and would be excited at infinity. Already an excitation at a finite position produces other modes in the vicinity of the excitation and thus invalidates the cTLT. The same is true for finite lines, where usually other modes occur close to the terminations. Also losses in the conductors destroy the exclusive TEM mode, because longitudinal components of the electric field are present.

Nonetheless for many applications, at least when the wavelength is much larger than the cross-sectional size of the line, the other modes decay exponentially, do not propagate and hence are negligible if compared to the dominant TEM mode. In this case one speaks of a quasi-TEM mode. The cTLT gives excellent results that are in very good agreement with experiments as well as with analytical calculations which take into account all modes.

The TEM mode allows the solution procedure to be split into two separate steps. First, the so-called per-unit-length parameters of the transmission line are determined. For a uniform line these parameters only depend on the cross-sectional geometry of the line. They describe the coupling between individual conductors as well as the propagation properties and thus fully characterize the transmission line. The second step is the solution of the governing equations, the telegrapher equations which are coupled partial first-order differential equations. The procedure for the computation of the results for these kinds of equations is very well known. For some important cases analytical solutions exist, for other cases usually numerical methods are used.

Things become much more complicated if the transmission lines are nonuniform, that is if the conductors are not in parallel anymore and the cross section changes along the line. One can imagine that the field distributions become much more involved than for uniform lines. In the literature there are many publications concerning nonuniform transmission lines and Heaviside made some contributions [30].

The majority of these papers discuss the solution procedure for the telegrapher equations with position-dependent parameters. Often the dependence is described by some simple mathematical function, for example linearly or exponentially varying parameters. Most papers do not clearly specify where these parameters come from, usually a TEM mode on the nonuniform transmission line is assumed. Recent publications on this topic include [31–34] where the product integral [35,36] is applied to formulate analytical solutions of the telegrapher equations for nonuniform multiconductor transmission lines.

There are several approaches to broaden the scope of the classical transmission-line theory (see [37] and references therein), however, most of them only work in a particular low-frequency range. Many works [22,38–45] are dedicated to the correction of the telegrapher equations when transmission lines with nonuniformities or discontinuities are under investigation. The models encompass abrupt bends, round bends, vertical risers or the cross coupling between two skewed lines. For low frequencies (quasi-static case) the additional effects are modeled with the aid of lumped elements (L,C) or additional transmission lines that are connected to the actual line. Sometimes nonuniform lines are treated as if they were built of uniform segments, that is within each section the line is regarded to be uniform [46].

There have been attempts to solve nonuniform transmission-line problems with higher mode effects using the cTLT. This is done by modeling a physical effect, for example radiation, which was previously excluded from the theory, by some additional term in the telegrapher equation [47].

Over the past few years considerable effort was put into the development of an extended transmission-line theory, which is not restricted to the TEM mode nor to any other mode. Instead of extending the existing theory, the preferred way to go is to derive generalized equations directly from Maxwell's theory. Some of the approaches deal with special geometric characteristics of the conductors like sharp bends, vertical termination wires or field excitations that generate non-TEM modes [48–55]. Although these works show great improvements compared with the cTLT, they do not provide a general procedure to handle arbitrary nonuniform transmission lines; they concentrate on one specific problem.

In [56] and [57] two different approaches are shown for the development of a generalized transmission-line theory, with interesting ideas. Unfortunately, only the theoretical derivation is considered and no practical examples are presented.

A different approach is shown in [58]. In this case a proof is given that the current distribution of a nonuniform transmission line is governed by a second-order differential equation. Then from known linearly independent solutions for the currents for different terminations it is possible to construct the coefficients of the differential equations which then become the per-unit-length parameters. The required solutions can be obtained numerically, for example with the method of moments (MoM).

The most recent contribution to this topic is [59], where a modal decomposition based on a spatial Fourier series is used to formulate modal telegrapher equations, that is one for every current component. After solving these equations one needs to sum up the infinite Fourier series to obtain the real current. In [60] these results are compared with the results obtained with the procedures introduced in this book and a very good agreement can be observed.

For a hybrid description of a complex system PEEC is a well-suited method to be combined with transmission-line theory, due to its circuit interpretation. Circuits are attached to interconnecting structures, for example transmission lines. Moreover, transmission lines may also be understood as a network of circuits in a topological model. Thus, both theories complement one another and provide a deep and clear physical understanding of relevant electromagnetic processes.

PEEC models can be developed both in the frequency domain and the time domain. Due to the circuit interpretation it is possible to include the circuit environment, which is defined as those circuits that are attached to the interconnection structure, into the analysis. In particular, time domain analysis and simulations allow nonlinear circuit environments to be included. This is of particular interest for the analysis of signal integrity and the analysis of transients in the context of microelectronic, electronic and power electronic systems. Within the framework of PEEC theory one gains a good understanding of the restrictions of classical transmission-line theory and the transition to generalized transmission-line theories which include radiation effects. These attractive features are the main reason the PEEC method has been chosen as a numerical method to complement the analysis of 'Radiating Nonuniform Transmission-Line Systems'.

The interpretation of electric, magnetic and electromagnetic field problems by means of electrical circuits has always been an important issue in electrical engineering. On the one hand, electric voltages and currents are more familiar to the electrical engineer than continuously distributed field strengths and flux densities; on the other hand, electrical circuit theory provides clearly laid out methods to establish the relevant system of equations and there are many well-established methods, realized as equation solvers and circuit solvers, to obtain corresponding solutions.

A circuit-oriented analysis of arbitrary three-dimensional interconnection structures first concentrates on the current–voltage relation of an accessible terminal pair which has been derived on a field-theoretical basis. The ohmic, inductive and capacitive part of the terminal impedance of a circular wire were determined in [61], taking into account the displacement current and retardation. Even the real part of this impedance contains, besides an ohmic contribution which is due to dissipation, an inductive and capacitive contribution which characterize radiation losses and vanish if retardation is neglected. In [62] and [63] this circumstance was formulated in a general form and it was noted that the ohmic, inductive and capacitive contributions to the terminal impedance depend in a complicated way on frequency. Again, it was noted that radiation losses vanish if retardation is neglected.

To characterize arbitrary three-dimensional structures, not only with respect to their terminal behavior but also with respect to the whole structure and its properties in a circuit-oriented way, it was necessary to extend the notion of inductance. This essential step was taken in [64] in 1946 based on earlier works [65,66]. Partial self and mutual inductances were introduced and formulas for their calculation were provided. Thus, it was possible to determine the inductive properties of pieces of wire and their interaction within the complete structure.

In 1965, exact equations were published [67] for the calculation of the partial self inductance of rectangular conductors and for the partial mutual inductance between combinations of parallel filaments, thin tapes and rectangular conductors. A general procedure was also given for calculating the self inductances of complicated geometries by dividing the geometry into simple elements whose self inductances can be calculated. In 1966 the notion of partial inductances, together with elementary calculation examples, had already found their way into a textbook on fundamentals of theoretical electrical engineering [68]. However, it must be stated that the important concept of partial inductances nowadays is not yet part of the standard literature.

In 1972, A. E. Ruehli published a very detailed and thorough paper on *Inductance Calculations in a Complex Integrated Circuit Environment* [69]. The major motivation for this work was the quasi-static analysis of inductive voltage drops and inductively coupled voltages for a large number of arbitrary loops of complex geometry. Using and elaborating the theory of partial inductances, complex conductor loops were subdivided into straight conductor segments with locally constant cross section and equations for an effective computerized calculation were derived. For straight conductor pieces with arbitrary cross sections, it was shown how a filament approximation of the partial self and mutual inductances could be calculated. On the way to improve the analysis and design of integrated circuits, the paper *Efficient Capacitance Calculation for Three-Dimensional Multiconductor Systems* [70] was published. In this work, an integral equation solution technique using the Galerkin method for capacitance calculations was developed. The derived expressions for the self and mutual potential coefficients of and between closely spaced conductor pieces both provided a high precision of calculation and minimized the computer storage without excessive computation times by complex capacitance calculations. After [69] and [70], the paper *Equivalent Circuit Models for Three-Dimensional Multiconductor Systems* [71] appeared as a logical consequence of well-aimed research work. In this paper the partial element equivalent circuit concept (PEEC), was developed as a numerical method for circuit-oriented modeling of the electromagnetic behavior of electric interconnection stuctures. This new approach was based on an integral equation formulation of Maxwell equations that used the Galerkin method. By taking into account retardation this approach provided full-wave models in the time and frequency domains and allowed conductor losses to be considered. The corresponding models lead to a flexible computer solution technique for the calculation of partial elements and for circuit analysis. Models of different complexity could be constructed to suit the application at hand. Because of this fundamental work and the substantial contributions for further development and applications, A. E. Ruehli is considered to be the father of the PEEC method. An important extension of the PEEC formulation to include arbitrary, finite, homogeneous dielectric regions was given in 1992 [72]. The key idea was to model the part of the displacement current for dielectrics with $\varepsilon_r > 1$ as a polarization current, caused by bound charges, in contrast to the conduction current based on free charges. The extended PEEC formulation was derived for the case of lossless dielectrics. However, the authors mentioned that the lossy dielectric case was given by the combination of a new term for lossless dielectrics and a term for finite conductivity that was already included in the PEEC concept [73]. In the paper *Circuit Models for 3D Structures with Incident Field* [74] the PEEC method was extended to three-dimensional structures that were illuminated by a nonhomogeneous incident electric field. Within a circuit interpretation, incident fields were represented as independent voltage sources. This enabled the PEEC method to model scattering problems in an adequate form. Despite the fact that interconnection structures are passive, physically stable systems, time domain solutions of their models that are derived from integral equations such as PEEC, may show instabilities. Since these instabilities can complicate or even prohibit the use of the PEEC method, much effort has been spent investigating and improving the stability of time domain solutions. As a general rule, the instability may be for two reasons: the numerical technique that is used for the time integration and the geometrical discretization that is required to obtain PEEC models.

In [75] it was shown by means of a very small problem that by changing the delay times the model can become unstable, that is the eigenvalues of the system are moved into the right half of the complex plane. It was stated that stable solutions could be obtained if an appropriate integration method was employed. In [76] the discretization issue was addressed and a circuit motivated technique to stabilize the time domain solution was suggested. The proposed stabilization scheme consisted in breaking the partial self inductances into two equal parts along the length with a delay between the two partial mutual inductances where the delay was used as a tuning parameter. In the papers [77] and [78] the stabilization scheme was improved. Cells obtained by a discretization criterion $\lambda/20$ were subdivided into a finite number of partitions. Equations for the calculation of partial inductances and potential coefficients were given. Despite the subdivision of cells the number of unknowns of the PEEC model was not increased. Additionally, a further stabilization measure was proposed: introducing a damping resistor in parallel to each self inductance of the PEEC model. In [79] the proposed stabilization scheme was investigated for a rectangular patch geometry in the time domain. The authors finished the paper with the statement for the unsatisfying situation: 'No systematic study of the general impact of the stabilization scheme has been done at this moment.' In [80] a computational procedure was presented for studying stability and passivity of PEEC models in specific regions of the complex plane. The proposed method could be used for studying all systems described by linear, time-invariant systems of delay-differential equations. In [81], and more detailed in [84], the authors examined the influence of the partial element accuracy on the stability of quasi-static PEEC models. As potential sources of instability the authors found out poor geometrical meshing and unsuitable partial element calculation routines. In [82] it was shown that, if a standard PEEC model was unstable, it was unstable independent of the excitation, and an unstable model cannot be made stable by refinement of discretization. As the main reason for the instability of PEEC models was revealed for the first time the replacement of the double integrals over the complex Green's function by the double integral over the static one with a separated retardation. Whereas the last integral lead to frequency independent partial elements, the first one showed a frequency dependence, namely a strong damping for frequencies higher than the maximum frequency of interest. Thus, on the basis of frequency dependent partial elements a generalized PEEC model in the frequency domain was built that could be used as a platform for developing stable time domain models. In the paper *Stable and effective full-wave PEEC models by full-spectrum convolution macromodeling* [83] stable time domain models were developed with higher precision as standard PEEC models and with a comparable runtime. The solution was based on an approximation of the frequency dependent partial elements by a complex function, an analytical inverse Fourier transformation of the function and using the convolution technique. The full-spectrum convolution macromodels obtained be realized by associated discrete circuits. The adequate generalized time domain PEEC models could be solved with SPICE-like solvers. In [84] an extension of standard PEEC models was proposed with low pass filters for improving stability. This was a more or less heuristic approach for taking into account the frequency dependence of partial elements and could be considered in a way as an extension of the damping resistor introduced in [78]. In paper [85] the frequency dependence of partial elements was taken into account by two terms: a delay term as in the standard PEEC and a remaining frequency dependent term approximated by a rational function represented in Foster's canonical form. The last term was the base for synthesizing linear *RLCG*-circuits. The broadband PEEC models obtained with improved stability and accuracy could be incorporated in a time domain MNA based solver.

An important concern for improving the versatility of the PEEC method in analyzing interconnection structures was the consideration of the skin effect. Although the concept of partial inductances provided the possibility of considering the skin and proximity effects by partitioning an arbitrary conductor cross section into filaments [69], the high number of filaments and of magnetic couplings between them made an analysis of real interconnection structures practically impossible with PEEC modeling.

In [86] an enhanced skin effect modeling for PEEC models was proposed. The method was based on the introduction of a global surface impedance (GSI) that accurately and efficiently modeled the quasi-static electromagnetic behavior of a two-dimensional lossy conductor (interior problem). Since the interior of the conductor cross section was discretized and incorporated into the formulation of the GSI model, it eliminated the need for a high-frequency volume filament approach. The GSI representation was to be integrated in the EFIE and, finally, in the PEEC model for solving the exterior problem. In [87] a modified PEEC method was proposed that allowed for approximately considering the skin effect for wire-like structures. This was achieved in time and frequency domains. The modification was based on replacing the partial resistance in the PEEC model by a mean surface impedance (MSI) representing the frequency dependent resistance and internal inductance of a wire part and by introducing external instead of the usual partial inductances. The MSI could be calculated analytically or numerically. Finally, its frequency dependence was approximated by *RL*-ladder circuits.

In the fundamental papers the PEEC method was derived for orthogonal meshing. Early in the history of PEEC an effort was made to improve the facilities of modeling by nonorthogonal meshing. The first extension, especially important for wire-like structures, was the modeling of rectangular bars in arbitrary orientation (e.g. [88]). Also in [88] the authors systematically and consistently extended the PEEC formulation to nonorthogonal geometries, especially to two-dimensional quadrilateral and three-dimensional hexahedral meshing. This more general formulation, using global and local coordinates, retained all the properties of the orthogonal PEEC method, in particular, the topologies of both models were exactly the same. In papers [89] and [90] PEEC models were developed using triangular cells and prisms as fundamental building blocks for modeling arbitrary conductor surfaces and conductor/dielectric volumes.

For the flexible triangular meshing Rao–Wilton–Glisson (RWG) expansion functions [91] were used.

Concerning the efforts for a systematic approach to forming circuit equations for PEEC models the papers [74] and [92] are worth mentioning. Generalized stamps for PEEC to minimize the number of unknowns in the so-called condensed MNA and MLA matrix formulations were developed. Most of the general circuit solvers like SPICE [93] or ASTAP [94] and their advancements were based on MNA. In [95] it was demonstrated which modifications were introduced to use Berkeley-SPICE for the calculation of the full-wave PEEC models with retardation in the time and frequency domains.

In [83] and in [85] it was shown how the MNA solver may be used to obtain time domain solutions for enhanced PEEC models with frequency dependent partial elements.

In the literature there are many publications dealing with single aspects of PEEC problems. However, up to date, there are no complete works where one can find a systematic and logical derivation and representation of the PEEC method. In view of this scientific and educational deficit, it is one of the tasks of this book to close this gap.

References

[1] Goerisch, A. 'Netzwerkorientierte Modellierung und Simulation elektrischer Verbindungsstrukturen mit der Methode der partiellen Elemente'. Otto-von-Guericke-Universität Magdeburg, Germany, Dissertation, May 2002.

[2] Kost, A. *Numerische Methoden in der Berechnung elektromagnetischer Felder*, Springer Verlag, Berlin–Heidelberg, Germany, 1994.

[3] Mittra, R. *Numerical and Asymptotic Techniques in Electromagnetics*, Springer Verlag, Germany, 1975.

[4] Mittra, R. (ed.) *Computer Techniques for Electromagnetics*, Springer Verlag, Germany, 1987.

[5] Singer, H., Brüns, H.-D., Mader, T. *et al. CONCEPT Manual of the Program System*, TU Hamburg-Harburg, Germany, 2000.

[6] Miller, E.K., Medgyesi-Mitschang, L.N. and Newman, E.H. (eds) *Computational Electromagnetics*, IEEE Press, New York, USA, 1992.

[7] Harrington, R.F. *Field Computation by Moment Method*, IEEE Press, New York, USA, 1993.

[8] Cristopoulos, C. *The Transmission-Line Modeling Method TLM*, IEEE Press, New York, USA, 1995.

[9] Sabath, F. 'Ein hybrides Verfahren zur Berechnung des abgestrahlten Feldes von Leiterplatten'. Universität Hannover, Germany, Dissertation, October 1998.

[10] Watson, W. 'Experiments and observations tending to illustrate the nature and properties of electricity'. In one letter to Martin Folkes and two letters to the Royal Society. The 2nd edn, London, Printed for C. Davis, 1746.

[11] Watson, W. 'In order to discover whether the electrical power would be sensible at great distances, with an experimental inquiry concerning the respective velocities of electricity and sound'. *Communicated to the Royal Society*. London, Printed for C. Davis, 1748.

[12] Kelvin, W.T. 'On the theory of the electric telegraph'. *Proc. R. Soc.*, **7**, 1855, 382–399.

[13] Kirchhoff, G. 'Ueber die Bewegung der Electricität in Drähten'. *Annalen der Physik (Poggendorfs Annalen)*, **100**, 1857, 193–217.

[14] Kirchhoff, G. 'Ueber die Bewegung der Electricität in Leitern'. *Annalen der Physik (Poggendorfs Annalen)*, **102**, 1857, 529–544.

[15] Kirchhoff, G. 'On the motion of electricity in wires'. *Philosophical Magazine*, **13**, 1857, 393–412.

[16] Graneau, P. and Assis, A.K.T. 'Kirchhoff on the motion of electricity in conductors'. *Apeiron*, **19**, 1994, 19–25.

[17] Heaviside, O. *Electromagnetic Theory*, E.& F.N. Spon Ltd, London, UK, 1951.

[18] Mie, G. 'Elektrische Wellen an zwei parallelen Drähten (Electrical waves along two parallel wires)'. *Annalen der Physik*, **2**, 1900, 201.

[19] Sommerfeld, A. *Electrodynamic*, Academic Press, New York, USA, 1964.

[20] Kuznetsov, P.I. and Stratonovich, R.L. *The Propagation of Electromagnetic Waves in Multiconductor Transmission Lines*, Pergamon Press, Oxford, London, New York, Paris, 1964 (translated from the Russian original).

[21] Schelkunoff, S.A. 'Conversion of Maxwell's equations into generalised telegraphists's equations'. *Bell Sust. Tech. J.*, **34**, 1955, 995–1043.
[22] King, R.P.W. *Transmission Line Theory*, McGraw-Hill Book Company, New York, USA, 1955.
[23] Djordjevic, A.R., Sarkar, T.R. and Harrington, R.F. 'Time domain response of multiconductor transmission lines's. *Proc. IEEE*, **75**, 1987, 743–764.
[24] Collin, R.E. *Field Theory of Guided Waves*, IEEE Press, New York, USA, 1991.
[25] Paul, C.R. *Analysis of Multiconductor Transmission Lines*, John Wiley & Sons, Inc., New York, USA, 1994.
[26] Unger, H.-G. *Elektromagnetische Wellen auf Leitungen*, 4th edn, Hüthig Buch Verlag, Heidelberg, Germany, 1996. ELTEX.
[27] Franceschetti, G. *Electromagnetics*, Plenum Press, New York, USA, 1997.
[28] Tesche, F.M., Ianoz, M.V. and Karlsson, T. *EMC Analysis Methods and Computational Models*, John Wiley & Sons, Inc., New York, USA, 1997.
[29] Reibiger, A. 'Field theoretic description of TEM waves in multiconductor transmission lines', Proceedings of the 6th IEEE Workshop on Signal Propagation on Interconnects, Torino, Politecnico di Torino, Italy, 2002, pp. 93–96.
[30] Heaviside, O. *Electromagnetic Theory*, volume II, Ernest Benn Ltd, London, UK, 1925. First published in 1899.
[31] Baum, C.E., Nitsch, J. and Sturm, R. 'Analytical solution for uniform and nonuniform multiconductor transmission lines with sources', in *Review of Radio Science*, (ed. W.R. Stone), Oxford University Press, Oxford, UK, 1996, pp. 433–464.
[32] Baum, C.E., Nitsch, J. and Sturm, R.J. 'Nonuniform multiconductor transmission lines and networks', in *Progress in Electromagnetic Research Symposium*, Hong Kong, January 1997.
[33] Nitsch, J. 'Exact analytical solution for nonuniform multiconductor transmission lines with the aid of the solution of the corresponding matrix Riccati equation'. *Electrical Engineering*, **81**, 1998, 117–120.
[34] Nitsch, J. and Gronwald, F. 'Analytical solutions in multiconductor transmission line theory'. *IEEE Transactions on Electromagnetic Compatibility*, **4**, 1999, 469–479.
[35] Gantmacher, F.R. *The Theory of Matrices*, volume 2. Chelsea Publishing Company, New York, USA, 1984.
[36] Dollard, J.D. and Friedman, C.N. *Product Integration with Application to Differential Equations*, Addison–Wesley Publishing Company, Reading, Massachusetts, USA, 1979.
[37] Haase, H., Nitsch, J. and Steinmetz, T. 'Transmission-Line Super Theory: A new Approach to an Effective Calculation of Electromagnetic Interactions'. *URSI Radio Science Bulletin (Review of Radio Science)*, **307**, 2003, 33–60.
[38] Lam, J. 'Equivalent lumped parameter for a bend in a two-wire transmission line: Part I. inductance'. *Interaction Note 303*, December 1976.
[39] Lam, J. 'Equivalent lumped parameter for a bend in a two-wire transmission line: Part II. capacitance'. *Interaction Note 304*, January 1977.
[40] Giri, D.V., Chang, S.K. and Tesche, F.M. 'A coupling model for a pair of skewed transmission lines'. *Interaction Note 349*, September 1978.
[41] Green, H.E. and Cashman, J.D. 'End effect in open-circuited two-wire transmission lines'. *IEEE Transactions on MTT*, **34**, 1986, 180.
[42] Getsinger, W.J. 'End-effects in quasi-TEM transmission lines'. *IEEE Transactions on MTT*, **41**, 1993, 666.
[43] Degauque, P. and Zeddam, A. 'Remarks on the transmission-line approach to determining the current induced on above-ground cables'. *IEEE Transactions on EMC*, **30**, 1988, 77.
[44] Tesche, F.M. and Brändli, B.R. 'Observations on the adequacy of transmission-line coupling models for long overhead cables'. Proceedings of the International Symposium on EMC, Rome, Italy, September 1994, p. 374.
[45] Liu, W. and Kami, Y. 'Vertical riser effects of a finite-length transmission line', in Proceedings of the International Symposium on EMC, Tokyo, Japan, 1999.
[46] Omid, M., Kami, Y. and Hayakawa, M. 'Field coupling to nonuniform and uniform transmission lines'. *IEEE Trans. EMC*, **39**, 1997, 201–211.
[47] Wendt, D.O. and ter Haseborg, J.L. 'Description of electromagnetic effects in the transmission line theory via concentrated and distributed linear elements', in Proceedings of the IEEE AP-S International Symposium and URSI North American Radion Science Meeting, volume 4, Montreal, Quebec, Canada, July 1997, pp. 2330–2333..
[48] Nakamura, T., Hayashi, N., Fukuda, H. and Yokokawa, S. 'Radiation from the transmission line with an acute bend', *IEEE Trans. EMC*, **37**, 1995, 317–325.

[49] Tkachenko, S., Rachidi, F. and Ianoz, M. 'On the theory of high-frequency wave propagation along nonuniform transmission lines', in Proceedings of the International Symposium on EMC, Magdeburg, Germany, October 1999.

[50] Tkachenko, S., Rachidi, F. and Ianoz, M. 'High-frequency electromagnetic field coupling to long terminated lines', *IEEE Trans. EMC*, **43**, May 2001, 117–129.

[51] Tkachenko, S. and Nitsch, J. 'Investigation of high-frequency coupling with uniform and non-uniform lines: comparison of exact analytical results with those of different approximations', in Proceedings of the 27th General Assembly of the International Union of Radio Science, Maastricht, Netherlands, August 2002.

[52] Tkachenko, S., Rachidi, F., Nitsch, J. and Steinmetz, T. 'Electromagnetic field coupling to nonuniform transmission lines: treatment of discontinuities', in Proceedings of the 15th International Symposium and Technical Exhibition on EMC, Zurich, Switzerland, February 2003, pp. 603–608.

[53] Nitsch, J. and Tkachenko, S. 'Source dependent transmission line parameters – plane wave vs tem excitation'. *Interaction Note 577*, October 2002.

[54] Nitsch, J. and Tkachenko, S. 'Complex-valued transmission-line parameters and their relation to the radiation resistance'. *IEEE Trans. EMC*, **46**, 2004, 477–487.

[55] Nitsch, J. and Tkachenko, S. 'Telegrapher equations for arbitrary frequencies and modes: radiation of an infinite, lossless transmission line'. *Radio Science*, **39**, 2004, RS2026, doi:10.1029/2002RS002817.

[56] Sturm, R.J. 'On the theory of nonuniform multiconductor transmission lines: an approach to irregular wire configurations', *Interaction Note 541*, June 1998.

[57] Sturm, R.J. 'The treatment of multiconductor nonuniform transmission lines; a comparison of high and low characteristic impedance lines', in Proceedings of the International Symposium on EMC, Magdeburg, Germany, October 1999, pp. 231–234. .

[58] Mei, K.K. 'Theory of Maxwellian circuits'. *Radio Sci. Bull.*, **305**, 2003, 6–13.

[59] Nitsch, J. and Tkachenko, S. 'Newest developments in transmissionline theory and applications'. *Interaction Note 592*, September 2004.

[60] Nitsch, J. and Tkachenko, S. 'Global and modal parameters in the generalized transmission line theory and their physical meaning'. *URSI Radio Sci. Bull.*, **312**, 2005, 21–31.

[61] Wessel, W. 'Über den Einfluss des Verschiebungsstromes auf den Wechselstromwiderstand'. *Zeitschrift Tech. Phys.*, **11**, 1936, 472–475.

[62] Plonsky, R. and Collin, R. E. *Principles and Applications of Electromagnetic Fields*, McGraw-Hill, New York, USA, 1961.

[63] Ramo, S., Whinnery, J. R. and Duzer, T. *Fields and Waves in Communication Electronics*. John Wiley & Sons, Inc., New York, USA, 1965.

[64] Grover, F. W. *Inductance Calculations*. Dover, New York, USA, 1946.

[65] Rosa, E. B. The self and mutual inductances of linear conductors. *Bull. Bur. Stand.*, **4**, No. 2, 1908.

[66] Rosa, E. B. and Grover, F. W. 'Formulas and tables for the calculation of mutual and self-induction'. *Bull. Bur. Stand.*, **8**, No. 1, 1911.

[67] Hoer, C. and Love, C. 'Exact inductance equations for rectangular conductors with applications to more complicated geometries'. *J. Res. Nat. Bur. Stand.*, **69C**, 1965, 127–137.

[68] Nejman, L. R. and Demircan, K. S. 'Teoreticeskie osnovy elektrotehniki, tom 2. Moskva-Leningrad, izdatelstvo' *Energia*, 1966.

[69] Ruehli, A. E. 'Inductance calculations in a complex integrated circuit environment'. *IBM Journal of Research and Development*, **16**, 1972, 470–481.

[70] Ruehli, A. E. and Brennan, P. A. 'Efficient Capacitance Calculations for Three-Dimensional Multiconductor Systems'. *IEEE Transactions on Microwave Theory and Techniques*, **21**, 1973, 76–82.

[71] Ruehli, A. E. 'Equivalent Circuit Models for Three-Dimensional Multiconductor Systems'. *IEEE Transactions on Microwave Theory and Techniques*, **22**, 1974, 216–221.

[72] Ruehli, A. E. and Heeb, H. 'Circuit Models for Three-Dimensional Geometries Including Dielectrics', *IEEE Trans. Microwave Theory and Techniques*, **40**, 1992, 1507–1516.

[73] Garrett, J. E. and Ruehli, A. E. 'PEEC-EFIE for Modeling 3D Geometries with Lossy Inhomogeneous Dielectrics and Incident Fields', *IBM Research Report RC 19245*, IBM T. J. Watson Research Center, Yorktown Heights, October 1993.

[74] Ruehli, A. E., Garrett, J. E. and Paul, C. R. 'Circuit Models for 3D Strucures with Incident Fields', Proceedings of the IEEE International Symposium on EMC, Dallas, Texas, USA, August 1993, pp. 28–32.

[75] Ruehli, A. E., Miekkala, U., Bellen, A. and Heeb, H. 'Stable Time Domain Solutions for EMC Problems using PEEC Circuit Models', Proceedings of the IEEE International Symposium on EMC, Chicago, Illinois, USA, 1994, 371–376.

[76] Ruehli, A. E., Miekkala, U. and Heeb, H. 'Stability of Discretized Partial Element Equivalent EFIE Circuit Models'. *IEEE Trans. Antennas and Propagation*, **43**, 1995, 553–559.

[77] Garrett, J., Ruehli, A. E. and Paul, C. 'Accuracy and Stability Advancements of the Partial Element Equivalent Circuit Model', Proceedings of the International Symposium on EMC, Zurich, Switzerland, March 1997, 529–534.

[78] Garrett, J., Ruehli, A. E. and Paul, C. 'Stability Improvements of Integral Equation Models'. *IBM Research Report RC 20701*, IBM T. J. Watson Research Division, January 1997.

[79] Pinello, W., Ruehli, A. E. and Cangellaris, A. 'Stabilization of Time Domain Solutions of EFIE based on Partial Element Equivalent Circuit Models'. *IBM Research Report RC 20700*, IBM T. J. Watson Research Division, January 1997.

[80] Cullum, J. and Ruehli, A. E. 'An Extension of Pseudospectral Analysis for Studying the Stability and Passivity of Models of VLSI Interconnects', *IBM Research Report RC 21016* T. J. Watson Research Center, York town Heights, January 1997.

[81] Ekman, J., Antonini, G., Orlandi, A. and Ruehli, A. E. 'Stability of PEEC models with respect to partial element accuracy', Proceedings of the IEEE International Symposium on EMC, Santa Clara, California, USA, 2004, Vol. 1, 271–276.

[82] Kochetov, S. V. and Wollenberg, G. 'Stability of full-wave PEEC models: reason for instabilities and approach for correction', *IEEE Trans. EMC*, **47**, 2005, 738–748.

[83] Kochetov, S. V. and Wollenberg, G. 'Stable and effective full-wave PEEC models by full-spectrum convolution macromodeling', *IEEE Trans. EMC*, **49**, 2007, 25–34.

[84] Ekman, J., Antonini, G. and Ruehli, A. E. 'Toward improved time domain stability and passivity for full-wave PEEC models', Proceedings of the IEEE International Symposium on EMC, Portland, Oregon, USA, 2006, 544–549.

[85] Antonini, G., Deschrijver, D. and Dhaene, T. 'Broadband macromodels for retarded partial element equivalent circuit (rPEEC) method', *IEEE Trans. EMC*, **49**, 2007, 35–48.

[86] Coperich, K. M., Ruehli, A. E. and Cangellaris, A. 'Enhanced skin effect for partial element equivalent circuit (PEEC) models'. *IEEE Trans. Microwave Theory and Technique*, **48**, 2000, 1435–1442.

[87] Wollenberg, G. and Kochetov, S. V. 'Modeling the skin effect in wire-like 3D interconnection structures with arbitrary cross section by a new modification of the PEEC method', Proceedings of the 15th International Symposium on EMC, Zurich, Switzerland, February 2003, pp. 609–614.

[88] Ruehli, A. E., Antonini, G., Esch, J. *et al.* 'Nonorthogonal PEEC formulation for time- and frequency-domain EM and circuit modeling', *IEEE Trans. EMC*, **45**, 2003, 167–176.

[89] Rong, A. and Cangellaris, A. C. 'Generalized PEEC models for three-dimensional interconnect stuctures and integrated passives of arbitrary shapes', Proceedings of the Digital Electrical Performance Electronic Packaging Conference, Vol. 10, Boston, Massachusetts, October 2001, pp. 225–228.

[90] Jandhyala, V., Wang, Y., Gope, D. and Shi, R. 'Coupled electromagnetic-circuit simulation of arbitraily-shaped conducting strucures using triangular meshes', Proceedings of the IEEE International Symposium on Quality Electronic Design, San Jose, California, USA, 2002, pp. 38–42.

[91] Rao, S. M., Wilton, D. R. and Glisson, A. W. 'Electromagnetic scattering by surfaces of arbitrary shape'. *IEEE Trans. Antennas and Propagation*, **AP-30**, 1982, 409–418.

[92] Garrett, J. E., Ruehli, A. E. and Paul, C. 'Efficient frequency domain solutions for sPEEC EFIE for modeling 3D geometries'. Proceedings of the International Symposium on EMC, Zurich, Switzerland, March 1995, pp. 179–184.

[93] Nagel, L. W. 'SPICE: a computer program to simulate semiconductor circuits', *Electr. Research Lab. Report ERL M520*, University of California, Berkeley, USA, March 1975.

[94] Weeks, W. T., Jimenez, A. J., Mahoney, G. W. *et al.* 'Algorithms for ASTAP – a network analysis program'. *IEEE Trans. Circuit Theory*, **20**, 1973, 628–634.

[95] Wollenberg, G. and Goerisch, A. 'Analysis of 3-D interconnect structures with PEEC using SPICE', *IEEE Trans. EMC*, **41**, 1999, 412–417.

1

Fundamentals of Electrodynamics

Most of electrical engineering rests upon classical electrodynamics. Classical electrodynamics summarizes the classical field theory which models the interaction between electric charges and the electromagentic field. The related field theoretical concepts are necessary to develop transmission-line models which describe in a general way the interaction between electric charges on transmission lines and the electromagnetic field.

Within this first chapter the fundamental concepts and results of classical electrodynamics which form the physical and mathematical basis of this book will be collected and developed. To begin with, an axiomatic approach is presented which derives the Maxwell equations, that is the equations of motion of electrodynamics, from conservation laws. This approach is followed by an alternative gauge field approach which also yields the Maxwell equations and, additionally, conveys profound insights into the concept of gauge symmetry. Both approaches will be compared and interrelated. Then the dynamical and nondynamical properties of the electromagnetic field are investigated in order to set up electromagnetic boundary value problems in terms of appropriate differential equations and integral equations.

1.1 Maxwell Equations Derived from Conservation Laws – an Axiomatic Approach

An axiomatic approach to classical electrodynamics is based on electric charge conservation, the Lorentz force, magnetic flux conservation and the existence of local and linear constitutive relations [1, 2]. The *inhomogeneous* Maxwell equations, expressed in terms of D^i and H_i , turn out to be a consequence of electric charge conservation, whereas the *homogeneous* Maxwell equations, expressed in terms of E_i and B^i, are derived from magnetic flux conservation. The excitations D^i and H_i , by means of constitutive relations, are linked to the field strengths E_i and B^i. Whereas the axiomatic approach has been presented in a relativistic framework [1, 2], basically no relativistic notions are used in this book. This is quite remarkable and requires, in particular for the derivation of the homogeneous Maxwell equations from magnetic flux conservation, some steps that are not necessary if the complete framework of relativity were available.

Radiating Nonuniform Transmission-Line Systems and the Partial Element Equivalent Circuit Method Jürgen Nitsch, Frank Gronwald, and Günter Wollenberg

The axiomatic approach is not only characterized by simplicity and beauty, but is also of appreciable pedagogical value. The more clearly a structure is presented, the easier it is to memorize. Moreover, an understanding of how the fundamental electromagnetic quantities D^i, H_i, E_i and B^i are related to each other facilitates the formulation and solution of actual electromagnetic problems.

As is appropriate for an axiomatic approach, as few prerequisites as possible are assumed. What will be needed is some elementary mathematical background that comprises differentiation and integration in the framework of tensor analysis in three-dimensional space. In particular, the concept of integration is necessary for introducing electromagnetic objects as integrands in a natural way. To this end, a tensor notation is used in which the components of mathematical quantities are explicitly indicated by means of upper (contravariant) or lower (covariant) indices [3]. The advantage of this notation is that it allows geometric properties to be represented clearly. In this way, the electromagnetic objects become more transparent and can be discussed more easily. For the formalism of differential forms, which provides similar conceptual advantages, the reader is referred to [2, 4].

Some mathematical material is compiled in Appendix A. This is helpful in order to become comfortable with the tensor notation. For a quick start the following conventions are introduced:

- Partial derivatives with respect to a spatial coordinate x^i (with $i, j, \ldots = 1, 2, 3$) or with respect to time t are abbreviated according to

$$\frac{\partial}{\partial x^i} \longrightarrow \partial_i \,, \qquad \frac{\partial}{\partial t} \longrightarrow \partial_t \,. \tag{1.1}$$

- The 'summation convention' is used. It states that a summation sign can be omitted if the same index occurs both in a lower and an upper position. That is, there is for example the correspondence

$$\sum_{i=1}^{3} \alpha_i \, \beta^i \longleftrightarrow \alpha_i \, \beta^i \,. \tag{1.2}$$

- The Levi-Civita symbols ϵ_{ijk} and ϵ^{ijk} are defined. They are antisymmetric with respect to all of their indices. Therefore, they vanish if two of their indices are equal. Their remaining components assume the values $+1$ or -1, depending on whether ijk is an even or an odd permutation of 123:

$$\epsilon_{ijk} = \epsilon^{ijk} = \begin{cases} 1\,, & \text{for } ijk = 123, 312, 231, \\ -1\,, & \text{for } ijk = 213, 321, 132. \end{cases} \tag{1.3}$$

With these conventions for the *gradient* of a function f the expression $\partial_i f$ is obtained. The *curl* of a (covariant) vector v_i is written according to $\epsilon^{ijk}\partial_j v_k$ and the *divergence* of a (contravariant) vector (density) w^i is given by $\partial_i w^i$.

In the next three subsections, classical electrodynamics will be established from electric charge conservation (axiom 1), the Lorentz force (axiom 2), magnetic flux conservation (axiom 3), and the existence of constitutive relations (axiom 4). This represents the core of classical electrodynamics: it results in the Maxwell equations together with the constitutive relations and the Lorentz force law.

In order to complete electrodynamics, one can require two more axioms, which are only mentioned briefly (see [2] for a detailed discussion). One can specify the energy–momentum distribution of the electromagnetic field (axiom 5) by means of its so-called energy–momentum tensor. This tensor yields the energy density $(D^i E_i + H_i B^i)/2$ and the energy flux density $\epsilon^{ijk} E_j H_k$ (the Poynting vector). Moreover, if one treats electromagnetic problems of materials in macrophysics, one needs a further axiom by means of which the total electric charge (and the current) is split (axiom 6) in a bound or material charge (and current), which is also conserved, and in a free or external charge (and current).

1.1.1 Charge Conservation

In classical electrodynamics, the electric charge is characterized by its density ρ. From a geometric point of view, the charge density ρ constitutes an integrand of a volume integral. This geometric identification is natural since, by definition, integration of ρ over a three-dimensional volume V yields the total charge Q enclosed in this volume

$$Q := \int_V \rho \, \mathrm{d}v \, . \tag{1.4}$$

It is noted that, in the SI system, electric charge is measured in units of 'ampere times second' or coulomb, $[Q] = \mathrm{As} = \mathrm{C}$. Therefore, the SI unit of charge density ρ is $[\rho] = \mathrm{As/m^3} = \mathrm{C/m^3}$.

It is instructive to invoke at this point the Poincaré lemma. There are different explicit versions of this lemma. Here the form (A.23) that is displayed in Appendix A is used. Then (if space fulfills suitable topological conditions) it is immediate to write the charge density ρ as the divergence of an integrand D^i of a surface integral. Thus,

$$\boxed{\partial_i D^i = \rho} \qquad\qquad (\nabla \cdot \boldsymbol{D} = \rho) \, . \tag{1.5}$$

This result already constitutes one inhomogeneous Maxwell equation, the Coulomb–Gauss law. The more conventional vector notation is displayed in parenthesis for comparison.

Electric charges often move. This motion is represented by a material velocity field u^i, that is, locally a velocity is assigned to each portion of charge in space. The product of electric charge density ρ and material velocity u^i defines[1] the electric current density J^i,

$$J^i = \rho u^i \, . \tag{1.6}$$

Geometrically, the electric current density constitutes an integrand of surface integrals since integration of J^i over a two-dimensional surface S yields the total electric current I that crosses this surface,

$$I = \int_S J^i \, \mathrm{d}a_i \, . \tag{1.7}$$

In SI units, $[I] = \mathrm{A}$ and $[J^i] = \mathrm{A/m^2}$.

[1] This definition is a microscopic one, since the movement of individual electric charges that constitute the electric charge density is considered. On a macroscopic 'averaged' level it is possible that the effective charge density vanishes while an electric current is present. An example is a configuration of an electric current that flows within a wire and exhibits no net charge density since negative charges of moving electrons are compensated by positive charges of atoms that constitute the wire.

Now electric charge conservation comes into play, the first axiom of the axiomatic approach. To this end it is determined how individual packets of charge change in time as they move with velocity u^i through space. A convenient way to describe this change is provided by the material derivative D_u/Dt which is also often called convective derivative [1, 5]. It allows the change of a physical quantity to be calculated as it appears to an observer or a probe that follows this quantity. Then electric charge conservation can be expressed as

$$\frac{D_u Q}{Dt} = 0\,, \tag{1.8}$$

where the material derivative is taken with respect to the velocity field u^i. It can be rewritten by means of the Reynold's transport theorem in the following way [5],

$$\begin{aligned} \frac{D_u Q}{Dt} &= \frac{D_u}{Dt}\int_{V(t)} \rho\,\mathrm{d}v \\ &= \int_{V(t)} \frac{\partial \rho}{\partial t}\,\mathrm{d}v + \oint_{\partial V(t)} \rho u^i\,\mathrm{d}a_i \\ &= \int_{V(t)} \left(\frac{\partial \rho}{\partial t} + \partial_i(\rho u^i)\right)\mathrm{d}v\,. \end{aligned} \tag{1.9}$$

In the last line Stokes' theorem in the form of (A.24) has been used. The volume $V(t)$ that is integrated over depends in general on time since it moves together with the electric charge that it contains. By means of Equations (1.6), (1.8) and (1.9) the axiom of electric charge conservation in the local form as continuity equation is obtained,

$$\partial_t \rho + \partial_i J^i = 0\,. \tag{1.10}$$

This result can also be obtained from the direct application of the material derivative D_u/Dt to the charge density. Noting that the material derivative is the sum of the partial time derivative and the Lie derivative with respect to the velocity field u^i [1],

$$\frac{D_u}{Dt} = \frac{\partial}{\partial t} + l_u\,, \tag{1.11}$$

it is found with (A.39) for the material time derivative of the scalar density ρ

$$\frac{D_u \rho}{Dt} = \frac{\partial \rho}{\partial t} + l_u \rho \tag{1.12}$$

$$= \frac{\partial \rho}{\partial t} + \partial_i(\rho u^i) \tag{1.13}$$

$$= \frac{\partial \rho}{\partial t} + \partial_i J^i\,. \tag{1.14}$$

In this way, the continuity Equation (1.10) follows from $D_u\rho/Dt = 0$.

Now the inhomogeneous Maxwell Equation (1.5) is used in order to replace within the continuity Equation (1.10) the charge density by the divergence of D^i. This yields

$$\partial_i\left(\partial_t D^i + J^i\right) = 0\,. \tag{1.15}$$

Again the Poincaré lemma is invoked, now in the form (A.22), and the sum $\partial_t D^i + J^i$ is written as the curl of the integrand of a line integral which is denoted by H_i. This yields

$$\boxed{\epsilon^{ijk}\partial_j H_k - \partial_t D^i = J^i} \qquad \left(\nabla \times \boldsymbol{H} - \frac{\partial \boldsymbol{D}}{\partial t} = \boldsymbol{J}\right). \tag{1.16}$$

Equation (1.16) constitutes the remaining inhomogeneous Maxwell equation, the Ampére–Maxwell law which, in this way, is derived from the axiom of electric charge conservation. The fields D^i and H_i are called electric excitation (historically, electric displacement) and magnetic excitation (historically, magnetic field), respectively. From Equations (1.5) and (1.16) it follows that their SI units are $[D^i] = \mathrm{As/m^2}$ and $[H_i] = \mathrm{A/m}$.

Some remarks are appropriate now. The excitations D^i and H_i are obtained from the Poincaré lemma and charge conservation, respectively, without introducing the concept of force. This is in contrast to other approaches that rely on the Coulomb and the Lorentz force laws [6]. Furthermore, since electric charge conservation is valid not only on macroscopic scales but also in microphysics,[2] the inhomogeneous Maxwell Equations (1.5) and (1.16) are microphysical equations as long as the source terms ρ and J^i are microscopically formulated as well. The same is valid for the excitations D^i and H_i. They are microphysical quantities – in contrast to what is often stated in textbooks (see [8] for example). It is finally remarked that the inhomogeneous Maxwell Equations (1.5) and (1.16) can be straightforwardly put into a relativistically invariant form. This is not self-evident but suggested by electric charge conservation in the form of the continuity Equation (1.10) since this fundamental equation can also be shown to be relativistically invariant.

1.1.2 Lorentz Force and Magnetic Flux Conservation

During the discovery of the electromagnetic field, the concept of force has played a major role. Electric and magnetic forces are directly accessible to experimental observation. Experimental evidence shows that, in general, an electric charge is subject to a force if an electromagnetic field acts on it. For a point charge q at position ${x_q}^i$ it follows $\rho(x^i) = q\delta(x^i - {x_q}^i)$. If it has the velocity u^i the Lorentz force

$$\boxed{F_i = q(E_i + \epsilon_{ijk}u^j B^k)} \tag{1.17}$$

is postulated as the second axiom. It introduces the electric field strength E_i and the magnetic field strength B^i. The Lorentz force already yields a prescription of how to measure E_i and B^i by means of the force that is experienced by an infinitesimally small test charge q which is either at rest or moving with velocity u^i. Turning to the dimensions, voltage is introduced as 'work per charge'. In SI, it is measured in volt (V). Then $[F_i] = \mathrm{VC/m}$ and, according to Equation (1.17), $[E_i] = \mathrm{V/m}$ and $[B^i] = \mathrm{Vs/m^2} = \mathrm{Wb/m^2} = \mathrm{T}$, with Wb an abbreviation for Weber and T for Tesla.

[2] Microphysics is commonly understood to be physics on a small scale that describes the interaction between single, elementary particles. The concept of an elementary particle does not necessarily involve quantum effects and is also useful and important for classical electrodynamics [7]. On microphysical scales electric charges and their related currents are often represented by distributions that reflect the physical model of a point particle.

From the axiom of the Lorentz force (Equation (1.17)), it follows that the electric and the magnetic field strengths are not independent of each other. The corresponding argument is based on the special relativity principle: according to the special relativity principle, the laws of physics are independent of the choice of an inertial system [6]. Different inertial systems move with constant velocities v^i relative to each other. The outcome of a physical experiment, as expressed by an empirical law, has to be independent of the inertial system where the experiment takes place.

It is assumed that a point charge q with a certain mass moves with velocity u^i in an electromagnetic field E_i and B^i. The velocity and the electromagnetic field are measured in an inertial laboratory frame. The point charge can also be observed from its instantaneous inertial *rest frame.* If quantities that are measured with respect to this rest frame are denoted by a prime (i.e. by u'^i, E'_i and B'^i) then it follows $u'^i = 0$. In the absence of an electric field in the rest frame (i.e. if additionally $E'_i = 0$) the charge experiences no Lorentz force and, therefore, no acceleration, so

$$F'_i = q(E'_i + \epsilon_{ijk} u'^j B'^k) = 0 . \tag{1.18}$$

The fact that the charge experiences no acceleration is also true in the laboratory frame. This is a consequence of the special relativity principle or, more precisely, of the fact that the square of the acceleration can be shown to form a relativistic invariant. Consequently,

$$F_i = q(E_i + \epsilon_{ijk} u^j B^k) = 0 . \tag{1.19}$$

Thus, in the laboratory frame, the electric and magnetic fields are related by

$$E_i = -\epsilon_{ijk} u^j B^k . \tag{1.20}$$

This situation is depicted in Figure 1.1. Accordingly, it is found that the electric and magnetic field strengths cannot be viewed as independent quantities. They are connected to each other by transformations between different inertial systems.

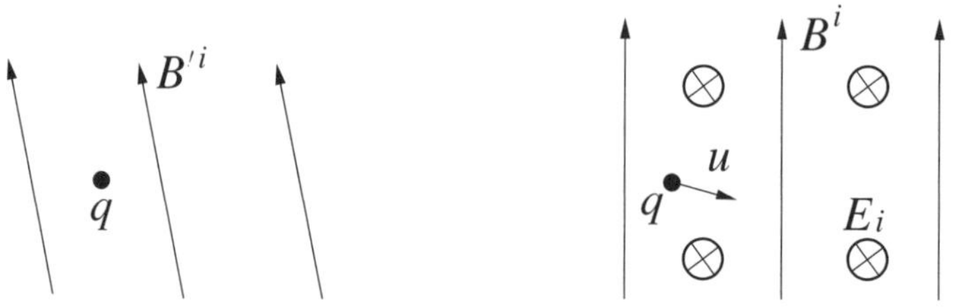

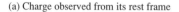

(a) Charge observed from its rest frame

(b) Charge observed from inertial frame moving with respect to q

Figure 1.1 (a) A charge which is, in some inertial frame, at rest and immersed in a purely magnetic field experiences no Lorentz force. The fact that there is no Lorentz force should be independent of the choice of the inertial system that is used to observe the charge. (b) A compensating electric field accompanies the magnetic field if viewed from an inertial laboratory system which is in motion relative to the charge.

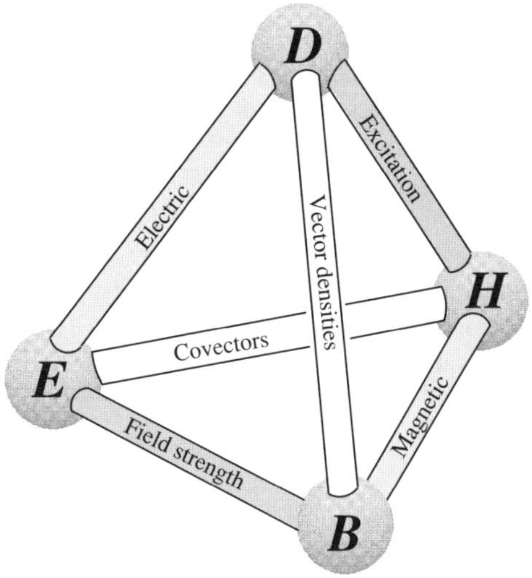

Figure 1.2 The tetrahedron of the electromagnetic field, according to Hehl [2]. The electric and magnetic excitations D^i and H_i and the electric and the magnetic field strengths E_i and B^i build up four-dimensional quantities in spacetime. These four fields describe the electromagnetic field completely. Of an electric nature are D^i and E_i, of a magnetic nature H_i and B^i.

So far, the four electromagnetic field quantities D^i, H_i and E_i, B^i have been introduced. These four quantities are interrelated by physical and mathematical properties. This is illustrated in Figure 1.2 by the 'tetrahedron of the electromagnetic field'.

To introduce the notion of magnetic flux it is instructive to digress for a moment and to turn to hydrodynamics. Helmholtz was one of the first who studied rotational or 'vortex' motion in hydrodynamics (see [9]). He derived theorems for vortex lines. An important consequence of his work was the conclusion that vortex lines are conserved. They may move or change orientation but they are never spontaneously created nor annihilated. The vortex lines that pierce through a two-dimensional surface can be integrated to yield a scalar quantity that is called circulation. The circulation in a perfect fluid, which satisfies certain conditions, is constant provided the loop enclosing the surface moves with the fluid [9].

There are certainly fundamental differences between electromagnetism and hydrodynamics, but some suggestive analogies exist. A vortex line in hydrodynamics seems analogous to a magnetic flux line. The magnetic flux Φ is determined from magnetic flux lines, represented by the magnetic field strength B^i, that pierce through a two-dimensional surface S,

$$\Phi := \int_S B^i \, \mathrm{d}a_i \, . \tag{1.21}$$

As the circulation in a perfect fluid is conserved, it is guessed that, in a similar way, the magnetic flux may be conserved. Of course, the consequences of such an axiom have to be borne out by experiment.

At first sight, to find the vortex lines of a fluid might be easier to visualize than magnetic flux lines. However, on a microscopic level, magnetic flux can occur in quanta.

The corresponding magnetic flux unit is called a flux quantum or fluxon and it carries $\Phi_0 = h/(2e) \approx 2.07 \times 10^{-15}$ Wb, with h the Planck constant and e the elementary charge. Single quantized magnetic flux lines have been observed in the interior of type II superconductors if exposed to a sufficiently strong magnetic field (see [2, p. 131]). They can even be counted. The corresponding experiments provide good evidence that magnetic flux is a conserved quantity.

But how can magnetic flux conservation be formulated mathematically? In Section 1.1.1 the material derivative D_u/Dt was applied with respect to a velocity field u^i to the total electric charge Q and, equivalently, to the electric charge density ρ. This yielded the continuity Equation (1.10) which expresses electric charge conservation. The same pattern is followed to express magnetic flux conservation and to write down the conservation law

$$\frac{D_u \Phi}{Dt} = 0 . \tag{1.22}$$

This expression has to be examined and this, in turn, requires the following two points to be clarified:

- What is the definition of a velocity field u^i with respect to a magnetic field B^i?
- What is a physically reasonable definition of the current of a magnetic flux?

To answer the first point it is necessary to know how to observe, in general, a magnetic field. The only means at our disposal is the Lorentz force law. With the Lorentz force law electric test charges can be used to measure the electric and magnetic field strength. It has already been noted in the last subsection that electric and magnetic field strengths are connected to each other by relativistic transformations. This makes it impossible to state in a relativistically invariant way which contribution to a Lorentz force is due to an electric field and which contribution is due to a magnetic field. *Any* observer who uses an electric test charge which is located in their rest frame might state that the Lorentz force on the test charge is 'purely electric' since for their test charge $u^i = 0$ in Equation (1.17) and, thus, $F_i = qE_i$. They might furthermore draw the conclusion that in their rest frame the velocity of the magnetic field vanishes as well. This is, of course, a wrong conclusion since the velocity which appears in the Lorenz force law is the relative velocity between an observer and a test charge but it is not, a priori, the relative velocity between a test charge and the magnetic field.

Nevertheless to associate a velocity field to a magnetic field it is noted that the *vanishing* of the Lorentz force on a test charge is relativistically invariant. If the Lorentz force on a test charge vanishes in one inertial system it will vanish in all inertial systems. In this case there will be exactly one inertial system where this test charge is at rest. In this inertial system,

$$F'_i = q(E'_i + \epsilon_{ijk} u'^j B'^k) = 0 \tag{1.23}$$

and it follows, since $u'^j = 0$, that $E'_i = 0$. Now this distinguished inertial system is *defined* to be, at the position considered, the rest frame of the magnetic field. This definition requires that an inertial system can always be found where $F'_i = 0$, that is that an inertial system can always be found where $E'_i = 0$. For an arbitrary electromagnetic field this will not be true, but this property is assigned to a *purely* magnetic field. A purely magnetic field is an electromagnetic field where, at any point in space and time, the electric field can be made to vanish in *one* inertial system. In the definition of magnetic flux conservation only electromagnetic fields which are purely magnetic will be considered. Otherwise it is not possible to associate a velocity field to a magnetic field in a relativistically invariant way. Therefore, the answer to the first point above is that the velocity u^i associated to a (purely) magnetic field is the velocity of a specified inertial system which moves with respect to a laboratory system with velocity u^i and where the Lorentz force on a test charge vanishes.

An answer to the second point needs to provide a physically meaningful definition of magnetic flux current. To this end the notion of electric charge is reconsidered,

$$Q = \int_V \rho \, \mathrm{d}v \,, \tag{1.24}$$

together with the corresponding conservation law

$$\partial_t Q + \int_{\partial V} J^i \, \mathrm{d}a_i = 0 \,. \tag{1.25}$$

It is seen from this representation that the rate of change of the electric charge within a specified volume V is balanced by the out- or inflowing charge across the surface ∂V. This charge transport is described by the electric charge current J^i that is integrated over the enveloping surface ∂V. By means of the Stokes' theorem in the form (A.24), Equation (1.25) yields the local continuity equation

$$\partial_t \rho + \partial_i J^i = 0 \,. \tag{1.26}$$

It is possible to follow the same pattern to define the current of a magnetic flux: starting with the definition (1.21) of the magnetic flux, the corresponding geometric conservation law, in analogy to (1.25), reads

$$\partial_t \Phi + \int_{\partial S} J_i^\Phi \, \mathrm{d}c^i = 0 \,, \tag{1.27}$$

where the magnetic flux current J_i^Φ is introduced. This is a covariant vector that is integrated along a line ∂S, that is along the curve bordering the two-dimensional surface S. The conservation law (1.27) shows that the rate of change of the magnetic flux within a specified area S is balanced by the magnetic flux current J_i^Φ that is integrated along the boundary ∂S. Then the Stokes' theorem in the form (A.25) yields the local continuity equation

$$\partial_t B^i + \epsilon^{ijk} \partial_j J_k^\Phi = 0 \,. \tag{1.28}$$

One interesting consequence is that the divergence of (1.28) reads

$$\partial_i (\partial_t B^i) = 0 \qquad \Longrightarrow \qquad \partial_i B^i = \rho_{\mathrm{mag}} \,, \quad \partial_t \rho_{\mathrm{mag}} = 0 \,. \tag{1.29}$$

Thus, a time-*in*dependent term ρ_{mag} is found which tentatively acquires the meaning of a magnetic charge density. Let a specific reference system be chosen in which ρ_{mag} is constant in time, that is $\partial_t \rho_{\mathrm{mag}} = 0$. Now an arbitrary reference system with time coordinate t' and spatial coordinates $x^{i'}$ is considered. Clearly, in general $\partial_{t'} \rho_{\mathrm{mag}} \neq 0$. The only way to evade a contradiction to (1.29) is to require $\rho_{\mathrm{mag}} = 0$, that is the magnetic field strength B^i has no sources, its divergence vanishes:

$$\boxed{\partial_i B^i = 0} \qquad (\nabla \cdot \boldsymbol{B} = 0)\,. \tag{1.30}$$

This is recognized as one of the homogeneous Maxwell equations.

To specify the magnetic flux current, magnetic flux conservation is explored, expressed by means of the material time derivative D_u/Dt with respect to a velocity field u^i which is associated to a purely magnetic field. It follows

$$\begin{aligned}\frac{D_u \Phi}{Dt} &= \frac{D_u}{Dt}\int_{S(t)} B^i \,\mathrm{d}a_i \\ &= \int_{S(t)} \left(\partial_t B^i - \epsilon^{ijk}\partial_j \epsilon_{klm} u^l B^m + u^i \partial_j B^j\right) \mathrm{d}a_i\,,\end{aligned} \tag{1.31}$$

where the Helmholtz transport theorem [5, p. 456] has been applied. Alternatively, it is possible to work with the local expression

$$\begin{aligned}\frac{D_u B^i}{Dt} &= \partial_t B^i + l_u B^i \\ &= \partial_t B^i + u^j \partial_j B^i - B^j \partial_j u^i + B^i \partial_j u^j\,,\end{aligned} \tag{1.32}$$

where the formula (A.36) for the Lie derivative of a contravariant vector density has been used. The Lie derivative of B^i can be rewritten according to[3]

$$u^j \partial_j B^i - B^j \partial_j u^i + B^i \partial_j u^j = -\epsilon^{ijk}\partial_j \epsilon_{klm} u^l B^m + u^i \partial_j B^j\,. \tag{1.33}$$

Hence, it follows

$$\frac{D_u B^i}{Dt} = \partial_t B^i - \epsilon^{ijk}\partial_j \epsilon_{klm} u^l B^m + u^i \partial_j B^j \tag{1.34}$$

and it is recognized that Equation (1.34) is the local version of Equation (1.31)

According to Equation (1.30) the divergence of the magnetic field strength B^i vanishes, $\partial_i B^i = 0$. Also, by virtue of Equation (1.20), the term $-\epsilon_{klm} u^l B^m$ can be locally identified with an electric field strength E_k. This is because it has been assumed that a purely magnetic field is presented, that is an electromagnetic field which does not exert a Lorentz force on an electric charge which moves with velocity u^i in the laboratory frame. Then magnetic flux conservation, $D_u B^i/Dt = 0$, yields

$$\boxed{\partial_t B^i + \epsilon^{ijk}\partial_j E_k = 0} \qquad \left(\frac{\partial \boldsymbol{B}}{\partial t} + \nabla \times \boldsymbol{E} = \boldsymbol{0}\right). \tag{1.35}$$

This equation reflects magnetic flux conservation. It constitutes the remaining homogeneous Maxwell equation, that is Faraday's induction law. This result is compared to the continuity

---

[3] In a more conventional notation this identity reads:

$$(\boldsymbol{u}\cdot\nabla)\boldsymbol{B} - \boldsymbol{B}(\nabla\cdot\boldsymbol{u}) + (\boldsymbol{B}\cdot\nabla)\boldsymbol{u} = -\nabla\times(\boldsymbol{u}\times\boldsymbol{B}) + \boldsymbol{u}(\nabla\cdot\boldsymbol{B})\,.$$

Equation (1.28) and it is deduced that the electric field that appears in the Faraday's induction law has to be interpreted as a magnetic flux current.

1.1.3 Constitutive Relations and the Properties of Spacetime

So far, $4 \times 3 = 12$ unknown electromagnetic field components D^i, H_i, E_i and B^i have been introduced. These components have to fulfill the Maxwell Equations (1.5), (1.16), (1.30) and (1.35), which represent $1 + 3 + 1 + 3 = 8$ partial differential equations. In fact, among the Maxwell Equations, only Equations (1.16) and (1.35) contain time derivatives and are dynamical. The remaining Equations (1.5) and (1.30) are so-called 'constraints'. They are, by virtue of the dynamical Maxwell equations, fulfilled at all times if they are fulfilled once. It follows that they do not contain information on the time evolution of the electromagnetic field. Therefore, there remain only six dynamical equations for 12 unknown field components. To make the Maxwell equations a determined set of partial differential equations it is also necessary to introduce the so-called 'constitutive relations' between the excitations D^i, H_i and the field strengths E_i, B^i.

The simplest case to begin with is to find constitutive relations for the case of electromagnetic fields in a vacuum. There are guiding principles that limit their structure. It is demanded that constitutive relations in a vacuum are invariant under translation and rotation, furthermore they should be local and linear, that is they should connect fields at the same position and at the same time. Finally, in a vacuum the constitutive relations should not mix electric and magnetic properties. These features characterize the vacuum and not the electromagnetic field itself. They cannot be proved but are postulated as the fourth axiom.

To relate the field strengths and the excitations it needs to be remembered that E_i, H_i are natural integrands of *line* integrals and D^i, B^i are natural integrands of *surface* integrals. Therefore, E_i, H_i transform under a change of coordinates as covariant vectors while D^i, B^i transform as contravariant vector densities. To compensate for these differences a symmetric metric field $g_{ij} = g_{ji}$ is introduced. The metric tensor determines spatial distances and introduces the notion of orthogonality. The determinant of the metric is denoted by g. It follows that $\sqrt{g}g^{ij}$ transforms like a density and maps a covariant vector into a contravariant vector density. Then the fourth axiom is given by the constitutive relations for vacuum,

$$\boxed{D^i = \varepsilon_0 \sqrt{g}\, g^{ij}\, E_j\,,} \tag{1.36}$$

$$\boxed{H_i = (\mu_0 \sqrt{g})^{-1} g_{ij}\, B^j\,.} \tag{1.37}$$

In flat spacetime and in Cartesian coordinates, $g = 1$, $g^{ii} = 1$, and $g^{ij} = 0$ for $i \neq j$, such that the familiar vacuum relations between field strengths and excitations are recognized. The electric constant ε_0 and the magnetic constant μ_0 characterize the vacuum. They acquire the SI units $[\varepsilon_0] = \text{As/Vm}$ and $[\mu_0] = \text{Vs/Am}$.

What seems to be conceptually important about the constitutive Equations (1.36) and (1.37) is that they not only provide relations between the excitations D^i, H_i and the field strengths E_i, B^i, but they also connect the electromagnetic field to the structure of spacetime, which here is represented by the metric tensor g_{ij}. The formulation of the first three axioms that were presented in the previous sections does not require information on this metric structure. The connection between the electromagnetic field and spacetime, as expressed by the constitutive relations, indicates that physical fields and spacetime are not independent of each other. The constitutive relations might suggest the point of view that the structure of spacetime determines the structure of the electromagnetic field. However, the opposite conclusion could also be true: it can be shown that the propagation properties of the electromagnetic field determine the metric structure of spacetime [2, 10].

Constitutive relations in matter usually assume a more complicated form than Equations (1.36) and (1.37). In this case it would be appropriate to derive the constitutive relations, after an averaging procedure, from a microscopic model of matter. Such procedures are the subject of solid state or plasma physics, for example. A discussion of these subjects is outside the scope of this work but, without going into details, the constitutive relations of a general linear *magnetoelectric* medium are simply quoted:

$$D^i = \left(\boldsymbol{\varepsilon}^{ij} - \epsilon^{ijk}\, n_k\right) E_j + \left(\boldsymbol{\gamma}^i{}_j + \tilde{s}_j{}^i\right) B^j + (\boldsymbol{\alpha} - s)\, B^i \,, \tag{1.38}$$

$$H_i = \left(\boldsymbol{\mu}^{-1}_{ij} - \epsilon_{ijk}\, m^k\right) B^j + \left(-\boldsymbol{\gamma}^j{}_i + \tilde{s}_i{}^j\right) E_j - (\boldsymbol{\alpha} + s)\, E_i \,. \tag{1.39}$$

This formulation is due to Hehl and Obukhov [2, 11, 12], an equivalent formulation was given by Lindell and Olyslager [4, 13]. Both matrices ε^{ij} and μ^{-1}_{ij} are symmetric and possess six independent components each, ε^{ij} is called the *permittivity* tensor and μ^{-1}_{ij} the *impermeability* tensor (reciprocal permeability tensor). The magnetoelectric cross-term $\gamma^i{}_j$, which is trace-free, $\gamma^k{}_k = 0$, has eight independent components. It is related to the Fresnel–Fizeau effects. Accordingly, these pieces altogether, which are printed in Equations (1.38) and (1.39) in bold type for better visibility, add up to $6 + 6 + 8 + 1 = 20 + 1 = 21$ independent components.

1.1.4 Remarks

An axiomatic approach to classical electrodynamics has been presented in which the Maxwell equations are derived from the conservation of electric charge and magnetic flux. In the context of the derivation of the inhomogeneous Maxwell equations, the electric and the magnetic excitation D^i and H_i, respectively, are introduced. The explicit calculation is rather simple because the continuity equation for electric charge is already relativistically invariant such that for the derivation of the inhomogeneous Maxwell equations no additional ingredients from special relativity are necessary. The situation is more complicated for the derivation of the homogeneous Maxwell equations from magnetic flux conservation since it is not immediately clear how to formulate magnetic flux conservation in a relativistic invariant way. It should be mentioned that if the complete framework of relativity were available, the derivation of the axiomatic approach could be done with considerably more ease and elegance [1, 2].

At this point it is appropriate to comment on a question that sometimes leads to controversial discussions, as summarized in [5], for example. This is the question of how the quantities E_i, D^i, B^i and H_i should be grouped in pairs, that is the question of 'which quantities belong together?'. Some people like to form the pairs (E_i, B^i) and (D^i, H_i), while others prefer to build (E_i, H_i) and (D^i, B^i). Already from a dimensional point of view, the answer to this question is obvious. Both E_i and B^i are *voltage*-related quantities, that is, related to the notions of force and work: in SI, the units are $[E_i] = \text{V/m}$, $[B^i] = \text{T=Vs/m}^2$, or $[B^i] = [E_i]/\text{velocity}$. Consequently, they belong together. Analogously, D^i and H_i are *current*-related quantities: $[D^i] = \text{C/m}^2 = \text{As/m}^2$, $[H_i] = \text{A/m}$, or $[D^i] = [H_i]/\text{velocity}$.

These conclusions are made irrefutable by relativity theory. Classical electrodynamics is a relativistic invariant theory and the implications of relativity have been proven to be correct on macro- and microscopic scales over and over again. Also relativity tells us that the electromagnetic field strengths E_i, B^i are inseparably intertwined by relativistic transformations, and the same is true for the electromagnetic excitations D^i, H_i. In the spacetime of relativity theory, the pair (E_i, B^i) forms one single quantity, the tensor of electromagnetic field strength, while the pair (D^i, H_i) forms another single quantity, the tensor of electromagnetic excitations. If compared to these facts, arguments in favor of the pairs (E_i, H_i), namely that they are both covectors, and (D^i, B^i) are both vector densities (see the tetrahedron in Figure 1.2), turn out to be of a secondary nature.

1.2 The Electromagnetic Field as a Gauge Field – a Gauge Field Approach

Modern descriptions of the fundamental interactions rely heavily on symmetry principles. In particular, this is true for the electromagnetic interaction which can be formulated as a gauge field theory that is based on a corresponding gauge symmetry. In recent articles this approach towards electromagnetism has been explained in an original and descriptive way [14, 15]. The gauge field approach is put next to the axiomatic approach since it furnishes further information that will complement the picture of classical electrodynamics. In particular, it allows the clarification of the concept of gauge invariance which often accompanies explicit calculations in the solution of electrodynamic boundary value problems. It also shows that the electromagnetic potentials, which are often viewed as mathematical auxiliary variables, are of major physical relevance. Furthermore, in the gauge field approach the inhomogeneous Maxwell equations turn out to be true equations of motion while the homogeneous Maxwell equations become a mere mathematical identity.

While it is rewarding to gain the additional insights that are provided by the gauge field approach it should be admitted that this approach, at first sight, extends on a rather abstract level. However, it only requires a small number of steps:

(1) Accept the fact that physical matter fields (which represent electrons, for example) are described microscopically by complex wave functions.
(2) Recognize that the absolute phase of these wave functions has no physical relevance. This arbitrariness of the absolute phase constitutes a one-dimensional rotational type symmetry $U(1)$ (the circle group). This is the gauge symmetry of electrodynamics.
(3) To derive observable physical quantities from the wave functions needs the definition of derivatives of wave functions with respect to space and time. These derivatives need to be invariant under the gauge symmetry. The construction of such 'gauge covariant' derivatives makes it necessary to introduce gauge fields. One gauge field, the vector potential A_i, defines gauge covariant derivatives D_i^A with respect to the three independent directions of space, while another gauge field, the scalar potential φ, defines a gauge covariant derivative D_t^φ with respect to time.
(4) Finally, the values of the gauge fields φ and A_i are obtained from equations of motion that turn out to be the inhomogeneous Maxwell equations. The gauge fields are related to the

gauge invariant electric and magnetic field strengths via

$$E_i = -\partial_i \varphi - \partial_t A_i \,, \tag{1.40}$$

$$B^i = \epsilon^{ijk} \partial_j A_k \,. \tag{1.41}$$

1.2.1 *Differences of Physical Fields that are Described by Reference Systems*

To introduce the concept of gauge symmetry a physical field is formally denoted by **F**, its components by F^i, and a reference frame with respects to these components by $\boldsymbol{e}_i$,

$$\mathbf{F} = F^i \boldsymbol{e}_i \,. \tag{1.42}$$

The field **F** might change in space or time. Its components F^i depend on the reference frame $\boldsymbol{e}_i$ and usually do not have an absolute significance since usually there is a certain freedom to choose $\boldsymbol{e}_i$. Therefore, a change of **F** involves both its components and the corresponding reference frame. Then it follows

$$\partial_i \mathbf{F} = (\partial_i F^j)\boldsymbol{e}_j + F^j(\partial_i \boldsymbol{e}_j)\,, \tag{1.43}$$

$$\partial_t \mathbf{F} = (\partial_t F^j)\boldsymbol{e}_j + F^j(\partial_t \boldsymbol{e}_j)\,. \tag{1.44}$$

This simple looking 'product rule for differentiation' poses severe difficulties: It might be possible to determine the differences $\partial_i F^j$ and $\partial_t F^j$, for example by measurements, but how is it possible to determine changes $\partial_i \boldsymbol{e}_j$, $\partial_t \boldsymbol{e}_j$ of reference frames? What is an *unchanged* reference frame? How is a reference frame *gauged*? The fact needs to be faced that a priori changes of reference frames are not defined as long as they involve the change between different points in space or time.

At this point the concept of 'interactions' enters the stage. If a physical field changes its value intuition indicates that this is due to an interaction. The gauge principle states that the information on the change of reference frames is contained in interaction fields. Mathematically, this is formulated as follows. The changes $\partial_i \boldsymbol{e}_j$, $\partial_t \boldsymbol{e}_j$ of a reference frame are determined by interaction fields ${A_{ij}}^k$ and ${\varphi_j}^k$ according to

$$\partial_i \boldsymbol{e}_j := {A_{ij}}^k \boldsymbol{e}_k \,, \tag{1.45}$$

$$\partial_t \boldsymbol{e}_j := {\varphi_j}^k \boldsymbol{e}_k \,. \tag{1.46}$$

With these definitions the interaction fields provide a gauging of reference system at different positions in space and time. This is the reason why they are called *gauge fields*. A corresponding mathematical term is *connection*, because gauge fields connect reference systems at different positions in space and time.

With the definitions (1.45) and (1.46) the differences $\partial_i \mathbf{F}$, $\partial_t \mathbf{F}$ are written as

$$\partial_i \mathbf{F} = (\partial_i F^j)\boldsymbol{e}_j + F^j {A_{ij}}^k \boldsymbol{e}_k \,, \tag{1.47}$$

$$\partial_t \mathbf{F} = (\partial_t F^j)\boldsymbol{e}_j + F^j {\varphi_j}^k \boldsymbol{e}_k \,. \tag{1.48}$$

So far it is not clear if the formal introduction of the gauge fields ${A_{ij}}^k$ and ${\varphi_j}^k$ is physically meaningful. However, it turns out that these gauge fields describe the fundamental interactions that are observed in nature correctly. In the following subsection this circumstance will be explained for the case of the electromagnetic interaction.

1.2.2 The Phase of Microscopic Matter Fields

As a first step, the physical field and its corresponding reference system which leads to the introduction of the electromagnetic field as a gauge field are identified. Then it is noted that microscopic matter fields, like electrons for example, are represented in the framework of quantum mechanics by *wave functions* $\Psi(x^i, t)$ [16, 17]. A wave function assumes complex values and depends on space coordinates x^i and a time coordinate t, $\Psi(x^i, t) \in \mathbb{C}$. It follows that a microscopic particle with specified momentum p_i and specified energy E is represented by a wave function of the form

$$\Psi(x^i, t) = \Psi_0 \, \mathrm{e}^{-\frac{j}{\hbar}(p_i x^i - Et)} , \tag{1.49}$$

with $\Psi_0 \in \mathbb{R}$ and $\hbar$ a fundamental constant which carries the dimension of an action, $\hbar = h/2\pi \approx 1.0546 \times 10^{-34}$ Js.

The plane wave in Equation (1.49) is a very special case of a wave function since the momentum and energy of a microscopic particle are usually not known exactly but are affected by an *uncertainty*. In this more general case microscopic particles are represented by *wave packets* that are obtained by the superposition of plane waves of the form (1.49) with different momenta and energies. The value $\Psi(x^i, t)$ has no *direct* physical significance, but the square of its absolute value yields a probability density. This is the real value

$$P(x^i, t) = |\Psi(x^i, t)|^2 \, \mathrm{d}v \tag{1.50}$$

which is the probability of finding a particle, which is described by $\Psi(x^i, t)$, at time t within a volume $\mathrm{d}v$.

The wave function $\Psi(x^i, t)$ of Equation (1.49) has the structure

$$\Psi(x^i, t) = \Psi_0 \, \mathrm{e}^{-j\theta} \tag{1.51}$$

with the phase

$$\theta(x^i, t) = (p_i x^i - Et)/\hbar . \tag{1.52}$$

It is seen that the characteristic quantities p_i and E of a free microscopic particle are included in its phase $\theta(x^i, t)$. They can be obtained explicitly by differentiation of the phase of the wave function,

$$p_i = \hbar \, \partial_i \theta , \tag{1.53}$$

$$E = -\hbar \, \partial_t \theta , \tag{1.54}$$

and these relations reveal that the momentum and the energy of a microscopic particle is determined by the phase differences $\partial_i \theta$ and $\partial_t \theta$. These phase differences depend on reference frames that, a priori, can be chosen arbitrarily, as will be seen next.

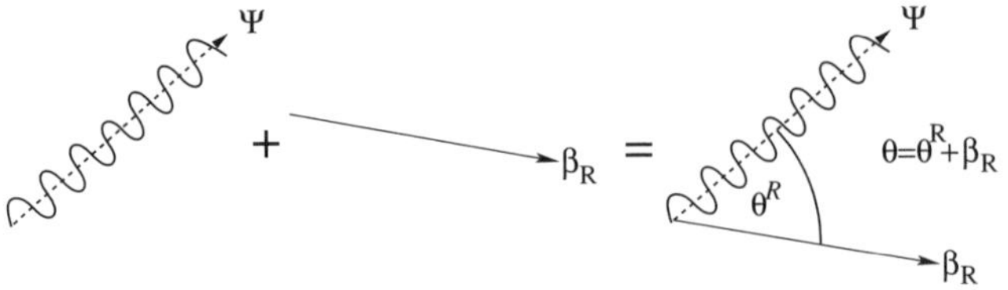

Figure 1.3 A reference system which determines a reference phase β_{R} allows the phase of a wave function to be determined according to $\theta = \theta^{\mathrm{R}} + \beta_{\mathrm{R}}$.

1.2.3 The Reference Frame of a Phase

In this subsection a wave function Ψ at *one* space point x^i at a *fixed* time t is considered. It is necessary to assign a specific phase θ to Ψ. The phase can be taken as a real number of the interval $[0, 2\pi]$, $0 \leq \theta < 2\pi$. In order to assign a fixed value to θ a reference system is necessary which determines a reference phase. This reference phase is denoted by β_{R}. The specification of θ is explained in Figure 1.3, where the wave function is displayed as a curly line. The arrow along this curly line indicates that the wave function is characterized by a certain phase. This phase can be thought of as a point on a circle and is given by an element of the interval $[0, 2\pi]$. The reference phase β_{R} is drawn as another arrow which also indicates a certain direction, i.e., represents an element of the interval $[0, 2\pi]$. This value is taken as reference phase β_{R}. The choice of β_{R} a priori is arbitrary. A convenient choice would be $\beta_{\mathrm{R}} = 0$, for example.

To read off the value θ of the wave function the angle between the wave function and the reference system is determined. This angle is denoted by θ^{R} and refers to the reference system β_{R}. Therefore, θ^{R} can be understood as a component with respect to the basis β_{R}. The component θ^{R} and the basis β_{R} yield the phase θ according to

$$\theta = \theta^{\mathrm{R}} + \beta_{\mathrm{R}} \,. \tag{1.55}$$

In terms of the wave function this is written as

$$\begin{aligned} \Psi &= \Psi_0\, \mathrm{e}^{-j\theta} = \Psi_0\, \mathrm{e}^{-j(\theta^{\mathrm{R}}+\beta_{\mathrm{R}})} \\ &= \Psi_0\, \mathrm{e}^{-j\theta^{\mathrm{R}}} \mathrm{e}^{-j\beta_{\mathrm{R}}} \,. \end{aligned} \tag{1.56}$$

It follows that the wave function is of the form (1.42),

$$\Psi = \Psi^{\mathrm{R}} \mathrm{e}_{\mathrm{R}} \,, \tag{1.57}$$

with the correspondences

$$\Psi^{\mathrm{R}} = \Psi_0\, \mathrm{e}^{-j\theta^{\mathrm{R}}} \qquad \text{(component)} \tag{1.58}$$

and

$$\mathrm{e}_{\mathrm{R}} = \mathrm{e}^{-j\beta_{\mathrm{R}}} \qquad \text{(reference system)}. \tag{1.59}$$

The choice of a reference system of a phase is not unique. This is clarified in Figure 1.4. There is the *gauge freedom* to choose between reference systems which differ from each other

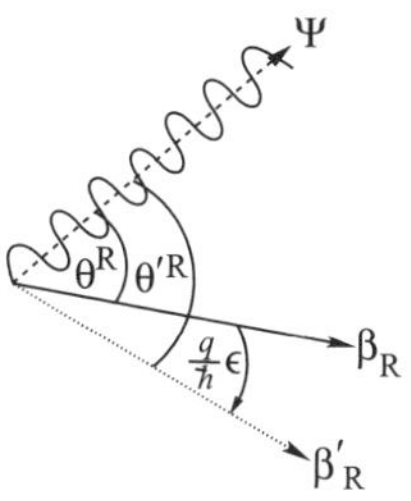

Figure 1.4 The phase θ^{R} of the wave function Ψ depends on the choice of a reference system. A gauge transformation corresponds to the transition from one reference system to another equivalent reference system. This transition is accomplished by a rotation about an angle $(q/\hbar)\epsilon$.

by a one-dimensional rotation. The choice of a fixed reference system is called the choice of a *gauge*. In this language the value θ^{R} is a *gauge dependent* quantity since it depends on the gauge, that is it depends on the choice of a reference system.

A gauge transformation is denoted by δ_ϵ. Its effect on the component θ^{R} of the phase is

$$\delta_\epsilon \theta^{\mathrm{R}} := \theta'^{\mathrm{R}} - \theta^{\mathrm{R}} = \frac{q}{\hbar}\epsilon\,. \tag{1.60}$$

Here, the difference $\theta'^{\mathrm{R}} - \theta^{\mathrm{R}}$ is assigned to the value $q\epsilon/\hbar$. This notation seems a bit awkward but it is in accordance with conventions that have their origin in quantum mechanics. The phase difference carries no physical dimension and q has the dimension of electric charge, $[q] = \mathrm{As} = \mathrm{C}$. Therefore, the parameter ϵ has the dimension of an action per charge, $[\epsilon] = \mathrm{J}/(\mathrm{As}^2)$. The interpretation of a gauge transformation (1.60) might appear to be rather trivial: if there is a shift between two reference systems that differ by an angle $(q/\hbar)\epsilon$ then the component θ^{R} of the phase changes by an angle $(q/\hbar)\epsilon$.

The reference system β_{R} can also be used to gauge the reference system β'_{R}. In order to show this the value of β'_{R} is set to

$$\beta'_{\mathrm{R}} = \beta_{\mathrm{R}} - \frac{q}{\hbar}\epsilon\,. \tag{1.61}$$

Then β_{R} transforms under a gauge transformation according to

$$\delta_\epsilon \beta_{\mathrm{R}} := \beta'_{\mathrm{R}} - \beta_{\mathrm{R}} = -\frac{q}{\hbar}\epsilon\,. \tag{1.62}$$

It follows that the phase θ is a gauge independent quantity,

$$\begin{aligned} \delta_\epsilon \theta &= \delta_\epsilon \theta^{\mathrm{R}} + \delta_\epsilon \beta_{\mathrm{R}} \\ &= \frac{q}{\hbar}\epsilon - \frac{q}{\hbar}\epsilon = 0\,. \end{aligned} \tag{1.63}$$

However, in spite of this gauge independence the value of θ has no absolute significance since it depends, by virtue of $\theta = \theta^{\mathrm{R}} + \beta_{\mathrm{R}}$, on a choice of a reference frame β_{R}.

The situation is different if the *difference* between two phases θ_1 and θ_2 is considered. Since both phases are defined at the same point x^i and to the same time t they can be characterized by a common reference system β_R,

$$\theta_1 = \theta_1^R + \beta_R \,, \tag{1.64}$$

$$\theta_2 = \theta_2^R + \beta_R \,. \tag{1.65}$$

Taking the difference yields

$$\theta_1 - \theta_2 = \theta_1^R - \theta_2^R \,, \tag{1.66}$$

that is, the difference $\theta_1 - \theta_2$ is both independent of the reference system β_R and, due to

$$\delta_\epsilon(\theta_1 - \theta_2) = \delta_\epsilon\theta_1 - \delta_\epsilon\theta_2 = 0 - 0 = 0 \,, \tag{1.67}$$

a gauge independent quantity. Therefore, the phase difference $\theta_1 - \theta_2$ at one point (x^i, t) in spacetime *has* an absolute significance.

1.2.4 The Gauge Fields of a Phase

So far, the phase θ at one point (x^i, t) in spacetime has been considered. In a next step, the change of the phase between *two different* points in spacetime is investigated. To this end, consider two points (x^i, t) and $(x^i + dx^i, t)$ that are separated by an infinitesimal spatial distance dx^i. Mathematically, from Equation (1.55) for the difference $\partial_i\theta$ between two points x^i and $x^i + dx^i$ the expression

$$\partial_i\theta = \partial_i\theta^R + \partial_i\beta_R \tag{1.68}$$

is found, that is the change of the phase θ is the sum of the change of its component and the change of its reference frame.

For a geometrical interpretation of Equation (1.68) it is useful to think of how to construct the difference $\partial_i\theta$. The construction is divided into several steps (see Figure 1.5):

1. According to the previous Section 1.2.3 first the phase $\theta(x^i, t)$ at the point (x^i, t) is determined by a reference system β_R according to $\theta(x^i, t) = \theta^R(x^i, t) + \beta_R(x^i, t)$.
2. At the point $(x^i + dx^i, t)$ an arbitrary reference frame β'_R is chosen. In Figure 1.5 this arbitrary reference frame is displayed by a dotted line. It can be used to read off the value

$$\theta'^R(x^i + dx^i, t) = \theta^R(x^i, t) + \partial_i\theta^R\, dx^i \,. \tag{1.69}$$

   However, this value has no immediate physical relevance since its corresponding reference system has been arbitrarily chosen.
3. At the point $(x^i + dx^i, t)$ there exists a unique reference system β_R which is defined to be *unchanged* if compared to the reference system β_R at (x^i, t). In mathematical terms, such an unchanged reference system is named a *parallel* reference system. It is obtained from the arbitrary reference system β'_R by a rotation about an angle $\partial_i\beta_R$,

$$\beta_R(x^i + dx^i, t) = \beta'_R(x^i + dx^i, t) - \partial_i\beta_R\, dx^i \tag{1.70}$$

   and has the *same* phase value as the reference system at the point (x^i, t),

$$\beta_R(x^i + dx^i, t) = \beta_R(x^i, t) \,. \tag{1.71}$$

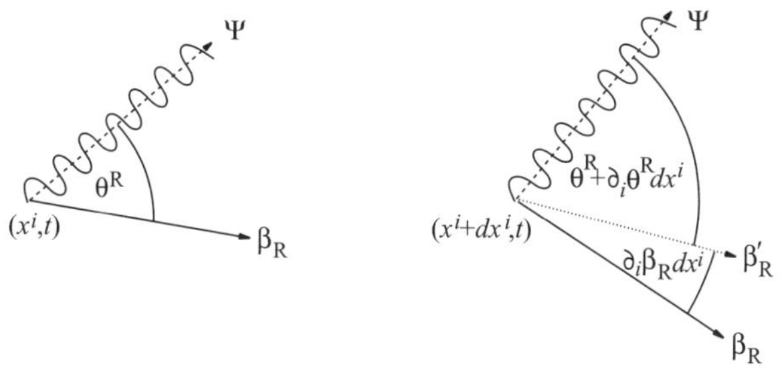

Figure 1.5 Determination of parallel reference systems at two spatially separated points (x^i, t) and $(x^i + dx^i, t)$. In the left part of the figure, at the point (x^i, t), the phase is given by $\theta = \theta^{\mathrm{R}} + \beta_{\mathrm{R}}$. In the right part of the figure, at the point $(x^i + dx^i, t)$, the phase is $\theta = \theta^{\mathrm{R}} + \partial_i\theta^{\mathrm{R}}\, dx^i + \beta_{\mathrm{R}} + \partial_i\beta_{\mathrm{R}}\, dx^i$.

With respect to this parallel reference system the component of the phase is given by

$$\theta^{\mathrm{R}}(x^i + dx^i, t) = \theta^{\mathrm{R}}(x^i, t) + \partial_i\theta^{\mathrm{R}}\, dx^i + \partial_i\beta_{\mathrm{R}}\, dx^i\,. \tag{1.72}$$

In summary, the relations are obtained:

$$\theta(x^i, t) = \theta^{\mathrm{R}}(x^i, t) + \beta_{\mathrm{R}}(x^i, t)\,, \tag{1.73}$$

$$\begin{aligned}\theta(x^i + dx^i, t) &= \theta'^{\mathrm{R}}(x^i + dx^i, t) + \beta'_{\mathrm{R}}(x^i + dx^i, t)\\ &= \theta^{\mathrm{R}}(x^i, t) + \partial_i\theta^{\mathrm{R}}\, dx^i + \beta_{\mathrm{R}}(x^i, t) + \partial_i\beta_{\mathrm{R}}\, dx^i\,,\end{aligned} \tag{1.74}$$

which leads back to Equation (1.68),

$$\partial_i\theta = \frac{\theta(x^i + \mathrm{d}x^i, t) - \theta(x^i, t)}{\mathrm{d}x^i} \tag{1.75}$$

$$= \partial_i\theta^{\mathrm{R}} + \partial_i\beta_{\mathrm{R}}\,. \tag{1.76}$$

Equation (1.76) determines the difference $\partial_i\theta$, but the contributions $\partial_i\theta^{\mathrm{R}}$ and $\partial_i\beta_{\mathrm{R}}$ need to be known explicitly and this leads to the conceptual problem that has been described at the end of Section 1.2.1: the change $\partial_i\theta^{\mathrm{R}}$ can be determined if the corresponding phases are properly read off but the difference $\partial_i\beta_{\mathrm{R}}$ a priori is not determined. It is simply not known which reference systems at different points are in parallel!

At this point the gauge fields come into play. In accordance with (1.45) and (1.46) the difference $\partial_i\beta_{\mathrm{R}}$ of the reference frame is determined from a gauge field A_i by

$$\partial_i\beta_{\mathrm{R}} := -\frac{q}{\hbar}A_i\,. \tag{1.77}$$

Similarly with (1.60) the factor $q/\hbar$ has been chosen to arrive at results that comply with quantum mechanics. Therefore,

$$\partial_i\theta = \partial_i\theta^{\mathrm{R}} - \frac{q}{\hbar}A_i\,. \tag{1.78}$$

In view of Equation (1.53) it is realized that the momentum p_i of a microscopic particle can only be defined by means of the gauge field A_i. It follows from Equation (1.78) that the SI unit of the gauge field A_i is $[A_i] = \mathrm{Vs/m}$.

So far, the difference $\partial_i\theta$ between two spatially separated points (x^i, t) and $(x^i + dx^i, t)$ has been considered. In the same way the difference $\partial_t\theta$ between two temporally separated points (x^i, t) and $(x^i, t + dt)$ can be considered. This leads to the relation

$$\partial_t\theta = \partial_t\theta^i + \partial_t\beta_i \,. \tag{1.79}$$

Since $\partial_t\beta_i$ is undetermined this requires the introduction of a gauge field φ,

$$\partial_t\beta_\mathrm{R} := \frac{q}{\hbar}\varphi \,, \tag{1.80}$$

and the result

$$\partial_t\theta = \partial_t\theta^\mathrm{R} + \frac{q}{\hbar}\varphi \tag{1.81}$$

is obtained. Similarly it can be shown in view of Equation (1.54), that the energy E of a microscopic particle can only be defined by means of the gauge field φ. Also it follows from Equation (1.81) that the SI unit of the gauge field φ is $[\varphi] = \mathrm{V}$.

A comparison between Equations (1.78) and (1.81) reveals that different signs in front of the fields A_i and Φ have been chosen. This is done in order to be able to merge A_i and φ in accordance to common conventions into a single relativistically covariant quantity with four components.

At the end of this subsection it is noted that the gauge fields A_i and φ are not gauge invariant. In fact, Equation (1.62) gives

$$\delta_\epsilon(\partial_i\beta_\mathrm{R}) = -\frac{q}{\hbar}\partial_i\epsilon \,, \tag{1.82}$$

$$\delta_\epsilon(\partial_t\beta_\mathrm{R}) = -\frac{q}{\hbar}\partial_t\epsilon \,, \tag{1.83}$$

and with the definitions (1.77) and (1.80) the behavior of A_i and φ under gauge transformations are obtained as

$$\delta_\epsilon A_i = \partial_i\epsilon \,, \tag{1.84}$$

$$\delta_\epsilon\varphi = -\partial_t\epsilon \,. \tag{1.85}$$

If the combinations

$$E_i := -\partial_i\varphi - \partial_t A_i \,, \tag{1.86}$$

$$B^i := \epsilon^{ijk}\partial_j A_k \tag{1.87}$$

are formed it can easily be verified that these are invariant under gauge transformations,

$$\delta_\epsilon E_i = -\partial_i\partial_t\epsilon + \partial_t\partial_i\epsilon = 0 \,. \tag{1.88}$$

$$\delta_\epsilon B^i = -\epsilon^{ijk}\partial_j\partial_k\epsilon = 0 \,. \tag{1.89}$$

It follows from (1.86) and (1.87) that the fields E_i and B^i have the same SI units as the electric and magnetic field strength, respectively, that have been introduced in Section 1.1.2 within the axiomatic approach, $[E_i] = \mathrm{V/m}$, $[B^i] = \mathrm{Vs/m^2}$.

More important information is obtained if the integrability conditions

$$\epsilon^{ijk}\partial_j E_k = -\underbrace{\epsilon^{ijk}\partial_j\partial_k\varphi}_{=0} - \partial_t\epsilon^{ijk}\partial_j A_k$$
$$= -\partial_t B^i \quad (1.90)$$
$$\partial_i B^i = \epsilon^{ijk}\partial_i\partial_j A_k$$
$$= 0 \quad (1.91)$$

are considered. These conditions exactly resemble the *homogeneous Maxwell Equations* (1.35) and (1.30).

1.2.5 Dynamics of the Gauge Field

In the previous section the gauge fields A_i and φ were introduced in order to define parallel reference frames at different points in spacetime. The approach was general and the values of A_i and φ are still unknown. How are these values obtained?

If it is assumed that A_i and φ are physical fields it can further be assumed that they are determined by equations of motion which can be constructed according to the guidelines of classical field theory. These guidelines imply that equations of motion can often (but not always) be concisely characterized by a Lagrangian density

$$\mathcal{L} = \mathcal{L}(\Psi, \partial_i\Psi, \partial_t\Psi) \quad (1.92)$$

which, in the standard case, is a function of the fields Ψ of the theory and their first derivatives. Integration of the Lagrangian density $\mathcal{L}$ over space yields the Lagrangian L,

$$L = \int \mathcal{L}(\Psi, \partial_i\Psi, \partial_t\Psi)\,\mathrm{d}v\,, \quad (1.93)$$

and further integration over time yields the action$S$,

$$S = \int L\,\mathrm{d}t\,. \quad (1.94)$$

There are guiding principles that show how to obtain an appropriate Lagrangian density for a given theory. Once an appropriate Lagrangian density is constructed, the properties of the fields Ψ can be conveniently derived. For example, the equations of motion which determine the dynamics of Ψ follow from extremization of the action S with respect to variations of Ψ,

$$\delta_\Psi S = 0 \qquad \Longrightarrow \qquad \text{equations of motion for } \Psi. \quad (1.95)$$

More explicitly, the equations of motion are given by the well-known *Euler–Lagrange equations*

$$\partial_j\left(\frac{\partial\mathcal{L}}{\partial_j\Psi}\right) - \partial_t\left(\frac{\partial\mathcal{L}}{\partial_t\Psi}\right) - \frac{\partial\mathcal{L}}{\partial\Psi} = 0. \quad (1.96)$$

The Lagrangian density $\mathcal{L}_{\text{gauge}}$ of the gauge fields A_i and φ has to fulfill a number of requirements:

- It should have the geometric character of a scalar density in order to be a proper volume integrand (compare section A.3).
- It should be gauge invariant.
- It should be relativistically invariant.
- It should have the SI unit $[\mathcal{L}] = \mathrm{VAs/m^3}$.
- It should be no more than quadratic in the fields A_i and φ.
- It should contain no higher-order derivatives than first-order derivatives to yield second-order equations of motion.

These requirements are quite stringent. In fact, with the fields A_i, φ and the gauge invariant quantities E_i, B^i alone it is not possible to construct a proper Lagrangian density. To construct a Lagrangian density it is necessary to introduce, as in Section 1.1.3, a metric structure $g_{ij} = g_{ji}$ that characterizes the geometry of spacetime. To get the dimensions right it is further necessary to introduce two constants ε_0 and μ_0 with SI units $[\varepsilon_0] = \mathrm{As/Vm}$ and $[\mu_0] = \mathrm{Vs/Am}$, respectively. Then the only meaningful combination of A_i and φ can be written in terms of E_i and B^i. It is given by the expression[4]

$$\mathcal{L}_{\text{gauge}} = \frac{1}{2}\left(\varepsilon_0\sqrt{g}g^{ij}E_iE_j - (\mu_0\sqrt{g})^{-1}g_{ij}B^iB^j\right). \tag{1.97}$$

So far, only the gauge fields have been taken into account. The Lagrangian density in Equation (1.97) corresponds to a free gauge field theory with no coupling to electrically charged matter fields. The inclusion of matter fields requires setting up a corresponding Lagrangian density $\mathcal{L}_{\text{matter}}$. As indicated in Equation (1.92), this will involve derivatives $\partial_i\Psi$, $\partial_t\Psi$ of the fields Ψ. In order to be gauge invariant, that is to be independent of a specific choice of reference frames, these derivatives are expected from Equations (1.47) and (1.48) to involve the gauge fields. Indeed, if the relations in Equations (1.57), (1.59) and (1.77) are recalled it is possible to write

$$\begin{aligned}\partial_i\Psi = (\partial_i\Psi)^{\mathrm{R}}e_{\mathrm{R}} &= (\partial_i\Psi^{\mathrm{R}})e_{\mathrm{R}} + \Psi^{\mathrm{R}}(\partial_i e_{\mathrm{R}})\\ &= (\partial_i\Psi^{\mathrm{R}})e_{\mathrm{R}} - j\partial_i\beta_{\mathrm{R}}\Psi^{\mathrm{R}}\,e_{\mathrm{R}}\\ &= \left(\partial_i\Psi^{\mathrm{R}} + j\frac{q}{\hbar}A_i\Psi^{\mathrm{R}}\right)e_{\mathrm{R}}\\ &= D_i^A\Psi^{\mathrm{R}}\,e_{\mathrm{R}}\end{aligned} \tag{1.98}$$

with the gauge covariant derivative

$$D_i^A := \partial_i + j\frac{q}{\hbar}A_i\,. \tag{1.99}$$

Analogously, it follows with Equation (1.80)

$$\begin{aligned}\partial_t\Psi = (\partial_t\Psi)^{\mathrm{R}}e_{\mathrm{R}} &= (\partial_t\Psi^{\mathrm{R}})e_{\mathrm{R}} + \Psi^{\mathrm{R}}(\partial_t e_{\mathrm{R}})\\ &= (\partial_t\Psi^{\mathrm{R}})e_{\mathrm{R}} - j\partial_t\beta_{\mathrm{R}}\Psi^{\mathrm{R}}\,e_{\mathrm{R}}\\ &= \left(\partial_t\Psi^{\mathrm{R}} - j\frac{q}{\hbar}\varphi\Psi^{\mathrm{R}}\right)e_{\mathrm{R}}\\ &= D_t^{\varphi}\Psi^{\mathrm{R}}\,e_{\mathrm{R}}\end{aligned} \tag{1.100}$$

[4] The factor $1/2$ is introduced to yield the correct Hamilton function (energy function) which can be obtained from the Lagrangian density by means of a Legendre transformation.

with the gauge covariant derivative

$$D_t^{\varphi} := \partial_t - j\frac{q}{\hbar}\varphi\,. \tag{1.101}$$

Therefore, in order to be gauge invariant, the matter Lagrangian density may only contain derivatives and gauge fields as combinations of gauge covariant derivatives. To explicitly obtain the matter Lagrangian density requires advanced knowledge of relativistic quantum mechanics. With reference to literature [18, 19] it is quoted as a result that the Lagrangian density of a certain class of electrically charged matter fields, like electrons, is given by[5]

$$\mathcal{L}_{\text{matter}} = -j\hbar c\overline{\Psi}^{\text{R}}\gamma^i\left(D_i^A - \frac{mc}{\hbar}\right)\Psi^{\text{R}} + j\hbar\overline{\Psi}^{\text{R}}\gamma^0\left(D_t^{\varphi} - \frac{mc^2}{\hbar}\right)\Psi^{\text{R}}\,. \tag{1.102}$$

The terms that couple the gauge fields to the matter fields can be written as

$$\mathcal{L}_{\text{coupling}} = -A_i J^i - \varphi\rho \tag{1.103}$$

with the definitions

$$J^i := qc\overline{\Psi}^{\text{R}}\gamma^i\Psi^{\text{R}}, \tag{1.104}$$

$$\rho := q\overline{\Psi}^{\text{R}}\gamma^0\Psi^{\text{R}}. \tag{1.105}$$

Then the dynamics of the gauge fields A_i and φ is determined from the combined Lagrangian density

$$\mathcal{L}_{\text{em}} = \mathcal{L}_{\text{gauge}} + \mathcal{L}_{\text{coupling}}\,. \tag{1.106}$$

The equations of motion that follow from this Lagrangian density are given by the Euler–Lagrange Equations (1.96). In the present case they acquire the form

$$\partial_j\left(\frac{\partial\mathcal{L}_{\text{em}}}{\partial_j A_i}\right) - \partial_t\left(\frac{\partial\mathcal{L}_{\text{em}}}{\partial_t A_i}\right) - \frac{\partial\mathcal{L}_{\text{em}}}{\partial A_i} = 0\,, \tag{1.107}$$

$$\partial_j\left(\frac{\partial\mathcal{L}_{\text{em}}}{\partial_j \varphi}\right) - \partial_t\left(\frac{\partial\mathcal{L}_{\text{em}}}{\partial_t \varphi}\right) - \frac{\partial\mathcal{L}_{\text{em}}}{\partial \varphi} = 0\,. \tag{1.108}$$

Inserting Equations (1.97) and (1.103) into Equation (1.106) yields

$$\epsilon^{ijk}\partial_j H_k - \partial_t D^i = J^i\,, \tag{1.109}$$

$$\partial_i D^i = \rho\,, \tag{1.110}$$

with the definitions

$$H_k := (\mu_0\sqrt{g})^{-1} g_{kl}\, B^l\,, \tag{1.111}$$

$$D^i := \varepsilon_0\sqrt{g}\, g^{ij}\, E_j\,. \tag{1.112}$$

Equations (1.109) and (1.110) are recognized as the inhomogeneous Maxwell Equations (1.16) and (1.5), respectively. It follows that the gauge fields A_i and φ, that have been introduced in the formal context of reference frames, do indeed constitute the electromagnetic potentials that are

[5] In this expression Ψ^{R} denotes a four-component spinor, $\overline{\Psi}^{\text{R}}$ is the adjoint spinor of Ψ^{R}, and γ^0, γ^i are 4×4 matrices [19].

familiar from classical electrodynamics. In this framework the homogeneous Maxwell equations turn out to be mathematical integrability conditions while the inhomogeneous Maxwell equations represent the equations of motion of the gauge fields A_i and φ.

1.3 The Relation Between the Axiomatic Approach and the Gauge Field Approach

In the following text some comments are made on the interrelation between the previously presented axiomatic approach and the gauge field approach. It is interesting to see how the axioms find their proper place within the gauge approach.

1.3.1 Noether Theorem and Electric Charge Conservation

In field theory there is a famous result which connects symmetries of laws of nature to conserved quantities. This is the Noether theorem [20] which has been proven to be of great importance in both classical and quantum contexts. It is, in particular, discussed in books on classical electrodynamics, (see [7, 21] for example). The Noether theorem connects the symmetry of a Lagrangian density $\mathcal{L}(\Psi, \partial_i\Psi, \partial_t\Psi)$, compare Equation (1.93), to conserved quantities. Suppose, for example, that $\mathcal{L}$ is invariant under time translations δ_t. From daily experience this assumption seems plausible since it is not expected that the laws of nature change in time. Then the Noether theorem implies a local conservation law which expresses the conservation of energy. Similarly, invariance under translations δ_{x^i} in space implies conservation of momentum, while invariance under rotations $\delta_{\omega_i{}^j}$ yields the conservation of angular momentum,

$$\delta_t\mathcal{L} = 0 \qquad \Longrightarrow \qquad \text{conservation of energy}\,, \tag{1.113}$$

$$\delta_{x^i}\mathcal{L} = 0 \qquad \Longrightarrow \qquad \text{conservation of momentum}\,, \tag{1.114}$$

$$\delta_{\omega_i{}^j}\mathcal{L} = 0 \qquad \Longrightarrow \qquad \text{conservation of angular momentum}\,. \tag{1.115}$$

These symmetries of spacetime are called *external* symmetries, but the Noether theorem also works for other types of symmetries that are called *internal* ones. Gauge symmetries often are internal symmetries. In this case, gauge invariance of the Lagrangian implies a conserved current with an associated charge. That is, if a gauge transformation is denoted by δ_ϵ it can be concluded that

$$\delta_\epsilon\mathcal{L} = 0 \qquad \Longrightarrow \qquad \text{charge conservation}\,. \tag{1.116}$$

If this conclusion is applied to electrodynamics it is necessary to specify the Lagrangian density to be the one of matter fields that represent electrically charged particles. Then, invariance of this Lagrangian density under the gauge symmetry of electrodynamics yields the conservation of electric charge. Thus, if the validity of the Lagrangian formalism is accepted, electric charge conservation is obtained from gauge invariance via the Noether theorem.

1.3.2 Minimal Coupling and the Lorentz Force

It has already been mentioned that, according to Equation (1.95), the equations of motion (1.96) of a physical theory can be derived from a Lagrangian density and its associated action.

This scheme can be used to derive the equations of motion of electrically charged particles. In this case, the corresponding Lagrangian density (that of the electrically charged particles) has to be gauge invariant.

For the electromagnetic case it has been demonstrated that the Lagrangian density will be gauge invariant if partial derivatives are converted to gauge covariant derivatives (1.99) and (1.101) according to

$$\partial_i \longrightarrow D_i^A := \partial_i + j\tfrac{q}{\hbar}A_i\,, \tag{1.117}$$
$$\partial_t \longrightarrow D_t^\varphi := \partial_t - j\tfrac{q}{\hbar}\varphi\,, \tag{1.118}$$

with q the electric charge of the particle under consideration. The substitutions of Equations (1.117) and (1.118) constitute the simplest way to ensure gauge invariance of the Lagrangian density of electrically charged particles. They constitute what is commonly called *minimal coupling*. Due to minimal coupling, electrically charged particles and the electromagnetic field are related in a natural way that is dictated by the requirement of gauge invariance.

If it is assumed, as in Section 1.2.2, that

$$\Psi^{\mathrm{R}} = \Psi_0\, \mathrm{e}^{-\frac{j}{\hbar}(p_i x^i - Et)}\,, \tag{1.119}$$

it follows that

$$j\hbar\partial_i \Psi^{\mathrm{R}} = p_i\, \Psi^{\mathrm{R}}, \tag{1.120}$$
$$-j\hbar\partial_t \Psi^{\mathrm{R}} = E\, \Psi^{\mathrm{R}}. \tag{1.121}$$

These are relations that indicate how to pass from quantum physics to classical physics, that is how to pass from the action of the differential operators $j\hbar\partial_i$ and $-j\hbar\partial_t$ on a wave function to the momentum and energy of a classical particle. It follows that the classical analogues of Equations (1.117) and (1.118) are given by

$$p_i \longrightarrow p_i - qA_i\,, \tag{1.122}$$
$$E \longrightarrow E - q\varphi\,. \tag{1.123}$$

That is, if electrically charged particles are represented by classical particles, rather than by wave functions, it is necessary to replace within the corresponding classical Lagrangian function the energy E and the momentum p_i of each particle according to Equations (1.122) and (1.123).

As a general example a nonrelativistic classical particle with mass m and charge q can be considered. In the absence of an electromagnetic field[6] the energy and momentum of the particle are related by

$$E = \frac{p_i p^i}{2m}\,. \tag{1.124}$$

In the presence of an electromagnetic field the replacements in Equations (1.122) and (1.123) lead to

$$E = \frac{(p_i - qA_i)(p^i - qA^i)}{2m} + q\varphi\,. \tag{1.125}$$

[6] Also the absence of a gravitational field is assumed.

It follows from this expression for the energy E that the Lagrange function L of the particle is given by [22, p. 167]

$$L = \frac{m}{2}(\partial_t x_i)(\partial_t x^i) + qA_i(\partial_t x^i) - q\varphi\,. \tag{1.126}$$

The equation of motion is obtained from the Lagrange function via

$$\frac{\mathrm{d}}{\mathrm{d}t}\frac{\partial L}{\partial(\partial_t x^i)} - \frac{\partial L}{\partial x^i} = 0\,. \tag{1.127}$$

This yields with Equation (1.126)

$$m(\partial_t^2 x_i) + q\big(\partial_t A_i + \partial_t x^j(\partial_j A_i)\big) - q\,\partial_t x^j(\partial_i A_j) + q\,\partial_i\varphi = 0\,. \tag{1.128}$$

The first term represents the force $F_i = m(\partial_t^2 x_i)$ that acts on the particle. Rearranging the other terms and noting, in particular, the identity

$$\partial_i A_j - \partial_j A_i = \epsilon_{ijk}\epsilon^{klm}\,\partial_l A_m \tag{1.129}$$

yields

$$F_i = q(-\partial_i\varphi - \partial_t A_i) + q(\epsilon_{ijk}\partial_t x^j\epsilon^{klm}\,\partial_l A_m)\,. \tag{1.130}$$

Finally, by means of Equations (1.86) and (1.87) the gauge fields A_i and φ are replaced by the field strengths E_i and B^i. The result is the Lorentz force law (1.17),

$$F_i = q(E_i + \epsilon_{ijk}\partial_t x^j B^k)\,. \tag{1.131}$$

Therefore, in the gauge field approach the Lorentz force is a consequence of the minimal coupling procedure which couples electrically charged particles to the electromagnetic potentials.

1.3.3 Bianchi Identity and Magnetic Flux Conservation

The electromagnetic gauge fields A_i and φ are often introduced as mathematical tools to facilitate the integration of the Maxwell equations. Indeed, Equations (1.90) and (1.91) have revealed that the homogeneous Maxwell equations reduce to mere integrability conditions that are automatically fulfilled if the electromagnetic field strengths are expressed in terms of the gauge potentials. This is an interesting observation since within the gauge approach the gauge potentials are fundamental physical quantities and are not only the outcome of a mathematical trick. Thus it can be stated that the mathematical structure of the gauge potentials already implies the homogeneous Maxwell equations and, in turn, magnetic flux conservation. In this light, magnetic flux conservation, within the gauge approach, appears as the consequence of a geometric identity. This is in contrast to electric charge conservation that can be viewed as the consequence of gauge invariance, that is via the Noether theorem as the consequence of a physical symmetry.

The integrability conditions that are reflected in the homogeneous Maxwell equations are special cases of *Bianchi identities*. Bianchi identities are the result of differentiating a potential twice. For example, in electrostatics the electric field strength E_i can be derived from a scalar potential φ according to

$$E_i = -\partial_i\varphi\,. \tag{1.132}$$

Differentiation reveals that the curl of E_i vanishes,

$$\epsilon^{ijk}\partial_j E_k = \epsilon^{ijk}\partial_j\partial_k\varphi = 0\,, \tag{1.133}$$

which is due to the antisymmetry of ϵ^{ijk}. Again, this equation is a mathematical identity, a simple example of a Bianchi identity.

1.3.4 Gauge Approach and Constitutive Relations

The gauge approach towards electrodynamics defines the properties of gauge fields, which represent the electromagnetic field, and depends on matter fields. However, it does not depend on a metric structure of spacetime. In contrast to this, the constitutive relations do depend on a metric structure of spacetime, as can already be seen from the constitutive relations of vacuum

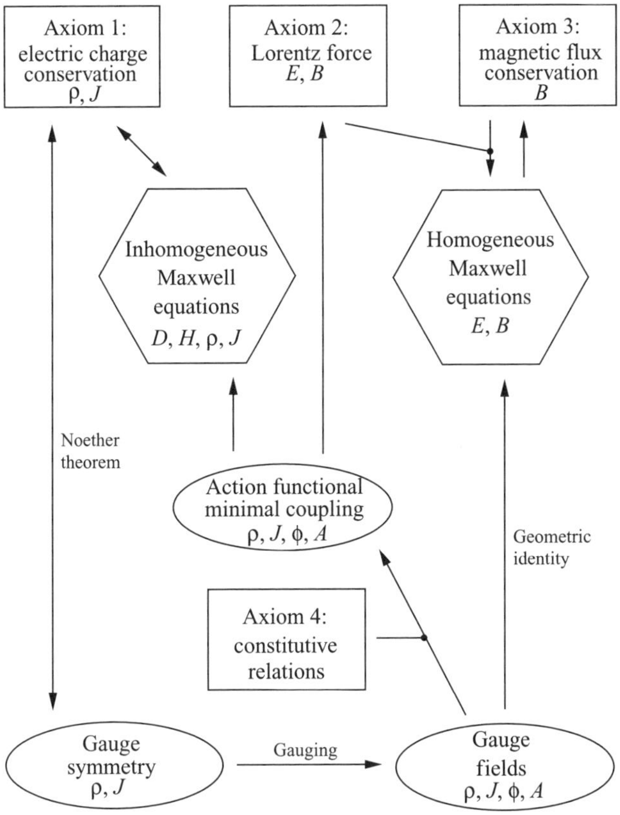

Figure 1.6 Interrelations between the axiomatic approach (rectangular frames) and the gauge field approach (elliptic frames) of classical electrodynamics. Both approaches yield the Maxwell equations. The gauge approach requires the knowledge of constitutive relations which represent the fourth axiom of the axiomatic approach. However, the first, second and third axiom of the axiomatic approach can be obtained from the gauge approach. Electric charge conservation, the first axiom of the gauge approach, represents the gauge symmetry of electrodynamics by means of the Noether theorem.

that involve the metric g_{ij}, compare Equations (1.36) and (1.37). Thus, also in the gauge approach the constitutive relations have to be postulated as an axiom in some way. One should note that, according to Equations (1.86) and (1.87), the gauge potentials are directly related to the field strengths E_i and B^i. The excitations D^i and H_i are part of the inhomogeneous Maxwell equations which, within the gauge approach, are derived as equations of motion from an action principle (see Equations (1.109) and (1.110)). Since the action itself involves the gauge potentials, one might wonder how it is possible to obtain equations of motion for the excitations rather than for the field strengths. The answer is that during the construction of the Lagrangian density (1.97) from the gauge potentials the constitutive relations are implicitly used. Figure 1.6 summarizes the interrelations between the axiomatic approach and the gauge approach.

1.4 Solutions of Maxwell Equations

From the axiomatic approach and from the gauge field approach to classical electrodynamics the Maxwell Equations (1.5), (1.16), (1.30) and (1.35) have been obtained. Appropriate constitutive relations of the form (1.36) and (1.37) or, more generally, (1.38) and (1.39) make Maxwell equations a set of determined partial differential equations. Within the limits of classical physics these equations model the interaction between electromagnetic sources ρ, J^i and the electromagnetic field, represented by (E_i, B^i) and (D^i, H_i). In order to explicitly formulate a specific problem it is necessary to impose physically meaningful initial and boundary conditions that lead to a well-defined boundary value problem. The solution of such a boundary value problem, in turn, determines a unique solution of Maxwell equations.

For the solution of an electromagnetic boundary value problem it is often advantageous to first rewrite Maxwell equations as (second-order) wave equations. This is straightforward as long as the constitutive relations are of a simple form.[7] Clearly, the solution of wave equations has been studied in many branches of physics and mathematics for a long time and a variety of corresponding solution procedures exist [23].

In the following, the tensor notation of the previous sections will no longer be used and it will be replaced by the more conventional vector notation of ordinary vector analysis that has been adopted by many authors of standard textbooks (see for example [1, 5, 6, 8, 24]). The widespread use of this notation is the reason it will be adopted here as well. Hence, covariant vectors E_i, H_i are replaced by ordinary vectors in three-dimensional space, $E_i \to \mathbf{E}$, $H_i \to \mathbf{H}$. Also the contravariant vector densities B^i, D^i and J^i are replaced by ordinary vectors, $B^i \to \mathbf{B}$, $D^i \to \mathbf{D}$ and $J^i \to \mathbf{J}$, while the notation for the scalar density ρ remains unchanged. It is obvious that due to this transition information on the geometric properties of the electromagnetic quantities is lost. For example, integration along lines or over surfaces is no longer defined in a natural way since line integrals $\int_c \mathbf{E} \cdot \mathrm{d}\boldsymbol{t}$ or surface integrals $\int_S \mathbf{B} \cdot \mathrm{d}\boldsymbol{a}$ now require a metric structure of space which is represented by the scalar product. Formally, also the expressions $\int_c \mathbf{D} \cdot \mathrm{d}\boldsymbol{t}$ or $\int_S \mathbf{H} \cdot \mathrm{d}\boldsymbol{a}$ can be formed but the physical interpretation of these integrals is not clear and certainly needs explanation since the electric excitation $\mathbf{D}$ or the magnetic excitation $\mathbf{H}$ are not natural integrands of line- or surface-integrals, respectively.

[7] Constitutive relations are usually *not* of a simple form if they introduce material parameters that are space or time dependent, if they mix electric and magnetic fields, or if they introduce nonlinearities, for example. In such cases analytic solutions of Maxwell equations often cannot be found and it is necessary to apply numerical methods directly to the Maxwell equations.

Table 1.1 Different fonts used to distinguish between the basic electromagnetic quantities in the time domain, frequency domain and reciprocal space

	Time domain	Frequency domain	Reciprocal space
Electric excitation	$\mathbf{D}(\boldsymbol{r}, t)$	$\boldsymbol{D}(\boldsymbol{r}, \omega)$	$\mathsf{D}(k, t)$
Magnetic excitation	$\mathbf{H}(\boldsymbol{r}, t)$	$\boldsymbol{H}(\boldsymbol{r}, \omega)$	$\mathsf{H}(k, t)$
Electric field strength	$\mathbf{E}(\boldsymbol{r}, t)$	$\boldsymbol{E}(\boldsymbol{r}, \omega)$	$\mathsf{E}(k, t)$
Magnetic field strength	$\mathbf{B}(\boldsymbol{r}, t)$	$\boldsymbol{B}(\boldsymbol{r}, \omega)$	$\mathsf{B}(k, t)$
Charge density	$\rho(\boldsymbol{r}, t)$	$\rho(\boldsymbol{r}, \omega)$	$\rho(k, t)$
Current density	$\mathbf{J}(\boldsymbol{r}, t)$	$\boldsymbol{J}(\boldsymbol{r}, \omega)$	$\mathsf{J}(k, t)$
Vector potential	$\mathbf{A}(\boldsymbol{r}, t)$	$\boldsymbol{A}(\boldsymbol{r}, \omega)$	$\mathsf{A}(k, t)$
Scalar potential	$\varphi(\boldsymbol{r}, t)$	$\varphi(\boldsymbol{r}, \omega)$	$\varphi(k, t)$

This indicates that it is important to keep the limitations of the vector notation in mind in order to construct mathematical expressions of physical objects in a meaningful way.

In the following text, electromagnetic quantities that are defined in the time domain will be considered, but these will quickly be transformed to the frequency domain (see Section 1.4.1.3) and to reciprocal space (see Section 1.4.1.4). It is convenient to print the corresponding Fourier transforms using different fonts. Then it is not necessary to always keep the arguments $\boldsymbol{r}$, t, ω or k in parentheses next to the symbols of the fields to indicate which domain or space they belong to. The letters of the different fonts are introduced in Table 1.1. Greek letters remain unaltered but this should not lead to confusion since usually they do not appear isolated in an equation.

1.4.1 Wave Equations

1.4.1.1 Decoupling of Maxwell Equations

Rewriting Maxwell Equations (1.5), (1.16), (1.30) and (1.35) in vector notation yields the familiar expressions

$$\nabla \cdot \mathbf{D}(\boldsymbol{r}, t) = \rho(\boldsymbol{r}, t) , \tag{1.134}$$

$$\nabla \times \mathbf{H}(\boldsymbol{r}, t) - \frac{\partial \mathbf{D}}{\partial t}(\boldsymbol{r}, t) = \mathbf{J}(\boldsymbol{r}, t) , \tag{1.135}$$

$$\nabla \cdot \mathbf{B}(\boldsymbol{r}, t) = 0 , \tag{1.136}$$

$$\nabla \times \mathbf{E}(\boldsymbol{r}, t) + \frac{\partial \mathbf{B}}{\partial t}(\boldsymbol{r}, t) = \mathbf{0} . \tag{1.137}$$

To decouple the Maxwell equations, constitutive relations are assumed which characterize a homogeneous, isotropic medium. These are of the form

$$\mathbf{D}(\boldsymbol{r}, t) = \varepsilon \mathbf{E}(\boldsymbol{r}, t) , \tag{1.138}$$

$$\mathbf{B}(\boldsymbol{r}, t) = \mu \mathbf{H}(\boldsymbol{r}, t) , \tag{1.139}$$

with constant parameters ε and μ. Then the curl operator $\nabla\times$ is applied to Equations (1.135) and (1.137) and the results are combined. This yields

$$\nabla \times \nabla \times \mathbf{E}(\boldsymbol{r},t) + \varepsilon\mu \frac{\partial^2}{\partial t^2}\mathbf{E}(\boldsymbol{r},t) = -\mu \frac{\partial \mathbf{J}}{\partial t}(\boldsymbol{r},t)\,, \tag{1.140}$$

$$\nabla \times \nabla \times \mathbf{B}(\boldsymbol{r},t) + \varepsilon\mu \frac{\partial^2}{\partial t^2}\mathbf{B}(\boldsymbol{r},t) = \mu \nabla \times \mathbf{J}(\boldsymbol{r},t)\,. \tag{1.141}$$

These equations can be transformed into standard wave equations if the identity (C.6) is used, together with the constitutive relations (1.138) and (1.139) and the Maxwell Equations (1.134) and (1.136). This yields

$$\Delta\mathbf{E}(\boldsymbol{r},t) - \varepsilon\mu \frac{\partial^2 \mathbf{E}}{\partial t^2}(\boldsymbol{r},t) = \frac{1}{\varepsilon}\nabla\rho(\boldsymbol{r},t) + \mu\frac{\partial \mathbf{J}}{\partial t}(\boldsymbol{r},t)\,, \tag{1.142}$$

$$\Delta\mathbf{B}(\boldsymbol{r},t) - \varepsilon\mu \frac{\partial^2 \mathbf{B}}{\partial t^2}(\boldsymbol{r},t) = -\mu\nabla \times \mathbf{J}(\boldsymbol{r},t)\,. \tag{1.143}$$

Due to the constitutive relations (1.138) and (1.139) two analogous equations are valid for $\mathbf{D}(\boldsymbol{r},t)$ and $\mathbf{H}(\boldsymbol{r},t)$ which furnish no additional information. Therefore, six scalar equations are obtained for six unknown field components. Equations (1.142) and (1.143) constitute inhomogeneous wave equations[8] with phase velocity

$$c = \frac{1}{\sqrt{\varepsilon\mu}}\,. \tag{1.144}$$

In a vacuum ($\rho(\boldsymbol{r},t) = 0,\ \mathbf{J}(\boldsymbol{r},t) = \mathbf{0}$) the inhomogeneous terms on the right-hand sides vanish. Then the homogeneous wave equations

$$\Delta\mathbf{E}(\boldsymbol{r},t) - \frac{1}{c^2}\frac{\partial^2 \mathbf{E}}{\partial t^2}(\boldsymbol{r},t) = \mathbf{0}\,, \tag{1.145}$$

$$\Delta\mathbf{B}(\boldsymbol{r},t) - \frac{1}{c^2}\frac{\partial^2 \mathbf{B}}{\partial t^2}(\boldsymbol{r},t) = \mathbf{0}\,. \tag{1.146}$$

are obtained

1.4.1.2 Equations of Motion for the Electromagnetic Potentials

It has already been seen that the field strengths $\mathbf{E}(\boldsymbol{r},t)$ and $\mathbf{B}(\boldsymbol{r},t)$ can be derived from the scalar potential $\varphi(\boldsymbol{r},t)$ and the vector potential $\mathbf{A}(\boldsymbol{r},t)$ via

$$\mathbf{E}(\boldsymbol{r},t) = -\nabla\varphi(\boldsymbol{r},t) - \frac{\partial \mathbf{A}}{\partial t}(\boldsymbol{r},t)\,, \tag{1.147}$$

$$\mathbf{B}(\boldsymbol{r},t) = \nabla \times \mathbf{A}(\boldsymbol{r},t)\,. \tag{1.148}$$

If the electromagnetic field is expressed by means of $\varphi(\boldsymbol{r},t)$ and $\mathbf{A}(\boldsymbol{r},t)$ the homogeneous Maxwell Equations (1.136) and (1.137) are recognized as geometric identities which are

[8] Since in Equations (1.142) and (1.143) the wave operator $\Delta - \varepsilon\mu\,\partial^2/\partial t^2$ acts on three-dimensional vectors it is clear that these equations are *vector wave equations*. However, equations of the type (1.140) and (1.141) are often also called vector wave equations, even though the differential operator $-(\nabla\times\nabla\times) - \varepsilon\mu\,\partial^2/\partial t^2$ does not always coincide with the wave operator. It does coincide with the wave operator if it acts on a divergence free vector field.

automatically fulfilled. Then the remaining inhomogeneous Maxwell Equations (1.134) and (1.135) determine the electromagnetic field. Replacing within the inhomogeneous Maxwell equations the excitations $\mathbf{D}(\boldsymbol{r},t)$, $\mathbf{H}(\boldsymbol{r},t)$ by means of the constitutive relations (1.138) and (1.139) and Equations (1.147) and (1.148) by $\varphi(\boldsymbol{r},t)$ and $\mathbf{A}(\boldsymbol{r},t)$ yields

$$\Delta\varphi(\boldsymbol{r},t) + \frac{\partial(\nabla\cdot\mathbf{A}(\boldsymbol{r},t))}{\partial t} = -\frac{\rho(\boldsymbol{r},t)}{\varepsilon}, \tag{1.149}$$

$$\Delta\mathbf{A}(\boldsymbol{r},t) - \frac{1}{c^2}\frac{\partial^2\mathbf{A}(\boldsymbol{r},t)}{\partial t^2} - \nabla\left(\nabla\cdot\mathbf{A}(\boldsymbol{r},t) + \frac{1}{c^2}\frac{\partial\varphi(\boldsymbol{r},t)}{\partial t}\right) = -\mu\mathbf{J}(\boldsymbol{r},t). \tag{1.150}$$

These are four scalar equations for the four unknown field components $\varphi(\boldsymbol{r},t)$ and $\mathbf{A}(\boldsymbol{r},t)$. Since electrodynamics is invariant under the gauge transformations

$$\delta_\epsilon\varphi(\boldsymbol{r},t) = -\frac{\partial\epsilon(\boldsymbol{r},t)}{\partial t}, \tag{1.151}$$

$$\delta_\epsilon\mathbf{A}(\boldsymbol{r},t) = \mathbf{A}'(\boldsymbol{r},t) - \mathbf{A}(\boldsymbol{r},t) = \nabla\epsilon(\boldsymbol{r},t), \tag{1.152}$$

with an arbitrary function $\epsilon(\boldsymbol{r},t)$ it is possible to simplify Equations (1.149) and (1.150) by the choice of a particular gauge. Common gauges are the Coulomb gauge

$$\nabla\cdot\mathbf{A}(\boldsymbol{r},t) = 0 \tag{1.153}$$

and the Lorenz gauge

$$\nabla\cdot\mathbf{A}(\boldsymbol{r},t) + \frac{1}{c^2}\frac{\partial\varphi(\boldsymbol{r},t)}{\partial t} = 0. \tag{1.154}$$

The Coulomb gauge leads to

$$\Delta\varphi(\boldsymbol{r},t) = -\frac{\rho(\boldsymbol{r},t)}{\varepsilon}, \quad \text{(Coulomb gauge)} \tag{1.155}$$

$$\Delta\mathbf{A}(\boldsymbol{r},t) - \frac{1}{c^2}\frac{\partial^2\mathbf{A}(\boldsymbol{r},t)}{\partial t^2} - \frac{1}{c^2}\frac{\partial(\nabla\varphi(\boldsymbol{r},t))}{\partial t} = -\mu\mathbf{J}(\boldsymbol{r},t), \quad \text{(Coulomb gauge)} \tag{1.156}$$

while the Lorenz gauge yields a scalar and a vector wave equation,

$$\Delta\varphi(\boldsymbol{r},t) - \frac{1}{c^2}\frac{\partial^2\varphi(\boldsymbol{r},t)}{\partial t^2} = -\frac{\rho(\boldsymbol{r},t)}{\varepsilon}, \quad \text{(Lorenz gauge)} \tag{1.157}$$

$$\Delta\mathbf{A}(\boldsymbol{r},t) - \frac{1}{c^2}\frac{\partial^2\mathbf{A}(\boldsymbol{r},t)}{\partial t^2} = -\mu\mathbf{J}(\boldsymbol{r},t). \quad \text{(Lorenz gauge)} \tag{1.158}$$

1.4.1.3 Maxwell Equations in the Frequency Domain and Helmholtz Equations

Time harmonic fields with sinusoidal time dependency can be expressed as

$$\mathbf{F}_{\text{sinus}}(\boldsymbol{r},t) = \text{Re}\left[\boldsymbol{F}(\boldsymbol{r},\omega)\mathrm{e}^{j\omega t}\right]. \tag{1.159}$$

This is a special case of the Fourier representation of a field with arbitrary time dependency,

$$\mathbf{F}(\boldsymbol{r},t) = \text{Re}\left[\frac{1}{\sqrt{2\pi}}\int_{-\infty}^{\infty}\boldsymbol{F}(\boldsymbol{r},\omega)\mathrm{e}^{j\omega t}\,\mathrm{d}\omega\right]. \tag{1.160}$$

In the time harmonic case it is straightforward to pass to the frequency domain and to write the Maxwell equations as

$$\nabla \cdot \boldsymbol{D}(\boldsymbol{r}, \omega) = \rho(\boldsymbol{r}, \omega), \tag{1.161}$$

$$\nabla \times \boldsymbol{H}(\boldsymbol{r}, \omega) - j\omega \boldsymbol{D}(\boldsymbol{r}, \omega) = \boldsymbol{J}(\boldsymbol{r}, \omega), \tag{1.162}$$

$$\nabla \cdot \boldsymbol{B}(\boldsymbol{r}, \omega) = 0, \tag{1.163}$$

$$\nabla \times \boldsymbol{E}(\boldsymbol{r}, \omega) + j\omega \boldsymbol{B}(\boldsymbol{r}, \omega) = \mathbf{0}. \tag{1.164}$$

The vector wave Equations (1.140) and (1.142) of the electric field **E**, for example, convert in the frequency domain to vector *Helmholtz equations*

$$\nabla \times \nabla \times \boldsymbol{E}(\boldsymbol{r}, \omega) - k^2 \boldsymbol{E}(\boldsymbol{r}, \omega) = -j\omega\mu \boldsymbol{J}(\boldsymbol{r}, \omega), \tag{1.165}$$

$$\Delta \boldsymbol{E}(\boldsymbol{r}, \omega) + k^2 \boldsymbol{E}(\boldsymbol{r}, \omega) = \frac{1}{\varepsilon} \nabla \rho(\boldsymbol{r}, \omega) + j\omega\mu \boldsymbol{J}(\boldsymbol{r}, \omega), \tag{1.166}$$

with $k = \omega / c$ the wave number. For the magnetic field **B** this yields

$$\nabla \times \nabla \times \boldsymbol{B}(\boldsymbol{r}, \omega) - k^2 \boldsymbol{B}(\boldsymbol{r}, \omega) = \mu \nabla \times \boldsymbol{J}(\boldsymbol{r}, \omega), \tag{1.167}$$

$$\Delta \mathbf{B}(\boldsymbol{r}, \omega) + k^2 \boldsymbol{B}(\boldsymbol{r}, \omega) = -\mu \nabla \times \boldsymbol{J}(\boldsymbol{r}, \omega). \tag{1.168}$$

The Helmholtz equations for the scalar and vector potential in the Lorenz gauge assume the form

$$\Delta \varphi(\boldsymbol{r}, \omega) + k^2 \varphi(\boldsymbol{r}, t) = -\frac{\rho(\boldsymbol{r}, \omega)}{\varepsilon}, \quad \text{(Lorenz gauge)} \tag{1.169}$$

$$\Delta \boldsymbol{A}(\boldsymbol{r}, \omega) + k^2 \boldsymbol{A}(\boldsymbol{r}, \omega) = -\mu \boldsymbol{J}(\boldsymbol{r}, \omega). \quad \text{(Lorenz gauge)} \tag{1.170}$$

If compared to Equations (1.166) and (1.168) these equations are more simple since now the source terms involve no derivatives.

1.4.1.4 Maxwell Equations in Reciprocal Space

Fields $\mathbf{F}(\boldsymbol{r}, t)$ that are defined in the time domain are transformed to reciprocal space by a spatial Fourier transform according to

$$\mathsf{F}(\boldsymbol{k}, t) = \frac{1}{(2\pi)^{3/2}} \int \mathbf{F}(\boldsymbol{r}, t) \mathrm{e}^{j\boldsymbol{k}\cdot\boldsymbol{r}} \, \mathrm{d}^3 r\,. \tag{1.171}$$

The inverse transform is given by

$$\mathbf{F}(\boldsymbol{r}, t) = \frac{1}{(2\pi)^{3/2}} \int \mathsf{F}(\boldsymbol{k}, t) \mathrm{e}^{-j\boldsymbol{k}\cdot\boldsymbol{r}} \, \mathrm{d}^3 k\,. \tag{1.172}$$

The operator ∇ transforms to multiplication by $-j\boldsymbol{k}$ in reciprocal space. Therefore, the Maxwell equations in reciprocal space become

$$-j\boldsymbol{k} \cdot \mathsf{D}(\boldsymbol{k}, t) = \rho(\boldsymbol{k}, t), \tag{1.173}$$

$$-j\boldsymbol{k} \times \mathsf{H}(\boldsymbol{k}, t) - \frac{\partial \mathsf{D}(\boldsymbol{k}, t)}{\partial t} = \mathsf{J}(\boldsymbol{k}, t), \tag{1.174}$$

$$-j\boldsymbol{k} \cdot \mathsf{B}(\boldsymbol{k}, t) = 0, \tag{1.175}$$

$$-j\boldsymbol{k} \times \mathsf{E}(\boldsymbol{k}, t) + \frac{\partial \mathsf{B}(\boldsymbol{k}, t)}{\partial t} = \mathsf{0}. \tag{1.176}$$

This representation of the Maxwell equations has the advantage that the fields and time derivatives all depend on the *same* point $\boldsymbol{k}$ in reciprocal space. Hence, the partial differential equations of real space become strictly local equations in reciprocal space.

1.4.2 Boundary Conditions at Interfaces

At the transition between two media with parameters $(\varepsilon_1, \mu_1, \sigma_1)$ and $(\varepsilon_2, \mu_2, \sigma_2)$ the boundary conditions can be derived from the Maxwell equations by means of integration and the application of Stokes' theorem. This is a standard procedure which is described in many textbooks (for example [25, Section 7.3.6.]). This yields

$$\nabla \times \mathbf{E} + \frac{\partial \mathbf{B}}{\partial t} = \mathbf{0} \quad \Longrightarrow \quad (\mathbf{E}_1 - \mathbf{E}_2) \times \boldsymbol{e}_n = \mathbf{0}\,, \tag{1.177}$$

$$\nabla \times \mathbf{H} - \frac{\partial \mathbf{D}}{\partial t} = \mathbf{J} \quad \Longrightarrow \quad (\mathbf{H}_1 - \mathbf{H}_2) \times \boldsymbol{e}_n = \boldsymbol{J}_{\mathrm{s}}\,, \tag{1.178}$$

$$\nabla \cdot \mathbf{B} = 0 \quad \Longrightarrow \quad (\mathbf{B}_2 - \mathbf{B}_1) \cdot \boldsymbol{e}_n = 0\,, \tag{1.179}$$

$$\nabla \cdot \mathbf{D} = \rho \quad \Longrightarrow \quad (\mathbf{D}_2 - \mathbf{D}_1) \cdot \boldsymbol{e}_n = \rho_{\mathrm{s}}\,. \tag{1.180}$$

These boundary conditions are valid both in the time and the frequency domain. The vector $\boldsymbol{e}_n$ denotes a normal unit vector that points on the interface between the different media from medium 1 to medium 2 and $\boldsymbol{J}_{\mathrm{s}}$ denotes a surface current that may flow on the interface between the media. Accordingly, ρ_{s} denotes a surface charge density.

Additionally, there is a boundary condition for the magnetic vector potential $\mathbf{A}$:

$$\mathbf{B} = \nabla \times \mathbf{A} \quad \Longrightarrow \quad (\mathbf{A}_1 - \mathbf{A}_2) \times \boldsymbol{e}_n = \mathbf{0}\,. \tag{1.181}$$

This boundary condition is gauge invariant since it is based on the rotational part of the magnetic vector potential.

1.4.3 Dynamical and Nondynamical Components of the Electromagnetic Field

1.4.3.1 Helmholtz's Vector Theorem, Longitudinal and Transverse Fields

In the study of vector fields $\mathbf{F}$ the *Helmholtz's vector theorem* is a useful tool [23]. It states that any vector field $\mathbf{F}$, which is finite, uniform, continuous and square integrable, may be split into a *longitudinal* or *irrotational* part $\mathbf{F}_{\parallel}$ and a *transverse* or *rotational* part $\mathbf{F}_{\perp}$,

$$\mathbf{F} = \mathbf{F}_{\parallel} + \mathbf{F}_{\perp}\,, \tag{1.182}$$

where $\mathbf{F}_{\parallel}$ and $\mathbf{F}_{\perp}$ are implicitly defined by

$$\nabla \times \mathbf{F}_{\parallel} = \mathbf{0}\,, \tag{1.183}$$

$$\nabla \cdot \mathbf{F}_{\perp} = 0\,. \tag{1.184}$$

This split is unique. A good discussion of the Helmholtz's vector theorem is contained in [26] where it is stressed that the theorem critically depends on the boundary conditions of the field $\mathbf{F}$. In the following discussion, it is assumed that the boundary conditions are such that the

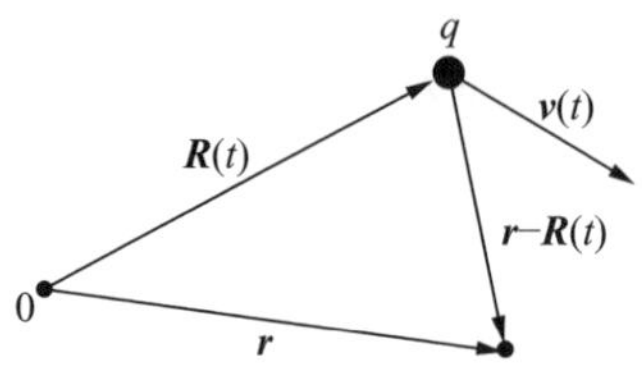

Figure 1.7 Coordinates of a moving point charge q. Position and velocity of the point charge are related by $\boldsymbol{v}(t) = \partial \boldsymbol{R}(t)/\partial t$.

Helmholtz's vector theorem can be applied.[9] The names *longitudinal* and *transverse* acquire a clear geometric interpretation in reciprocal space where Equations (1.183) and (1.184) become

$$-j\boldsymbol{k} \times \mathsf{F}_{\parallel} = \mathbf{0}\,, \tag{1.185}$$

$$-j\boldsymbol{k} \cdot \mathsf{F}_{\perp} = 0\,, \tag{1.186}$$

that is, $\mathsf{F}_{\parallel}$ is parallel to $\boldsymbol{k}$ and $\mathsf{F}_{\perp}$ is perpendicular to $\boldsymbol{k}$.

By means of the vector identity (C.6) and the relation

$$\Delta\left(\frac{1}{|\boldsymbol{r}-\boldsymbol{r}'|}\right) = -4\pi\delta(\boldsymbol{r}-\boldsymbol{r}') \tag{1.187}$$

it can be shown that in real space the longitudinal and transverse part of a vector field $\mathbf{F}(\boldsymbol{r}, t)$ are given by

$$\mathbf{F}_{\parallel}(\boldsymbol{r}, t) = -\frac{1}{4\pi}\nabla \int \frac{\nabla' \cdot \mathbf{F}(\boldsymbol{r}', t)}{|\boldsymbol{r}-\boldsymbol{r}'|}\,\mathrm{d}^3 r'\,, \tag{1.188}$$

$$\mathbf{F}_{\perp}(\boldsymbol{r}, t) = \frac{1}{4\pi}\nabla \times \nabla \times \int \frac{\mathbf{F}(\boldsymbol{r}', t)}{|\boldsymbol{r}-\boldsymbol{r}'|}\,\mathrm{d}^3 r'\,, \tag{1.189}$$

respectively. These explicit formulas show that the split $\mathbf{F} = \mathbf{F}_{\parallel} + \mathbf{F}_{\perp}$ introduces nonlocal effects: both $\mathbf{F}_{\parallel}(\boldsymbol{r}, t)$ and $\mathbf{F}_{\perp}(\boldsymbol{r}, t)$, considered at a fixed time t and at a specific point $\boldsymbol{r}$, depend on the values of $\mathbf{F}(\boldsymbol{r}', t)$ at the *same* time and at *all* points $\boldsymbol{r}'$ in space. Conversely, even if $\mathbf{F}(\boldsymbol{r}, t)$ is localized in space, that is if it vanishes outside a compact region, the parts $\mathbf{F}_{\parallel}(\boldsymbol{r}, t)$ and $\mathbf{F}_{\perp}(\boldsymbol{r}, t)$ will generally extend over the whole space.

Example 1.1: Consider a point charge q at a position $\boldsymbol{R}(t)$ which moves with velocity $\boldsymbol{v}(t)$ and is observed from a position $\boldsymbol{r}$, (see Figure 1.7).

The charge density ρ and current density $\boldsymbol{J}$ of this point charge are given by

$$\rho(\boldsymbol{r}, t) = q\delta(\boldsymbol{r} - \boldsymbol{R}(t))\,, \tag{1.190}$$

[9] From a microscopic but still classical point of view macroscopic boundary conditions are the result of the interaction between the electromagnetic field and electrically charged particles, like electrons or protons. Therefore, if fundamental properties of the electromagnetic field are considered, it is possible to focus on these microscopic interactions and there is no need to separately consider macroscopic boundary conditions. For example, the boundary conditions that are imposed by a perfect conductor can be replaced by the interaction between the electromagnetic field and the electrons and protons that represent the electrically charged particles of the perfect conductor.

and

$$\boldsymbol{J}(\boldsymbol{r},t) = q\boldsymbol{v}(t)\delta(\boldsymbol{r}-\boldsymbol{R}(t)), \tag{1.191}$$

respectively. To calculate the corresponding longitudinal current $\boldsymbol{J}_{\|}(\boldsymbol{r},t)$, Equation (1.188) is used and the continuity equation

$$\nabla' \cdot \boldsymbol{J}(\boldsymbol{r}',t) = -\frac{\partial \rho}{\partial t}(\boldsymbol{r}',t) \tag{1.192}$$

is applied. This yields

$$\boldsymbol{J}_{\|}(\boldsymbol{r},t) = \frac{q}{4\pi}\frac{\partial}{\partial t}\nabla \int \frac{\delta(\boldsymbol{r}'-\boldsymbol{R}(t))}{|\boldsymbol{r}-\boldsymbol{r}'|}\,\mathrm{d}^3 r' \tag{1.193}$$

$$= \frac{q}{4\pi}\frac{\partial}{\partial t}\nabla\left(\frac{1}{|\boldsymbol{r}-\boldsymbol{R}(t)|}\right) \tag{1.194}$$

$$= -\frac{q}{4\pi}\frac{\partial}{\partial t}\left(\frac{\boldsymbol{r}-\boldsymbol{R}(t)}{|\boldsymbol{r}-\boldsymbol{R}(t)|^3}\right) \tag{1.195}$$

$$= \frac{q}{4\pi}\left[\frac{\boldsymbol{v}(t)}{|\boldsymbol{r}-\boldsymbol{R}(t)|^3} - \frac{3(\boldsymbol{r}-\boldsymbol{R}(t))[(\boldsymbol{r}-\boldsymbol{R}(t))\cdot\boldsymbol{v}(t)]}{|\boldsymbol{r}-\boldsymbol{R}(t)|^5}\right]. \tag{1.196}$$

Accordingly, due to $\boldsymbol{J}_{\perp}(\boldsymbol{r},t) = \boldsymbol{J}(\boldsymbol{r},t) - \boldsymbol{J}_{\|}(\boldsymbol{r},t)$, there is also

$$\boldsymbol{J}_{\|}(\boldsymbol{r},t) = \frac{q}{4\pi}\left[4\pi\,\boldsymbol{v}(t)\delta(\boldsymbol{r}-\boldsymbol{R}(t)) - \frac{\boldsymbol{v}(t)}{|\boldsymbol{r}-\boldsymbol{R}(t)|^3} + \frac{3(\boldsymbol{r}-\boldsymbol{R}(t))[(\boldsymbol{r}-\boldsymbol{R}(t))\cdot\boldsymbol{v}(t)]}{|\boldsymbol{r}-\boldsymbol{R}(t)|^5}\right] \tag{1.197}$$

and it is clearly seen that both $\boldsymbol{J}_{\|}(\boldsymbol{r},t)$ and $\boldsymbol{J}_{\perp}(\boldsymbol{r},t)$ extend over the whole space.

1.4.3.2 Nondynamical Maxwell Equations as Boundary Conditions in Time

Among the complete set of Maxwell Equations (1.134) to (1.137) the Equations (1.134) and (1.136) are not dynamical equations but rather so-called *boundary conditions* that determine appropriate initial conditions of the fields. By virtue of the remaining dynamical Maxwell equations they are fulfilled at all times if they are fulfilled at one time. To illustrate this circumstance for the boundary condition (1.136) it is assumed that at some initial time t_0

$$\nabla \cdot \mathbf{B}|_{t_0} = 0. \tag{1.198}$$

It then needs to be shown that at an infinitesimally later time $t_0 + dt$

$$\nabla \cdot \mathbf{B}|_{t_0+dt} = 0, \tag{1.199}$$

that is

$$\left.\frac{\partial(\nabla\cdot\mathbf{B})}{\partial t}\right|_{t_0} = 0. \tag{1.200}$$

However, this condition immediately follows if the divergence of the dynamical Maxwell Equation (1.137) is taken. Similarly, it is found from Equation (1.135)

$$\left.\frac{\partial}{\partial t}(\nabla\cdot\mathbf{D}-\rho)\right|_{t_0} = -\left.\left(\nabla\cdot\mathbf{J}+\tfrac{\partial\rho}{\partial t}\right)\right|_{t_0} \tag{1.201}$$

$$= 0, \tag{1.202}$$

where in the second step the continuity Equation (1.10) has been employed. Therefore, it is sufficient to calculate the solutions of Equations (1.134) and (1.136) at an initial time t_0 and then solve with these solutions as boundary conditions the dynamical Maxwell Equations (1.135) and (1.137) to obtain the time evolution of the electromagnetic field.

1.4.3.3 Longitudinal Part of the Maxwell Equations

The Maxwell Equation (1.134) can be written as

$$\nabla \cdot \mathbf{D}_{\parallel}(\boldsymbol{r}, t) = \rho(\boldsymbol{r}, t) \tag{1.203}$$

and relates the longitudinal electric excitation $\mathbf{D}_{\parallel}$ to the charge density ρ. In reciprocal space this relation becomes

$$-j\boldsymbol{k} \cdot \mathsf{D}_{\parallel}(\boldsymbol{k}, t) = \rho(\boldsymbol{k}, t) \tag{1.204}$$

and can easily be solved for $\mathsf{D}_{\parallel}(\boldsymbol{k}, t)$ to yield

$$\mathsf{D}_{\parallel}(\boldsymbol{k}, t) = j\rho(\boldsymbol{k}, t)\frac{\boldsymbol{k}}{k^2}\,. \tag{1.205}$$

An inverse Fourier transform to real space gives the result

$$\mathbf{D}_{\parallel}(\boldsymbol{r}, t) = \frac{1}{4\pi}\int \rho(\boldsymbol{r}', t)\frac{\boldsymbol{r} - \boldsymbol{r}'}{|\boldsymbol{r} - \boldsymbol{r}'|^3}\,\mathrm{d}^3r'\,. \tag{1.206}$$

This result is quite remarkable since it turns out that the longitudinal electric displacement is completely determined from the *instantaneous* Coulomb field of the charge distribution. With the constitutive relation of Equation (1.138) the same is true for the longitudinal electric field strength,

$$\mathbf{E}_{\parallel}(\boldsymbol{r}, t) = \frac{1}{4\pi\varepsilon}\int \rho(\boldsymbol{r}', t)\frac{\boldsymbol{r} - \boldsymbol{r}'}{|\boldsymbol{r} - \boldsymbol{r}'|^3}\,\mathrm{d}^3r'\,. \tag{1.207}$$

The fact that $\mathbf{D}_{\parallel}(\boldsymbol{r}, t)$ and $\mathbf{E}_{\parallel}(\boldsymbol{r}, t)$ instantly respond to a change of the charge density does, at this point, not necessarily imply that causality is violated since it is required that the *complete* fields $\mathbf{D}(\boldsymbol{r}, t)$, $\mathbf{E}(\boldsymbol{r}, t)$ are causal.

The longitudinal part of the second inhomogeneous Maxwell Equation (1.135) is given by

$$-\frac{\partial \mathbf{D}_{\parallel}(\boldsymbol{r}, t)}{\partial t} = \mathbf{J}_{\parallel}(\boldsymbol{r}, t). \tag{1.208}$$

Taking the divergence of this equation reveals that it reduces to the continuity equation

$$\frac{\partial \rho(\boldsymbol{r}, t)}{\partial t} + \nabla \cdot \mathbf{J}_{\parallel}(\boldsymbol{r}, t) = 0. \tag{1.209}$$

Therefore, Equation (1.208) conveys no additional information.

In summary, the longitudinal components of the electromagnetic field are determined from the instantaneous Coulomb field of the electric charge density. It follows that the longitudinal components do not have their own degrees of freedom, they are tied to the degrees of freedom of the electric charge density.

1.4.3.4 Transverse Part of the Maxwell Equations

What is left to investigate are the transverse parts of the Maxwell Equations (1.135) and (1.137),

$$\nabla \times \mathbf{H}_\perp(\boldsymbol{r},t) - \frac{\partial \mathbf{D}_\perp}{\partial t}(\boldsymbol{r},t) = \mathbf{J}_\perp(\boldsymbol{r},t)\,, \tag{1.210}$$

$$\nabla \times \mathbf{E}_\perp(\boldsymbol{r},t) + \frac{\partial \mathbf{B}_\perp}{\partial t}(\boldsymbol{r},t) = \mathbf{0}\,. \tag{1.211}$$

With simple constitutive relations of the form in Equations (1.138) and (1.139) these equations are easily decoupled, arriving at the transverse part of the wave Equations (1.142) and (1.143),

$$\Delta \mathbf{E}_\perp(\boldsymbol{r},t) - \varepsilon\mu \frac{\partial^2 \mathbf{E}_\perp}{\partial t^2}(\boldsymbol{r},t) = \mu \frac{\partial \mathbf{J}_\perp}{\partial t}(\boldsymbol{r},t)\,, \tag{1.212}$$

$$\Delta \mathbf{B}_\perp(\boldsymbol{r},t) - \varepsilon\mu \frac{\partial^2 \mathbf{B}_\perp}{\partial t^2}(\boldsymbol{r},t) = -\mu \nabla \times \mathbf{J}_\perp(\boldsymbol{r},t)\,, \tag{1.213}$$

since

$$\mathbf{B}_\perp = \mathbf{B}. \tag{1.214}$$

In the following the transverse index $\perp$ of the magnetic field strength will be dropped.

From the wave Equations (1.212) and (1.213) it appears that $\mathbf{E}_\perp$ and $\mathbf{B}$ are the dynamical quantities of the electromagnetic field with two independent components each. However, one needs to note that $\mathbf{E}_\perp$ and $\mathbf{B}$ are not independent of each other. To explicitly show how both quantities are interrelated the dynamical Maxwell Equations (1.210) and (1.211) are rewritten in reciprocal space. With the constitutive relations (1.138) and (1.139), the relations $c^2 = 1/(\varepsilon\mu)$, $\omega = ck$ and the notation $\hat{\boldsymbol{k}} = \boldsymbol{k}/k$ one obtains, similar to Equations (1.174) and (1.176), the equations

$$\frac{\partial \mathsf{E}_\perp(\boldsymbol{k},t)}{\partial t} = -j\omega c\hat{\boldsymbol{k}} \times \mathsf{B}(\boldsymbol{k},t) - \frac{\mathsf{J}_\perp(\boldsymbol{k},t)}{\varepsilon_0}, \tag{1.215}$$

$$c\hat{\boldsymbol{k}} \times \frac{\partial \mathsf{B}(\boldsymbol{k},t)}{\partial t} = -j\omega \mathsf{E}_\perp(\boldsymbol{k},t). \tag{1.216}$$

In the sourceless case with $\mathsf{J}_\perp = 0$ it is recognized from these equations by addition and subtraction that eigenfunctions of this system are determined from

$$\frac{\partial}{\partial t}\left(\mathsf{E}_\perp(\boldsymbol{k},t) - c\hat{\boldsymbol{k}} \times \mathsf{B}(\boldsymbol{k},t)\right) = j\omega\left(\mathsf{E}_\perp(\boldsymbol{k},t) - c\hat{\boldsymbol{k}} \times \mathsf{B}(\boldsymbol{k},t)\right), \tag{1.217}$$

$$\frac{\partial}{\partial t}\left(\mathsf{E}_\perp(\boldsymbol{k},t) + c\hat{\boldsymbol{k}} \times \mathsf{B}(\boldsymbol{k},t)\right) = -j\omega\left(\mathsf{E}_\perp(\boldsymbol{k},t) + c\hat{\boldsymbol{k}} \times \mathsf{B}(\boldsymbol{k},t)\right). \tag{1.218}$$

To label these eigenfunctions variables $\mathsf{a}(\boldsymbol{k},t)$ and $\mathsf{b}(\boldsymbol{k},t)$ are introduced by

$$\mathsf{a}(\boldsymbol{k},t) := \tfrac{j}{2\mathsf{N}(k)}\left[\mathsf{E}_\perp(\boldsymbol{k},t) - c\hat{\boldsymbol{k}} \times \mathsf{B}(\boldsymbol{k},t)\right], \tag{1.219}$$

$$\mathsf{b}(\boldsymbol{k},t) := \tfrac{j}{2\mathsf{N}(k)}\left[\mathsf{E}_\perp(\boldsymbol{k},t) + c\hat{\boldsymbol{k}} \times \mathsf{B}(\boldsymbol{k},t)\right]. \tag{1.220}$$

The factor $j/2\mathsf{N}(k)$ denotes a normalization coefficient and is in accordance with a common notation that is used in the context of the quantization of the electromagnetic field [27]. In this context the function $\mathsf{N}(k)$ is related to the energy of a quantum state of the electromagnetic field. For classical purposes the explicit form of $\mathsf{N}(k)$ is not important. Within expressions of the electromagnetic field in real space the function $\mathsf{N}(k)$ will cancel and drop out. It could also be absorbed in the definition of $\mathsf{a}(\boldsymbol{k}, t)$ and $\mathsf{b}(\boldsymbol{k}, t)$.

It is straightforward to solve Equations (1.219) and (1.220) for $\mathsf{E}_\perp(\boldsymbol{k}, t)$ and $\mathsf{B}(\boldsymbol{k}, t)$. Since both quantities have to be real it follows that

$$\mathsf{b}(\boldsymbol{k}, t) = -\mathsf{a}^*(-\boldsymbol{k}, t)\,, \tag{1.221}$$

where the asterisk * denotes complex conjugation. Then,

$$\mathsf{E}_\perp(\boldsymbol{k}, t) = -j\mathsf{N}(k)\Big[\mathsf{a}(\boldsymbol{k}, t) - \mathsf{a}^*(-\boldsymbol{k}, t)\Big], \tag{1.222}$$

$$\mathsf{B}(\boldsymbol{k}, t) = -\frac{j\mathsf{N}(k)}{c}\Big[\hat{\boldsymbol{k}} \times \mathsf{a}(\boldsymbol{k}, t) + \hat{\boldsymbol{k}} \times \mathsf{a}^*(-\boldsymbol{k}, t)\Big]\,. \tag{1.223}$$

Therefore, the transverse electromagnetic field is completely specified by the function $\mathsf{a}(\boldsymbol{k}, t)$. Since $\mathsf{E}_\perp(\boldsymbol{k}, t)$ and $\mathsf{B}(\boldsymbol{k}, t)$ are transverse functions it follows that $\mathsf{a}(\boldsymbol{k}, t)$ is a transverse function, too. Hence, it is concluded that $\mathsf{a}(\boldsymbol{k}, t)$ exhibits two degrees of freedom which are the two dynamical components of the electromagnetic field. The function $\mathsf{a}(\boldsymbol{k}, t)$ is said to represent the *normal modes* of the electromagnetic field. This term indicates that $\mathsf{a}(\boldsymbol{k}, t)$ represents a whole class of electromagnetic excitations which is parameterized by a discrete or continuous set of values for the wave number $\boldsymbol{k}$.

Inserting Equations (1.222) and (1.223) in the Maxwell equations yields for the time evolution of $\mathsf{a}(\boldsymbol{k}, t)$ the equation

$$\frac{\partial \mathsf{a}(\boldsymbol{k}, t)}{\partial t} - j\omega \mathsf{a}(\boldsymbol{k}, t) = -\frac{j}{2\varepsilon \mathsf{N}(k)}\mathsf{J}_\perp(\boldsymbol{k}, t)\,. \tag{1.224}$$

This equation of motion for the normal modes represents, in fact, the motion of harmonic oscillation: if a new variable $\mathsf{c}(\boldsymbol{k}, t)$ is implicitly introduced via

$$\mathsf{a}(\boldsymbol{k}, t) = \mathsf{c}(\boldsymbol{k}, t) - \frac{j}{\omega}\frac{\partial \mathsf{c}(\boldsymbol{k}, t)}{\partial t} \tag{1.225}$$

the familiar equation of motion of a harmonic oscillator follows from Equation (1.224),

$$\frac{\partial^2 \mathsf{c}(\boldsymbol{k}, t)}{\partial t^2} + \omega^2 \mathsf{c}(\boldsymbol{k}, t) = \frac{\omega}{2\varepsilon \mathsf{N}(k)}\mathsf{J}_\perp(\boldsymbol{k}, t)\,. \tag{1.226}$$

From Equations (1.222) and (1.223) one obtains for the fields $\mathbf{E}_\perp(\boldsymbol{r}, t)$ and $\mathbf{B}(\boldsymbol{t}, t)$ from a Fourier transformation the expansions

$$\mathbf{E}_\perp(\boldsymbol{r}, t) = -\frac{j}{(2\pi)^{3/2}}\int \mathsf{N}(k)\Big[\mathsf{a}(\boldsymbol{k}, t)\mathrm{e}^{-j\boldsymbol{k}\cdot\boldsymbol{r}} - \mathsf{a}^*(\boldsymbol{k}, t)\mathrm{e}^{j\boldsymbol{k}\cdot\boldsymbol{r}}\Big]\,\mathrm{d}^3k\,, \tag{1.227}$$

$$\mathbf{B}(\boldsymbol{r}, t) = -\frac{j}{(2\pi)^{3/2}}\int \frac{\mathsf{N}(k)}{c}\Big[\hat{\boldsymbol{k}} \times \mathsf{a}(\boldsymbol{k}, t)\mathrm{e}^{-j\boldsymbol{k}\cdot\boldsymbol{r}} - \hat{\boldsymbol{k}} \times \mathsf{a}^*(\boldsymbol{k}, t)\mathrm{e}^{j\boldsymbol{k}\cdot\boldsymbol{r}}\Big]\,\mathrm{d}^3k\,. \tag{1.228}$$

It is recalled that $\mathsf{a}(\boldsymbol{k}, t)$ and $\mathsf{a}^*(\boldsymbol{k}, t)$ are purely complex quantities such that the fields $\mathbf{E}_\perp(\boldsymbol{r}, t)$ and $\mathbf{B}(\boldsymbol{r}, t)$ are real.

In the absence of sources, $\mathsf{J}_\perp(\boldsymbol{k}, t) = 0$. Then the equation of motion (1.224) yields the solution

$$\mathsf{a}(\boldsymbol{k}, t) = \mathsf{a}(\boldsymbol{k})\mathrm{e}^{j\omega t} \tag{1.229}$$

and the expansions of Equations (1.227) and (1.228) turn into expansions in traveling plane waves,

$$\mathbf{E}_{\perp\mathrm{free}}(\boldsymbol{r}, t) = -\frac{j}{(2\pi)^{3/2}} \int \mathsf{N}(k)\Big[\mathsf{a}(\boldsymbol{k})\mathrm{e}^{j(\omega t - \boldsymbol{k}\cdot\boldsymbol{r})} - \mathsf{a}^*(\boldsymbol{k})\mathrm{e}^{-j(\omega t - \boldsymbol{k}\cdot\boldsymbol{r})}\Big]\,\mathrm{d}^3k\,, \tag{1.230}$$

$$\mathbf{B}_{\mathrm{free}}(\boldsymbol{r}, t) = -\frac{j}{(2\pi)^{3/2}} \int \frac{\mathsf{N}(k)}{c}\Big[\hat{\boldsymbol{k}} \times \mathsf{a}(\boldsymbol{k})\mathrm{e}^{j(\omega t - \boldsymbol{k}\cdot\boldsymbol{r})} - \hat{\boldsymbol{k}} \times \mathsf{a}^*(\boldsymbol{k})\mathrm{e}^{-j(\omega t - \boldsymbol{k}\cdot\boldsymbol{r})}\Big]\,\mathrm{d}^3k\,. \tag{1.231}$$

In these expansions of the free fields the functions $\mathsf{a}(\boldsymbol{k}, t) = \mathsf{a}(\boldsymbol{k})e^{j\omega t}$ corresponding to different $\boldsymbol{k}$ are completely decoupled. This also holds true if the electromagnetic sources, represented by $\mathsf{J}_\perp(\boldsymbol{k}, t)$, are independent of $\mathsf{a}(\boldsymbol{k}, t)$, that is independent of the electromagnetic field. However, if the electromagnetic sources do interact with the electromagnetic field the time evolution of $\mathsf{J}_\perp(\boldsymbol{k}, t)$ will depend on $\mathsf{a}(\boldsymbol{k}, t)$ and, in general, lead to a coupling between $\mathsf{a}(\boldsymbol{k}, t)$ with different $\boldsymbol{k}$.

1.4.4 Electromagnetic Energy and the Singularities of the Electromagnetic Field

The energy density $w(\boldsymbol{r}, t)$ of the electromagnetic field is given by

$$w_{\mathrm{em}}(\boldsymbol{r}, t) = \frac{1}{2}\Big(\mathbf{E}(\boldsymbol{r}, t)\cdot\mathbf{D}(\boldsymbol{r}, t) + \mathbf{B}(\boldsymbol{r}, t)\cdot\mathbf{H}(\boldsymbol{r}, t)\Big) \tag{1.232}$$

and the corresponding energy $W(t)$ is

$$W_{\mathrm{em}}(t) = \int w_{\mathrm{em}}(\boldsymbol{r}, t)\,\mathrm{d}^3r \tag{1.233}$$

$$= \frac{1}{2}\int \big(\mathbf{E}(\boldsymbol{r}, t)\cdot\mathbf{D}(\boldsymbol{r}, t) + \mathbf{B}(\boldsymbol{r}, t)\cdot\mathbf{H}(\boldsymbol{r}, t)\big)\,\mathrm{d}^3r\,. \tag{1.234}$$

This energy, in general, is not a constant of motion if the electromagnetic field interacts with electromagnetic sources.

For further analysis it is useful to split the electromagnetic field energy into a contribution of the longitudinal fields $\mathbf{E}_\parallel$, $\mathbf{D}_\parallel$ and a contribution of the transverse fields $\mathbf{E}_\perp$, $\mathbf{D}_\perp$, $\mathbf{B}$ and $\mathbf{H}$. Clearly, the magnetic part of $W_{\mathrm{em}}(t)$ only involves the transverse fields $\mathbf{B}$ and $\mathbf{H}$. The electric part can be written as

$$\frac{1}{2}\int \mathbf{E}(\boldsymbol{r}, t)\cdot\mathbf{D}(\boldsymbol{r}, t)\,\mathrm{d}^3r$$

$$= \frac{1}{2}\int \mathsf{E}^*(\boldsymbol{k}, t)\cdot\mathsf{D}(\boldsymbol{k}, t)\,\mathrm{d}^3k \tag{1.235}$$

$$= \frac{1}{2}\int \big(\mathsf{E}_\parallel{}^*(\boldsymbol{k}, t) + \mathsf{E}_\perp{}^*(\boldsymbol{k}, t)\big)\cdot\big(\mathsf{D}_\parallel(\boldsymbol{k}, t) + \mathsf{D}_\perp(\boldsymbol{k}, t)\big)\,\mathrm{d}^3k \tag{1.236}$$

$$= \frac{1}{2}\int \mathsf{E}_{\parallel}{}^{*}(\boldsymbol{k},t)\cdot\mathsf{D}_{\parallel}(\boldsymbol{k},t)\,\mathrm{d}^3k + \frac{1}{2}\int \mathsf{E}_{\perp}{}^{*}(\boldsymbol{k},t)\cdot\mathsf{D}_{\perp}(\boldsymbol{k},t)\,\mathrm{d}^3k \tag{1.237}$$

$$= \frac{1}{2}\int \mathbf{E}_{\parallel}(\boldsymbol{r},t)\cdot\mathbf{D}_{\parallel}(\boldsymbol{r},t)\,\mathrm{d}^3r + \frac{1}{2}\int \mathbf{E}_{\perp}(\boldsymbol{r},t)\cdot\mathbf{D}_{\perp}(\boldsymbol{r},t)\,\mathrm{d}^3r\,. \tag{1.238}$$

The result is the desired split into longitudinal and transverse contributions to $W_{\mathrm{em}}(t)$.

If constitutive relations of the form are assumed Equation (1.138) and the results (1.206) and (1.207) are taken into account it is easy to see that the longitudinal contribution $W_{\mathrm{em}\parallel}$ to the electromagnetic field energy W_{em} is just given by the electrostatic Coulomb energy,

$$\begin{aligned} W_{\mathrm{em}\parallel}(t) &= \frac{1}{2}\int \mathbf{E}_{\parallel}(\boldsymbol{r},t)\cdot\mathbf{D}_{\parallel}(\boldsymbol{r},t)\,\mathrm{d}^3r \\ &= \frac{1}{8\pi\varepsilon}\iint \frac{\rho(\boldsymbol{r},t)\rho(\boldsymbol{r}',t)}{|\boldsymbol{r}-\boldsymbol{r}'|}\,\mathrm{d}^3r\,\mathrm{d}^3r'\,. \end{aligned} \tag{1.239}$$

For the transverse contribution

$$W_{\mathrm{em}\perp}(t) = \frac{1}{2}\int \big(\mathbf{E}_{\perp}(\boldsymbol{r},t)\cdot\mathbf{D}_{\perp}(\boldsymbol{r},t) + \mathbf{B}(\boldsymbol{r},t)\cdot\mathbf{H}(\boldsymbol{r},t)\big)\,\mathrm{d}^3r \tag{1.240}$$

it is not possible to find in real space a result of similar simplicity. However, if the validity of the constitutive relations (1.138) and (1.139) is assumed and a shift to reciprocal space is performed one can write

$$W_{\mathrm{em}\perp}(t) = \frac{\varepsilon}{2}\int \big(\mathsf{E}_{\perp}{}^{*}(\boldsymbol{k},t)\cdot\mathsf{E}_{\perp}(\boldsymbol{k},t) + \mathsf{B}^{*}(\boldsymbol{k},t)\cdot\mathsf{B}(\boldsymbol{k},t)\big)\,\mathrm{d}^3k\,. \tag{1.241}$$

Inserting into this expression the relations (1.222) and (1.223) it then follows from simple vector algebra that

$$W_{\mathrm{em}\perp}(t) = \varepsilon\int \mathsf{N}^2(k)\left[\mathsf{a}^{*}(\boldsymbol{k},t)\cdot\mathsf{a}(\boldsymbol{k},t) + \mathsf{a}(-\boldsymbol{k},t)\cdot\mathsf{a}^{*}(-\boldsymbol{k},t)\right]\mathrm{d}^3k. \tag{1.242}$$

Therefore, the transverse contribution to the electromagnetic energy is completely determined from the normal modes $\mathsf{a}(\boldsymbol{k},t)$.

Next it is investigated under which conditions the electromagnetic energy of a system becomes divergent: from Equation (1.239) for the electrostatic Coulomb energy of the longitudinal fields it is seen that this energy depends on the relative position of the electric sources within a system. It diverges in the limit $|\boldsymbol{r}\to\boldsymbol{r}'|$ in which case two sources become arbitrarily close. This behavior characterizes the *Coulomb singularity* of classical electrodynamics and is related to the infinite amount of Coulomb energy that is carried by any electric source. It is independent of the kinematical or dynamical state of a source, that is it is independent of velocity and acceleration. In particular, in the time-harmonic case it is independent of frequency. The expression (1.242) for the energy of the transverse fields exhibits no spatial singularity. It is seen from Equations (1.224) and (1.226) that the normal modes $\mathsf{a}(\boldsymbol{k},t)$ represent oscillations of the electromagnetic field that are driven by the transverse part $\mathsf{J}_{\perp}(\boldsymbol{k},t)$ of the electric current. Dominant contributions to the energy given by Equation (1.242) occur if the excitation is such that it operates at an eigenfrequency ω of the system. In this case of *electromagnetic resonance* it is known from the elementary solution of the equation of motion of a forced harmonic oscillator that the amplitude of the resulting oscillation will tend to infinity if no loss mechanism is present [22]. Then the energy in Equation (1.242) will tend to infinity as well. It follows that in the lossless case the oscillations of the electromagnetic field become singular at resonance. This is the second type of electromagnetic singularity which leads to diverging electromagnetic energy. Since forced harmonic motion is the only solution of the motion Equation (1.226) this type of singularity is the only one that is contained in Equation (1.242).

As a result it can be stated that the electromagnetic field exhibits two types of singularities that lead to diverging electromagnetic energy densities. These are:

- Coulomb singularities that are related to the mutual position of electric sources;
- electromagnetic resonances that are related to forced oscillations of the electromagnetic field.

1.4.5 Coulomb Fields and Radiation Fields

The split of the electromagnetic fields **D**, **H**, **E** and **B** and the electric current **J** in longitudinal and transverse parts has already been proven to be useful in the study of basic electromagnetic field properties. Unfortunately, it also has some disadvantages. Two major disadvantages which, in fact, are interrelated, are the following:

- The split in longitudinal and transverse parts is not relativistically invariant. A field which is purely longitudinal if observed from a first reference system will contain transverse contributions if observed from a second reference system which is in relative motion to the first one.
- The decomposition in longitudinal and transverse parts introduces nonlocal effects. This has been noted in Section 1.4.3.1 where the explicit form of the decomposition is given by Equations (1.188) and (1.189).

The physical relevance of these disadvantages becomes apparent if one returns to the analysis of the longitudinal and transverse parts of the Maxwell equations, as outlined in Sections 1.4.3.3 and 1.4.3.4. It was found that the longitudinal electric field $\mathbf{E}_{\parallel}$ represents the instantaneous Coulomb field of the electric charge density ρ. Since the complete electric field **E**, according to Equation (1.142), fulfills a proper wave equation which leads to causal solutions it follows that the transverse electric part $\mathbf{E}_{\perp}$ must contain an instantaneous contribution which exactly cancels the instantaneous contribution of $\mathbf{E}_{\parallel}$. Indeed, it can be shown that this is the case [27]. At first sight this is surprising since $\mathbf{E}_{\perp}$ fulfills the wave Equation (1.212). However, the solution of this wave equation contains instantaneous contributions since the source term $\mathbf{J}_{\perp}$ is nonlocal and contains instantaneous contributions as well (compare the short discussion for the general transverse field $\mathbf{F}_{\perp}$ after Equation (1.189)). Therefore, the split in longitudinal and transverse electromagnetic fields does not separate physically independent electromagnetic field contributions since longitudinal and transverse electromagnetic fields are inseparably connected to each other.

To illustrate this circumstance, the electromagnetic field of a moving point charge is considered where the validity of the constitutive relations in Equations (1.36) and (1.37) is assumed. At a fixed time the point charge is located at a position $\boldsymbol{r}'$ where it moves with velocity $\boldsymbol{v}$, and an observer is positioned at a position $\boldsymbol{r}$. The unit vector $\boldsymbol{e}_{r',r} := (\boldsymbol{r} - \boldsymbol{r}')/|\boldsymbol{r} - \boldsymbol{r}'|$ points from the charge to the observer (see Figure 1.8) and also the abbreviation $\boldsymbol{\beta} := \boldsymbol{v}/c$ is introduced.

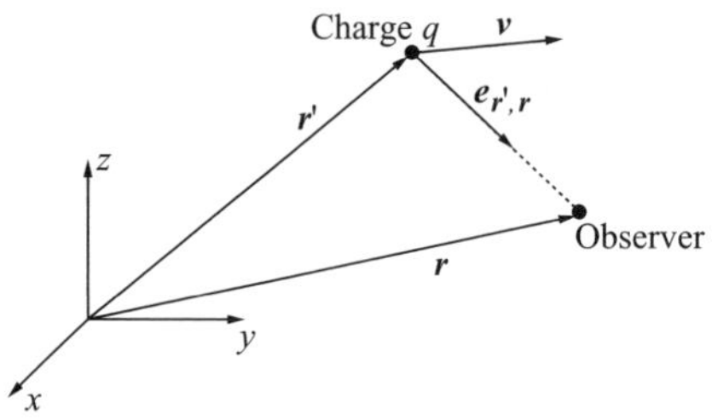

Figure 1.8 A charge q moves with velocity $\boldsymbol{v}$ in the presence of an observer. It is assumed that the observer does not move with respect to the inertial system xyz. The electromagnetic field that is generated by the electric charge requires the time $|\boldsymbol{r} - \boldsymbol{r}'|/c$ to reach the observer. Therefore, the electromagnetic field that is noticed by the observer at a time t has been generated by the electric charge at the earlier, *retarded time* $t_{\text{ret}} = t - |\boldsymbol{r} - \boldsymbol{r}'|/c$.

Then the electromagnetic field, expressed in terms of the field strengths **E** and **B**, that is noticed by the observer is given by [8]

$$\mathbf{E}(\boldsymbol{r}, t) = \underbrace{\frac{q}{4\pi\varepsilon_0}\left[\frac{(\boldsymbol{e}_{r',r} - \boldsymbol{\beta})(1-\beta^2)}{(1-\boldsymbol{\beta}\cdot\boldsymbol{e}_{r',r})^3\,|\boldsymbol{r}-\boldsymbol{r}'|^2}\right]_{\text{ret}}}_{\text{velocity field (Coulomb field)}} + \underbrace{\frac{q}{4\pi\varepsilon_0}\left[\frac{\boldsymbol{e}_{r',r}\times\left((\boldsymbol{e}_{r',r}-\boldsymbol{\beta})\times\frac{\partial\boldsymbol{\beta}}{\partial t}\right)}{c(1-\boldsymbol{\beta}\cdot\boldsymbol{e}_{r',r})^3\,|\boldsymbol{r}-\boldsymbol{r}'|}\right]_{\text{ret}}}_{\text{acceleration field (radiation field)}}, \tag{1.243}$$

$$\mathbf{B}(\boldsymbol{r}, t) = \frac{1}{c}\boldsymbol{e}_{r',r} \times \mathbf{E}(\boldsymbol{r}, t) \tag{1.244}$$

$$= \underbrace{\frac{q}{4\pi\varepsilon_0}\left[\frac{(\boldsymbol{\beta}\times\boldsymbol{e}_{r',r})(1-\beta^2)}{c(1-\boldsymbol{\beta}\cdot\boldsymbol{e}_{r',r})^3\,|\boldsymbol{r}-\boldsymbol{r}'|^2}\right]_{\text{ret}}}_{\text{velocity field (Coulomb field)}} + \underbrace{\frac{q}{4\pi\varepsilon_0}\left[\frac{\boldsymbol{e}_{r',r}\times\left(\boldsymbol{e}_{r',r}\times\left((\boldsymbol{e}_{r',r}-\boldsymbol{\beta})\times\frac{\partial\boldsymbol{\beta}}{\partial t}\right)\right)}{c^2(1-\boldsymbol{\beta}\cdot\boldsymbol{e}_{r',r})^3\,|\boldsymbol{r}-\boldsymbol{r}'|}\right]_{\text{ret}}}_{\text{radiation field (acceleration field)}}. \tag{1.245}$$

The brackets $[\ ]_{\text{ret}}$ indicate that the enclosed quantities have to be taken at the retarded time t_{ret} that is introduced in the caption of Figure 1.8. Each of the fields in Equations (1.243) and (1.244) splits nicely into a first part which depends on the velocity of the charge and a second part which depends on both the velocity and the acceleration of the charge.

The velocity fields can also be obtained by a Lorentz transformation of the static fields

$$\mathbf{E}_{\text{static}}(\boldsymbol{r}) = \frac{q}{4\pi\varepsilon}\frac{\boldsymbol{r}-\boldsymbol{r}'}{|\boldsymbol{r}-\boldsymbol{r}'|^3}, \tag{1.246}$$

$$\mathbf{B}_{\text{static}}(\boldsymbol{r}) = \mathbf{0}. \tag{1.247}$$

This suggests the name *Coulomb fields* for the electromagnetic velocity fields. They constitute the static Coulomb field of a point charge as noticed by an observer which is in relative motion to the charge. The remaining acceleration fields are commonly called *radiation fields*.

The fields in Equations (1.243) and (1.244) can be split into longitudinal and transverse parts. Since the magnetic field already is purely transverse only the electric field is considered. The longitudinal component $\mathbf{E}_{\|}$ is given by the instantaneous Coulomb field (see Equation (1.207)) and it follows

$$\mathbf{E}_{\|}(\boldsymbol{r}, t) = \underbrace{\frac{q}{4\pi\varepsilon_0}\frac{\boldsymbol{r}-\boldsymbol{r}'(t)}{|\boldsymbol{r}-\boldsymbol{r}'(t)|^3}}_{\text{instantaneous Coulomb field}}, \tag{1.248}$$

$$\begin{aligned}\mathbf{E}_{\perp}(\boldsymbol{r}, t) = &\underbrace{\frac{q}{4\pi\varepsilon_0}\left(\left[\frac{(\boldsymbol{e}_{r',r}-\boldsymbol{\beta})(1-\beta^2)}{(1-\boldsymbol{\beta}\cdot\boldsymbol{e}_{r',r})^3\,|\boldsymbol{r}-\boldsymbol{r}'|^2}\right]_{\text{ret}} - \frac{\boldsymbol{r}-\boldsymbol{r}'(t)}{|\boldsymbol{r}-\boldsymbol{r}'(t)|^3}\right)}_{\text{remaining part of the Coulomb field}} \\ &+ \underbrace{\frac{q}{4\pi\varepsilon_0}\left[\frac{\boldsymbol{e}_{r',r}\times\left((\boldsymbol{e}_{r',r}-\boldsymbol{\beta})\times\frac{\partial\boldsymbol{\beta}}{\partial t}\right)}{c(1-\boldsymbol{\beta}\cdot\boldsymbol{e}_{r',r})^3\,|\boldsymbol{r}-\boldsymbol{r}'|}\right]_{\text{ret}}}_{\text{radiation field}}.\end{aligned} \tag{1.249}$$

Therefore, the split of the electric field into longitudinal and transverse parts also splits the Coulomb field into two parts and assigns the aforementioned instantaneous field contributions to both the longitudinal and the transverse component of the electric field.

In many electrical engineering applications of classical electrodynamics the microscopic picture of single electric charges that generate an electromagnetic field is not available. Then electromagnetic sources are expressed by the charge and current density ρ and $\mathbf{J}$, respectively. In this case it is not possible to characterize a resulting electromagnetic field by the velocities and accelerations of microscopic charges and, accordingly, we no longer have the notion of velocity fields and acceleration fields. This implies that, in general, it is no longer possible to split an electromagnetic field into its Coulomb part and its radiation part. There still is the split into longitudinal and transverse parts, but the transverse part contains contributions of the Coulomb part and the complete radiation part.

In principle, the radiation part of the electromagnetic field could be isolated if, at a particular time, it were possible to switch off the coupling between electric charges and the electromagnetic field, that is if it were possible to switch off electric charges and their accompanying Coulomb fields. Then the remaining radiation field would be the solution of the sourceless Maxwell equations with $\rho = 0$, $\boldsymbol{J} = \mathbf{0}$ and nontrivial initial conditions. A solution of the sourceless Maxwell equations commonly is called *free electromagnetic field*. It is a pure radiation field and fulfills the homogeneous wave Equations (1.145) and (1.146). A free electromagnetic field has no longitudinal components since its longitudinal electric field component in Equation (1.207) vanishes by virtue of $\rho = 0$. The solutions for the transverse components are characterized by oscillatory motion, as is recognized from the solutions in Equations (1.230) and (1.231) which, in turn, reflect the solutions of the harmonic oscillator Equations (1.224) and (1.226) for a vanishing transverse current. The solutions in Equations (1.230) and (1.231) also exhibit that for a radiation field the wave vector $\boldsymbol{k}$, the electric field $\mathbf{E} = \mathbf{E}_{\perp}$ and the magnetic field $\mathbf{B}$ are always mutually orthogonal to each other. It is in this way that a radiation field propagates electromagnetic energy with phase velocity $c = 1/\sqrt{\varepsilon\mu}$ through space.

However, in practice electric charges cannot simply be switched off to neglect the coupling between electric charges and the electromagnetic field. Instead, it is necessary to consider the nonlocal transverse electric current $\mathbf{J}_\perp$ which drives the transverse electromagnetic field components and, in general, extends through the whole of space. Then the concept of a free electromagnetic field turns to an ideal which, nevertheless, is often a useful one. An example is given by the far field of an antenna in free space. In free space the transverse electric current falls off faster in intensity than the electromagnetic field does. Then the electromagnetic field becomes asymptotically free at large distances where it constitutes the common radiation field.

The fact that generally it is not possible to split a given electromagnetic field into a Coulomb part and a radiation part indicates that there are some conceptual difficulties in classical electrodynamics which cannot be resolved. These do not necessarily have to be a matter of concern. If an electromagnetic boundary value problem is solved it is usually solved for the complete fields and it might be of no practical interest to know which part of the solution constitutes a Coulomb field and which part represents a radiation field. However, the fact that there are two different categories of electric fields with different, and often complementary, properties is the reason for many difficulties that are present in the solution of practical problems in electrical engineering.

1.4.6 The Green's Function Method

The construction of solutions to linear differential equations with specified sources and given boundary conditions belongs to the fundamental problems of the Maxwell theory and other physical field theories. The Green's function method provides a technique to find these solutions and in this section this method is introduced. For a proper understanding of the Green's function method it is necessary to be familiar with a number of functional analytic notions such as a linear operator which is self-adjoint and acts between Hilbert spaces with specified inner products. For those readers who are not familiar with these notions, Appendix B provides the necessary functional analytic background.

Formally, a linear differential equation is often expressed as an operator equation of the form

$$\mathcal{L}_{\mathrm{D}} f = g \tag{1.250}$$

with a linear differential operator $\mathcal{L}_{\mathrm{D}}$, a source function g which is assumed to be known, and an unknown function f. The Green's function method consists of finding a Green's function G such that the unknown function f is expressed as an integral over the source function g, weighted with the Green's function. From a physical point of view the Green's function method is a representation of the superposition principle: the Green's function is the solution of the given linear differential equation with respect to a unit source which is placed at a specific position. Then the solution with respect to a general source is obtained from the superposition of known solutions of individual unit sources at various positions.

1.4.6.1 Basic Ideas

To introduce the Green's function method a real, self-adjoint differential operator $\mathcal{L}_\mathrm{D}$ is considered with

$$(\mathcal{L}_\mathrm{D} f)(\boldsymbol{r}) = g(\boldsymbol{r}) \,. \tag{1.251}$$

In this case, it is explicitly indicated that the functions depend on a variable $\boldsymbol{r}$ which usually represents a position in space. It is also common to have the time t as an additional parameter.

A *Green's function* is implicitly defined by

$$\mathcal{L}_\mathrm{D} G(\boldsymbol{r}, \boldsymbol{r}') = \delta(\boldsymbol{r} - \boldsymbol{r}') \tag{1.252}$$

with

$$\delta_{\boldsymbol{r}} := \delta(\boldsymbol{r} - \boldsymbol{r}') \tag{1.253}$$

the *Dirac delta function* which is a generalized function that is defined in the distributional sense [28, 29]. In a Hilbert space H with an inner product $\langle \ , \ \rangle$ it can be introduced via the relationship

$$\langle f, \delta_{\boldsymbol{r}} \rangle = f(\boldsymbol{r}) \,. \tag{1.254}$$

Suppose that a Green's function has been constructed that fulfills Equation (1.252). Then the expression

$$\langle \mathcal{L}_\mathrm{D} f, G^* \rangle = \langle f, \mathcal{L}_\mathrm{D} G^* \rangle \tag{1.255}$$

$$= \langle f, (\mathcal{L}_\mathrm{D} G)^* \rangle \tag{1.256}$$

may be considered. Here the fact that $\mathcal{L}_\mathrm{D}$ is self-adjoint and real has been used. Applying Equation (1.252) and noting that the delta function is a real function yields

$$\langle \mathcal{L}_\mathrm{D} f, G^* \rangle = \langle f, \delta \rangle \tag{1.257}$$

$$= f \,. \tag{1.258}$$

With Equation (1.251) the solution of the original problem is found as

$$f = \langle g, G^* \rangle \,. \tag{1.259}$$

This establishes the Green's function method for solving differential equations that are represented by a linear, self-adjoint and real differential operator $\mathcal{L}_\mathrm{D}$.

Example 1.2: Consider as an example the Hilbert space $\boldsymbol{L}^2(\Omega)^m$ with inner product (B.27) and $\boldsymbol{f}, \boldsymbol{g} \in \boldsymbol{L}^2(\Omega)^m$ (see Appendix B). In an analogy to Equations (1.251) and (1.252) a linear differential equation is assumed,

$$(\mathcal{L}_\mathrm{D} \boldsymbol{f})(\boldsymbol{r}) = \boldsymbol{g}(\boldsymbol{r}) \,. \tag{1.260}$$

For the corresponding *Green's function* the ansatz

$$\mathcal{L}_\mathrm{D} \overline{\boldsymbol{G}}(\boldsymbol{r}, \boldsymbol{r}') = \delta(\boldsymbol{r} - \boldsymbol{r}') \overline{\boldsymbol{I}} \tag{1.261}$$

is considered where now the Green's function $\overline{\boldsymbol{G}}(\boldsymbol{r}, \boldsymbol{r}')$ is represented as a dyadic and $\overline{\boldsymbol{I}}$ denotes the unit dyad [30]. With the inner product (B.27) the delta function acts according to

$$\langle \boldsymbol{f}, \delta_{\boldsymbol{r}} \rangle = \int_{\Omega} \delta(\boldsymbol{r} - \boldsymbol{r}') \boldsymbol{f}(\boldsymbol{r}') \, \mathrm{d}\Omega' \tag{1.262}$$

$$= \boldsymbol{f}(\boldsymbol{r}) . \tag{1.263}$$

Repeating the steps that led from Equations (1.255) to (1.259) yields the solution of Equation (1.260) in the form

$$\boldsymbol{f}(\boldsymbol{r}) = \int_{\Omega} \boldsymbol{g}(\boldsymbol{r}') \cdot \overline{\boldsymbol{G}}(\boldsymbol{r}, \boldsymbol{r}') \, \mathrm{d}\Omega' . \tag{1.264}$$

1.4.6.2 Self-Adjointness of Differential Operators and Boundary Conditions

The general solutions given in Equation (1.259) and (1.264) that have been obtained by means of the Green's function method look deceptively simple because they only involve weighting the source function g with the Green's function G. However, from the theory of differential equations it is known that boundary conditions play a fundamental role in the determination of a unique solution. Therefore, the information on boundary conditions must have been incorporated in the derivation of Equations (1.259) and (1.264). Indeed, self-adjointness of the real differential operator $\mathcal{L}_{\mathrm{D}}$ has been presupposed,

$$\langle \mathcal{L}_{\mathrm{D}} f, g \rangle = \langle f, \mathcal{L}_{\mathrm{D}} g \rangle . \tag{1.265}$$

Since in most function spaces of physical interest the inner product is represented by means of integration it follows that self-adjointness is closely connected to 'generalized partial integration', that is to the generalized Green's identity

$$\int_{\Omega} (\mathcal{L}_{\mathrm{D}} f) g^* \, \mathrm{d}\Omega = \int_{\Omega} f (\mathcal{L}_{\mathrm{D}} g)^* \, \mathrm{d}\Omega + \int_{\Gamma = \partial\Omega} J(f, g) \, \mathrm{d}\Gamma \tag{1.266}$$

where Ω is the integration volume and $\partial\Omega = \Gamma$ is its boundary. This identity can be written in terms of the inner product as

$$\langle \mathcal{L}_{\mathrm{D}} f, g \rangle = \langle f, \mathcal{L}_{\mathrm{D}} g \rangle + \int_{\Gamma = \partial\Omega} J(f, g) \, \mathrm{d}\Gamma . \tag{1.267}$$

It follows that $\mathcal{L}_{\mathrm{D}}$ is self-adjoint if and only if the integral on the right-hand side of this equation vanishes. This requirement will pose restrictions on the a priori unknown function f and the Green's function G.

This circumstance is demonstrated by the following general application of the Green's function method. Consider again a differential equation of the form

$$\mathcal{L}_{\mathrm{D}} f = g \tag{1.268}$$

and a corresponding Green's function G which satisfies

$$\mathcal{L}_{\mathrm{D}} G = \delta . \tag{1.269}$$

At this point it is not required for $\mathcal{L}_{\mathrm{D}}$ to be self-adjoint. Taking the inner product of Equation (1.268) with G^* and the inner product of the complex conjugate of Equation (1.269) with

f yields

$$\langle \mathcal{L}_D f, G^* \rangle = \langle g, G^* \rangle , \tag{1.270}$$

$$\langle f, (\mathcal{L}_D G)^* \rangle = \langle f, \delta^* \rangle = f . \tag{1.271}$$

Forming the differences of both equations leads to

$$\langle \mathcal{L}_D f, G^* \rangle - \langle f, (\mathcal{L}_D G)^* \rangle = \langle g, G^* \rangle - f. \tag{1.272}$$

In the notation of Equation (1.267) it is found that

$$\langle g, G^* \rangle - f = \int_{\Gamma=\partial\Omega} J(f, G)\, \mathrm{d}\Gamma , \tag{1.273}$$

that is the simple solution of Equation (1.259) is obtained if and only if the boundary integral, which involves the so-called *conjunct* $J(f, G)$, vanishes. This explicitly shows the relation between self-adjointness of the differential operator $\mathcal{L}_D$ and boundary conditions of f and G.

Example 1.3: Consider the Helmholtz Equation (1.169) for the magnetic vector potential $\boldsymbol{A}$ in the Lorenz gauge,

$$\Delta \boldsymbol{A}(\boldsymbol{r}) + k^2 \boldsymbol{A}(\boldsymbol{r}) = -\mu \boldsymbol{J}(\boldsymbol{r}) . \quad \text{(Lorenz gauge)} \tag{1.274}$$

Up to a factor μ the dyadic Green's function of this equation has to fulfill

$$\Delta \overline{\boldsymbol{G}}^A(\boldsymbol{r}, \boldsymbol{r}') + k^2 \overline{\boldsymbol{G}}^A(\boldsymbol{r}, \boldsymbol{r}') = -\overline{\boldsymbol{I}} \delta(\boldsymbol{r} - \boldsymbol{r}') . \tag{1.275}$$

With the inner product (B.27) of $\boldsymbol{L}^2(\Omega)^3$ the general Equation (1.272) yields

$$\begin{aligned} \int_\Omega \left[(\Delta \boldsymbol{A}(\boldsymbol{r}')) \cdot \overline{\boldsymbol{G}}^A(\boldsymbol{r}, \boldsymbol{r}') - \boldsymbol{A}(\boldsymbol{r}') \cdot \Delta \overline{\boldsymbol{G}}^A(\boldsymbol{r}, \boldsymbol{r}') \right] \mathrm{d}^3 r' = \\ -\mu \int_\Omega \boldsymbol{J}(\boldsymbol{r}') \cdot \overline{\boldsymbol{G}}^A(\boldsymbol{r}, \boldsymbol{r}')\, \mathrm{d}^3 r' + \boldsymbol{A}(\boldsymbol{r}) . \end{aligned} \tag{1.276}$$

By means of the second vector-dyadic Green's second theorem (C.21) the integral on the left-hand side can be transformed to a boundary integral. Then,

$$\begin{aligned} \boldsymbol{A}(\boldsymbol{r}) = \mu \int_\Omega \boldsymbol{J}(\boldsymbol{r}') \cdot \overline{\boldsymbol{G}}^A(\boldsymbol{r}, \boldsymbol{r}')\, \mathrm{d}^3 r' \\ + \oint_\Gamma \Big[(\boldsymbol{e}_n \times \boldsymbol{A}(\boldsymbol{r}')) \cdot (\nabla' \times \overline{\boldsymbol{G}}^A(\boldsymbol{r}, \boldsymbol{r}')) - (\nabla \times \boldsymbol{A}(\boldsymbol{r}')) \cdot (\boldsymbol{e}_n \times \overline{\boldsymbol{G}}^A(\boldsymbol{r}, \boldsymbol{r}')) \\ + \boldsymbol{e}_n \cdot \boldsymbol{A}(\boldsymbol{r}')(\nabla' \cdot \overline{\boldsymbol{G}}^A(\boldsymbol{r}, \boldsymbol{r}')) - \boldsymbol{e}_n \cdot \overline{\boldsymbol{G}}^A(\boldsymbol{r}, \boldsymbol{r}')(\nabla' \cdot \boldsymbol{A}(\boldsymbol{r}')) \Big] \mathrm{d}^2 r' . \end{aligned} \tag{1.277}$$

Since the boundary integral should vanish it is necessary to think about appropriate boundary conditions. If it is supposed that the boundary is perfectly conducting it follows that

$$\boldsymbol{e}_n \times \boldsymbol{A}(\boldsymbol{r})|_{\boldsymbol{r}\in\Gamma} = \boldsymbol{0} , \tag{1.278}$$

$$\nabla \cdot \boldsymbol{A}(\boldsymbol{r})|_{\boldsymbol{r}\in\Gamma} = 0 , \tag{1.279}$$

such that the first and last term within the surface integral will vanish. There are also the corresponding boundary conditions

$$\boldsymbol{e}_n \times \overline{\boldsymbol{G}}^A(\boldsymbol{r}, \boldsymbol{r}')\Big|_{\boldsymbol{r}\in\Gamma} = \boldsymbol{0}, \tag{1.280}$$

$$\nabla \cdot \overline{\boldsymbol{G}}^A(\boldsymbol{r}, \boldsymbol{r}')\Big|_{\boldsymbol{r}\in\Gamma} = 0 \tag{1.281}$$

of the dyadic Green's function such that the surface integral completely vanishes. Therefore, the magnetic vector potential can be calculated from the expression

$$\boldsymbol{A}(\boldsymbol{r}) = \mu \int \boldsymbol{J}(\boldsymbol{r}') \cdot \overline{\boldsymbol{G}}^A(\boldsymbol{r}, \boldsymbol{r}')\, \mathrm{d}^3 r', \tag{1.282}$$

which involves no boundary terms.

Example 1.4: Consider the vector Helmholtz Equation (1.165) for the electric field $\boldsymbol{E}$,

$$\nabla \times \nabla \times \boldsymbol{E}(\boldsymbol{r}) - k^2 \boldsymbol{E}(\boldsymbol{r}) = -j\omega\mu \boldsymbol{J}(\boldsymbol{r}). \tag{1.283}$$

Up to a factor $-j\omega\mu$ the corresponding Green's function needs to fulfill

$$\nabla \times \nabla \times \overline{\boldsymbol{G}}^E(\boldsymbol{r}, \boldsymbol{r}') - k^2 \overline{\boldsymbol{G}}^E(\boldsymbol{r}, \boldsymbol{r}') = \overline{\boldsymbol{I}}\delta(\boldsymbol{r}, \boldsymbol{r}'). \tag{1.284}$$

The general Equation (1.272) yields

$$\begin{aligned} &\int_\Omega \left[(\nabla \times \nabla \times \boldsymbol{E}(\boldsymbol{r}')) \cdot \overline{\boldsymbol{G}}^E(\boldsymbol{r}, \boldsymbol{r}') - \boldsymbol{E}(\boldsymbol{r}') \cdot (\nabla \times \nabla \times \overline{\boldsymbol{G}}^E(\boldsymbol{r}, \boldsymbol{r}')) \right] \mathrm{d}^3 r' \\ &\qquad = -j\omega\mu \int_\Omega \boldsymbol{J}(\boldsymbol{r}') \cdot \overline{\boldsymbol{G}}^E(\boldsymbol{r}, \boldsymbol{r}')\, \mathrm{d}^3 r' - \boldsymbol{E}(\boldsymbol{r}). \end{aligned} \tag{1.285}$$

From application of the Green's theorem (C.20) and the identity (C.15) it is found that

$$\begin{aligned} \boldsymbol{E}(\boldsymbol{r}) = &-j\omega\mu \int_\Omega \boldsymbol{J}(\boldsymbol{r}') \cdot \overline{\boldsymbol{G}}^E(\boldsymbol{r}, \boldsymbol{r}')\, \mathrm{d}^3 r' \\ &+ \oint_\Gamma \left[(\boldsymbol{e}_n \times \boldsymbol{E}(\boldsymbol{r}')) \cdot (\nabla \times \overline{\boldsymbol{G}}^E(\boldsymbol{r}, \boldsymbol{r}')) - (\nabla \times \boldsymbol{E}(\boldsymbol{r}')) \cdot (\boldsymbol{e}_n \times \overline{\boldsymbol{G}}^E(\boldsymbol{r}, \boldsymbol{r}')) \right] \mathrm{d}^2 r'. \end{aligned} \tag{1.286}$$

If it is supposed again that the interior of the boundary Γ is perfectly conducting the boundary condition

$$\boldsymbol{e}_n \times \boldsymbol{E}(\boldsymbol{r})|_{\boldsymbol{r}\in\Gamma} = \boldsymbol{0} \tag{1.287}$$

is valid. The corresponding boundary condition of the Green's function is

$$\boldsymbol{e}_n \times \overline{\boldsymbol{G}}^E(\boldsymbol{r}, \boldsymbol{r}')\Big|_{\boldsymbol{r}\in\Gamma} = \boldsymbol{0} \tag{1.288}$$

and, as a consequence, the surface integral vanishes such that the electric field can be calculated according to

$$\boldsymbol{E}(\boldsymbol{r}) = -j\omega\mu \int_\Omega \boldsymbol{J}(\boldsymbol{r}') \cdot \overline{\boldsymbol{G}}^E(\boldsymbol{r}, \boldsymbol{r}')\, \mathrm{d}^3 r'. \tag{1.289}$$

Examples 1.3 and 1.4 show that the Helmholtz equations for the magnetic vector potential and the electric field strength, respectively, form self-adjoint boundary value problems if the fields are defined in a finite domain which is enclosed by a perfectly conducting boundary.

1.4.6.3 General Solutions of Maxwell Equations

In Sections 1.4.1.1 and 1.4.1.3 it has been seen that for homogeneous and isotropic media it is straightforward to decouple the Maxwell equations and to rewrite them in the form of wave equations or, in the time harmonic case, as Helmholtz equations. In the absence of boundaries, that is in free space, and within a homogeneous medium the general solution of the Maxwell equations is given in terms of the solution of the scalar Helmholtz equation. This follows from Equations (1.169) and (1.170) which constitute in free space four independent scalar Helmholtz equations. These have the general structure

$$(\Delta + k^2) f(\boldsymbol{r}, \omega) = -g(\boldsymbol{r}, \omega) \tag{1.290}$$

and the appropriate Green's function $G_0(\boldsymbol{r}, \boldsymbol{r}')$ needs to satisfy

$$(\Delta + k^2) G_0(\boldsymbol{r}, \boldsymbol{r}') = -\delta(\boldsymbol{r} - \boldsymbol{r}') . \tag{1.291}$$

The solution for $G_0(\boldsymbol{r}, \boldsymbol{r}')$ is most easily obtained in spherical coordinates, taking advantage of the symmetries of free space. This yields the retarded solution [8, p. 243]

$$G_0(\boldsymbol{r}, \boldsymbol{r}') = \frac{1}{4\pi} \frac{\mathrm{e}^{-jk|\boldsymbol{r}-\boldsymbol{r}'|}}{|\boldsymbol{r} - \boldsymbol{r}'|} . \tag{1.292}$$

Therefore, in free space the general solution of the Maxwell equations is represented by the equations

$$\begin{aligned} \varphi(\boldsymbol{r}) &= \int G_0(\boldsymbol{r}, \boldsymbol{r}') \rho(\boldsymbol{r}') \, \mathrm{d}^3 r' \\ &= \frac{1}{4\pi\varepsilon} \int \frac{\mathrm{e}^{-jk|\boldsymbol{r}-\boldsymbol{r}'|}}{|\boldsymbol{r} - \boldsymbol{r}'|} \rho(\boldsymbol{r}') \, \mathrm{d}^3 r' , \end{aligned} \tag{1.293}$$

$$\begin{aligned} \boldsymbol{A}(\boldsymbol{r}) &= \int G_0(\boldsymbol{r}, \boldsymbol{r}') \boldsymbol{J}(\boldsymbol{r}') \, \mathrm{d}^3 r' \\ &= \frac{\mu}{4\pi} \int \frac{\mathrm{e}^{-jk|\boldsymbol{r}-\boldsymbol{r}'|}}{|\boldsymbol{r} - \boldsymbol{r}'|} \boldsymbol{J}(\boldsymbol{r}') \, \mathrm{d}^3 r' , \end{aligned} \tag{1.294}$$

which relate the electromagnetic sources ρ, $\boldsymbol{J}$ to the electromagnetic field, expressed by φ and $\boldsymbol{A}$ in the Lorenz gauge.

1.4.6.4 Basic Relations Between Electromagnetic Green's Functions

In general, there are some basic relations between the various electromagnetic Green's functions. It is recalled from Section 1.4.1.3 that in a linear, isotropic and homogeneous medium the Maxwell equations in the frequency domain can be reduced to Helmholtz equations. For the vector potential $\boldsymbol{A}(\boldsymbol{r})$ in the Lorenz gauge, the electric field $\boldsymbol{E}(\boldsymbol{r})$ and the magnetic field

$\boldsymbol{B}(\boldsymbol{r})$, respectively, the (vector) Helmholtz equations

$$\Delta \boldsymbol{A}(\boldsymbol{r},\omega) + k^2 \boldsymbol{A}(\boldsymbol{r},\omega) = -\mu \boldsymbol{J}(\boldsymbol{r},\omega)\,, \tag{1.295}$$

$$\nabla \times \nabla \times \boldsymbol{E}(\boldsymbol{r},\omega) - k^2 \boldsymbol{E}(\boldsymbol{r},\omega) = -j\omega\mu \boldsymbol{J}(\boldsymbol{r},\omega)\,, \tag{1.296}$$

$$\nabla \times \nabla \times \boldsymbol{B}(\boldsymbol{r},\omega) - k^2 \boldsymbol{B}(\boldsymbol{r},\omega) = \mu \nabla \times \boldsymbol{J}(\boldsymbol{r},\omega) \tag{1.297}$$

were found. The corresponding dyadic Green's functions obey the differential equations[10]

$$\Delta \overline{\boldsymbol{G}}^{A}(\boldsymbol{r},\boldsymbol{r}') + k^2 \overline{\boldsymbol{G}}^{A}(\boldsymbol{r},\boldsymbol{r}') = -\overline{\boldsymbol{I}}\delta(\boldsymbol{r}-\boldsymbol{r}')\,, \tag{1.298}$$

$$\nabla \times \nabla \times \overline{\boldsymbol{G}}^{E}(\boldsymbol{r},\boldsymbol{r}') - k^2 \overline{\boldsymbol{G}}^{E}(\boldsymbol{r},\boldsymbol{r}') = \overline{\boldsymbol{I}}\delta(\boldsymbol{r}-\boldsymbol{r}')\,, \tag{1.299}$$

$$\nabla \times \nabla \times \overline{\boldsymbol{G}}^{B} - k^2 \overline{\boldsymbol{G}}^{B}(\boldsymbol{r},\boldsymbol{r}') = \nabla\delta(\boldsymbol{r}-\boldsymbol{r}') \times \overline{\boldsymbol{I}}\,. \tag{1.300}$$

Equations (1.298) and (1.299) are already familiar from the examples of Section 1.4.6.2.

Suppose that it is possible to construct $\overline{\boldsymbol{G}}^{A}(\boldsymbol{r},\boldsymbol{r}')$. Then $\overline{\boldsymbol{G}}^{E}(\boldsymbol{r},\boldsymbol{r}')$ and $\overline{\boldsymbol{G}}^{B}(\boldsymbol{r},\boldsymbol{r}')$ are obtained via

$$\overline{\boldsymbol{G}}^{E}(\boldsymbol{r},\boldsymbol{r}') = \left(\overline{\boldsymbol{I}} + \frac{1}{k^2}\nabla\nabla\right)\overline{\boldsymbol{G}}^{A}(\boldsymbol{r},\boldsymbol{r}')\,, \tag{1.301}$$

$$\overline{\boldsymbol{G}}^{B}(\boldsymbol{r},\boldsymbol{r}') = \nabla \times \overline{\boldsymbol{G}}^{A}(\boldsymbol{r},\boldsymbol{r}')\,. \tag{1.302}$$

Clearly, this is an immediate consequence of the relations

$$\boldsymbol{E}(\boldsymbol{r}) = -j\omega\left(1 + \frac{1}{k^2}\nabla\nabla\cdot\right)\boldsymbol{A}(\boldsymbol{r})\,, \tag{1.303}$$

$$\boldsymbol{B}(\boldsymbol{r}) = \nabla \times \boldsymbol{A}(\boldsymbol{r})\,. \tag{1.304}$$

If $\overline{\boldsymbol{G}}^{E}(\boldsymbol{r},\boldsymbol{r}')$ is constructed rather than $\overline{\boldsymbol{G}}^{A}(\boldsymbol{r},\boldsymbol{r}')$ then $\overline{\boldsymbol{G}}^{B}(\boldsymbol{r},\boldsymbol{r}')$ is obtained from

$$\overline{\boldsymbol{G}}^{B}(\boldsymbol{r},\boldsymbol{r}') = \nabla \times \overline{\boldsymbol{G}}^{E}(\boldsymbol{r},\boldsymbol{r}') \tag{1.305}$$

since

$$\boldsymbol{B}(\boldsymbol{r}) = -\frac{1}{j\omega}\nabla \times \boldsymbol{E}(\boldsymbol{r})\,. \tag{1.306}$$

1.5 Boundary Value Problems and Integral Equations

1.5.1 Surface Integral Equations in Short

The problem of calculating the electric current on a scatterer is a standard problem in electromagnetic theory. A current is generated by primary sources that produce incident electromagnetic fields $\boldsymbol{E}^{\mathrm{inc}}$, $\boldsymbol{H}^{\mathrm{inc}}$. Then the total fields $\boldsymbol{E}$, $\boldsymbol{H}$ in the presence of the scatterer are a superposition of the incident fields and scattered fields $\boldsymbol{E}^{\mathrm{sca}}$, $\boldsymbol{H}^{\mathrm{sca}}$,

$$\boldsymbol{E} = \boldsymbol{E}^{\mathrm{inc}} + \boldsymbol{E}^{\mathrm{sca}}\,, \tag{1.307}$$

$$\boldsymbol{H} = \boldsymbol{H}^{\mathrm{inc}} + \boldsymbol{H}^{\mathrm{sca}}\,. \tag{1.308}$$

[10] It is a convention to skip in the definitions of the Green's functions the factors μ and $-j\omega\mu$ that appear on the right-hand sides of the corresponding Helmholtz equations.

A general method to find from this decomposition an equation for the unknown current on the scatterer consists of three steps which require the surface equivalence principle, the source–field relationships that follow from the solution of the Maxwell equations and the boundary conditions for the total electromagnetic fields [31, 1]:

(1) By virtue of the surface equivalence principle the scatterer is replaced by equivalent electromagnetic sources which, a priori, are unknown. If the scatterer is assumed to be perfectly conducting the equivalent electromagnetic sources are represented by an electric current. In general, the equivalent electromagnetic sources will be determined from the incident electromagnetic fields.
(2) The equivalent electromagnetic sources generate a scattered electromagnetic field according to the source–field relationships that express a field by the integral over a source, weighted with the appropriate Green's function. This allows the unknown scattered electromagnetic field, which is usually defined within an entire volume, to be replaced by the unknown equivalent sources, which are usually defined on a boundary surface.
(3) On the boundary surface, where the unknown equivalent sources are defined, the boundary conditions of Equations (1.307) and (1.308) for the total fields must be enforced. This relates on the boundary surface the known incident electromagnetic field to integrals over the unknown equivalent sources.

These three steps comprise, in short, how to construct surface integral equations for unknown electromagnetic sources that are induced by primary electromagnetic fields.

1.5.2 The Standard Electric Field Integral Equations of Antenna Theory and Radiating Nonuniform Transmission-Line Systems

Linear antennas constitute the classical antenna prototype and generalize the concept of an idealized, mathematical electric dipole to an actual engineering device. In the development of antenna theory linear antennas have played a dominant role because many of their properties can be modeled by analytic methods. However, even in the simplest realistic cases simplifying approximations have to be made in order to arrive at analytic results [32–35].

There are four standard electric field integral equations which can be used to determine the electric current on cylindrical thin-wire antennas [36–38]. These equations also constitute the basis of generalized transmission-line theories and are summarized below.

1.5.2.1 Pocklington's Equation

It is remarkable that 10 years after the discovery of electromagnetic radiation by Hertz [39] in 1887 an integral equation for the current distribution along cylindrical wire dipole antennas was published by Pocklington [40] in 1897. Pocklington's equation constitutes an electric field integral equation that is adapted to cylindrical, thin-wire antennas. To formulate this integral equation one can follow the three-step procedure of the previous section and first introduce a surface current $\boldsymbol{J}_s$ which is related to a scattered electric field $\boldsymbol{E}^{\text{sca}}$ via the electric Green's function $\overline{\boldsymbol{G}}^E$,

$$\boldsymbol{E}^{\text{sca}}(\boldsymbol{r}) = -j\omega\mu \int_\Gamma \overline{\boldsymbol{G}}^E(\boldsymbol{r}, \boldsymbol{r}')\, \boldsymbol{J}_s(\boldsymbol{r}')\, \mathrm{d}^2r' \,. \tag{1.309}$$

Here the antenna surface is denoted by Γ. The boundary condition for the total electric field $\boldsymbol{E}$ on a perfectly conducting surface is $\boldsymbol{e}_n \times \boldsymbol{E} = \boldsymbol{0}$ or, alternatively, $E_t = 0$, with $E_t = \boldsymbol{E} \cdot \boldsymbol{e}_t$ the projection of $\boldsymbol{E}$ on a given tangential vector $\boldsymbol{e}_t$. With this boundary condition and Equation (1.307) an electric field integral equation is obtained,

$$j\omega\mu \left[\int_\Gamma \overline{\boldsymbol{G}}^E(\boldsymbol{r}, \boldsymbol{r}') \boldsymbol{J}_s(\boldsymbol{r}')\, \mathrm{d}^2r' \right] \cdot \boldsymbol{e}_t(\boldsymbol{r}) = E_t^{\text{inc}}(\boldsymbol{r}) \,. \tag{1.310}$$

This equation simplifies if the antenna geometry is that of a thin cylindrical wire. Then a thin-wire approximation can be performed where azimuthal currents are neglected and the surface current $\boldsymbol{J}_s$ turns to a filamentary current I that flows along the cylinder axis [41, 42]. Furthermore, if the wire is assumed to be straight and, in Cartesian coordinates, directed along the z-axis one obtains from Equation (1.310)

$$j\omega\mu \int_{-L/2}^{L/2} G_{zz}^E(z, z') I(z')\, \mathrm{d}z' = E_z^{\text{inc}}(z) \tag{1.311}$$

with L the length of the antenna. Finally, Equation (1.301) is used to replace the zz-component of the electric dyadic Green's function by the zz-component of the dyadic Green's function for the magnetic vector potential in the Lorenz gauge. This yields Pocklington's equation in the form

$$-\frac{1}{j\omega\varepsilon} \int_{-L/2}^{L/2} \left(\frac{\partial^2}{\partial z^2} + k^2 \right) G_{zz}^A(z, z') I(z')\, \mathrm{d}z' = E_z^{\text{inc}}(z) \,. \tag{1.312}$$

1.5.2.2 Hallén's Equation

Not until 40 years after the publication of Pocklington's integral equation was a different integral equation for the calculation of antenna currents proposed by Hallén [43]. This integral equation is derived in close analogy to Pocklington's equation. First, an equivalent surface current $\boldsymbol{J}_s$ is introduced on the boundary of the antenna surface. This current is related to the scattered magnetic vector potential via

$$\boldsymbol{A}^{\text{sca}}(\boldsymbol{r}) = \mu \int_\Gamma \overline{\boldsymbol{G}}^A(\boldsymbol{r}, \boldsymbol{r}')\, \boldsymbol{J}_s(\boldsymbol{r}')\, \mathrm{d}^2r' \,. \tag{1.313}$$

The boundary condition for the total magnetic vector potential $\boldsymbol{A}$ on a perfectly conducting surface is $\boldsymbol{e}_n \times \boldsymbol{A} = \boldsymbol{0}$ or, equivalently, $A_t = 0$. With this boundary condition and the relation

$$\boldsymbol{A} = \boldsymbol{A}^{\text{inc}} + \boldsymbol{A}^{\text{sca}} \tag{1.314}$$

one obtains

$$\mu \left[\int_\Gamma \overline{\boldsymbol{G}}^A(\boldsymbol{r}, \boldsymbol{r}') \boldsymbol{J}_s(\boldsymbol{r}')\, \mathrm{d}^2r' \right] \cdot \boldsymbol{e}_t(\boldsymbol{r}) = -A_t^{\text{inc}}(\boldsymbol{r}) \,. \tag{1.315}$$

Again, the special case of a thin, straight cylindrical wire which is directed along the z-axis is considered. Then, similar to Equation (1.311),

$$\mu \int_{-L/2}^{L/2} G_{zz}^{A}(z, z') I(z') \, \mathrm{d}z' = -A_z^{\mathrm{inc}}(z) \,. \tag{1.316}$$

This result looks rather simple but it must be noted that, in practice, the incident electromagnetic field will usually be given in terms of the electric field strength $\boldsymbol{E}^{\mathrm{inc}}$ rather than in terms of the magnetic vector potential $\boldsymbol{A}^{\mathrm{inc}}$. In the Lorenz gauge,

$$\boldsymbol{E}^{\mathrm{inc}}(\boldsymbol{r}) = -\frac{j\omega}{k^2} \left(\nabla(\nabla \cdot \boldsymbol{A}^{\mathrm{inc}}(z)) + k^2 \boldsymbol{A}^{\mathrm{inc}}(z) \right) \tag{1.317}$$

and this second-order partial differential equation needs to be solved in order to obtain $\boldsymbol{A}^{\mathrm{inc}}$ from $\boldsymbol{E}^{\mathrm{inc}}$. In the case of a z-directed thin-wire antenna Equation (1.317) simplifies to

$$E_z^{\mathrm{inc}}(z) = -\frac{j\omega}{k^2} \left(\frac{\partial^2}{\partial z^2} + k^2 \right) A_z^{\mathrm{inc}}(z) \,. \tag{1.318}$$

This ordinary differential equation has well-known solutions that are given by the sum of a general solution of the homogeneous problem and a special solution of the inhomogeneous problem. It follows that

$$-A_z^{\mathrm{inc}}(z) = C_1 \mathrm{e}^{jkz} + C_2 \mathrm{e}^{-jkz} + \frac{k}{2j\omega} \int_{-L/2}^{L/2} \sin(k|z - z'|) E_z^{\mathrm{inc}}(z') \, \mathrm{d}z' \,, \tag{1.319}$$

where C_1, C_2 denote two integration constants that need to be determined from the boundary condition that the antenna current vanishes at the antenna ends. The function $G(z) = \sin(k|z|)$ that appears in the special solution of the inhomogeneous problem is, up to a constant factor, a Green's function for the differential Equation (1.318). It fulfills $(\partial^2/\partial z^2 + k^2)G(z) = 2k\delta(z)$. With Equation (1.319) one obtains from Equation (1.316) Hallén's equation in the form

$$\mu \int_{-L/2}^{L/2} G_{zz}^{A}(z, z') I(z') \, \mathrm{d}z' = C_1 \mathrm{e}^{jkz} + C_2 \mathrm{e}^{-jkz} + \frac{k}{2j\omega} \int_{-L/2}^{L/2} \sin(k|z - z'|) E_z^{\mathrm{inc}}(z') \, \mathrm{d}z' \,. \tag{1.320}$$

Compared with Pocklington's Equation (1.312) the integral kernel of Hallén's equation is less singular and thus preferable for numerical evaluations. The integral kernel of Pocklington's equation exhibits a spatial singularity that is proportional to $1/|\boldsymbol{r} - \boldsymbol{r}'|^3$ while in the case of Hallén's equation the spatial singularity is proportional to $1/|\boldsymbol{r} - \boldsymbol{r}'|$. However, to determine the integration constants C_1, C_2 in Hallén's integral equation can be cumbersome and in such cases Pocklington's equation might be the more practical choice.

1.5.2.3 Mixed-Potential Integral Equation

The mixed-potential integral equation represents another version of an electric-field integral equation. It is often used in numerical calculations since its integral kernel is proportional to $1/|\boldsymbol{r} - \boldsymbol{r}'|^2$, that is, the singularity of the integral kernel is weaker than in the case of Pocklington's equation. Additionally, the mixed-potential integral equation does not require integration constants to be determined as in the case of Hallén's equation.

To derive the mixed-potential integral equation the relation (1.147) for the scattered electromagnetic field in the frequency domain is considered:

$$\boldsymbol{E}^{\rm sca}(\boldsymbol{r}) = -\nabla\varphi^{\rm sca} - j\omega\boldsymbol{A}^{\rm sca}\,. \tag{1.321}$$

If one replaces by means of the Lorenz gauge the scalar potential $\varphi^{\rm sca}$ by the vector potential $\boldsymbol{A}^{\rm sca}$ and uses Equation (1.313) one will be led back to Equation (1.309) and obtain nothing new. Alternatively, it is possible to consider, besides Equation (1.313), the source–field relation

$$\varphi^{\rm sca}(\boldsymbol{r}) = \mu\int_{\Gamma} G^{\varphi}(\boldsymbol{r},\boldsymbol{r}')\,\rho_{\rm s}(\boldsymbol{r}')\,{\rm d}^2r' \tag{1.322}$$

with $G^{\varphi}(\boldsymbol{r},\boldsymbol{r}')$ the scalar Green's function of the scalar Helmholtz equation, (compare Equation (1.169)), and $\rho_{\rm s}$ a surface charge density. This surface charge density is related to a surface current $\boldsymbol{J}_{\rm s}$ by a continuity equation which, in integral form, reads

$$j\omega\int_{\Gamma}\rho_s\,{\rm d}A + \int_{\partial\Gamma}\boldsymbol{J}_{\rm s}\cdot{\rm d}\boldsymbol{A} = 0\,. \tag{1.323}$$

If the surface Γ is simply connected Stokes' theorem (A.25) can be applied to yield the local continuity equation

$$j\omega\rho_{\rm s}(\boldsymbol{r}) + (\nabla\times\boldsymbol{J}_{\rm s}(\boldsymbol{r}))\cdot\boldsymbol{e}_{\rm n} = 0 \tag{1.324}$$

with $\boldsymbol{e}_{\rm n}$ a normal vector on Γ. This is a special case of the fundamental continuity Equation (1.10). It follows that Equation (1.321) can be rewritten in the form

$$\boldsymbol{E}^{\rm sca}(\boldsymbol{r}) = \frac{1}{j\omega\varepsilon}\int_{\Gamma}\Big(\big(\nabla G^{\varphi}(\boldsymbol{r},\boldsymbol{r}')\big)\big(\nabla'\times\boldsymbol{J}_{\rm s}(\boldsymbol{r}')\big)\cdot\boldsymbol{e}_{\rm n} + k^2\overline{\boldsymbol{G}}^{A}(\boldsymbol{r},\boldsymbol{r}')\boldsymbol{J}_{\rm s}(\boldsymbol{r}')\Big)\,{\rm d}^2r'\,. \tag{1.325}$$

This yields the electric field integral equation

$$\frac{1}{j\omega\varepsilon}\left[\int_{\Gamma}\Big(\big(\nabla G^{\varphi}(\boldsymbol{r},\boldsymbol{r}')\big)\big(\nabla'\times\boldsymbol{J}_{\rm s}(\boldsymbol{r}')\big)\cdot\boldsymbol{e}_{\rm n} + k^2\overline{\boldsymbol{G}}^{A}(\boldsymbol{r},\boldsymbol{r}')\boldsymbol{J}_{\rm s}(\boldsymbol{r}')\Big)\,{\rm d}^2r'\right]\cdot\boldsymbol{e}_t(\boldsymbol{r}) = -E_t^{\rm inc}(\boldsymbol{r})\,. \tag{1.326}$$

Consider again the special case of a z-directed, cylindrical and straight antenna. In this case Equation (1.324) is not valid since the surface of a cylinder is not simply connected. However, after a thin-wire approximation Equation (1.324) can be replaced by

$$j\omega q' + \frac{{\rm d}I}{{\rm d}z} = 0\,, \tag{1.327}$$

with q' the electric charge per unit length and I the total electric current on the antenna, and one finds in analogy to Equation (1.326) the mixed-potential integral equation

$$\frac{1}{j\omega\varepsilon}\int_{-L/2}^{L/2}\left[\frac{\partial G^{\varphi}(z,z')}{\partial z}\frac{\partial I(z')}{\partial z'} + k^2G^{A}_{zz}(z,z')I(z')\right]{\rm d}z' = -E_z^{\rm inc}(z)\,. \tag{1.328}$$

In free space $G^{\varphi}(z,z')$ and $G^{A}(z,z')$ are the same functions and the mixed-potential integral equation simplifies further.

1.5.2.4 Schelkunoff's Equation

For completeness another electric field integral equation is mentioned that is known as Schelkunoff's equation. It requires the condition

$$\frac{\partial G^{\varphi}(z, z')}{\partial z} = -\frac{\partial G^{\varphi}(z, z')}{\partial z'} \tag{1.329}$$

which implies translational invariance of the Green's function, $G^{\varphi}(z, z') = G^{\varphi}(|z - z'|)$. This condition is fulfilled for the Green's function of free space (compare Equation (1.293)), but it will not be valid in general.

If Equation (1.329) holds, the first term in the integral of Equation (1.328) can be integrated by parts,

$$\int_{-L/2}^{L/2} \frac{\partial G^{\varphi}(z, z')}{\partial z'} \frac{\partial I(z')}{\partial z'} \, dz' = G^{\varphi}(z, z') \frac{\partial I(z')}{\partial z'} \bigg|_{z'=-L/2}^{z'=L/2} - \int_{-L/2}^{L/2} G^{\varphi}(z, z') \frac{\partial^2 I(z')}{\partial z'^2} \, . \tag{1.330}$$

Then the mixed-potential integral equation can be rewritten to yield Schelkunoff's equation,

$$\frac{1}{j\omega\epsilon} \int_{-L/2}^{L/2} \left[G^{\varphi}(z, z') \frac{\partial^2 I(z')}{\partial z'^2} + k^2 G_{zz}^{A}(z, z') I(z') \right] dz'$$
$$- \frac{1}{j\omega\varepsilon} \, G^{\varphi}(z, z') \frac{\partial I(z')}{\partial z'} \bigg|_{z'=-L/2}^{z'=L/2} = -E_z^{\mathrm{inc}}(z) \, . \tag{1.331}$$

As in the case of Hallén's equation the integral kernel of Schelkunoff's equation is proportional to $1/|\boldsymbol{r} - \boldsymbol{r}'|$ and, thus, advantageous for numerical evaluation.

References

[1] Truesdell, C. and Toupin, R.A. 'The classical field theories', in *Handbuch der Physik, vol. III/1*, (ed. S. Flügge Springer, Berlin, Germany), 1960, 226–793.

[2] Hehl, F.W. and Obukhov, Y. *Foundations of Classical Electrodynamics: Charge, Flux and Metric*, Birkhäuser, Boston, USA, 2003.

[3] Schouten, J.A. *Tensor Analysis for Physicists*, 2nd edn. Dover, New York, USA, 1989.

[4] Lindell, I.V. *Differential Forms in Electromagnetics*, IEEE Press, Piscataway, NJ, USA, and John Wiley & Sons, New York, USA, 2004.

[5] Rothwell, E.J. and Cloud, M.J. *Electromagnetics*, CRC Press, Boca Raton, USA, 2001, p. 457.

[6] Elliott, R.S. *Electromagnetics – History, Theory and Applications*, IEEE Press, New York, USA, 1992.

[7] Rohrlich, F. *Classical Charged Particles*, Addison-Wesley, Reading, Massachusetts, USA, 1965.

[8] Jackson, J.D. *Classical Electrodynamics*, 3rd edn John Wiley & Sons, Inc., New York, USA, 1998.

[9] Lamb, H. *Hydrodynamics*, 6th edn., Cambridge University Press, Cambridge, UK, 1936 and Dover, New York, USA, 1993.

[10] Lämmerzahl, C. and Hehl, F.W. 'Riemannian light cone from vanishing birefringence in premetric vacuum electrodynamics,' *Phys. Rev. D*, **70**, 2004, 105022.

[11] Hehl, F.W. and Obukhov, Yu.N. 'Linear media in classical electrodynamics and the Post constraint,' *Phys. Lett. A*, **334**, 2005, 249–259.

[12] Obukhov, Yu.N. and Hehl, F.W. 'Measuring a piecewise constant axion field in classical electrodynamics,' *Phys. Lett. A*, **341**, 2005, 357–365.

[13] Olyslager, F. and Lindell, V. 'Electromagnetics and exotic media: A quest for the holy grail,' *IEEE Antennas and Propagation Magazine*, **44**, 2002, 48–58.

[14] Gronwald, F. and Nitsch, J. 'The physical origin of gauge invariance in electrodynamics,' *Electrical Engineering*, **81**, 1999, 363–367.

[15] Gronwald, F. and Nitsch, J. 'The structure of the electrodynamic field as derived from first principles,' *IEEE Antennas and Propagation Magazine*, **43**, 2001, 64–79.
[16] Bohm, D. *Quantum Theory*, Prentice Hall, New York, USA, 1951.
[17] Gilmore R. *Alice in Quantumland*, Copernicus Books, New York, USA, 1995.
[18] Chen, T.-P. and Li, L.-F. *Gauge Theory of Elementary Particle Physics*, Clarendon Press, Oxford, 1984.
[19] Ryder, L. *Quantum Field Theory*, 2nd edn, Cambridge University Press, Cambridge, UK, 1996.
[20] Noether, E. 'Invariante Variationsprobleme', *Nachr. König. Gesell. Wiss. Göttingen, Math.-Phys. Kl.* 1918, 235–257; English translation in: *Transport Theory and Stat. Phys.*, **1**, 1971, 186–207.
[21] Schwinger, J., DeRaad Jr., L.L., Milton, K.A., and Tsai, W. *Classical Electrodynamics*, Perseus Books, Reading, MA, USA, 1998.
[22] Budó, A. *Theoretische Mechanik*, 9th edn., VEB Verlag, Berlin, Germany, 1978.
[23] Morse, P.M. and Feshbach, H. *Methods of Theoretical Physics I+II*, McGraw Hill, New York, USA, 1953.
[24] Collin, R.E. *Field Theory of Guided Waves*, 2nd edn, IEEE Press, New York, USA, 1997.
[25] Griffiths, D.J. *Introduction to Electrodynamics*, 3rd edn, Prentice Hall, New Jersey, USA, 1999.
[26] Sommerfeld, A. *Mechanics of Deformable Bodies*, Academic Press, New York, USA, 1950.
[27] Cohen-Tannoudji, C., Dupont-Roc, J. and Grynberg G. *Photons and Atoms – Introduction to Quantum Electrodynamics*, John Wiley & Sons, Inc., New York, USA, 1997.
[28] Gelfand, I.M. and Shilov, G.E. *Generalized Functions, Vols. 1-4*, Academic Press, New York, USA, 1964–1968.
[29] Schwartz, L. *Mathematics for the Physical Sciences*, Addison-Wesley, Reading, Massachusetts, USA, 1966.
[30] Tai, C.-T. *Dyadic Green Functions in Electromagnetic Theory*, IEEE Press, New York, USA, 1994.
[31] Peterson, A.F., Ray, S.L. and Mittra, R. *Computational Methods for Electromagnetism*, IEEE Press, New York, USA, 1998.
[32] King, R.W.P. *The Theory of Linear Antennas*, Harvard University Press, Cambridge, MA, USA, 1956.
[33] Schelkunoff, S.A. and Friis, H.T. *Antennas – Theory and Practice*, 3rd edn., John Wiley & Sons, Inc., New York, USA, 1966.
[34] Collin, R.E. and Zucker, F.J. *Antenna Theory*, Part I and II, McGraw-Hill, New York, USA, 1969.
[35] King, R.W.P. and Harrison, C.W. *Antennas and Waves: A Modern Approach*, The MIT Press, Cambridge, USA, 1969.
[36] Popović, B.D., Dragović, M.B. and Djordjecić, A.R. *Analysis and Synthesis of Wire Antennas*, Research Studies Press, Chichester, UK, 1982.
[37] Popović , B.D. and Kolundžija B.M. *Analysis of Metallic Antennas and Scatterers*, IEE, London, UK, 1994.
[38] Sarkar, T.K., Djordjević, A.R., and Kolundžija, B.M. 'Method of Moments Applied to Antennas', in *The Handbook of Antennas in Wireless Communications*, (eds L. Godara, and V. Barroso), CRC Press, Boca Raton, USA, 2001.
[39] Hertz, H. 'Ueber sehr schnelle elektrische Schwingungen' and 'Nachtrag zu der Abhandlung ueber sehr schnelle elektrische Schwingungen', *Annalen der Physik und Chemie*, **31**, 1887, 421–448 and 543–544.
[40] Pocklington, H.C. 'Electrical oscillations in wires', *Cambridge Philos. Soc. Proc.*, **9**, 1897, 324–332.
[41] Butler, C.M. and Wilton, D.R. 'Analysis of Various Numerical Techniques Applied to Thin-Wire Scatterers', *IEEE Trans. Antennas Propagat.*, **23**, 1975, 534–540.
[42] Werner, D.H., Huffman, J.A. and Werner, P.L. 'Techniques for Evaluating the Uniform Current Vector Potential at the Isolated Singularity of the Cylindrical Wire Kernel', *IEEE Trans. Antennas Propagat.*, **42**, 1994, 1549–1553.
[43] Hallén, E. 'Theoretical investigations into the transmitting and receiving qualities of antennae', *Nova Acta Regiae Soc. Sci. Upsaliensis*, **4**, 1938, 1–44.

2

Nonuniform Transmission-Line Systems

Transmission lines consist of metallic structures that transmit electromagnetic signals and energy. In this respect they are similar to the systems of transmitting and receiving antennas. However, the physical mechanisms that govern the electromagnetic transmission along transmission lines are quite different from the electromagnetic transmission between pairs of antennas. This is illustrated in Figure 2.1.

Between a pair of antennas the electromagnetic transmission results from a propagating electromagnetic field which, for practical purposes, can often be approximated by a radiation field. This does not mean that in such a situation no Coulomb field is present. A Coulomb field will be related to the electric charges that move along the antennas and will constitute their near-field. However, in many cases the transmitting and receiving antennas are sufficiently far apart so that the main coupling is mediated by the radiation field which resembles a freely propagating electromagnetic field.[1] Electric charges are *not* involved in the actual electromagnetic transmission that occurs between the antennas. They are only required at the beginning and at the end of the transmission in order to generate and receive the transmitting electromagnetic field respectively.

The electromagnetic transmission along a transmission line *does* involve electric charges. These charges are located on the transmission line which normally consists of a highly conducting material. They are accompanied by a Coulomb field which dominates their mutual electromagnetic interaction at short distances. While the electric charges are accelerated, a radiation field will be produced. In particular, this will happen at high frequencies or if the transmission-line is strongly curved or bent. Normally, the generation of a radiation field by electric charges on a transmission line is an unwanted effect which may affect the properties of a transmission line. For many situations this effect is small and negligible.

[1] It should be recalled at this point that in the presence of moving electric charges there will be an extended nonvanishing transverse current $\mathbf{J}_\perp$ which, in turn, will couple to the transverse parts of the electromagnetic field. In this general case there is no free propagation of an electromagnetic field since the wave Equations (1.212) and (1.213) will always contain a nonvanishing source term. The propagation can only be *approximately* free at large distances to moving charges.

Radiating Nonuniform Transmission-Line Systems and the Partial Element Equivalent Circuit Method Jürgen Nitsch, Frank Gronwald, and Günter Wollenberg

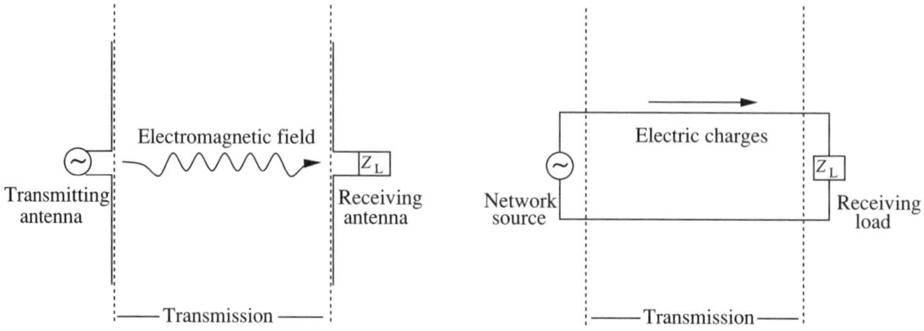

Figure 2.1 Electromagnetic transmission by means of a pair of antennas (left) and a transmission line (right). Between the antennas an electromagnetic field mediates the actual transmission while the transmission line provides electric charges that mediate the transmission between the source and the load.

It follows that the dynamics along a transmission line is determined from the motion of electric charges and not from the degrees of freedom of an electromagnetic field. The fact that for the electromagnetic transmission along a transmission line the degrees of freedom of the electromagnetic field can often be neglected implies that the *classical transmission-line theory* (classical TLT) has a much simpler structure than the Maxwell theory. The classical TLT is a limiting case of the Maxwell theory and contains the electric current $I(z)$, representing electric charges, and the electric voltage $V(z)$, representing the associated Coulomb fields, as main physical quantities [1–4]. Clearly, these two quantities are not independent of each other. They are related by the *telegrapher equations* which, for a two-wire transmission line and in the frequency domain, are of the form

$$\frac{\partial V(z)}{\partial z} + (j\omega L' + R')I(z) = V_s'(z), \tag{2.1}$$

$$\frac{\partial I(z)}{\partial z} + (j\omega C' + G')V(z) = I_s'(z). \tag{2.2}$$

The prime quantities denote *per-unit-length parameters*. Explicitly, the quantities L', R', C' and G' are the per-unit-length inductance, resistance, capacitance and conductance, respectively. They represent geometric and material properties of the transmission line. The quantities V_s' and I_s' denote *distributed voltage and current sources*, respectively, and represent the electromagnetic excitation of the line. For a given exciting electromagnetic field their explicit form depends on the choice of one of several but equivalent coupling models which is used to calculate V_s' or I_s' [5–7].

Even though the classical TLT is approximate it is often preferred over the exact Maxwell theory because the telegrapher Equations (2.1) and (2.2), which constitute a set of coupled first-order differential equations, are much easier to solve than the complete set of the Maxwell equations. The price that has to be paid for this simplification is the limited scope of the classical TLT. It is obvious that prior to the application of classical TLT its limitations need to be understood.

If the conditions for an application of the classical TLT are not met it is suggested that the range of applicability of the telegrapher equations should be increased while keeping their mathematical structure. This leads to the subject of a *generalized transmission-line theory* (generalized TLT) which will be developed in the following. There are the relations

$$\text{classical TLT} \quad \subseteq \quad \text{generalized TLT} \quad \subset \quad \text{Maxwell theory}.$$

A generalized TLT, in particular, is required to take into account radiation effects properly. In order to arrive at a generalized TLT, two meaningful possibilities come to mind:

(1) supplement the classical TLT with corrective terms that take into account additional effects;
(2) derive more general Telegrapher equations from the exact Maxwell theory.

The second possibility appears to be the more systematic and logical one and will be pursued further in this chapter.[2]

2.1 Multiconductor Transmission Lines: General Equations

2.1.1 Geometric Representation of Nonuniform Transmission Lines

In order to derive a generalized TLT it is necessary to characterize mathematically the geometry of nonuniform transmission lines. This requires some techniques from differential geometry, as oulined, for example, in [10].

The position of a conductor i is given by a general space curve. The vector $\mathbf{C}_i(\zeta)$ points to that curve, which must be smooth and two-times differentiable. For all conductors the curve parameter is chosen to be the same. One side of the conductors, for example the beginning of the line, corresponds to the curve parameter ζ_0 and the other side coincides with ζ_l. The conductors may have different lengths, they can be bent or have loops. There is no restriction on the shape of the conductor, as long as it can be described by a space curve.

One may raise the question, how to determine the space curve of a given conductor. There is no unique way, in fact there are infinite possibilities. The most favorable way is to find a circular tube that tightly surrounds the conductor and choose the center line of this tube.

2.1.1.1 Local Coordinate System

A local coordinate system (see Figure 2.2) can be constructed using the so-called *Frenet* frame. This frame is composed of three vectors, the tangential unit vector $\mathbf{T}_i^{\mathrm{u}}(\zeta)$, the normal unit vector $\mathbf{N}_i^{\mathrm{u}}(\zeta)$ and the binormal unit vector $\mathbf{B}_i^{\mathrm{u}}(\zeta)$. The tangential vector can be calculated by differentiating $\mathbf{C}_i(\zeta)$ with respect to ζ:

$$\mathbf{T}_i = \frac{\partial \mathbf{C}_i}{\partial \zeta}. \tag{2.3}$$

[2] With the kind permission of H. Haase and T. Steinmetz, significant portions of their dissertations [8,9] are adapted in this chapter.

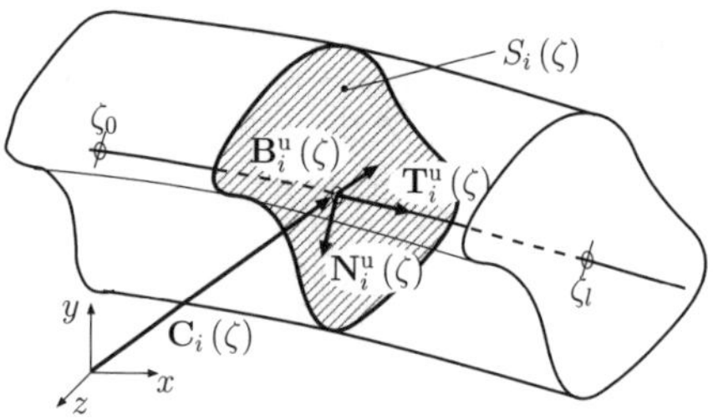

Figure 2.2 A conductor with the central space curve and the tangential, normal and binormal vectors.

This vector is not a unit vector because the curve parameter is not the arc length, instead the unit vector $\mathbf{T}_i^{\mathrm{u}}$ is

$$\mathbf{T}_i^{\mathrm{u}}(\zeta) = \frac{\mathbf{T}_i(\zeta)}{u_{ii}(\zeta)}, \tag{2.4}$$

where u_{ii} is the length of the tangential vector at the location ζ:

$$u_{ii}(\zeta) = |\mathbf{T}_i(\zeta)|. \tag{2.5}$$

The normal and binormal unit vectors are given by

$$\mathbf{N}_i^{\mathrm{u}} = \frac{1}{\kappa_i u_{ii}} \frac{\partial \mathbf{T}_i^{\mathrm{u}}}{\partial \zeta} \tag{2.6}$$

and

$$\mathbf{B}_i^{\mathrm{u}} = \mathbf{T}_i^{\mathrm{u}} \times \mathbf{N}_i^{\mathrm{u}}, \tag{2.7}$$

with the quantity κ_i being the curvature

$$\kappa_i = \left|\frac{\partial \mathbf{T}_i^{\mathrm{u}}}{\partial \zeta}\right| \frac{1}{u_{ii}}. \tag{2.8}$$

It is noted that for straight conductors the derivative of the tangential vector vanishes and it is thus not possible to define a normal vector. Under this circumstances an alternative direction is chosen which is more convenient.

Next, a polar coordinate system in the plane spanned by $\mathbf{N}_i^{\mathrm{u}}$ and $\mathbf{B}_i^{\mathrm{u}}$ is introduced. Every point $\mathbf{x}_i$ inside the conductor volume V_i can then be characterized by coordinates (ζ, r, α) (see Figure 2.3):

$$\mathbf{x}_i(\zeta, r, \alpha) = \mathbf{C}_i(\zeta) + \mathbf{N}_i^u(\zeta)\, r\cos\alpha + \mathbf{B}_i^u(\zeta)\, r\sin\alpha. \tag{2.9}$$

For all conductors the coordinate ζ is in the interval $[\zeta_0, \zeta_l]$. The upper and lower bounds coincide with the two ends of the conductors. The angle α can have values between 0 and 2π. The radius r is bounded by $\hat{r}_i(\zeta, \alpha)$, which describes the surface of the conductor. This way of specifying the conductor volume does not allow all surface shapes, but the most interesting shapes are covered.

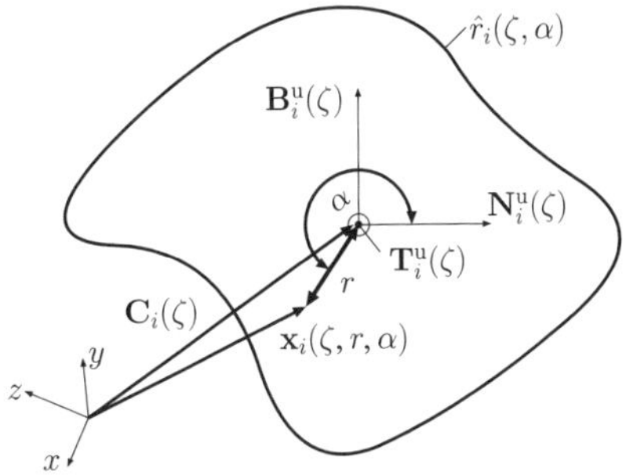

Figure 2.3 The conductor cross section with the polar coordinate system.

2.1.1.2 Tangential Surface Vector

In order to apply a boundary condition for the electric field the tangential vector on the surface of a conductor must be known. With the above coordinates the surface of the conductor is given by

$$\hat{\mathbf{x}}_i(\zeta,\alpha) = \mathbf{C}_i(\zeta) + \mathbf{N}_i^{\mathrm{u}}(\zeta)\,\hat{r}(\zeta,\alpha)\cos\alpha + \mathbf{B}_i^{\mathrm{u}}(\zeta)\,\hat{r}(\zeta,\alpha)\sin\alpha. \tag{2.10}$$

The tangential vector on the surface in the direction of the conductor is then

$$\frac{\partial\hat{\mathbf{x}}_i(\zeta,\alpha)}{\partial\zeta} = \mathbf{T}_i(\zeta) + \frac{\partial\left(\mathbf{N}_i^{\mathrm{u}}(\zeta)\,\hat{r}(\zeta,\alpha)\right)}{\partial\zeta}\cos\alpha + \frac{\partial\left(\mathbf{B}_i^{\mathrm{u}}(\zeta)\,\hat{r}(\zeta,\alpha)\right)}{\partial\zeta}\sin\alpha. \tag{2.11}$$

To simplify matters, it is required that the length of the last two vectors in this sum should be small compared with the first one. Thus if

$$|\mathbf{T}_i(\zeta)| \gg \left|\frac{\partial\left(\mathbf{N}_i^{\mathrm{u}}(\zeta)\,\hat{r}(\zeta,\alpha)\right)}{\partial\zeta}\cos\alpha + \frac{\partial\left(\mathbf{B}_i^{\mathrm{u}}(\zeta)\,\hat{r}(\zeta,\alpha)\right)}{\partial\zeta}\sin\alpha\right| \tag{2.12}$$

one can write

$$\frac{\partial\hat{\mathbf{x}}_i(\zeta,\alpha)}{\partial\zeta} \approx \mathbf{T}_i(\zeta). \tag{2.13}$$

This restricts the theory to conductors where the cross-sectional shape varies very slowly along ζ, which is true for many configurations. For a constant cross section the last two terms even vanish. If the above condition is not fulfilled the correct tangential vector must be taken into account. This does not influence the derivation, but it would make the equations more complicated.

2.1.1.3 Volume and Surface Integrals

It will be necessary to integrate a function $F(\ldots,\mathbf{x}')$ over the conductor volume V_i, for example

$$\int_{V_i} F(\ldots,\mathbf{x}')\mathrm{d}v'. \tag{2.14}$$

With the coordinate system just established the volume element $\mathrm{d}v'$ becomes

$$\mathrm{d}v' = \left|\frac{\partial(x, y, z)}{\partial(\zeta, r, \alpha)}\right| \mathrm{d}r\,\mathrm{d}\alpha\,\mathrm{d}\zeta \tag{2.15}$$

$$= (1 - r\kappa_i(\zeta)\cos\alpha)\, r\,\mathrm{d}r\,\mathrm{d}\alpha\, u_{ii}(\zeta)\,\mathrm{d}\zeta \tag{2.16}$$

$$= f_{c_i}(\zeta, r, \alpha)\, r\,\mathrm{d}r\,\mathrm{d}\alpha\, u_{ii}(\zeta)\,\mathrm{d}\zeta, \tag{2.17}$$

where

$$f_{c_i}(\zeta, r, \alpha) = (1 - r\,\kappa_i(\zeta)\cos\alpha) \tag{2.18}$$

and the integral can be written as

$$\int_{V_i} F(\ldots, \mathbf{x}')\,\mathrm{d}v' = \int_{\zeta_0}^{\zeta_l}\int_0^{2\pi}\int_0^{\hat{r}_i(\zeta,\alpha)} F(\ldots, \mathbf{x}'_i)\, f_{c_i} r'\,\mathrm{d}r'\,\mathrm{d}\alpha'\, u_{ii}\,\mathrm{d}\zeta'. \tag{2.19}$$

The two inner integrals correspond to an integration over the cross section of the conductor. This will be abbreviated as

$$\int_0^{2\pi}\int_0^{\hat{r}_i(\zeta',\alpha')} F(\ldots, \mathbf{x}'_i) f_{c_i} r'\,\mathrm{d}r'\,\mathrm{d}\alpha' = \int_{S_i(\zeta')} F(\ldots, \mathbf{x}'_i)\, f_{c_i}\,\mathrm{d}a'. \tag{2.20}$$

2.1.2 Derivation of Generalized Transmission-Line Equations

Based on the continuity equation the first of the two generalized telegrapher equations is derived next. This step is rather simple because it only involves the adaption of the continuity equation to the conductor geometry.

The second telegrapher equation is then derived from an electric field integral equation. This step requires some more advanced techniques. The basic idea is to formulate a trial function for the current which corresponds to a solution of a second-order differential equation. If this trial function is inserted into the integral equation it becomes the second telegrapher equation.

2.1.2.1 Continuity Equation

The continuity equation connects the charge density with the current density (see Section 1.1.1). To adapt this equation to the conductor geometry, one can define a conductor current and a per-unit-length charge.

First, the integral form of Equation (1.10) is applied to a volume enclosing part of the conductor i. The two cross-sectional surfaces $S_i(\zeta_0)$ at the beginning of the conductor and $S_i(\zeta)$ at position ζ somewhere on the conductor, must belong to this volume surface. Then,

$$\oint_{\partial V_{i_{\zeta_0\ldots\zeta}}} \mathbf{J}\,\mathrm{d}\mathbf{a} + \mathrm{j}\omega\int_{V_{i_{\zeta_0\ldots\zeta}}} \rho\,\mathrm{d}v = 0. \tag{2.21}$$

Because there is no current through the surface of the conductor, except at the cross sections, this equation can be written as

$$\int_{S_i(\zeta)} \mathbf{J}\cdot\mathbf{T}_i^{\mathrm{u}}(\zeta)\,\mathrm{d}a - \int_{S_i(\zeta_0)} \mathbf{J}\cdot\mathbf{T}_i^{\mathrm{u}}(\zeta_0)\,\mathrm{d}a + j\omega\int_{\zeta_0}^{\zeta} u_{ii}\int_{S_i(\zeta')} \rho\, f_{c_i}\,\mathrm{d}a\,\mathrm{d}\zeta' = 0. \tag{2.22}$$

This expression is differentiated with respect to ζ resulting in

$$\frac{\partial}{\partial \zeta}\int_{S_i(\zeta)} \mathbf{J}\cdot\mathbf{T}_i^u(\zeta)\,\mathrm{d}a + j\omega u_{ii}\int_{S_i(\zeta)} \rho\, f_{c_i}\,\mathrm{d}a = 0. \tag{2.23}$$

The first term is the derivative of the conductor current and the second term is a per-unit-length charge multiplied with $j\omega$. With the definitions

$$i_i(\zeta) := \int_{S_i(\zeta)} \mathbf{J}(\mathbf{x}_i)\cdot\mathbf{T}_i^{\mathrm{u}}(\zeta)\,\mathrm{d}a \tag{2.24}$$

and

$$q_i(\zeta) := \int_{S_i(\zeta)} \rho(\mathbf{x}_i)\, f_{c_i}\,\mathrm{d}a\, u_{ii}(\zeta) \tag{2.25}$$

Equation (2.23) becomes

$$\frac{\partial i_i}{\partial \zeta} + j\omega q_i = 0, \tag{2.26}$$

which is the continuity equation tailored for the current and the per-unit-length charge in a conductor, and is called the first extended telegrapher equation.

2.1.2.2 Reconstruction of the Densities

For the calculation of electromagnetic fields the current density and the charge density are required. With a given current and per-unit-length charge, respectively, the densities can be reconstructed with the aid of distribution functions as follows:

$$\mathbf{J}(\mathbf{x}_i) = \mathbf{T}_i^{\mathrm{u}}(\zeta)\, i_i(\zeta)\, d_{\mathbf{J}_i}(\mathbf{x}_i) \tag{2.27}$$

$$\rho(\mathbf{x}_i) = q_i(\zeta)\,\frac{d_{\rho_i}(\mathbf{x}_i)}{u_{ii}(\zeta)}. \tag{2.28}$$

The distribution functions are defined as

$$d_{\mathbf{J}_i}(\mathbf{x}_i) := \frac{\mathbf{J}(\mathbf{x}_i)\cdot\mathbf{T}_i^{\mathrm{u}}(\zeta)}{i_i(\zeta)} \tag{2.29}$$

and

$$d_{\rho_i}(\mathbf{x}_i) := \frac{\rho(\mathbf{x}_i)}{q_i(\zeta)}\, u_{ii}(\zeta). \tag{2.30}$$

The present theory is not able to predict these functions, rather it is necessary to specify them before any calculations are made. In many cases it is sufficient either to select some simple function or to 'guess' the current and charge distribution. For thin wires, for instance, the thin-wire approximation can be used (see below).

In general, this approach takes into account current and charge displacement effects, such as the skin effect; however, most of the time the solutions for the current and charge distribution inside the conductors are unknown.

There are also several publications dealing with the solution for the fields and source distributions inside straight wires, for example [11–13]. Unfortunately, it is not possible to transfer the results directly to the nonuniform lines that are discussed in the present context.

In [14] a Fourier series approach was used to calculate the azimuthal current distribution in a straight thick wire above ground. This seems to be one possible solution to this problem. One could expand the distribution functions in a combination of a Fourier (for α) and a Taylor series (for r). Then a system of telegrapher equations for every coefficient of these series expansions would be obtained.

2.1.3 Mixed Potential Integral Equation

The second equation arises from the mixed potential integral equation (see also Equation (1.328))

$$\nabla \frac{1}{\epsilon} \int_V G\left(\mathbf{x}, \mathbf{x}'\right) \rho\left(\mathbf{x}'\right) \mathrm{d}v' + j\omega\mu \int_V G\left(\mathbf{x}, \mathbf{x}'\right) \mathbf{J}\left(\mathbf{x}'\right) \mathrm{d}v' + \frac{\mathbf{J}(\mathbf{x})}{\sigma} = \mathbf{E}^{(i)}(\mathbf{x}). \tag{2.31}$$

If this equation is applied to the geometry of a multiconductor transmission line the volume integrals can be split into a sum of volume integrals over the conductor volumes. The observation point $\mathbf{x}$ is on the surface of the conductor j and thus becomes $\hat{\mathbf{x}}_j$. Furthermore, Equations (2.27) and (2.28) are substituted for $\mathbf{J}$ and ρ. Now the equation is multiplied by the tangential vector $\mathbf{T}_j^{\mathrm{u}}$, to extract the tangential component of the current and fields along the conductors. Eventually, the average on the surface is computed by performing the operation

$$\frac{1}{2\pi} \int_0^{2\pi} f(\ldots, \alpha)\, \mathrm{d}\alpha. \tag{2.32}$$

This yields an adapted version of the integral Equation (2.31) for a nonuniform multiconductor transmission line. A detailed derivation is shown in Appendix D, resulting in the equation

$$\frac{\partial}{\partial \zeta} \sum_{i=1}^{N} \int_{\zeta_0}^{\zeta_l} k_{c_{ji}}\left(\zeta, \zeta'\right) q_i\left(\zeta'\right) \mathrm{d}\zeta' + j\omega \sum_{i=1}^{N} \int_{\zeta_0}^{\zeta_l} k_{l_{ji}}\left(\zeta, \zeta'\right) i_i\left(\zeta'\right) \mathrm{d}\zeta' + z_{jj}(\zeta) i_j(\zeta) = v_j^{(i)'}(\zeta), \tag{2.33}$$

where the source term

$$v_j^{(i)'}(\zeta) := \frac{1}{2\pi} \int_0^{2\pi} \mathbf{T}_j(\zeta) \cdot \mathbf{E}^{(i)}\left(\hat{\mathbf{x}}_j\right) \mathrm{d}\alpha \tag{2.34}$$

is the averaged incident electric field. The variable z_{jj} represents the per-unit-length surface impedance of conductor j and is calculated from

$$z_{jj}(\zeta) := \frac{u_{jj}}{2\pi} \int_0^{2\pi} \frac{d_{\mathbf{J}_i}\left(\hat{\mathbf{x}}_i\right)}{\sigma} \mathrm{d}\alpha. \tag{2.35}$$

The integral kernels are composed of Green's functions, current or charge distribution functions and geometrical factors and read

$$k_{l_{ji}}\left(\zeta, \zeta'\right) = \tfrac{\mu}{2\pi} \mathbf{T}_j(\zeta) \cdot \mathbf{T}_i\left(\zeta'\right) \int_0^{2\pi} \int_{S_i(\zeta')} G\left(\hat{\mathbf{x}}_j, \mathbf{x}_i'\right) d_{\mathbf{J}_i}\left(\mathbf{x}_i\right) f_{c_i} da'\, \mathrm{d}\alpha \tag{2.36}$$

and

$$k_{c_{ji}}\left(\zeta, \zeta'\right) = \tfrac{1}{2\pi\epsilon} \int_0^{2\pi} \int_{S_i(\zeta')} G\left(\hat{\mathbf{x}}_j, \mathbf{x}_i'\right) d_{\rho_i}\left(\mathbf{x}_i\right) f_{c_i} \mathrm{d}a'\, \mathrm{d}\alpha. \tag{2.37}$$

2.1.3.1 Thin-Wire Approximation

Very often it is sufficient to think of the conductors as thin wires with a circular cross section. The thin-wire approximation then assumes that the currents and charges are concentrated at the center of the wires. In this situation the integral kernels can be simplified significantly:

$$k_{l_{ji}}\left(\zeta,\zeta'\right) := \mu \mathbf{T}_j(\zeta)\cdot\mathbf{T}_i\left(\zeta'\right) G\left(\mathbf{C}_j(\zeta),\mathbf{C}_i\left(\zeta'\right)\right), \tag{2.38}$$

$$k_{c_{ji}}\left(\zeta,\zeta'\right) := \frac{1}{\epsilon} G\left(\mathbf{C}_j(\zeta),\mathbf{C}_i\left(\zeta'\right)\right). \tag{2.39}$$

If $\mathbf{C}_j$ and $\mathbf{C}_i$ point to the center of the same conductor, $\mathbf{C}_i$ must be moved from the center to the surface.

With this approximation many problems can be solved. However, the current and charge distributions inside the conductors are fixed and effects such as the skin effect or the proximity effect cannot be taken into account.

2.1.3.2 Representation as Matrix Equations

For a transmission line with N conductors there are now $2N$ equations (N equations of type (2.26) and N equations of type (2.33)). There are also $2N$ unknowns ($q_1 \ldots q_N, i_1 \ldots i_N$). This system of equations can be combined into two matrix equations. The first set of telegrapher Equations (2.26) may then be written as

$$\frac{\partial}{\partial\zeta}\mathbf{i} + j\omega\mathbf{q} = \mathbf{0}. \tag{2.40}$$

The column vectors $\mathbf{i} = [i_1 \ldots i_N]^{\mathsf{T}}$ and $\mathbf{q} = [q_1 \ldots q_N]^{\mathsf{T}}$ contain the conductor currents and the per-unit-length charges, respectively. The second Equation (2.33) becomes

$$\frac{\partial}{\partial\zeta}\int_{\zeta_0}^{\zeta_l} \mathbf{k}_c\left(\zeta,\zeta'\right)\mathbf{q}\left(\zeta'\right)\mathrm{d}\zeta' + j\omega\int_{\zeta_0}^{\zeta_l} \mathbf{k}_l\left(\zeta,\zeta'\right)\mathbf{i}\left(\zeta'\right)\mathrm{d}\zeta' + \mathbf{z}(\zeta)\mathbf{i}(\zeta) = \mathbf{v}^{(i)'}(\zeta). \tag{2.41}$$

The elements of the matrix functions $\mathbf{k}_c$ and $\mathbf{k}_l$ are the integral kernels of Equations (2.36) and (2.37), while the column vector $\mathbf{v}^{(i)'}$ contains the distributed sources. The diagonal matrix $\mathbf{z}$ accommodates the per-unit-length surface impedances.

2.1.3.3 Current and Charge Trial Function

Equation (2.41) already shows some similarities to the telegrapher equation. However, the unknown quantities $\mathbf{i}$ and $\mathbf{q}$ are inside integrals. It is necessary to pull the current and the per-unit-length charge out of the integrals. This can be achieved by introducing trial functions for the current and the per-unit-length charge.

King presented similar equations for uniform lines in [1]. His trial function is a Taylor series expansion of the current, truncated after the linear term. He shows that only these terms are important, which is due to the TEM mode. Since there is no exclusive TEM mode anymore, the trial function is more complicated.

On an excited transmission line there are usually forward and backward traveling waves. These waves are governed by wave equations. Also the currents in the conductors obey wave equations. Thus it is obvious to use the solution of a wave equation as a trial function for the current. In the frequency domain a wave equation is usually a second-order ordinary differential equation (ODE) with, in the most general case, nonconstant coefficients. For the N-conductor system this yields N coupled ODEs, which are most conveniently written as a matrix equation:

$$\left(\frac{\partial^2}{\partial\zeta^2} + j\omega\mathbf{P}_{t_{11}}(\zeta)\frac{\partial}{\partial\zeta} + \omega^2\mathbf{P}_{t_{12}}(\zeta)\right)\mathbf{i}(\zeta) = -j\omega\mathbf{q}'_t(\zeta). \tag{2.42}$$

The matrices $\mathbf{P}_{t_{11}}$ and $\mathbf{P}_{t_{12}}$ are the coefficient matrices, and the column vector $\mathbf{q}'_t$ is the excitation. With the aid of Equation (2.40) this equation can be converted into a system of coupled first-order differential equations. In supermatrix notation this becomes

$$\frac{\partial}{\partial\zeta}\begin{bmatrix}\mathbf{q}(\zeta)\\ \mathbf{i}(\zeta)\end{bmatrix} + j\omega\underbrace{\begin{bmatrix}\mathbf{P}_{t_{11}} & \mathbf{P}_{t_{12}}\\ \mathbf{1} & \mathbf{0}\end{bmatrix}}_{\overline{\mathbf{P}}_t}\begin{bmatrix}\mathbf{q}(\zeta)\\ \mathbf{i}(\zeta)\end{bmatrix} = \begin{bmatrix}\mathbf{q}'_t(\zeta)\\ \mathbf{0}\end{bmatrix}. \tag{2.43}$$

The general solution of Equation (2.43) can be written in a very compact form:

$$\begin{bmatrix}\mathbf{q}(\zeta')\\ \mathbf{i}(\zeta')\end{bmatrix} = \mathcal{M}_{\zeta}^{\zeta'}\{-j\omega\overline{\mathbf{P}}_t\}\begin{bmatrix}\mathbf{q}(\zeta)\\ \mathbf{i}(\zeta)\end{bmatrix} + \int_{\zeta}^{\zeta'}\mathcal{M}_{\xi}^{\zeta'}\{-j\omega\overline{\mathbf{P}}_t\}\begin{bmatrix}\mathbf{q}'_t(\xi)\\ \mathbf{0}\end{bmatrix}\mathrm{d}\xi, \tag{2.44}$$

where $\mathcal{M}_{\zeta}^{\zeta'}\{-j\omega\overline{\mathbf{P}}_t\}$ is the product integral [15, 16] (see also Appendix E). This expression is the desired trial function. Although, the supermatrix $\overline{\mathbf{P}}_t$ and the excitation vector $\mathbf{q}'_t$ are unknown, this expression gives us a very general relationship for the currents and per-unit-length charges at two different positions along the transmission lines.

2.1.3.4 Generalized Telegrapher Equations and TLST

In the next step to the generalized telegrapher equations, this trial function (Equation (2.44)) is substituted into Equation (2.41). For this Equation (2.41) may be rearranged to

$$\left[\mathbf{1} j\omega\,\mathbf{1}\frac{\partial}{\partial\zeta}\right]_{(N,2N)}\int_{\zeta_0}^{\zeta_l}\overline{\mathbf{K}}(\zeta,\zeta')\begin{bmatrix}\mathbf{q}(\zeta')\\ \mathbf{i}(\zeta')\end{bmatrix}\mathrm{d}\zeta' + \mathbf{z}(\zeta)\mathbf{i}(\zeta) = \mathbf{v}^{(i)'}(\zeta), \tag{2.45}$$

where the supermatrix $\overline{\mathbf{K}}$ is given by

$$\overline{\mathbf{K}} := \begin{bmatrix}\mathbf{0} & \mathbf{k}_l\\ \mathbf{k}_c & \mathbf{0}\end{bmatrix}. \tag{2.46}$$

Now Equation (2.44) can be directly inserted and after some minor rearrangements the result becomes

$$\mathbf{I}_{21}\frac{\partial\mathbf{q}}{\partial\zeta} + \mathbf{I}_{22}\frac{\partial\mathbf{i}}{\partial\zeta} + \left(j\omega\mathbf{I}_{11} + \frac{\partial\mathbf{I}_{21}}{\partial\zeta}\right)\mathbf{q} + \left(j\omega\mathbf{I}_{12} + \frac{\partial\mathbf{I}_{22}}{\partial\zeta} + \mathbf{z}\right)\mathbf{i} + j\omega\mathbf{I}_{01} + \frac{\partial\mathbf{I}_{02}}{\partial\zeta} = \mathbf{v}^{(i)'}. \tag{2.47}$$

The new matrices and vectors $\mathbf{I}_{ij}$ are the blocks of the supermatrix

$$\overline{\mathbf{I}}(\zeta) = \int_{\zeta_0}^{\zeta_l}\overline{\mathbf{K}}(\zeta,\zeta')\,\mathcal{M}_{\zeta}^{\zeta'}\{-j\omega\overline{\mathbf{P}}_t\}\,\mathrm{d}\zeta' \tag{2.48}$$

and of the supervector

$$\begin{bmatrix} \mathbf{I}_{10}(\zeta) \\ \mathbf{I}_{20}(\zeta) \end{bmatrix} = \int_{\zeta_0}^{\zeta_l} \overline{\mathbf{K}}(\zeta, \zeta') \int_{\zeta}^{\zeta'} \mathcal{M}_{\xi}^{\zeta'} \{-j\omega\overline{\mathbf{P}}_t\} \begin{bmatrix} \mathbf{q}'_t(\xi) \\ \mathbf{0} \end{bmatrix} \mathrm{d}\xi \mathrm{d}\zeta'. \tag{2.49}$$

Equations (2.40) and (2.47) are now combined into a supermatrix equation

$$\frac{\partial}{\partial \zeta} \begin{bmatrix} \mathbf{q}(\zeta) \\ \mathbf{i}(\zeta) \end{bmatrix} + j\omega\overline{\mathbf{P}}(\zeta) \begin{bmatrix} \mathbf{q}(\zeta) \\ \mathbf{i}(\zeta) \end{bmatrix} = \begin{bmatrix} \mathbf{q}'_s(\zeta) \\ \mathbf{0} \end{bmatrix} \tag{2.50}$$

where

$$\overline{\mathbf{P}} = \begin{bmatrix} \mathbf{I}_{21} & \mathbf{I}_{22} \\ \mathbf{0} & \mathbf{1} \end{bmatrix}^{-1} \begin{bmatrix} \mathbf{I}_{11} + \dfrac{1}{j\omega} \dfrac{\partial \mathbf{I}_{21}}{\partial \zeta} & \mathbf{I}_{12} + \dfrac{1}{j\omega} \left(\dfrac{\partial \mathbf{I}_{22}}{\partial \zeta} + \mathbf{z} \right) \\ \mathbf{1} & \mathbf{0} \end{bmatrix} \tag{2.51}$$

and

$$\mathbf{q}'_s = \mathbf{I}_{21}{}^{-1} \left(\mathbf{v}^{(i)'} - j\omega\mathbf{I}_{10} - \frac{\partial \mathbf{I}_{20}}{\partial \zeta} \right). \tag{2.52}$$

This is an important result which constitutes the desired generalized transmission-line theory. With the introduction of the trial function the mixed potential integral equation together with the continuity relation is cast into a system of first-order ODEs. These equations are identical to the ODEs, the solution of which is the trial function. Thus the current is indeed governed by a second-order ODE.

Equation (2.50) has the same structure as the telegrapher equations, although it encompasses the full integral equations which come directly from Maxwell's theory. No restricting assumptions are made, apart from assumptions about the current and charge distribution inside the conductors. The equations are valid for a nonuniform multiconductor transmission line and take into account all field modes and physical effects that might occur, including radiation losses. The corresponding framework of a generalized transmission-line theory will be called 'Transmission-Line Super Theory' (TLST) in the following.

2.1.4 Computation of Generalized Transmission-Line Parameters

Before the above generalized telegrapher equations can be solved, the parameters and source terms must be determined. This is the most complicated part because it involves the solution of integral equations.

2.1.4.1 Parameters

During the derivation of the generalized telegrapher equations a formula was obtained for the determination of the parameters (see Equation (2.51)). For evaluation it is necessary to compute $\overline{\mathbf{I}}$ which depends on the parameters of the trial function $\overline{\mathbf{P}}_t$. Unfortunately these parameters are unknown. However, Equations (2.43) and (2.50) contain the same physical quantities, and thus the coefficients as well as the source terms must be identical. Therefore it is meaningful to set

$$\overline{\mathbf{P}}_t = \overline{\mathbf{P}} = \begin{bmatrix} \mathbf{I}_{21}^{-1}\left(\mathbf{I}_{11} - \mathbf{I}_{22} + \frac{1}{j\omega}\frac{\partial \mathbf{I}_{21}}{\partial \xi}\right) & \mathbf{I}_{21}^{-1}\left(\mathbf{I}_{12} + \frac{1}{j\omega}\left(\mathbf{z} + \frac{\partial \mathbf{I}_{22}}{\partial \xi}\right)\right) \\ \mathbf{1} & \mathbf{0} \end{bmatrix}. \quad (2.53)$$

If Equation (2.53) is inserted into Equation (2.48) an integral equation for $\overline{\mathbf{I}}$ is obtained:

$$\overline{\mathbf{I}}(\zeta) = \int_{\zeta_0}^{\zeta_l} \overline{\mathbf{K}}(\zeta, \zeta') \mathcal{M}_\zeta^{\zeta'} \{-j\omega\overline{\mathbf{P}}\} \, \mathrm{d}\zeta'. \quad (2.54)$$

There seems to be no analytic way to explicitly solve for $\overline{\mathbf{I}}$ in the general case. However, an iterative procedure can be applied to determine $\overline{\mathbf{I}}$:

$$\overline{\mathbf{I}}^{(k+1)} = \mathcal{L}_\mathrm{p}\left\{\overline{\mathbf{I}}^{(k)}\right\}. \quad (2.55)$$

The operation $\mathcal{L}_\mathrm{p}$ is given by Equations (2.53) and (2.54). $\overline{\mathbf{I}}^{(k)}$ is the approximation for $\overline{\mathbf{I}}$ after the kth iteration. The final result of this operation is then used to calculate the parameters with the aid of Equations (2.51) or (2.53).

Concerning convergence of the procedure, for several examples it has been shown that one can obtain very good results from only one iteration. Additionally, for line setups, where analytical solutions are possible, the iterative approach yields the same results as other methods (see Section 2.3). However, there is no general proof for the general convergence of the iteration procedure.

To start the iteration a zeroth order term $\overline{\mathbf{I}}^{(0)}$ is chosen to begin with. In this context it is advantageous to obtain as much information out of Equation (2.54) as possible. One can expand $\overline{\mathbf{I}}$ into a series:

$$\overline{\mathbf{I}} = \overline{\mathbf{I}}_0 + j\omega\overline{\mathbf{I}}_1 + \dots. \quad (2.56)$$

For a pure TEM mode the parameters are frequency independent. Consequently all but the constant term of this series will vanish. If there are additional modes on the line the higher-order terms occur. However, for many cases the constant term will still be dominant. Thus this term is an appropriate starting value for the iteration:

$$\overline{\mathbf{I}}^{(0)} = \overline{\mathbf{I}}_0. \quad (2.57)$$

To calculate $\overline{\mathbf{I}}^{(0)}$ it is necessary to evaluate Equation (2.54) for $\omega = 0$. With some additional approximations this can be done explicitly:

$$\overline{\mathbf{I}}^{(0)}(\zeta) = \int_{\zeta_0}^{\zeta_l} \overline{\mathbf{K}}(\zeta, \zeta')\big|_{\omega=0} \mathcal{M}_\zeta^{\zeta'} \left\{ - \begin{bmatrix} \mathbf{I}_{21}^{-1}\frac{\partial \mathbf{I}_{21}}{\partial \xi} & \mathbf{I}_{21}^{-1}\left(\mathbf{z} + \frac{\partial \mathbf{I}_{22}}{\partial \xi}\right) \\ \mathbf{1} & \mathbf{0} \end{bmatrix} \right\} \mathrm{d}\zeta'. \quad (2.58)$$

The product integral is then replaced by its series representation

$$\overline{\mathbf{I}}^{(0)}(\zeta) = \int_{\zeta_0}^{\zeta_l} \overline{\mathbf{K}}(\zeta, \zeta')\big|_{\omega=0} \left(\overline{\mathbf{1}} - \int_{\zeta}^{\zeta'} \begin{bmatrix} \mathbf{I}_{21}^{-1}\frac{\partial \mathbf{I}_{21}}{\partial \xi} & \mathbf{I}_{21}^{-1}\left(\mathbf{z} + \frac{\partial \mathbf{I}_{22}}{\partial \xi}\right) \\ \mathbf{1} & \mathbf{0} \end{bmatrix} \mathrm{d}\xi + \dots \right) \mathrm{d}\zeta'. \quad (2.59)$$

It is now argued that the major contribution to the integral arises if $\zeta \approx \zeta'$. Physically this means, that the Coulomb interaction of regions of the transmission line that are very close to each other is much stronger than the interaction of more distant regions. In this case, the second term of the series representation of the product integral is smaller than the constant term and can be neglected. This argument is not necessarily true for a general nonuniform transmission line, but it is a reasonable way to obtain good starting values. Thus the starting values for the iteration are set to

$$\overline{\mathbf{I}}^{(0)}(\zeta) = \int_{\zeta_0}^{\zeta_l} \overline{\mathbf{K}}(\zeta, \zeta')\big|_{\omega=0} \mathrm{d}\zeta'. \tag{2.60}$$

Note that this expression is frequency independent because it is the constant term of the Taylor expansion of the actual frequency-dependent function. Nonetheless, this result is used as the starting values for the iteration at all frequencies.

2.1.4.2 Source Terms

Once the parameters are determined the source terms $\mathbf{q}_s'$ are calculated, which are given by Equation (2.52). For the evaluation $\mathbf{I}_{10}$ and $\mathbf{I}_{20}$ are needed. Both depend on the already known $\overline{\mathbf{P}}_t$ and on $\mathbf{q}_t'$. This second quantity is equal to the unknown source term because Equations (2.43) and (2.50) are identical:

$$\mathbf{q}_t' = \mathbf{q}_s'. \tag{2.61}$$

With this relation an implicit expression for the determination of $\mathbf{q}_s'$ is obtained. Because an explicit expression is not known, an iterative approach is used to calculate the $\mathbf{q}_s'$:

$$\mathbf{q}_s^{(k+1)'} = \mathcal{L}_q \left\{ \mathbf{q}_s^{(k)'} \right\}, \tag{2.62}$$

where $\mathcal{L}_q$ consists of Equations (2.52) and (2.49). The iteration is started with vanishing sources, that is $\mathbf{q}_s^{(0)'} = \mathbf{0}$. It follows from Equation (2.49)

$$\mathbf{I}_{10}^{(0)} = \mathbf{0} \tag{2.63}$$

$$\mathbf{I}_{20}^{(0)} = \mathbf{0} \tag{2.64}$$

and, with Equation (2.52),

$$\mathbf{q}_s^{(1)'} = \mathbf{I}_{21}^{-1} \mathbf{v}^{(i)'}. \tag{2.65}$$

The result can be used to calculate $\mathbf{I}_{10}^{(1)}$ and $\mathbf{I}_{20}^{(1)}$ which, in turn, are used to calculate $\mathbf{q}_s^{(2)'}$, and so on. For practical applications the first iteration already often yields very satisfying results.

2.1.4.3 Solution of the Extended Telegrapher Equations

The extended telegrapher equations in the frequency domain constitute a system of first-order ordinary differential equations with nonconstant (i.e. position-dependent) coefficients. Solution methods for this kind of equation are very well known and the subject of many books and publications, for example from the mathematical point of view [15, 16], from the transmission-line-theory point of view [17, 18] or from the numerical point of view [19]. As was anticipated in the previous section the general solution is

$$\begin{bmatrix} \mathbf{q}(\zeta) \\ \mathbf{i}(\zeta) \end{bmatrix} = \mathcal{M}_{\zeta_0}^{\zeta}\{-j\omega\mathbf{P}\} \begin{bmatrix} \mathbf{q}(\zeta_0) \\ \mathbf{i}(\zeta_0) \end{bmatrix} + \int_{\zeta_0}^{\zeta} \mathcal{M}_{\xi}^{\zeta}\{-j\omega\mathbf{P}\} \begin{bmatrix} \mathbf{q}_s'(\xi) \\ \mathbf{0} \end{bmatrix} \mathrm{d}\xi. \tag{2.66}$$

Here $\mathcal{M}_{\zeta_0}^{\zeta}\{-j\omega\mathbf{P}\}$ is the product integral, which is often written as

$$\mathcal{M}_{\zeta_0}^{\zeta}\{\mathbf{X}\} = \prod_{\zeta_0}^{\zeta} \mathrm{e}^{\mathbf{X}(\eta)\mathrm{d}\eta}. \tag{2.67}$$

In the next Section 2.2 and Appendix E more details on the product integral and its evaluation are provided. The column vectors $\mathbf{q}(\zeta_0)$ and $\mathbf{i}(\zeta_0)$ are integration constants and are determined by the boundary conditions at the terminals of the transmission lines.

2.1.4.4 Returning to Voltages?

In many practical cases it is more convenient to deal with a voltage instead of a per-unit-length charge. For instance, if the transmission line is embedded into a circuit and other elements are connected to the line the voltages at the terminals must be known to perform the calculations.

The voltage v_{12} between the two points $\mathbf{x}_1$ and $\mathbf{x}_2$ is defined as the integral of the electric field strength $\mathbf{E}$ along a path $\mathcal{C}$ which connects the two points:

$$v_{12} := -\int_{\mathcal{C}_{\mathbf{x}_1 \to \mathbf{x}_2}} \mathbf{E} \cdot \mathrm{d}\mathbf{s}. \tag{2.68}$$

If $\mathbf{E}$ is represented with the aid of potentials one can also write

$$v_{12} = \int_{\mathcal{C}_{\mathbf{x}_1 \to \mathbf{x}_2}} (\nabla\varphi + j\omega\mathbf{A}) \cdot \mathrm{d}\mathbf{s} \tag{2.69}$$

$$= \varphi(\mathbf{x}_2) - \varphi(\mathbf{x}_1) + j\omega \int_{\mathcal{C}_{\mathbf{x}_1 \to \mathbf{x}_2}} \mathbf{A} \cdot \mathrm{d}\mathbf{s}. \tag{2.70}$$

In general, the remaining integral depends on the path that is chosen and thus the voltage is different for different paths. For a TEM-mode line, in a plane perpendicular to the conductors, the second integral becomes zero because $\mathbf{A} \perp \mathrm{d}\mathbf{s}$ and the voltage can be uniquely defined as $v_{12} := \varphi(\mathbf{x}_2) - \varphi(\mathbf{x}_1)$. For a general nonuniform line this is not the case. Thus, in order to compute the voltage, the vector potential along that path must be known. This can be computed but it is rather cumbersome to do this calculation along a transmission line.

If the distance between the conductors at the terminals, that is the lengths of the integration paths for the voltage computation, is much smaller than the smallest wavelength λ, this region of the transmission line can be regarded to be in the 'static' regime. Fortunately, this is quite often the case and then the voltage is given by the difference of potentials at the corresponding conductor terminals. For the potential the average potential at the surface of the conductors can be used, which is calculated by

$$\varphi_i(\zeta) = \frac{1}{2\pi} \int_0^{2\pi} \varphi(\hat{\mathbf{x}}_i(\zeta, \alpha))\, \mathrm{d}\alpha. \tag{2.71}$$

It is suggested that this potential should be called the quasi-voltage. Now the relation

$$\varphi(\mathbf{x}) = \frac{1}{\epsilon} \int_V G(\mathbf{x}, \mathbf{x}') \rho(\mathbf{x}') \, \mathrm{d}v' \tag{2.72}$$

is inserted into this equation. Then, with some more manipulations, in fact the same as in the derivation of Equation (2.41) which contains the same expression, one arrives at

$$\boldsymbol{\varphi}(\zeta) = \int_{\zeta_0}^{\zeta_l} \mathbf{k}_c(\zeta, \zeta') \mathbf{q}(\zeta') \, \mathrm{d}\zeta'. \tag{2.73}$$

Here $\boldsymbol{\varphi}$ is the column vector of the average conductor potentials. With the aid of Equations (2.44), (2.24) and (2.49) this expression becomes

$$\boldsymbol{\varphi} = \mathbf{I}_{21}\mathbf{q} + \mathbf{I}_{22}\mathbf{i} + \mathbf{I}_{20}, \tag{2.74}$$

which allows us to formulate the following transformation:

$$\begin{bmatrix} \boldsymbol{\varphi} \\ \mathbf{i} \end{bmatrix} = \begin{bmatrix} \mathbf{I}_{21} & \mathbf{I}_{22} \\ \mathbf{0} & \mathbf{1} \end{bmatrix} \begin{bmatrix} \mathbf{q} \\ \mathbf{i} \end{bmatrix} + \begin{bmatrix} \mathbf{I}_{20} \\ \mathbf{0} \end{bmatrix}. \tag{2.75}$$

Because this relation is valid everywhere on the line, not only at the beginning or the end, the telegrapher equations are easily converted into the quasivoltage–current representation

$$\frac{\partial}{\partial \zeta} \begin{bmatrix} \boldsymbol{\varphi}(\zeta) \\ \mathbf{i}(\zeta) \end{bmatrix} + j\omega \overline{\mathbf{P}}^{*}(\zeta) \begin{bmatrix} \boldsymbol{\varphi}(\zeta) \\ \mathbf{i}(\zeta) \end{bmatrix} = \begin{bmatrix} \boldsymbol{\varphi}'_{\mathrm{s}}(\zeta) \\ \mathbf{i}'_{\mathrm{s}}(\zeta) \end{bmatrix}, \tag{2.76}$$

where

$$\overline{\mathbf{P}}^{*} = \begin{bmatrix} \mathbf{I}_{11}\mathbf{I}_{21}^{-1} & \mathbf{I}_{12} - \mathbf{I}_{11}\mathbf{I}_{21}^{-1}\mathbf{I}_{22} + \frac{\mathbf{z}}{j\omega} \\ \mathbf{I}_{21}^{-1} & -\mathbf{I}_{21}^{-1}\mathbf{I}_{22} \end{bmatrix} \tag{2.77}$$

and

$$\begin{bmatrix} \boldsymbol{\varphi}'_{\mathrm{s}} \\ \mathbf{i}'_{\mathrm{s}} \end{bmatrix} = \begin{bmatrix} \mathbf{v}^{(i)\prime} + j\omega \left(\mathbf{I}_{11}\mathbf{I}_{21}^{-1}\mathbf{I}_{20} - \mathbf{I}_{10} \right) \\ j\omega \mathbf{I}_{21}^{-1}\mathbf{I}_{20} \end{bmatrix}. \tag{2.78}$$

Remember, the quasivoltage along the line is still a gauge-dependent quantity with no physical meaning. Only when the prescribed conditions are met, the quasivoltage becomes a real, physical voltage.

In the above equations, $j\omega\mathbf{P}^{*}_{12}$ can be identified as the parameter matrix which is equivalent to the per-unit-length impedance matrix of the classical transmission-line theory. Moreover, the matrix $j\omega\mathbf{P}^{*}_{21}$ is equivalent to the per-unit-length admittance matrix. The two other block matrices $j\omega\mathbf{P}^{*}_{11}$ and $j\omega\mathbf{P}^{*}_{22}$ vanish in the classical theory and are only nonzero for nonuniform lines. However, when interpreting these parameters, one should keep in mind that they connect the current, which is a physical quantity, to the quasivoltage, which, in general, has no physical meaning.

Also it is noted that, unlike the parameters in Equation (2.50), these parameters do not contain any derivatives. This makes the numerical evaluation of the product integral much simpler than for Equation (2.50).

2.1.4.5 Discussion of the New Parameters

Wave Equation

In the previous sections a new set of differential equations for nonuniform transmission lines has been derived. The equations contain novel coefficients, the per-unit-length parameters of the TLST. In order to give some meaning to these parameters, a different representation is considered next. First, it is instructive to convert the system of $2N$ first-order differential equations in q–i representation into a system of N second-order differential equations, involving the conductor currents and its first and second derivatives.

$$\left(\frac{\partial^2}{\partial\zeta^2} - \mathbf{D}\frac{\partial}{\partial\zeta} - \mathbf{\Gamma}^2\right)\mathbf{i} = \mathbf{i}_s \tag{2.79}$$

where

$$\mathbf{\Gamma}^2 = -\omega^2\mathbf{P}_{12} \tag{2.80}$$

$$= -\omega^2\mathbf{I}_{21}^{-1}\mathbf{I}_{12} + j\omega\mathbf{I}_{21}^{-1}\left(\frac{\partial}{\partial\zeta}\mathbf{I}_{22} + \mathbf{z}\right) \tag{2.81}$$

$$\mathbf{D} = -j\omega\mathbf{P}_{11} \tag{2.82}$$

$$= -\mathbf{I}_{21}^{-1}\frac{\partial}{\partial\zeta}\mathbf{I}_{21} - j\omega\mathbf{I}_{21}^{-1}\left(\mathbf{I}_{11} - \mathbf{I}_{22}\right) \tag{2.83}$$

$$\mathbf{i}_s = -j\omega\mathbf{q}_s. \tag{2.84}$$

The propagation function $\mathbf{\Gamma}$, in general, is a complex valued matrix, incorporating losses, not only caused by conductor losses but also caused by radiation. The matrix $\mathbf{D}$ describes the damping along the line due to the reflections caused by the nonuniformity.

Radiation Losses

To further investigate the radiation losses, in [20] it was suggested that a power balance for the transmission line should be established. Here a simple model will be used, consisting only of a one-conductor transmission line above a ground plane. Equation (2.76) is used to describe this transmission line and it is assumed that all conditions for the validity of this equation are met. At position ζ_0 the line is fed with a source and there is a load at ζ_l. The real power generated by the source is then given by

$$p(\zeta_0) = \frac{1}{2}\left(\varphi(\zeta_0)\overline{i(\zeta_0)} + \overline{\varphi(\zeta_0)}i(\zeta_0)\right) \tag{2.85}$$

where the line above the variables indicates the complex conjugate. Likewise, the power dissipated in the load is given by

$$p(\zeta_1) = \frac{1}{2}\left(\varphi(\zeta_1)\overline{i(\zeta_1)} + \overline{\varphi(\zeta_1)}i(\zeta_1)\right). \tag{2.86}$$

Then the power balance is $p(\zeta_0) = p_{\mathrm{TL}} + p(\zeta_1)$ or rearranged

$$p_{\mathrm{TL}} = -\left(p(\zeta_1) - p(\zeta_0)\right), \tag{2.87}$$

where p_{TL} is the power dissipated in the transmission line. The expression inside the parentheses can be regarded as an integral of the form

$$p_{\mathrm{TL}} = -\int_{\zeta_0}^{\zeta_l} \frac{\mathrm{d}p(\zeta)}{\mathrm{d}\zeta}\mathrm{d}\zeta. \tag{2.88}$$

where

$$p(\zeta) = \frac{1}{2}\left(\varphi(\zeta)\overline{i(\zeta)} + \overline{\varphi(\zeta)}i(\zeta)\right). \tag{2.89}$$

After some longer manipulations [20] one obtains the relation

$$p_{\mathrm{TL}} = -\int_{\zeta_0}^{\zeta_l} \omega\left(\Im\left(P_{12}^{*}\right)|i|^2 + \Im\left(P_{21}^{*}\right)|\varphi|^2 + \Im\left(\left(P_{11}^{*} - \overline{P_{22}^{*}}\right)\varphi\bar{i}\right)\right)\mathrm{d}\zeta \tag{2.90}$$

which describes the real power dissipated by the transmission line. This includes the power which is dissipated in the conductors as well as the radiated power. As can be seen, the imaginary parts of the per-unit-length parameters $\mathbf{P}_{12}^{*}$ and $\mathbf{P}_{21}^{*}$ are important coefficients. Multiplied with $j\omega$ these imaginary parts become real and hence, act as a series resistance and a shunt conductance. The above expression can also be obtained for other representations of the generalized telegrapher equations. However, the imaginary parts of the $\overline{\mathbf{P}}^{*}$ parameters, which are responsible for losses and radiation, are 'hidden' in the real and imaginary parts of the corresponding parameters of the other representations and cannot be isolated as in Equation (2.90).

2.1.4.6 Asymmetric Parameter Matrices

In the classical transmission-line theory, reciprocity implies symmetric parameter matrices. Unlike these matrices, the parameter matrices of the supertheory are not necessarily symmetric.

For instance, imagine a transmission line, such as shown in Figure 2.4. Each of the parameter matrices will contain one coefficient describing the coupling from conductor 1 to conductor 2 and one for the coupling from conductor 2 to conductor 1. In the classical theory, because of reciprocity, these two coefficients are identical and thus the corresponding matrix is symmetric.

In the TLST the first coefficient describes the coupling from all parts of conductor 1 to the point ζ on conductor 2, it is a nonlocal interaction. The same is valid for the other parameter. It is very reasonable that only in rare cases are these couplings and thus the coefficients the same. Hence, the coefficient matrix is not symmetric.

Figure 2.4 Nonuniform transmission lines and asymmetric parameter matrices.

2.1.5 *Numerical Evaluation of the Parameters*

It is now appropriate to discuss methods to calculate the parameters of the generalized telegrapher equations. This mainly involves the evaluation of Equations (2.53) and (2.54). The following treatment will be somewhat specialized; however, more general methods to evaluate the matrizant/product integral will be presented in the next Section 2.2.

Here, for an efficient processing, some assumptions must be made which require a careful preparation of the problems to be calculated. The setup should be in free space, an ideal ground plane or a corner can be taken into account with the aid of the method of images [21–23]. Also all conductors should be thin wires. This is not strictly necessary for the following derivations, but some integrals can only be solved in closed form if the thin-wire approximation is used.

The transmission line is divided into M segments. Within each segment, the conductors should be straight, but not necessarily in parallel to the reference conductor. Thus their position can be approximated by a linear function. The parameters are determined at the centers of the segments and are interpolated for positions in between. Therefore, segment length should reflect the nonuniformity of the line, for strongly nonuniform lines a finer segmentation is required.

2.1.5.1 Starting Values for the Iteration

The first step is the calculation of $\overline{\mathbf{I}}^{(0)}$ using Equation (2.60). For the segmented transmission line this equation becomes

$$\overline{\mathbf{I}}^{(0)}(\zeta) = \sum_{m=0}^{M} \int_{\zeta_m}^{\zeta_{m+1}} \overline{\mathbf{K}}\left(\zeta, \zeta'\right)\Big|_{\omega=0} \mathrm{d}\zeta'. \tag{2.91}$$

The integration of the matrix function is performed by the integration of the single elements resulting in

$$\overline{\mathbf{I}}^{(0)}(\zeta) = \sum_{m=0}^{M} \begin{bmatrix} \mathbf{0} & \mu\left[\mathbf{T}_j(\zeta) \cdot \mathbf{T}_i\left(\zeta_{c_m}\right) I_{m_{ij}}^{(0)}\right] \\ \frac{1}{\epsilon}\left[I_{m_{ij}}^{(0)}\right] & \mathbf{0} \end{bmatrix} \tag{2.92}$$

where

$$I_{m_{ij}}^{(0)}(\zeta) = \int_{\zeta_m}^{\zeta_{m+1}} G\left(\mathbf{C}_j(\zeta), \mathbf{C}_i\left(\zeta'\right)\right)\Big|_{\omega=0} \mathrm{d}\zeta'. \tag{2.93}$$

Here the new quantity

$$\zeta_{c_m} = \frac{\zeta_{m+1} + \zeta_m}{2} \tag{2.94}$$

was introduced to indicate the center of the segments. The above integral can be evaluated in closed form.

For a line in free space one can write (see also Figure 2.5)

$$I_{m_{ij}}^{(0)} = \int_{\zeta_m}^{\zeta_{m+1}} \frac{1}{4\pi\left|\mathbf{C}_j(\zeta) - \mathbf{C}_i\left(\zeta'\right)\right|} \mathrm{d}\zeta' \tag{2.95}$$

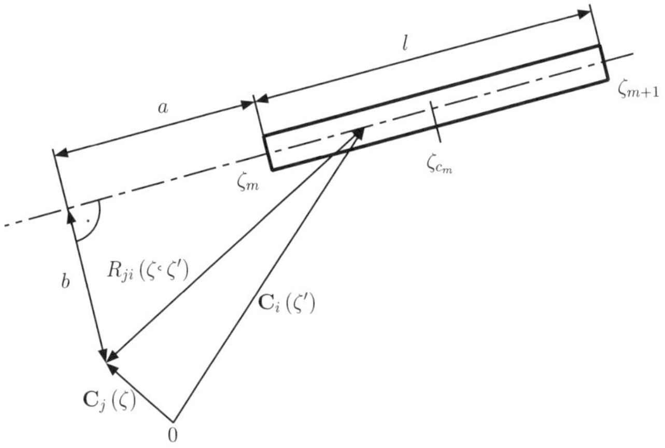

Figure 2.5 One segment of a conductor of a nonuniform transmission line.

$$= \int_{\zeta_m}^{\zeta_{m+1}} \frac{1}{4\pi R_{ji}(\zeta, \zeta')} \mathrm{d}\zeta' \tag{2.96}$$

with $x = a + l \dfrac{\zeta - \zeta_m}{\zeta_{m+1} - \zeta_m}$

$$I_{m_{ij}}^{(0)} = \frac{\zeta_{m+1} - \zeta_m}{4\pi l} \int_a^{a+l} \frac{1}{\sqrt{x^2 + b^2}} \mathrm{d}x \tag{2.97}$$

$$= \frac{\zeta_{m+1} - \zeta_m}{4\pi l} \begin{cases} \operatorname{arcsinh} \frac{a+l}{b} - \operatorname{arcsinh} \frac{a}{b} & b > 0 \\ \ln \frac{a+l}{a} & b = 0. \end{cases} \tag{2.98}$$

If there is a ground plane or a corner the additional images have to be taken into account. The integrals can be solved in a similar way to the one for free space. These results are inserted in Equation (2.53) to obtain the starting parameters $\overline{\mathbf{P}}^{(0)}(\zeta)$ for the first iteration. Note that the parameters are only evaluated at the centers $\zeta = \zeta_{c_m}, m = 1 \ldots M$ of the segments.

2.1.5.2 First Iteration

In the next step Equation (2.54) has to be calculated with the previously determined zeroth-order parameters. Again, the line is divided into segments giving:

$$\overline{\mathbf{I}}^{(1)}(\zeta) = \sum_{m=1}^{M} \overline{\mathbf{I}}_m^{(1)}(\zeta) \tag{2.99}$$

and

$$\overline{\mathbf{I}}_m^{(1)}(\zeta) = \int_{\zeta_m}^{\zeta_{m+1}} \overline{\mathbf{K}}(\zeta, \zeta') \mathcal{M}_{\zeta}^{\zeta'} \left\{ -j\omega \overline{\mathbf{P}}^{(0)} \right\} \mathrm{d}\zeta'. \tag{2.100}$$

The product integral is split into two factors (see Equation (E.15)). One of these factors is independent of the integration variable and can be pulled out of the integral:

$$\overline{\mathbf{I}}_m^{(1)}(\zeta) = \int_{\zeta_m}^{\zeta_{m+1}} \overline{\mathbf{K}}(\zeta, \zeta') \mathcal{M}_{\zeta_{cm}}^{\zeta'} \left\{-j\omega\overline{\mathbf{P}}^{(0)}\right\} \mathrm{d}\zeta' \, \mathcal{M}_{\zeta}^{\zeta_{cm}} \left\{-j\omega\overline{\mathbf{P}}^{(0)}\right\}. \tag{2.101}$$

The remaining product integral inside the integral is expressed as a sum

$$\mathcal{M}_{\zeta_{cm}}^{\zeta'} \left\{-j\omega\overline{\mathbf{P}}^{(0)}\right\} = \sum_{n=0}^{\infty} \lambda_{m_n}(\zeta') \overline{\mathbf{M}}_{m_n}, \tag{2.102}$$

where $\lambda_{m_n}(\zeta')$ is a scalar function and $\overline{\mathbf{M}}_{m_n}$ a possibly frequency-dependent supermatrix, which is, however, independent of ζ'. There are several possibilities for the expansion of the product integral. Below, two of these will be discussed. The first possibility takes advantage of a Taylor series and the second is based on an eigenvalue/eigenvector decomposition of the product integral. Equation (2.101) may now be written as

$$\overline{\mathbf{I}}_m^{(1)}(\zeta) = \left(\sum_{n=0}^{\infty} \int_{\zeta_m}^{\zeta_{m+1}} \overline{\mathbf{K}}(\zeta, \zeta') \lambda_{m_n}(\zeta') \mathrm{d}\zeta' \, \overline{\mathbf{M}}_{m_n}\right) \mathcal{M}_{\zeta}^{\zeta_{cm}} \left\{-j\omega\overline{\mathbf{P}}^{(0)}\right\}. \tag{2.103}$$

With the previous manipulations the problem of integrating the product of two matrix functions in Equation (2.100) has been reduced to the problem of integrating the product of a matrix function and a scalar function in Equation (2.103). This integration can be carried out elementwise, yielding

$$\overline{\mathbf{I}}_m^{(1)}(\zeta) = \left(\sum_{n=0}^{\infty} \begin{bmatrix} \mathbf{0} & \left[\mu \mathbf{T}_j(\zeta) \cdot \mathbf{T}_i(\zeta_{m_c}) I_{m_{n\,ji}}^{(1)}\right] \\ \left[\frac{1}{\epsilon} I_{m_{n\,ji}}^{(1)}\right] & \mathbf{0} \end{bmatrix} \overline{\mathbf{M}}_{m_n}\right) \mathcal{M}_{\zeta}^{\zeta_{cm}} \left\{-j\omega\overline{\mathbf{P}}^{(0)}\right\}, \tag{2.104}$$

with

$$I_{m_{n\,ji}}^{(1)} = \int_{\zeta_m}^{\zeta_{m+1}} G\left(\mathbf{C}_j(\zeta), \mathbf{C}_i(\zeta')\right) \lambda_{m_n}(\zeta') \mathrm{d}\zeta'. \tag{2.105}$$

Depending on the decomposition of the product integral in Equation (2.102) there are several ways to evaluate the last expression. Because of the singularity of the Green's function a numerical integration is rather complicated. In a few cases analytical closed-form formulas can be obtained. However, for the majority of the possible cases a series expansion of the Green's function and λ_{m_n} or of the product of these two is suggested. Some techniques can be adapted from other numerical methods like the MoM. For instance, in [24] some techniques for the evaluation of terms involving this Green's function are shown.

2.1.5.3 Taylor Series Expansion of the Product Integral

The Taylor series expansion of the product integral is given in Appendix E.2. The parameters $\overline{\mathbf{P}}^{(0)}(\zeta)$ are linearly interpolated between the segment centers, and hence are given in segment m by

$$\overline{\mathbf{P}}^{(0)}(\zeta) = \overline{\mathbf{P}}_{m_0} + (\zeta - \zeta_{c_m}) \overline{\mathbf{P}}_{m_1}, \quad \text{for} \quad \zeta_m \le \zeta \le \zeta_{m+1}, \tag{2.106}$$

where

$$\overline{\mathbf{P}}_{m_0} = \overline{\mathbf{P}}^{(0)}\left(\zeta_{c_m}\right) \tag{2.107}$$

$$\overline{\mathbf{P}}_{m_1} = \begin{cases} \frac{\overline{\mathbf{P}}^{(0)}(\zeta_{cm})-\overline{\mathbf{P}}^{(0)}\left(\zeta_{c_{m-1}}\right)}{\zeta_{cm}-\zeta_{c_{m-1}}} & \zeta < \zeta_{c_m} \\ \frac{\overline{\mathbf{P}}^{(0)}\left(\zeta_{c_{m+1}}\right)-\overline{\mathbf{P}}^{(0)}(\zeta_{cm})}{\zeta_{c_{m+1}}-\zeta_{cm}} & \zeta > \zeta_{c_m} \end{cases}. \tag{2.108}$$

Because a linear interpolation is used, the second-and all higher-order derivatives vanish. This makes the evaluation of the coefficients of the Taylor series (cf. Equation (E.11)) rather simple and we obtain

$$\lambda_{m_n} = \left(\zeta' - \zeta_{c_m}\right)^n \tag{2.109}$$

and

$$\overline{\mathbf{M}}_{m_n} = -\frac{j\omega}{n!}\left((n-1)\,\overline{\mathbf{P}}_{m_1}\overline{\mathbf{M}}_{m_{n-2}} + \overline{\mathbf{P}}_{m_0}\overline{\mathbf{M}}_{m_{n-1}}\right) \tag{2.110}$$

where

$$\overline{\mathbf{M}}_{m_0} = \overline{\mathbf{1}} \quad \text{and} \quad \overline{\mathbf{M}}_{m_1} = -j\omega\overline{\mathbf{P}}_{m_0}\overline{\mathbf{M}}_{m_0}. \tag{2.111}$$

Now, for a line in free space Equation (2.105) becomes

$$I^{(1)}_{m_{n\,ji}} = \int_{\zeta_m}^{\zeta_{m+1}} \frac{\mathrm{e}^{-jkR_{ji}(\zeta,\zeta')}\left(\zeta' - \zeta_{c_m}\right)^n}{4\pi R_{ji}(\zeta,\zeta')}\mathrm{d}\zeta', \tag{2.112}$$

which has no closed-form solution. In a similar way to the product integral above, the exponential can be split into two factors, where one is independent of the integration:

$$I^{(1)}_{m_{n\,ji}} = \frac{\mathrm{e}^{-jkR_{ji}(\zeta,\zeta_{cm})}}{4\pi}\int_{\zeta_m}^{\zeta_{m+1}} \frac{\mathrm{e}^{-jk(R_{ji}(\zeta,\zeta')-R_{ji}(\zeta,\zeta_{cm}))}(\zeta' - \zeta_{c_m})^n}{R_{ji}(\zeta,\zeta')}\mathrm{d}\zeta'. \tag{2.113}$$

The argument of the remaining exponential function is small if the conductor length within a segment is small compared with the wavelength:

$$\left|R_{ji}\left(\zeta,\zeta'\right) - R_{ji}\left(\zeta,\zeta_{c_m}\right)\right| \leq |\mathbf{C}_i(\zeta_{m+1}) - \mathbf{C}_i(\zeta_m)| \ll \lambda. \tag{2.114}$$

Thus this function can be expanded into a Taylor series

$$I^{(1)}_{m_{n\,ji}} = \frac{\mathrm{e}^{-jkR_{ji}(\zeta,\zeta_{cm})}}{4\pi}\sum_{p=0}^{\infty}\frac{(-jk)^p}{p!}\int_{\zeta_m}^{\zeta_{m+1}} \frac{(R_{ji}(\zeta,\zeta') - R_{ji}(\zeta,\zeta_{c_m}))^p(\zeta' - \zeta_{c_m})^n}{R_{ji}(\zeta,\zeta')}\mathrm{d}\zeta' \tag{2.115}$$

and truncated after a few terms. The integral can then be modified, similar to Equation (2.97), and becomes

$$\int_{\zeta_m}^{\zeta_{m+1}} \frac{\left(R_{ji}\left(\zeta,\zeta'\right) - R_{ji}\left(\zeta,\zeta_{c_m}\right)\right)^p\left(\zeta' - \zeta_{c_m}\right)^n}{R_{ji}\left(\zeta,\zeta'\right)}\mathrm{d}\zeta'$$
$$= \left(\frac{\zeta_{m+1} - \zeta_m}{l}\right)^{n+1}\int_a^{a+l}\frac{\left(\sqrt{x^2+b^2} - \sqrt{x_c^2+b^2}\right)^p}{\sqrt{x^2+b^2}}(x - x_c)^n\mathrm{d}x \tag{2.116}$$

which has a closed-form solution. For small numbers p and n the solutions are given in Appendix F.1. Now the parameters can be calculated after the first iteration.

2.1.5.4 Eigenvalue Decomposition

Instead of representing the product integral as a Taylor expansion one could also use an eigenvalue decomposition. This method is especially useful for slowly varying parameters. To this end, one chooses a segment size such that one can assume constant parameters within a segment, that is

$$\overline{\mathbf{P}}^{(0)}(\zeta) = \overline{\mathbf{P}}^{(0)}\left(\zeta_{c_m}\right) \quad \text{for} \quad \zeta_m \leq \zeta < \zeta_{m+1}. \tag{2.117}$$

Then the product integral in Equation (2.101) becomes the matrix exponential:

$$\mathcal{M}_{\zeta_{cm}}^{\zeta'}\left\{-j\omega\overline{\mathbf{P}}^{(0)}\right\} = \mathrm{e}^{-j\omega(\zeta'-\zeta_{cm})\overline{\mathbf{P}}^{(0)}(\zeta_{cm})}. \tag{2.118}$$

The evaluation of this expression can be carried out by diagonalizing[3] the matrix $\overline{\mathbf{P}}^{(0)}(\zeta_m)$:

$$\overline{\mathbf{P}}^{(0)}(\zeta_m) = \overline{\mathbf{w}}_m^{-1}\overline{\mathbf{D}}_m\overline{\mathbf{w}}_m \tag{2.119}$$

$$= \sum_{n=1}^{2N} d_{m_n}\left(\overline{\mathbf{w}}_m^{-1}\right)_n \overline{\mathbf{w}}_{m_n}^{\mathbf{T}} \tag{2.120}$$

$$= \sum_{n=1}^{2N} d_{m_n}\overline{\mathbf{W}}_{m_n}. \tag{2.121}$$

The diagonal matrix $\overline{\mathbf{D}}_m$ contains the eigenvalues of $\overline{\mathbf{P}}^{(0)}(\zeta_m)$, the columns of $\overline{\mathbf{w}}_m$ are the corresponding eigenvectors, $\overline{\mathbf{w}}_{m_n}$ and $(\overline{\mathbf{w}}_m^{-1})_n$ are the nth columns of this matrix and its inverse, respectively. $\overline{\mathbf{W}}_{m_n}$ is the dyadic product of these two vectors, that is a matrix, and d_{m_n} is the nth eigenvalue.

For the matrix exponential one can write

$$\mathcal{M}_{\zeta_{cm}}^{\zeta'}\left\{-j\omega\overline{\mathbf{P}}^{(0)}\right\} = \sum_{n=1}^{2N} \mathrm{e}^{-j\omega(\zeta'-\zeta_{cm})d_{mn}}\overline{\mathbf{W}}_{m_n} \tag{2.122}$$

and thus

$$\lambda_{m_n} = \begin{cases} \mathrm{e}^{-j\omega(\zeta'-\zeta_{cm})d_{mn}} & 1 \leq n \leq 2N \\ 0 & \text{otherwise} \end{cases} \tag{2.123}$$

$$\overline{\mathbf{M}}_{m_n} = \begin{cases} \overline{\mathbf{W}}_{m_n} & 1 \leq n \leq 2N \\ \overline{\mathbf{0}} & \text{otherwise.} \end{cases} \tag{2.124}$$

[3] The diagonalization of a matrix is not always possible. However, here it is assumed that it is possible for our matrix. Otherwise, a different approach to calculate the integral must be used.

Equation (2.105), for a transmission line in free space, becomes

$$I_{m_n ji}^{(1)} = \int_{\zeta_m}^{\zeta_{m+1}} \frac{e^{-jkR_{ji}(\zeta,\zeta')} e^{-jk_{m_n}(\zeta'-\zeta_{cm})}}{4\pi R_{ji}(\zeta,\zeta')} d\zeta', \tag{2.125}$$

with $k_{m_n} = \omega d_{m_n}$. Unfortunately this expression cannot be integrated in closed form; however, when $k \approx k_{m_n}$ a series expansion can be used:

$$e^{-jk_{m_n}(\zeta'-\zeta_{cm})} = e^{-jk_{m_n}(\zeta'-\zeta)} e^{-jk_{m_n}(\zeta-\zeta_{cm})} \tag{2.126}$$

$$= e^{-jk(\zeta'-\zeta)} e^{j(k-k_{m_n})(\zeta'-\zeta)} e^{-jk_{m_n}(\zeta-\zeta_{cm})} \tag{2.127}$$

$$= e^{-jk(\zeta'-\zeta)} \left(\sum_{p=0}^{\infty} \frac{j^p}{p!} \left(k - k_{m_n}\right)^p \left(\zeta' - \zeta\right)^p \right) e^{-jk_{m_n}(\zeta-\zeta_{cm})}. \tag{2.128}$$

Inserting this into the integral yields

$$I_{m_n ji}^{(1)} = \frac{1}{4\pi} e^{-jk_{m_n}(\zeta-\zeta_{cm})} \sum_{p=0}^{\infty} \frac{j^p}{p!} \left(k - k_{m_n}\right)^p \int_{\zeta_m}^{\zeta_{m+1}} \frac{e^{-jkR_{ji}(\zeta,\zeta')} e^{-jk(\zeta'-\zeta)}}{R_{ji}(\zeta,\zeta')} \left(\zeta' - \zeta\right)^p d\zeta', \tag{2.129}$$

which has closed-form solutions, see Appendix F.2. For $k \approx k_{m_n}$ this series can be truncated after a few terms.

2.1.5.5 Discussion of the Numerical Methods

Two different methods for the calculation of the parameters have been discussed so far. The first is based on a Taylor series expansion and the second is based on an eigenvalue decomposition. In the first approach the truncation of the Taylor series after a few terms requires the segment length to be small. The number of terms and the segment length are directly related to each other. The more terms one takes into account the longer the segment can be. In applications it turned out that terms up to a power of three and a segment length, such that the conductors are not longer than $\lambda/10$, produced very satisfying results.

Unlike the Taylor series, the eigenvalue expansion is independent from the segment length. However, it requires almost constant parameters within a segment. Thus it is especially useful for long, almost uniform line sections where it allows an efficient determination of the line parameters. An analytical variant of this technique will be used for the first examples in Section 2.3.

2.2 General Calculation Methods for the Product Integral/Matrizant

In view of the general solution of the extended telegrapher equations it is obvious that it is of the utmost importance to calculate the product integral–which is often also called the *matrizant*. Unfortunately, in general cases there are no closed-form analytic solutions. Therefore, this section provides a detailed discussion on different semi-analytical and numerical methods to compute the matrizant and, thus, to obtain explicit results for the extended telegrapher equations.

Quite generally, telegrapher equations for nonuniform transmission-line systems are described by the equation

$$\frac{\mathrm{d}\overline{\mathbf{x}}(\xi)}{\mathrm{d}\xi} = \overline{\mathbf{R}}^{x}(\xi)\overline{\mathbf{x}}(\xi) + \overline{\mathbf{x}}^{s}(\xi). \tag{2.130}$$

The super vector $\overline{\mathbf{x}}$ represents the unknown quantities that need to be determined, the matrix function $\overline{\mathbf{R}}^{x}(\xi)$ contains the parameters that characterize the transmission-line system and $\overline{\mathbf{x}}^{s}(\xi)$ denotes the distributed excitation, that is the source terms.

From a mathematical perspective, the telegrapher equations for nonuniform transmission-line systems reduce to a linear, inhomogeneous system of first-order differential equations with a non-constant coefficient matrix. The dimension of this system is twice the number of conductors, excluding the reference conductor. The super matrix $\overline{\mathbf{R}}^{x}(\xi)$ is defined over an interval $[\xi_0, \xi]$ where it constitutes a finite, complex-valued matrix function, that is all elements $R^{x}_{i,k}(\xi)$ of the matrix $\overline{\mathbf{R}}^{x}(\xi)$ are finite, real or complex function of the real argument ξ.

Quite generally, computational methods for the solution of a linear, inhomogeneous system of first-order differential equations with nonconstant coefficients can be divided into four different classes:

- iteration methods and successive approximation,
- analytic methods, based on series expansion and general matrix methods,
- diagonalization and eigenvalue methods and
- numerical integration.

2.2.1 *Picard Iteration*

The system of Equation (2.130) can be solved by iteration [25] where for the calculation of the homogeneous solution the method of successive approximation can be used. This yields an approximation of the solution by means of the recursion formula

$$\frac{\mathrm{d}\overline{\mathbf{X}}^{(k+1)}(\xi)}{\mathrm{d}\xi} = \overline{\mathbf{R}}^{x}(\xi)\overline{\mathbf{X}}^{(k)}(\xi) \qquad k = 0, 1, 2, \cdots, \tag{2.131}$$

where $\overline{\mathbf{X}}^{(0)}(\xi)$ is given by the unit matrix $\overline{\mathbf{E}}$. If, furthermore, $\overline{\mathbf{X}}^{(k)}(\xi_0) = \overline{\mathbf{E}}$ for all k the form

$$\overline{\mathbf{X}}^{(k+1)}(\xi) = \overline{\mathbf{E}} + \int_{\xi_0}^{\xi} \overline{\mathbf{R}}^{x}(\xi)\overline{\mathbf{X}}^{(k)}(\xi)\mathrm{d}\xi \qquad k = 0, 1, 2, \cdots. \tag{2.132}$$

is obtained. Collecting all terms results in a matrix series which is known as *Picard iteration*,

$$\overline{\mathbf{X}}(\xi) = \overline{\mathbf{E}} + \int_{\xi_0}^{\xi} \overline{\mathbf{R}}^{x}(\zeta)\mathrm{d}\zeta + \int_{\xi_0}^{\xi} \overline{\mathbf{R}}^{x}(\zeta) \int_{\xi_0}^{\zeta} \overline{\mathbf{R}}^{x}(\eta)\mathrm{d}\eta\mathrm{d}\zeta + \cdots \equiv \mathcal{M}_{\xi_0}^{\xi}\left\{\overline{\mathbf{R}}^{x}\right\}. \tag{2.133}$$

In the literature, the result of this formula is known as a propagator, a matrizant or a product integral. A proof of absolute and uniform convergence of the solution in Equation (2.133) can be found in [25].

For a numerical evaluation of the matrizant one discretizes the interval $[\xi_0, \xi]$ to obtain a one-dimensional grid. The discretization is not required to be equidistant, it can also be adapted to the characteristics of the matrix function $\overline{\mathbf{R}}^{x}(\xi)$. The matrizant along the interval is then given by

$$\mathcal{M}_{\xi_0}^{\xi}\left\{\overline{\mathbf{R}}^{x}\right\} = \overline{\mathbf{E}} + \overline{\mathbf{X}}^{(1a)}(\xi) + \overline{\mathbf{X}}^{(2a)}(\xi) + \overline{\mathbf{X}}^{(3a)}(\xi) + \cdots. \tag{2.134}$$

with the following intermediate values at the grid points:

$$\begin{aligned}
\overline{\mathbf{X}}^{(1a)}(\xi_\kappa) &= \int_{\xi_0}^{\xi_\kappa} \overline{\mathbf{R}}^{x}(\zeta)\mathrm{d}\zeta &\longrightarrow\quad& \overline{\mathbf{X}}^{(1b)}(\xi_\kappa) = \overline{\mathbf{R}}^{x}(\xi_\kappa)\overline{\mathbf{X}}^{(1a)}(\xi_\kappa)\\
\overline{\mathbf{X}}^{(2a)}(\xi_\kappa) &= \int_{\xi_0}^{\xi_\kappa} \overline{\mathbf{X}}^{(1b)}(\zeta)\mathrm{d}\zeta &\longrightarrow\quad& \overline{\mathbf{X}}^{(2b)}(\xi_\kappa) = \overline{\mathbf{R}}^{x}(\xi_\kappa)\overline{\mathbf{X}}^{(2a)}(\xi_\kappa)\\
\overline{\mathbf{X}}^{(3a)}(\xi_\kappa) &= \int_{\xi_0}^{\xi_\kappa} \overline{\mathbf{X}}^{(2b)}(\zeta)\mathrm{d}\zeta &\longrightarrow\quad& \overline{\mathbf{X}}^{(3b)}(\xi_\kappa) = \overline{\mathbf{R}}^{x}(\xi_\kappa)\overline{\mathbf{X}}^{(3a)}(\xi_\kappa).\\
&\qquad\qquad\cdots & & \qquad\qquad .
\end{aligned} \tag{2.135}$$

Once the term $\overline{\mathbf{X}}^{(na)}(\xi)$ yields no more significant values the series is truncated. For constant $\overline{\mathbf{R}}^{x}(\xi) = \overline{\mathbf{R}}^{x}$ the series reduces to the Taylor series of $\exp\overline{\mathbf{R}}^{x}(\xi - \xi_0)$. The actual numerical effort for Picard iteration is rather high. It is necessary to store matrix values for any grid point and, besides the actual integration, it is also necessary to perform matrix multiplications for any grid point. Therefore the method is only suitable if the iteration converges quickly .

Once a homogeneous solution of the matrizant has been obtained by iteration it is possible [25] to solve along the grid a corresponding inhomogeneous problem (see Equation (2.130)) by application of

$$\overline{\mathbf{x}}(\xi) = \mathcal{M}_{\xi_0}^{\xi}\left\{\overline{\mathbf{R}}^{x}\right\}\overline{\mathbf{x}}(\xi_0) + \int_{\xi_0}^{\xi} \mathcal{M}_{\zeta}^{\xi}\left\{\overline{\mathbf{R}}^{x}\right\}\overline{\mathbf{x}}^{\mathrm{s}}(\zeta)\mathrm{d}\zeta \tag{2.136}$$

and (compare Appendix E)

$$\mathcal{M}_{\zeta}^{\xi}\left\{\overline{\mathbf{R}}^{x}\right\} = \mathcal{M}_{\xi_0}^{\xi}\left\{\overline{\mathbf{R}}^{x}\right\}\left(\mathcal{M}_{\xi_0}^{\zeta}\left\{\overline{\mathbf{R}}^{x}\right\}\right)^{-1}. \tag{2.137}$$

2.2.2 Volterra's Method and the Product Integral

The interval $[\xi_0, \xi]$ can be subdivided in n subintervals. Then, according to Equation (E.15), the matrizant can be written as the ordered matrix product of the matrizants of the subintervals,

$$\mathcal{M}_{\xi_0}^{\xi}\left\{\overline{\mathbf{R}}^{x}\right\} = \mathcal{M}_{\xi_{n-1}}^{\xi}\left\{\overline{\mathbf{R}}^{x}\right\}\cdots\mathcal{M}_{\xi_1}^{\xi_2}\left\{\overline{\mathbf{R}}^{x}\right\}\mathcal{M}_{\xi_0}^{\xi_1}\left\{\overline{\mathbf{R}}^{x}\right\}. \tag{2.138}$$

Under the assumption of small $\Delta\xi_i = \xi_{i+1} - \xi_i$ it may be assumed for the calculation of the matrizant that the argument $\overline{\mathbf{R}}^{x}(\xi_i \le \xi \le \xi_{i+1}) = \overline{\mathbf{R}}_i^{x}$ is constant within the subinterval i. It then follows

$$\mathcal{M}_{\xi_i}^{\xi_{i+1}}\left\{\overline{\mathbf{R}}^{x}\right\} \approx \exp\overline{\mathbf{R}}_i^{x}\Delta\xi_i\,. \tag{2.139}$$

This corresponds to a step approximation of the matrix function $\overline{\mathbf{R}}^x$. Performing the limit $\Delta\xi_i \to 0$ leads to

$$\mathcal{M}_{\xi_0}^{\xi}\left\{\overline{\mathbf{R}}^x\right\} = \lim_{\Delta\xi_i \to 0}\left[\exp\overline{\mathbf{R}}_i^x\Delta\xi_i \cdots \exp\overline{\mathbf{R}}_2^x\Delta\xi_2 \exp\overline{\mathbf{R}}_1^x\Delta\xi_1\right] = \prod_{\xi_0}^{\xi}\exp\overline{\mathbf{R}}^x(\zeta)\mathrm{d}\zeta \;. \quad (2.140)$$

This limiting value is analogous to the approximation of a usual Riemannian integral by the infinite sum of rectangular segments and, thus, often called the product integral. It was first introduced by Volterra [26].

If the values of the matrix function $\overline{\mathbf{R}}(\xi)$ commute, that is

$$\overline{\mathbf{R}}^x(\xi_1)\overline{\mathbf{R}}^x(\xi_2) = \overline{\mathbf{R}}^x(\xi_2)\overline{\mathbf{R}}^x(\xi_1) \quad (2.141)$$

for arbitrary ξ_1 and ξ_2 of the interval (ξ_0, ξ), the limit simplifies to

$$\mathcal{M}_{\xi_0}^{\xi}\left\{\overline{\mathbf{R}}^x\right\} = \exp\int_{\xi_0}^{\xi}\overline{\mathbf{R}}^x(\zeta)\mathrm{d}\zeta \;, \quad (2.142)$$

as is easily deduced from Equation (2.140) and the relation $\exp\mathbf{A}\exp\mathbf{B} = \exp(\mathbf{A}+\mathbf{B})$ which is valid for commutative matrices **A** and **B**. For nonuniform transmission lines this simplifying condition is usually not fulfilled.

The matrix exponential functions that appear in Equations (2.138) and (2.140) can be approximated to various degrees. If a Taylor approximation is truncated after the linear term, for example, one obtains the approximation

$$\mathcal{M}_{\xi_0}^{\xi}\left\{\overline{\mathbf{R}}^x\right\} = \lim_{\Delta\xi_i \to 0}\left[\left(\overline{\mathbf{E}} + \overline{\mathbf{R}}_i^x\Delta\xi_i\right)\cdots\left(\overline{\mathbf{E}} + \overline{\mathbf{R}}_2^x\Delta\xi_2\right)\left(\overline{\mathbf{E}} + \overline{\mathbf{R}}_1^x\Delta\xi_1\right)\right] . \quad (2.143)$$

Further possible approximations and resulting matrizants $\mathcal{M}_{\xi_i}^{\xi_{i+1}}\{\overline{\mathbf{R}}_i^x\}$ along a subinterval are listed in Table 2.1.

Table 2.1 Different approximations of the matrix exponential function within the product integral

Approximation	Matrizant within segment	Equivalent integration method
Taylor approximation, truncation after the linear term	$\overline{\mathbf{E}} + \overline{\mathbf{R}}_i^x\Delta\xi_i$	Euler, explicit
Pade approximation of order $\begin{pmatrix}0\\1\end{pmatrix}$	$\left(\overline{\mathbf{E}} - \overline{\mathbf{R}}_i^x\Delta\xi_i\right)^{-1}$	Euler, implicit
Pade approximation of order $\begin{pmatrix}1\\1\end{pmatrix}$	$\left(\overline{\mathbf{E}} - \frac{1}{2}\overline{\mathbf{R}}_i^x\Delta\xi_i\right)^{-1}\cdot\left(\overline{\mathbf{E}} + \frac{1}{2}\overline{\mathbf{R}}_i^x\Delta\xi_i\right)$	Crank–Nicolson, implicit
Taylor approximation, truncation after the third term	$\overline{\mathbf{E}} + \overline{\mathbf{R}}_i^x\Delta\xi_i + \frac{1}{2}\overline{\mathbf{R}}_i^{x\,2}\Delta\xi_i^2$	Modified Euler, explicit
Taylor approximation, truncation after the fifth term	$\overline{\mathbf{E}} + \overline{\mathbf{R}}_i^x\Delta\xi_i + \frac{1}{2}\overline{\mathbf{R}}_i^{x\,2}\Delta\xi_i^2 + \frac{1}{6}\overline{\mathbf{R}}_i^{x\,3}\Delta\xi_i^3 + \frac{1}{24}\overline{\mathbf{R}}_i^{x\,4}\Delta\xi_i^4$	Runge–Kutta, explicit, fourth order

Here it is distinguished, quite generally, between two approximations. The first one is the stepwise approximation of the matrix function $\overline{\mathbf{R}}^{x}(\zeta)$ and the second one is the approximation of the matrix exponential function for the constant argument $\overline{\mathbf{R}}_{i}^{x}$. The second step corresponds to a numerical integration of a constant matrix function along one segment. A disadvantage of this method is the artificial scattering at the transition between adjacent segments of the transmission line. This scattering is due to the stepwise approximation and only disappears during the transition $\Delta\xi_i \to 0$.

2.2.3 Recursion Formulas for Linear Interpolation

In the following the Taylor expansion of the solution at position ξ is used to derive a calculation method of the matrizant. If for the matrix function $\overline{\mathbf{R}}^{x}(\xi)$ and the vector function $\overline{\mathbf{x}}^{\mathrm{s}}(\xi)$ a linear interpolation inbetween a segment within the intervall $[\xi_0, \xi_1]$ is assumed, that is if

$$\overline{\mathbf{R}}^{x}(\xi) = \overline{\mathbf{R}}_0^{x} + \frac{\Delta\overline{\mathbf{R}}^{x}}{\Delta\xi}(\xi - \xi_0) \tag{2.144}$$

and

$$\overline{\mathbf{x}}^{\mathrm{s}}(\xi) = \overline{\mathbf{x}}_0^{\mathrm{s}} + \frac{\Delta\overline{\mathbf{x}}^{\mathrm{s}}}{\Delta\xi}(\xi - \xi_0) \tag{2.145}$$

with $\overline{\mathbf{R}}_0^{x} = \overline{\mathbf{R}}^{x}(\xi_0)$, $\Delta\overline{\mathbf{R}}^{x} = \overline{\mathbf{R}}^{x}(\xi_1) - \overline{\mathbf{R}}^{x}(\xi_0)$, $\overline{\mathbf{x}}_0^{\mathrm{s}} = \overline{\mathbf{x}}^{\mathrm{s}}(\xi_0)$, $\Delta\overline{\mathbf{x}}^{\mathrm{s}} = \overline{\mathbf{x}}^{\mathrm{s}}(\xi_1) - \overline{\mathbf{x}}^{\mathrm{s}}(\xi_0)$ and $\Delta\xi = \xi_1 - \xi_0$, then one finds in $[\xi_0, \xi_1]$ for the first derivative of the matrix function

$$\frac{\mathrm{d}\overline{\mathbf{R}}^{x}(\xi)}{\mathrm{d}\xi} = \frac{\Delta\overline{\mathbf{R}}^{x}}{\Delta\xi}, \tag{2.146}$$

and for the first derivative of the inhomogeneous term

$$\frac{\mathrm{d}\overline{\mathbf{x}}^{\mathrm{s}}(\xi)}{\mathrm{d}\xi} = \frac{\Delta\overline{\mathbf{x}}^{\mathrm{s}}}{\Delta\xi}. \tag{2.147}$$

Higher derivatives vanish and thus do not need to be considered.

In the following, use will be made of the Taylor approximation of the matrizant around the position ξ_0,

$$\overline{\mathbf{x}}(\xi_1) = \overline{\mathbf{x}}(\xi_0) + \Delta\xi\frac{\mathrm{d}}{\mathrm{d}\xi}\overline{\mathbf{x}}(\xi_0) + \frac{\Delta\xi^2}{2}\frac{\mathrm{d}^2}{\mathrm{d}\xi^2}\overline{\mathbf{x}}(\xi_0) + \cdots + \frac{\Delta\xi^n}{n!}\frac{\mathrm{d}^n}{\mathrm{d}\xi^n}\overline{\mathbf{x}}(\xi_0). \tag{2.148}$$

To eliminate the derivatives of the matrizant within the Taylor series, the initial differential Equation (2.130) is differentiated various times and the results are inserted into Equation (2.148). Taking into account Equations (2.144) and (2.146) this yields

$$\frac{\mathrm{d}}{\mathrm{d}\xi}\overline{\mathbf{x}}(\xi_0) = \overline{\mathbf{R}}^{x}(\xi_0)\overline{\mathbf{x}}(\xi_0) + \overline{\mathbf{x}}^{\mathrm{s}}(\xi_0) = \overline{\mathbf{Q}}_1\overline{\mathbf{x}}(\xi_0) + \overline{\mathbf{x}}^{\mathrm{s}}{}_1 \tag{2.149}$$

$$\frac{\mathrm{d}^2}{\mathrm{d}\xi^2}\overline{\mathbf{x}}(\xi_0) = \frac{\mathrm{d}\overline{\mathbf{R}}^{x}(\xi_0)}{\mathrm{d}\xi}\overline{\mathbf{x}}(\xi_0) + \overline{\mathbf{R}}^{x}(\xi_0)\frac{\mathrm{d}\overline{\mathbf{x}}(\xi_0)}{\mathrm{d}\xi} + \frac{\mathrm{d}\overline{\mathbf{x}}^{\mathrm{s}}(\xi_0)}{\mathrm{d}\xi} \tag{2.150}$$

$$= \underbrace{\left(\frac{\Delta\overline{\mathbf{R}}^x}{\Delta\xi} + \overline{\mathbf{R}}_0^x\overline{\mathbf{Q}}_1\right)}_{\overline{\mathbf{Q}}_2}\overline{\mathbf{x}}(\xi_0) + \underbrace{\frac{\Delta\overline{\mathbf{x}}^s}{\Delta\xi} + \overline{\mathbf{R}}_0^x\overline{\mathbf{x}}_1^s}_{\overline{\mathbf{x}}^s{}_2}$$

$$\frac{d^3}{d\xi^3}\overline{\mathbf{x}}(\xi_0) = \frac{d^2\overline{\mathbf{R}}^x(\xi_0)}{d\xi^2}\overline{\mathbf{x}}(\xi_0) + 2\frac{d\overline{\mathbf{R}}^x(\xi_0)}{d\xi}\frac{d\overline{\mathbf{x}}(\xi_0)}{d\xi} + \overline{\mathbf{R}}^x(\xi_0)\frac{d^2\overline{\mathbf{x}}(\xi_0)}{d\xi^2} + \frac{d^2\overline{\mathbf{x}}^s(\xi_0)}{d\xi^2} \tag{2.151}$$

$$= 2\frac{d\overline{\mathbf{R}}^x(\xi_0)}{d\xi}\frac{d\overline{\mathbf{x}}(\xi_0)}{d\xi} + \overline{\mathbf{R}}^x(\xi_0)\frac{d^2\overline{\mathbf{x}}(\xi_0)}{d\xi^2}$$

$$= \underbrace{\left(2\frac{\Delta\overline{\mathbf{R}}^x}{\Delta\xi}\overline{\mathbf{Q}}_1 + \overline{\mathbf{R}}_0^x\overline{\mathbf{Q}}_2\right)}_{\overline{\mathbf{Q}}_3}\overline{\mathbf{x}}(\xi_0) + \underbrace{2\frac{\Delta\overline{\mathbf{R}}^x}{\Delta\xi}\overline{\mathbf{x}}_1^s + \overline{\mathbf{R}}_0^x\overline{\mathbf{x}}_2^s}_{\overline{\mathbf{x}}_3^s}$$

$$\ldots$$

$$\frac{d^n\overline{\mathbf{x}}(\xi_0)}{d\xi^n} = \underbrace{\left((n-1)\frac{\Delta\overline{\mathbf{R}}^x}{\Delta\xi}\overline{\mathbf{Q}}_{n-2} + \overline{\mathbf{R}}_0^x\overline{\mathbf{Q}}_{n-1}\right)}_{\overline{\mathbf{Q}}_n}\overline{\mathbf{x}}(\xi_0) + \underbrace{(n-1)\frac{\Delta\overline{\mathbf{R}}^x}{\Delta\xi}\overline{\mathbf{x}}_{n-2}^s + \overline{\mathbf{R}}_0^x\overline{\mathbf{x}}_{n-1}^s}_{\overline{\mathbf{x}}_n^s}, \tag{2.152}$$

and the result of the Taylor series of Equation (2.148) can be written in the form

$$\overline{\mathbf{x}}(\xi) = \left(\overline{\mathbf{E}} + \sum_{k=1}^{\infty}\frac{\overline{\mathbf{Q}}_k(\xi-\xi_0)^k}{k!}\right)\overline{\mathbf{x}}(\xi_0) + \sum_{k=1}^{\infty}\frac{\overline{\mathbf{x}}_k^s(\xi-\xi_0)^k}{k!} \tag{2.153}$$

$$\overline{\mathbf{x}}(\xi) = \mathcal{M}_{\xi_0}^{\xi}\left\{\overline{\mathbf{R}}^x\right\}\overline{\mathbf{x}}(\xi_0) + \overline{\mathbf{x}}_{eq}^s(\xi)\,.$$

The elements $\overline{\mathbf{Q}}_n$ and $\overline{\mathbf{x}}_n^s$ can be determined recursively from Equation (2.152).

For a numerical implementation it is advantageous not to calculate $\overline{\mathbf{Q}}_n$ or $\overline{\mathbf{x}}_n^s$ but rather to evaluate the expressions

$$\hat{\overline{\mathbf{Q}}}_n = \overline{\mathbf{Q}}_n\frac{(\xi-\xi_0)^n}{n!} \qquad \text{and} \qquad \hat{\overline{\mathbf{x}}}^s{}_n = \overline{\mathbf{x}}_n^s\frac{(\xi-\xi_0)^n}{n!}. \tag{2.154}$$

Substitution of these expressions into Equations (2.149) to (2.152) yields expressions for $\hat{\overline{\mathbf{Q}}}_n$

$$\hat{\overline{\mathbf{Q}}}_0 = \overline{\mathbf{E}} \tag{2.155}$$

$$\hat{\overline{\mathbf{Q}}}_1 = \overline{\mathbf{R}}_0^x\,\Delta\xi \tag{2.156}$$

$$\hat{\overline{\mathbf{Q}}}_2 = \frac{\Delta\xi}{2}\left[\Delta\overline{\mathbf{R}}^x\hat{\overline{\mathbf{Q}}}_0 + \overline{\mathbf{R}}_0^x\hat{\overline{\mathbf{Q}}}_1\right] = \frac{\Delta\xi}{2}\left[\Delta\overline{\mathbf{R}}^x + \frac{\overline{\mathbf{R}}_0^{x\,2}}{\Delta\xi}\right] \tag{2.157}$$

$$\ldots$$

$$\hat{\overline{\mathbf{Q}}}_n = \frac{\Delta\xi}{n}\left[\Delta\overline{\mathbf{R}}^x\hat{\overline{\mathbf{Q}}}_{n-2} + \overline{\mathbf{R}}_0^x\hat{\overline{\mathbf{Q}}}_{n-1}\right], \tag{2.158}$$

which are more simple and stable for numerical evaluation. It follows that the numerical effort reduces to two matrix multiplications, one matrix addition and one scaling of a matrix per series term. Also the division by large numbers due to the factorial is avoided. The inhomogeneous parts $\hat{\overline{\mathbf{x}}}^{\mathrm{s}}{}_n$ can be treated in the same way. The terms $\hat{\overline{\mathbf{Q}}}_n$ and $\hat{\overline{\mathbf{x}}}^{\mathrm{s}}{}_n$ are calculated and summed over until a previously defined accuracy is achieved.

2.2.4 Approximation by Power Series

Further analytical solution formulas are obtained by the approximation of the matrix functions $\overline{\mathbf{R}}^x(\xi)$ and $\overline{\mathbf{x}}^{\mathrm{s}}(\xi)$ by power series. Without loss of generality $\xi_0 = 0$ can be assumed since a substitution $\hat{\xi} = \xi - \xi_0$ always allows this choice. With this assumption $\overline{\mathbf{R}}^x(\xi)$ and $\overline{\mathbf{x}}^{\mathrm{s}}(\xi)$ can be expressed as

$$\overline{\mathbf{R}}^x(\xi) = \sum_{k=0}^{\infty} \overline{\mathbf{A}}_k \xi^k \tag{2.159}$$

$$\overline{\mathbf{x}}^{\mathrm{s}}(\xi) = \sum_{k=0}^{\infty} \overline{\mathbf{b}}_k \xi^k \,. \tag{2.160}$$

Assuming that the series of Equations (2.159) and (2.160) converge for $\xi < R_{\mathrm{conv}}$, it is possible to approximate the solution for an initial value $\overline{\mathbf{x}}(0)$ as a power series:

$$\overline{\mathbf{x}}(\xi) = \sum_{k=0}^{\infty} \overline{\mathbf{c}}_k \xi^k = \overline{\mathbf{c}}_0 + \overline{\mathbf{c}}_1 \xi + \overline{\mathbf{c}}_2 \xi^2 + \cdots + \overline{\mathbf{c}}_n \xi^n. \tag{2.161}$$

In the following, this will be shown and also equations for the determination of $\overline{\mathbf{c}}_k$ will be given.

Starting from Equation (2.161) the derivative of $\overline{\mathbf{x}}(\xi)$ can be written as

$$\frac{\mathrm{d}}{\mathrm{d}\xi}\overline{\mathbf{x}}(\xi) = \sum_{k=0}^{\infty}(k+1)\,\overline{\mathbf{c}}_{k+1}\xi^k = \overline{\mathbf{c}}_1 + 2\,\overline{\mathbf{c}}_2\xi + 3\,\overline{\mathbf{c}}_3\xi^2 + \cdots + (n+1)\,\overline{\mathbf{c}}_n\xi^{(n+1)} \,. \tag{2.162}$$

If Equations (2.159) to (2.162) are inserted into the initial differential Equation (2.130) it follows that

$$\sum_{k=0}^{\infty}(k+1)\,\overline{\mathbf{c}}_{k+1}\xi^k = \left(\sum_{k=0}^{\infty}\overline{\mathbf{A}}_k\xi^k\right)\left(\sum_{k=0}^{\infty}\overline{\mathbf{c}}_k\xi^k\right) + \sum_{k=0}^{\infty}\overline{\mathbf{b}}_k\xi^k \,. \tag{2.163}$$

Expanding the sums and comparison of coefficients yields determining equations for the coefficients $\overline{\mathbf{c}}_n$

$$\overline{\mathbf{c}}_0 = \overline{\mathbf{x}}(0) \tag{2.164}$$

$$\overline{\mathbf{c}}_1 = \overline{\mathbf{A}}_0\overline{\mathbf{c}}_0 + \overline{\mathbf{b}}_0 \tag{2.165}$$

$$2\,\overline{\mathbf{c}}_2 = \overline{\mathbf{A}}_1\overline{\mathbf{c}}_0 + \overline{\mathbf{A}}_0\overline{\mathbf{c}}_1 + \overline{\mathbf{b}}_1 \tag{2.166}$$

$$3\,\overline{\mathbf{c}}_3 = \overline{\mathbf{A}}_2\overline{\mathbf{c}}_0 + \overline{\mathbf{A}}_1\overline{\mathbf{c}}_1 + \overline{\mathbf{A}}_0\overline{\mathbf{c}}_2 + \overline{\mathbf{b}}_2 \tag{2.167}$$

$$\cdots$$

$$(n+1)\,\overline{\mathbf{c}}_{n+1} = \sum_{k=0}^{n}\overline{\mathbf{A}}_{n-k}\overline{\mathbf{c}}_k + \overline{\mathbf{b}}_n \,. \tag{2.168}$$

The convergence of the series solution still needs to be clarified. If it is assumed that the approximation series of Equations (2.159) and (2.160) are convergent for $0 \le \rho \le R_{\mathrm{conv}}$, then, with a constant and finite bound $S > 0$, it follows

$$\rho^n \left\|\overline{\mathbf{A}}_n\right\| \leq S \qquad \text{and} \qquad \rho^n \left\|\overline{\mathbf{b}}_n\right\| \leq S\,, \tag{2.169}$$

and furthermore, by means of Equation (2.168),

$$(n+1)\left\|\overline{\mathbf{c}}_{n+1}\right\| \leq \sum_{k=0}^{n} S\rho^{k-n}\left\|\overline{\mathbf{c}}_k\right\| + S\rho^{-n}\,. \tag{2.170}$$

From the further definitions of upper bounds of the norms $\|\overline{\mathbf{c}}_n\| \leq s_n$ it follows with Equation (2.170) that

$$(n+1)\,s_{n+1} \leq \frac{S}{\rho^n}\left(s_0 + s_1\rho^1 + s_2\rho^2 + \cdots s_n\rho^n + 1\right) \tag{2.171}$$

$$\leq S s_n + \frac{1}{\rho}\underbrace{\frac{S}{\rho^{n-1}}\left(1 + s_0 + s_1\rho + s_2\rho^2 + \cdots + s_{n-1}\rho^{n-1}\right)}_{ns_n} \tag{2.172}$$

$$s_{n+1} \leq \left(\frac{S}{n+1} + \frac{n}{n+1}\frac{1}{\rho}\right)s_n\,. \tag{2.173}$$

The convergence of the series in Equation (2.161) is assured if $\sum_{n=0}^{\infty} s_n\xi^n$ converges. The quotient criterion of d'Alembert

$$\lim_{n\to\infty}\left|\frac{s_{n+1}\xi^{n+1}}{s_n\xi^n}\right| \leq \lim_{n\to\infty}\left(\frac{S}{n+1} + \frac{n}{n+1}\frac{1}{\rho}\right)|\xi| \tag{2.174}$$

$$\lim_{n\to\infty}\left|\frac{s_{n+1}\xi^{n+1}}{s_n\xi^n}\right| \leq \frac{|\xi|}{\rho} \tag{2.175}$$

shows the convergence of the solution series for $|\xi| < \rho$. Since by assumption ρ is arbitrary within the interval $0 \leq \rho \leq R_{\text{conv}}$ it follows that the solution series converges for all $|\xi| \leq R_{\text{conv}}$.

To arrive at a representation which involves the matrizant from Equation (2.168) is possible if the super vector $\overline{\mathbf{x}}(0)$ is separated from the coefficients $\overline{\mathbf{c}}_n$,

$$\overline{\mathbf{c}}_0 = \underbrace{\overline{\mathbf{E}}}_{\hat{\overline{\mathbf{C}}}_0}\,\overline{\mathbf{x}}(0) \tag{2.176}$$

$$\begin{aligned}\overline{\mathbf{c}}_1 &= \overline{\mathbf{A}}_0\overline{\mathbf{c}}_0 + \overline{\mathbf{b}}_0 \\ &= \underbrace{\overline{\mathbf{A}}_0\hat{\overline{\mathbf{C}}}_0}_{\hat{\overline{\mathbf{C}}}_1}\,\overline{\mathbf{x}}(0) + \underbrace{\overline{\mathbf{b}}_0}_{\hat{\overline{\mathbf{b}}}_0}\end{aligned} \tag{2.177}$$

$$\begin{aligned}\overline{\mathbf{c}}_2 &= \frac{1}{2}\left(\overline{\mathbf{A}}_1\overline{\mathbf{c}}_0 + \overline{\mathbf{A}}_0\overline{\mathbf{c}}_1 + \overline{\mathbf{b}}_1\right) \\ &= \underbrace{\frac{1}{2}\left(\overline{\mathbf{A}}_1\hat{\overline{\mathbf{C}}}_0 + \overline{\mathbf{A}}_0\hat{\overline{\mathbf{C}}}_1\right)}_{\hat{\overline{\mathbf{C}}}_2}\overline{\mathbf{x}}(0) + \underbrace{\frac{1}{2}\left(\overline{\mathbf{A}}_0\hat{\overline{\mathbf{b}}}_0 + \overline{\mathbf{b}}_1\right)}_{\hat{\overline{\mathbf{b}}}_1}\end{aligned} \tag{2.178}$$

$$\overline{\mathbf{c}}_3 = \frac{1}{3}\left(\overline{\mathbf{A}}_2\overline{\mathbf{c}}_0 + \overline{\mathbf{A}}_1\overline{\mathbf{c}}_1 + \overline{\mathbf{A}}_0\overline{\mathbf{c}}_2 + \overline{\mathbf{b}}_2\right) \tag{2.179}$$

$$= \underbrace{\frac{1}{3}\left(\overline{\mathbf{A}}_2\hat{\overline{\mathbf{C}}}_0 + \overline{\mathbf{A}}_1\hat{\overline{\mathbf{C}}}_1 + \overline{\mathbf{A}}_0\hat{\overline{\mathbf{C}}}_2\right)}_{\hat{\overline{\mathbf{C}}}_3}\overline{\mathbf{x}}(0) + \underbrace{\frac{1}{3}\left(\overline{\mathbf{A}}_1\hat{\overline{\mathbf{b}}}_0 + \overline{\mathbf{A}}_0\hat{\overline{\mathbf{b}}}_1 + \overline{\mathbf{b}}_2\right)}_{\hat{\overline{\mathbf{b}}}_2}$$

...

$$\overline{\mathbf{c}}_{n+1} = \underbrace{\frac{1}{n+1}\left(\sum_{k=0}^{n}\overline{\mathbf{A}}_{n-k}\hat{\overline{\mathbf{C}}}_k\right)}_{\hat{\overline{\mathbf{C}}}_n}\overline{\mathbf{x}}(0) + \underbrace{\frac{1}{n+1}\left(\sum_{k=0}^{n-2}\overline{\mathbf{A}}_{n-2-k}\,\hat{\overline{\mathbf{b}}}_k + \overline{\mathbf{b}}_{n-1}\right)}_{\hat{\overline{\mathbf{b}}}_{n-1}} . \tag{2.180}$$

It then follows for the matrizant or, likewise, for the source vector

$$\mathcal{M}_0^{\xi}\left\{\overline{\mathbf{R}}^x\right\} = \sum_{k=0}^{\infty}\hat{\overline{\mathbf{C}}}_k\xi^k , \qquad \overline{\mathbf{x}}^{\mathrm{s}}_{\mathrm{eq}}(\xi) = \sum_{k=0}^{\infty}\hat{\overline{\mathbf{b}}}_k\xi^k . \tag{2.181}$$

For the simple case of a constant matrix $\overline{\mathbf{R}}^x(\xi) = \overline{\mathbf{A}}_0$ it is recognized that, according to Equations (2.176) to (2.180) and (2.181),

$$\hat{\overline{\mathbf{C}}}_0 = \overline{\mathbf{E}} \tag{2.182}$$

$$\hat{\overline{\mathbf{C}}}_1 = \overline{\mathbf{A}}_0\hat{\overline{\mathbf{C}}}_0 = \overline{\mathbf{A}}_0 \tag{2.183}$$

$$\hat{\overline{\mathbf{C}}}_2 = \frac{1}{2}\overline{\mathbf{A}}_0\hat{\overline{\mathbf{C}}}_1 = \frac{1}{2}\overline{\mathbf{A}}_0^2 \tag{2.184}$$

...

$$\hat{\overline{\mathbf{C}}}_n = \frac{1}{n}\overline{\mathbf{A}}_0\hat{\overline{\mathbf{C}}}_{n-1} = \frac{1}{n!}\overline{\mathbf{A}}_0^n . \tag{2.185}$$

Thus, for the matrizant the exact analytical solution of the series expansion of the matrix exponential function is retrieved,

$$\mathcal{M}_0^{\xi}\left\{\overline{\mathbf{R}}^x\right\} = \sum_{k=0}^{\infty}\frac{1}{n!}\overline{\mathbf{A}}_0^n\xi^n = \exp\overline{\mathbf{R}}^x\xi . \tag{2.186}$$

If a linear function is used for the approximation of $\overline{\mathbf{R}}^x(\xi) = \overline{\mathbf{A}}_0 + \overline{\mathbf{A}}_1\xi$ one finds for the terms $\hat{\overline{\mathbf{C}}}_n$

$$\hat{\overline{\mathbf{C}}}_0 = \overline{\mathbf{E}} \tag{2.187}$$

$$\hat{\overline{\mathbf{C}}}_1 = \overline{\mathbf{A}}_0\hat{\overline{\mathbf{C}}}_0 = \overline{\mathbf{A}}_0 \tag{2.188}$$

$$\hat{\overline{\mathbf{C}}}_2 = \frac{1}{2}\left(\overline{\mathbf{A}}_1 + \overline{\mathbf{A}}_0^2\right) \tag{2.189}$$

...

$$\hat{\overline{\mathbf{C}}}_n = \frac{1}{n}\left(\overline{\mathbf{A}}_0\hat{\overline{\mathbf{C}}}_{n-1} + \overline{\mathbf{A}}_1\hat{\overline{\mathbf{C}}}_{n-2}\right) , \tag{2.190}$$

and this yields for the matrizant

$$\mathcal{M}_0^{\xi}\left\{\overline{\mathbf{R}}^x\right\} = \overline{\mathbf{E}} + \xi\overline{\mathbf{A}}_0 + \tfrac{\xi^2}{2!}\left(\overline{\mathbf{A}}_1 + \overline{\mathbf{A}}_0^2\right) + \tfrac{\xi^3}{3!}\left[\overline{\mathbf{A}}_0\left(\overline{\mathbf{A}}_1 + \overline{\mathbf{A}}_0^2\right) + 2\overline{\mathbf{A}}_1\overline{\mathbf{A}}_0\right] + \cdots . \tag{2.191}$$

This result is identical to the formulas of Section 2.2.3 for $\xi_0 = 0$.

As a third example, a quadratic approximation of $\overline{\mathbf{R}}^x(\xi) = \overline{\mathbf{A}}_0 + \overline{\mathbf{A}}_1\xi + \overline{\mathbf{A}}_2\xi^2$ is examined. In this case the coefficients are calculated to

$$\hat{\overline{\mathbf{C}}}_0 = \overline{\mathbf{E}} \tag{2.192}$$

$$\hat{\overline{\mathbf{C}}}_1 = \overline{\mathbf{A}}_0\hat{\overline{\mathbf{C}}}_0 = \overline{\mathbf{A}}_0 \tag{2.193}$$

$$\hat{\overline{\mathbf{C}}}_2 = \frac{1}{2}\left(\overline{\mathbf{A}}_1 + \overline{\mathbf{A}}_0^2\right) \tag{2.194}$$

$$\hat{\overline{\mathbf{C}}}_3 = \frac{1}{3!}\left[\overline{\mathbf{A}}_0\left(\overline{\mathbf{A}}_1 + \overline{\mathbf{A}}_0^2\right) + 2\overline{\mathbf{A}}_1\overline{\mathbf{A}}_0 + 2\overline{\mathbf{A}}_2\right] \tag{2.195}$$

$$\cdots$$

$$\hat{\overline{\mathbf{C}}}_n = \frac{1}{n}\left(\overline{\mathbf{A}}_0\hat{\overline{\mathbf{C}}}_{n-1} + \overline{\mathbf{A}}_1\hat{\overline{\mathbf{C}}}_{n-2} + \overline{\mathbf{A}}_2\hat{\overline{\mathbf{C}}}_{n-3}\right). \tag{2.196}$$

It is an advantage of the power series method that it yields a general and simple algorithm to construct formulas for the matrizant and the equivalent source vector for an arbitrary order of approximation of $\overline{\mathbf{R}}^x(\xi)$ and $\bar{x}^s(\xi)$. Within the recursive formula for $\hat{\overline{\mathbf{C}}}_n$ the number of terms is identical to the highest power of ξ in the approximation.

2.2.5 *Interpolation from Diagonalization*

In the literature several different solution methods can be found which take advantage of diagonal transformations by means of eigenvalues and eigenvectors in order to find solutions for special types of nonuniform transmission lines, such as cyclic transmission lines, for example [18, 27, 28]. This section discusses an interpolation scheme which is applicable to general nonuniform multiconductor transmission lines. To this end, the matrix function $\overline{\mathbf{R}}^x(\xi)$ along an interval $[\xi_0, \xi_1]$ of length $\Delta\xi$ is divided into two terms,

$$\overline{\mathbf{R}}^x(\xi) = \overline{\mathbf{R}}^x_m + \overline{\mathbf{R}}^x_\Delta(\xi). \tag{2.197}$$

The first term, that is the reference term $\overline{\mathbf{R}}^x_m$, is independent of the coordinate ξ which, in the following, is an element of the interval $[\xi_0, \xi_1]$. The reference term could be chosen as a mean value with respect to the chosen interval, for example. The second term describes the deviation from the reference value and usually will vary along the interval. With this decomposition the sum rule of the product integral of Equation (E.17) yields

$$\mathcal{M}_{\xi_0}^{\xi_1}\left\{\overline{\mathbf{R}}^x\right\} = \overline{\mathbf{G}}^{(0)}(\xi_1)\,\mathcal{M}_{\xi_0}^{\xi_1}\{\overline{\mathbf{H}}\} = \overline{\mathbf{G}}^{(0)}(\xi_1)\,\overline{\mathbf{G}}^{(1)}(\xi_1) \tag{2.198}$$

with

$$\overline{\mathbf{H}}(\xi) = \{\overline{\mathbf{G}}^{(0)}(\xi)\}^{-1}\,\overline{\mathbf{R}}^x_\Delta(\xi)\,\overline{\mathbf{G}}^{(0)}(\xi) \tag{2.199}$$

and

$$\overline{\mathbf{G}}^{(0)}(\xi) = \mathcal{M}_{\xi_0}^{\xi}\left\{\overline{\mathbf{R}}^x_m\right\} = \exp\overline{\mathbf{R}}^x_m(\xi - \xi_0). \tag{2.200}$$

From a similarity transformation of the reference term $\overline{\mathbf{R}}^x_m$ into diagonal form[4]

$$\overline{\mathbf{R}}^x_m = \overline{\mathbf{U}}\,\overline{\Lambda}\,\overline{\mathbf{V}} = \overline{\mathbf{U}}\,\overline{\Lambda}\,\overline{\mathbf{U}}^{-1} = \sum_{n=1}^{N} \lambda_n \overline{\boldsymbol{u}}_n \overline{\boldsymbol{v}}_n^{\mathsf{T}}, \tag{2.201}$$

one finds

$$\overline{\mathbf{G}}^{(0)}(\xi) = \sum_{n=1}^{N} \exp\left(\lambda_n(\xi - \xi_0)\right)\overline{\boldsymbol{u}}_n \overline{\boldsymbol{v}}_n^{\mathsf{T}} \tag{2.202}$$

and

$$\{\overline{\mathbf{G}}^{(0)}(\xi)\}^{-1} = \sum_{n=1}^{N} \exp\left(-\lambda_n(\xi - \xi_0)\right)\overline{\boldsymbol{u}}_n \overline{\boldsymbol{v}}_n^{\mathsf{T}}. \tag{2.203}$$

The sum of Equation (2.201) is a representation of the diagonal transformation in terms of dyadic products where the vectors $\overline{\boldsymbol{v}}_n^{\mathsf{T}}$ and $\overline{\boldsymbol{u}}_n$ represent the rows and columns of the transformation matrix $\overline{\mathbf{U}}$ and its corresponding inverse $\overline{\mathbf{V}}$ to obtain the diagonal matrix $\overline{\Lambda}$. It is clear that the transformation matrix $\overline{\mathbf{U}}$ may not be singular. The value N is the dimension of the matrix function $\overline{\mathbf{R}}^x(\xi)$.

If the reference term $\overline{\mathbf{R}}^x_m$ of Equation (2.197) is chosen as the mean value with respect to the interval (ξ_0, ξ_1)

$$\overline{\mathbf{R}}^x_m = \frac{1}{2}\left(\overline{\mathbf{R}}^x(\xi_0) + \overline{\mathbf{R}}^x(\xi_1)\right), \tag{2.204}$$

then the varying term $\overline{\mathbf{R}}^x_\Delta(\xi)$ can be written as a constant difference matrix $\overline{\mathbf{C}}$ multiplied with a varying function $f(\xi)$

$$\overline{\mathbf{R}}^x_\Delta(\xi) = f(\xi)\frac{1}{2}\left(\overline{\mathbf{R}}^x(\xi_1) - \overline{\mathbf{R}}^x(\xi_0)\right) = f(\xi)\overline{\mathbf{C}}. \tag{2.205}$$

Here, $f(\xi)$ is a monotonically increasing function which is chosen such that $f(\xi_0) = -1$ and $f(\xi_1) = +1$. With this choice it is assured that the matrix function $\overline{\mathbf{R}}^x(\xi)$ assumes its correct values at the boundaries of the segment. The simplest choice of $f(\xi)$ is a linear function

$$f(\xi) = \frac{2\xi - \xi_1 - \xi_0}{\xi_1 - \xi_0} = \frac{2\xi - 2\xi_0 - \Delta\xi}{\Delta\xi} = \frac{2}{\Delta\xi}(\xi - \xi_m) \tag{2.206}$$

with

$$\xi_m = \frac{\xi_0 + \xi_1}{2} = \xi_0 + \frac{\Delta\xi}{2}. \tag{2.207}$$

This ansatz resembles the linear term of a truncated Taylor expansion of $\overline{\mathbf{R}}^x(\xi)$. The reference term $\overline{\mathbf{R}}^x_m$ corresponds to the constant term which is obtained if the expansion is done around the mean value and, furthermore, a linear term emerges which is a constant matrix, multiplied with the variable ξ. However, this Taylor expansion does not guarantee that the matrix function $\overline{\mathbf{R}}^x(\xi)$ is properly modeled at the boundaries of the segment. Rather, discontinuities are

[4] Since the reference term is arbitrary it can always be chosen such that it can be transformed into diagonal form.

possible. However, if the function $f(\xi)$ is chosen to be of higher order it is possible to have continuous derivatives of the matrix function $\overline{\mathbf{R}}^{x}(\xi)$ at the boundaries of the segment.

Once the terms $\overline{\mathbf{G}}^{(0)}(\xi)$ and $\{\overline{\mathbf{G}}^{(0)}(\xi)\}^{-1}$ are determined from the diagonal transformation according to Equations (2.202) and (2.203) the matrix function $\overline{\mathbf{H}}(\xi)$ and the corresponding matrizant have to be computed. The matrix function $\overline{\mathbf{H}}(\xi)$ turns out to be

$$\overline{\mathbf{H}}(\xi) = f(\xi)\{\overline{\mathbf{G}}^{(0)}(\xi)\}^{-1}\overline{\mathbf{C}}\,\overline{\mathbf{G}}^{(0)}(\xi) \tag{2.208}$$

$$= f(\xi)\sum_{k=1}^{N}\sum_{n=1}^{N}\left\{\exp\left(\lambda_k - \lambda_n\right)(\xi - \xi_0)\overline{\boldsymbol{u}}_n\overline{\boldsymbol{v}}_n^{\mathsf{T}}\,\overline{\mathbf{C}}\,\overline{\boldsymbol{u}}_k\overline{\boldsymbol{v}}_k^{\mathsf{T}}\right\} \tag{2.209}$$

$$= f(\xi)\sum_{k=1}^{N}\sum_{n=1}^{N}\left\{\exp\left(\lambda_k - \lambda_n\right)(\xi - \xi_0)\overline{\boldsymbol{u}}_n\overline{\boldsymbol{v}}_k^{\mathsf{T}}\alpha_{nk}\right\}, \tag{2.210}$$

with the scalar

$$\alpha_{nk} = \overline{\boldsymbol{v}}_n^{\mathsf{T}}\,\overline{\mathbf{C}}\,\overline{\boldsymbol{u}}_k\,. \tag{2.211}$$

The matrix function $\overline{\mathbf{H}}(\xi)$ is a similarity transformation of the variance matrix $f(\xi)\overline{\mathbf{C}}$ and possesses the same eigenvalues. If $\overline{\mathbf{C}}$ is 'small' in comparison to $\overline{\mathbf{R}}_m^{x}$ then the matrizant of $\overline{\mathbf{H}}(\xi)$ can be calculated numerically with comparatively large steps if compared to the equivalent calculation of the matrizant of $\overline{\mathbf{R}}^{x}(\xi)$. By means of Equation (2.210) and the diagonalization of $\overline{\mathbf{R}}_m^{x}$ the matrizant of $\overline{\mathbf{H}}(\xi)$ can be calculated with the numerical methods that are part of the next section.

Another computational method of the matrix $\overline{\mathbf{G}}^{(1)}(\xi_1)$ can be used if the variance matrix $\overline{\mathbf{C}}$ is rather small. Then an evaluation of the first two terms of the Picard iteration of Equation (2.133) might be sufficient,

$$\overline{\mathbf{G}}^{(1)}(\xi_1) = \overline{\mathbf{E}} + \int_{\xi_0}^{\xi_1}\overline{\mathbf{H}}(\zeta)\mathrm{d}\zeta + O(|\lambda_{\max}^{(C)}|\Delta\xi^2) \tag{2.212}$$

$$= \overline{\mathbf{E}} + \int_{\xi_0}^{\xi_1} f(\zeta)\sum_{k=1}^{N}\sum_{n=1}^{N}\exp\left((\lambda_k - \lambda_n)(\zeta - \xi_0)\right)\overline{\boldsymbol{u}}_n\overline{\boldsymbol{v}}_k^{\mathsf{T}}\alpha_{nk}\mathrm{d}\zeta + O(|\lambda_{\max}^{(C)}|\Delta\xi^2)\,. \tag{2.213}$$

Here, $|\lambda_{\max}^{(C)}|$ is the absolute value of the largest eigenvalue of the matrix $\overline{\mathbf{C}}$. Decreasing the step size $\Delta\xi$ also leads to a decrease of $|\lambda_{\max}^{(C)}|$ such that the error approaches the order $O(\Delta\xi^3)$. Since the rows $\overline{\boldsymbol{u}}_n$ and columns $\overline{\boldsymbol{v}}_k^{\mathsf{T}}$ of the transformation matrix $\overline{\mathbf{U}}$ and the scalar coefficient α_{nk} do not depend on the integration variable ζ they can be pulled out of the integral

$$\overline{\mathbf{G}}^{(1)}(\xi_1) = \overline{\mathbf{E}} + \sum_{k=1}^{N}\sum_{n=1}^{N}\overline{\boldsymbol{u}}_n\overline{\boldsymbol{v}}_k^{\mathsf{T}}\alpha_{nk}\int_{\xi_0}^{\xi_1} f(\zeta)\exp\left(\lambda_k - \lambda_n\right)(\zeta - \xi_0)\mathrm{d}\zeta + O(|\lambda_{\max}^{(C)}|\Delta\xi^2)\,. \tag{2.214}$$

If the interpolation function $f(\xi)$, according to Equation (2.206), is inserted one obtains for $\lambda_k \neq \lambda_n$ the integral of Equation (2.214) in the form

$$F_{nk} = \int_{\xi_0}^{\xi_1} f(\zeta)\exp\left((\lambda_k - \lambda_n)(\zeta - \xi_0)\right)\mathrm{d}\zeta \tag{2.215}$$

$$= \int_{\xi_0}^{\xi_1} \frac{2}{\Delta\xi}(\zeta - \xi_m) \exp\left((\lambda_k - \lambda_n)(\zeta - \xi_0)\right) d\zeta \tag{2.216}$$

$$= \frac{2\exp\left((\lambda_n - \lambda_k)\,\xi_0\right)}{\Delta\xi} \int_{\xi_0}^{\xi_1} (\zeta - \xi_m) \exp(\lambda_k - \lambda_n)\,\zeta\, d\zeta \tag{2.217}$$

$$= \frac{\exp(\lambda_n - \lambda_k)\,\xi_0}{(\lambda_k - \lambda_n)^2\,\Delta\xi} \Big\{ [(\lambda_k - \lambda_n)\,\Delta\xi + 2] \exp(\lambda_k - \lambda_n)\,\xi_0$$

$$+ [(\lambda_k - \lambda_n)\,\Delta\xi - 2] \exp(\lambda_k - \lambda_n)\xi_1 \Big\} \tag{2.218}$$

$$= \frac{1}{(\lambda_k - \lambda_n)^2\,\Delta\xi} \Big\{ [(\lambda_k - \lambda_n)\,\Delta\xi + 2]$$

$$+ [(\lambda_k - \lambda_n)\,\Delta\xi - 2] \exp(\lambda_k - \lambda_n)\,\Delta\xi \Big\}. \tag{2.219}$$

If the eigenvalues are equal the integral turns out to be

$$F_{nk} = \int_{\xi_0}^{\xi_1} f(\xi) \exp\left((\lambda_k - \lambda_n)(\zeta - \xi_0)\right) d\zeta \tag{2.220}$$

$$= \int_{\xi_0}^{\xi_1} \frac{2}{\Delta\xi}(\zeta - \xi_m) d\zeta \tag{2.221}$$

$$= \frac{2}{\Delta\xi}\left[\frac{1}{2}\zeta^2 - \xi_m \zeta\right]_{\zeta=\xi_0}^{\xi_1} = 0\,. \tag{2.222}$$

With these results the approximation of the matrizant within the segment $[\xi_0, \xi_1]$ can be written as

$$\mathcal{M}_{\xi_0}^{\xi_1}\left\{\overline{\mathbf{R}}^x\right\} \approx \left(\sum_{n=1}^{N} \exp \lambda_n \Delta\xi \overline{\boldsymbol{u}}_n \overline{\boldsymbol{v}}_n^{\mathsf{T}}\right)\left(\overline{\mathbf{E}} + \sum_{k=1}^{N}\sum_{n=1}^{N} \overline{\boldsymbol{u}}_n \overline{\boldsymbol{v}}_k^{\mathsf{T}} \alpha_{nk} F_{nk}\right) \tag{2.223}$$

$$\approx \exp \overline{\mathbf{R}}_m^x \Delta\xi + \exp \overline{\mathbf{R}}_m^x \Delta\xi \sum_{k=1}^{N}\sum_{n=1}^{N} \overline{\boldsymbol{u}}_n \overline{\boldsymbol{v}}_k^{\mathsf{T}} \alpha_{nk} F_{nk}\,. \tag{2.224}$$

This result can be interpreted as the sum of a matrizant of a constant matrix function, that is the matrizant of a uniform transmission line with the properties of the matrix $\overline{\mathbf{R}}_m^x$, plus a correction term. For a constant matrix function the correction term vanishes since in this case the scalars α_{nk} vanish.

2.2.6 *Numerical Integration*

Besides the analytical and semianalytical solution methods it is also possible to apply purely numerical integration methods in order to calculate the matrizant. In principle, it is possible to use numerical integration n times with different initial conditions $\overline{\mathbf{x}}(\xi_0)$ for the calculation. However, it turns out to be more advantageous to derive general integration formulas for one segment first by transforming the integration methods into the framework of matrix calculus.

For the classification of integration methods the number of processed intervals per integration step can be utilized. In this context one distinguishes between one-step and multiple-step methods. In the following focus is put on one-step methods since these can be applied straightforwardly to the required matrix calculus. Multiple-step methods have the disadvanatge that they require, depending on the number of steps, various matrix inversions and, thus, quickly become inefficient.

For the application of integration methods the differential Equation (2.130) is transformed into an integral equation by integration over the interval $[\xi_i, \xi_{i+1}]$,

$$\overline{\mathbf{x}}(\xi_{i+1}) = \overline{\mathbf{x}}(\xi_i) + \int_{\xi_i}^{\xi_{i+1}} \overline{\mathbf{R}}^x(\zeta)\overline{\mathbf{x}}(\zeta)\mathrm{d}\zeta + \int_{\xi_i}^{\xi_{i+1}} \overline{\mathbf{x}}^{\mathrm{s}}(\zeta)\mathrm{d}\zeta. \tag{2.225}$$

Then a numerical integration method is applied in order to put the integral equation into the form

$$\overline{\mathbf{x}}(\xi_{i+1}) = \mathcal{F}\left(\Delta\xi_i, \overline{\mathbf{R}}^x(\zeta)\right)\overline{\mathbf{x}}(\xi_i) + \mathcal{F}_{\mathrm{S}}\left(\Delta\xi_i, \overline{\mathbf{R}}^x(\zeta), \overline{\mathbf{x}}^{\mathrm{s}}(\zeta)\right) \tag{2.226}$$

$$= \mathcal{M}_{\xi_i}^{\xi_{i+1}}\left\{\overline{\mathbf{R}}^x\right\}\overline{\mathbf{x}}(\xi_i) + \overline{\mathbf{x}}^{\mathrm{s}}_{\mathrm{eq},i} \tag{2.227}$$

in order to obtain a closed formula for the matrizant and the source vector at the end of the segment. The source vector $\overline{\mathbf{x}}^{\mathrm{s}}_{\mathrm{eq},i}$ represents the effect of the distributed excitation along the segment $\Delta\xi_i$ on the end of the segment and also represents an initial condition for the subsequent segment $\Delta\xi_{i+1}$. The matrizant of the complete transmission line results from chain multiplication of subsequent segments. For n segments the equivalent source vector at the end of the transmission line is given by

$$\overline{\mathbf{x}}^{\mathrm{s}}_{\mathrm{eq}}(\xi) = \sum_{i=1}^{n} \mathcal{M}_{\xi_{i+1}}^{\xi}\left\{\overline{\mathbf{R}}^x\right\}\overline{\mathbf{x}}^{\mathrm{s}}_{\mathrm{eq},i}\,. \tag{2.228}$$

As abbreviation in the following the index i and $i+1$ is used to indicate function values at the beginning and the end, respectively, of the ith segment. If function values are required within a segment $[\xi_i, \xi_{i+1}]$ these are denoted by noninteger indices, for example $\overline{\mathbf{R}}^x_{i+\frac{1}{2}}$ corresponds to $\overline{\mathbf{R}}^x(\xi_i + \frac{1}{2}\Delta\xi_i)$. The value $\Delta\xi_i$ denotes the step size of the ith segment, defined as $\Delta\xi_i = \xi_{i+1} - \xi_i$.

For the transmission-line equations the existence and uniqueness of a solution is guaranteed if the Lipschitz-condition

$$\left\|\overline{\mathbf{R}}(\xi)\overline{\boldsymbol{u}} - \overline{\mathbf{R}}(\xi)\overline{\boldsymbol{v}}\right\| \leq L\ \|\overline{\boldsymbol{u}} - \overline{\boldsymbol{v}}\| \text{ with } L > 0 \tag{2.229}$$

is fulfilled for all ξ [29]. It then follows that within the interval $[\xi_i, \xi_{i+1}]$ a unique solution exists if the Lipschitz-constant

$$L = \sup_{\xi_i \leq \xi \leq \xi_{i+1}} \left\|\overline{\mathbf{R}}(\xi)\right\| \tag{2.230}$$

is finite, that is if the parameter matrix $\overline{\mathbf{R}}(\xi)$ is finite for all ξ. For meaningful transmission line geometries this condition should be fulfilled.

2.2.6.1 Euler–Cauchy Method

At the beginning ξ_i of a segment the slope $\frac{\mathrm{d}\overline{\mathbf{x}}_i}{\mathrm{d}\xi} = \overline{\mathbf{R}}_i^x \overline{\mathbf{x}}_i + \overline{\mathbf{x}}_i^{\mathrm{s}}$ is known exactly. Under the assumption of a constant slope it allows the solution at the end the segment to be determined approximately according to

$$\overline{\mathbf{x}}_{i+1} = \overline{\mathbf{x}}_i + \frac{\mathrm{d}\overline{\mathbf{x}}_i}{\mathrm{d}\xi} \Delta\xi_i \tag{2.231}$$

$$= \underbrace{\left(\overline{\mathbf{E}} + \overline{\mathbf{R}}_i^x \Delta\xi_i\right)}_{\mathcal{M}_{\xi_i}^{\xi_{i+1}}\left\{\overline{\mathbf{R}}^x\right\}} \overline{\mathbf{x}}_i + \underbrace{\overline{\mathbf{x}}_i^{\mathrm{s}} \Delta\xi_i}_{\overline{\mathbf{x}}_{\mathrm{eq},i}^{\mathrm{s}}} . \tag{2.232}$$

Thus, in this approximation the matrizant within a segment interval $[\xi_i, \xi_{i+1}]$ results in

$$\mathcal{M}_{\xi_i}^{\xi_{i+1}}\left\{\overline{\mathbf{R}}^x\right\} = \overline{\mathbf{E}} + \overline{\mathbf{R}}_i^x \Delta\xi_i . \tag{2.233}$$

Since the matrix function $\overline{\mathbf{R}}^x(\xi_i)$ is evaluated at the beginning of the segment only it yields identical computation rules for constant and varying matrix functions within a segment. Therefore, it is obvious that the result is in accordance with the Volterra method of Section 2.2.2.

Equation (2.233), in essence, represents the *explicit Euler method* which relies on the slope at the beginning of the segment. Likewise, a formula for the *implicit Euler method* is obtained which uses the slope at the end of the segment for the approximation:

$$\frac{\mathrm{d}\overline{\mathbf{x}}}{\mathrm{d}\xi} \approx \frac{\overline{\mathbf{x}}_{i+1} - \overline{\mathbf{x}}_i}{\Delta\xi_i} = \overline{\mathbf{R}}_{i+1}^x \overline{\mathbf{x}}_{i+1} + \overline{\mathbf{x}}_{i+1}^{\mathrm{s}}$$

$$\overline{\mathbf{x}}_{i+1} - \Delta\xi_i \overline{\mathbf{R}}_{i+1}^x \overline{\mathbf{x}}_{i+1} = \overline{\mathbf{x}}_i + \Delta\xi_i \overline{\mathbf{x}}_{i+1}^{\mathrm{s}}$$

$$\overline{\mathbf{x}}_{i+1} = \underbrace{\left(\overline{\mathbf{E}} - \overline{\mathbf{R}}_{i+1}^x \Delta\xi_i\right)^{-1}}_{\mathcal{M}_{\xi_i}^{\xi_{i+1}}\left\{\overline{\mathbf{R}}^x\right\}} \overline{\mathbf{x}}_i + \underbrace{\left(\overline{\mathbf{E}} - \overline{\mathbf{R}}_{i+1}^x \Delta\xi_i\right)^{-1} \Delta\xi_i \overline{\mathbf{x}}_{i+1}^{\mathrm{s}}}_{\overline{\mathbf{x}}_{\mathrm{eq},i}^{\mathrm{s}}} . \tag{2.234}$$

Here, the matrix function is evaluated at the end of the segment only.

The explicit Euler method can also be obtained on the basis of the Taylor series

$$\overline{\mathbf{x}}_{i+1} = \overline{\mathbf{x}}_i + \overline{\mathbf{x}}_i' \Delta\xi_i + \overline{\mathbf{x}}_i'' \frac{\Delta\xi_i^2}{2} + \cdots . \tag{2.235}$$

The explicit Euler method follows if the series is truncated after the second term and the substitution $\overline{\mathbf{x}}_i' = \overline{\mathbf{R}}_i^x \overline{\mathbf{x}}_i + \overline{\mathbf{x}}_i^{\mathrm{s}}$ is made. It is thus clear that the error of the explicit Euler method is of the order $O(\Delta\xi_i^2)$.

2.2.6.2 Integration by trapezoidal rule

The rather crude Euler method can be improved by means of the trapezoidal rule: from the matrix function values at the beginning and the end of the segment an average slope is calculated and used to determine the value $\overline{\mathbf{x}}$ at the end of the segment. This yields

$$\overline{\mathbf{x}}_{i+1} = \overline{\mathbf{x}}_i + \left(\overline{\mathbf{R}}_i^x \overline{\mathbf{x}}_i + \overline{\mathbf{R}}_{i+1}^x \overline{\mathbf{x}}_{i+1}\right) \frac{\Delta\xi_i}{2} + \left(\overline{\mathbf{x}}_i^s + \overline{\mathbf{x}}_{i+1}^s\right) \frac{\Delta\xi_i}{2} \tag{2.236}$$

$$\overline{\mathbf{x}}_{i+1} = \underbrace{\left(\overline{\mathbf{E}} - \frac{\Delta\xi_i}{2}\overline{\mathbf{R}}_{i+1}^x\right)^{-1} \left(\overline{\mathbf{E}} + \frac{\Delta\xi_i}{2}\overline{\mathbf{R}}_i^x\right)}_{\mathcal{M}_{\xi_i}^{\xi_{i+1}}\{\overline{\mathbf{R}}^x\}} \overline{\mathbf{x}}_i + \underbrace{\left(\overline{\mathbf{E}} - \frac{\Delta\xi_i}{2}\overline{\mathbf{R}}_{i+1}^x\right)^{-1} \left(\overline{\mathbf{x}}_i^s + \overline{\mathbf{x}}_{i+1}^s\right) \frac{\Delta\xi_i}{2}}_{\overline{\mathbf{x}}_{\mathrm{eq},i}^s} . \tag{2.237}$$

Since the slope at both the beginning *and* at the end of the segment is evaluated it follows that, in contrast to the Euler method, a variation of the slope, that is of the parameter matrix $\overline{\mathbf{R}}^x(\xi)$) along the segment, is taken into account. This will also be true for the numerical integration methods that will be discussed below.

The trapezoidal rule requires a matrix inversion. If this operation is to be avoided it is also possible to estimate in a first step the value ($\overline{\mathbf{x}}_{i+1}$) and improve this value in a second step by means of the trapezoidal rule. A simple estimate is the choice ($\overline{\mathbf{x}}_{i+1} = \overline{\mathbf{x}}_i$). Obviously, this choice will usually be incorrect but it leads to the integration formula

$$\overline{\mathbf{x}}_{i+1} = \underbrace{\left(\overline{\mathbf{E}} + \Delta\xi_i \frac{\overline{\mathbf{R}}_i^x + \overline{\mathbf{R}}_{i+1}^x}{2}\right)}_{\mathcal{M}_{\xi_i}^{\xi_{i+1}}\{\overline{\mathbf{R}}^x\}} \overline{\mathbf{x}}_i + \underbrace{\left(\overline{\mathbf{x}}_i^s + \overline{\mathbf{x}}_{i+1}^s\right) \frac{\Delta\xi_i}{2}}_{\overline{\mathbf{x}}_{\mathrm{eq},i}^s} , \tag{2.238}$$

which is of first order and thus of the same quality as the Euler method. A better estimate for the solution at the end of the segment is provided by the explicit Euler method with $\overline{\mathbf{x}}_{i+1} = \overline{\mathbf{x}}_i + \Delta\xi_i \overline{\mathbf{R}}_i^x \overline{\mathbf{x}}_i + \Delta\xi_i \overline{\mathbf{x}}_i^s$. This leads to the following modified trapezoidal integration method,

$$\overline{\mathbf{x}}_{i+1} = \underbrace{\left(\overline{\mathbf{E}} + \Delta\xi_i \frac{\overline{\mathbf{R}}_i^x + \overline{\mathbf{R}}_{i+1}^x + \Delta\xi_i \overline{\mathbf{R}}_{i+1}^x \overline{\mathbf{R}}_i^x}{2}\right)}_{\mathcal{M}_{\xi_i}^{\xi_{i+1}}\{\overline{\mathbf{R}}^x\}} \overline{\mathbf{x}}_i + \underbrace{\left(\overline{\mathbf{x}}_i^s + \overline{\mathbf{x}}_{i+1}^s\right) \frac{\Delta\xi_i}{2} + \overline{\mathbf{R}}_{i+1}^x \overline{\mathbf{x}}_i^s \frac{\Delta\xi_i^2}{2}}_{\overline{\mathbf{x}}_{eq,i}^s} . \tag{2.239}$$

The error of the trapezoidal and the modified trapezoidal method is of the order of $(O(\Delta\xi_i^3))$.

2.2.6.3 Explicit Runge–Kutta Method

Nowadays, the most important class of one-step method is given by the explicit Runge–Kutta methods. For a derivation of this class one can consult, for the scalar case, the relevant literature, (e.g. see, [29, 30]). Generally, a one-step method of the form

$$\overline{\mathbf{x}}_{i+1} = \overline{\mathbf{x}}_i + \sum_{m=1}^{s} b_m \overline{\mathbf{K}}_m \tag{2.240}$$

$$\overline{\mathbf{K}}_m = \Delta\xi_i\, \overline{\mathbf{f}}\left(\xi_{i+c_m}, \overline{\mathbf{x}}_i + \sum_{n=1}^{m-1} a_{m,n} \overline{\mathbf{K}}_n\right) \tag{2.241}$$

Table 2.2 Butcher scheme for an s-level and for the classical Runge–Kutta method

$$
\begin{array}{c|cccc}
0 & & & & \\
c_2 & a_{2,1} & & & \\
c_3 & a_{3,1} & a_{3,2} & & \\
\vdots & \vdots & \vdots & \ddots & \\
c_s & a_{s,1} & a_{s,2} & \cdots & a_{s,s-1} \\
\hline
 & b_1 & b_2 & \cdots & b_s
\end{array}
\qquad
\begin{array}{c|cccc}
0 & & & & \\
1/2 & 1/2 & & & \\
1/2 & 0 & 1/2 & & \\
1 & 0 & 0 & 1 & \\
\hline
 & 1/6 & 1/3 & 1/3 & 1/6
\end{array}
$$

$$= \Delta\xi_i \, \overline{\mathbf{R}}^x_{i+c_m} \left(\overline{\mathbf{x}}_i + \sum_{n=1}^{m-1} a_{m,n} \overline{\mathbf{K}}_n \right) + \Delta\xi_i \overline{\mathbf{x}}^s_{i+c_m} \tag{2.242}$$

is called an s-level Runge–Kutta method. Substitution of $\overline{\mathbf{K}}_m = \hat{\overline{\mathbf{K}}}_m \overline{\mathbf{x}}_i + \overline{\mathbf{x}}^s_{K_m}$, yields a general, recursive formula for the matrizant,

$$\mathcal{M}_{\xi_i}^{\xi_{i+1}} \left\{ \overline{\mathbf{R}}^x \right\} = \overline{\mathbf{E}} + \sum_{m=1}^{s} b_m \hat{\overline{\mathbf{K}}}_m \tag{2.243}$$

$$\hat{\overline{\mathbf{K}}}_m = \Delta\xi_i \, \overline{\mathbf{R}}^x_{i+c_m} \left(\overline{\mathbf{E}} + \sum_{n=1}^{m-1} a_{m,n} \hat{\overline{\mathbf{K}}}_n \right) \tag{2.244}$$

and for the inhomogeneous term

$$\overline{\mathbf{x}}^s_{eq,i} = \sum_{m=1}^{s} b_m \overline{\mathbf{x}}^s_{K_m} \tag{2.245}$$

$$\overline{\mathbf{x}}^s_{K_m} = \Delta\xi_i \, \overline{\mathbf{x}}^s_{i+c_m} + \Delta\xi_i \, \overline{\mathbf{R}}^x_{i+c_m} \sum_{n=1}^{m-1} a_{m,n} \overline{\mathbf{x}}^s_{K_n} \,. \tag{2.246}$$

The parameters $a_{n,m}$, b_m and c_m are suitably chosen and can be ordered according to the so-called Butcher scheme [29] (see Table 2.2). With the given Butcher scheme for the classical four-level Runge–Kutta method, the solution at the end of the segment follows from the solution at the beginning of the segment according to [19]

$$\overline{\mathbf{x}}_{i+1} = \overline{\mathbf{x}}_i + \left(\frac{1}{6}\overline{\mathbf{K}}_1 + \frac{1}{3}\overline{\mathbf{K}}_2 + \frac{1}{3}\overline{\mathbf{K}}_3 + \frac{1}{6}\overline{\mathbf{K}}_4 \right) . \tag{2.247}$$

The intermediate values are

$$\overline{\mathbf{K}}_1 = \Delta\xi_i \, \overline{\mathbf{R}}^x_i \overline{\mathbf{x}}_i + \Delta\xi_i \, \overline{\mathbf{x}}^s_i \tag{2.248}$$

$$\overline{\mathbf{K}}_2 = \Delta\xi_i \, \overline{\mathbf{R}}^x_{i+\frac{1}{2}} \left(\overline{\mathbf{x}}_i + \frac{1}{2}\overline{\mathbf{K}}_1 \right) + \Delta\xi_i \, \overline{\mathbf{x}}^s_{i+\frac{1}{2}} \tag{2.249}$$

$$\overline{\mathbf{K}}_3 = \Delta\xi_i \, \overline{\mathbf{R}}^x_{i+\frac{1}{2}} \left(\overline{\mathbf{x}}_i + \frac{1}{2}\overline{\mathbf{K}}_2 \right) + \Delta\xi_i \, \overline{\mathbf{x}}^s_{i+\frac{1}{2}} \tag{2.250}$$

$$\overline{\mathbf{K}}_4 = \Delta\xi_i\, \overline{\mathbf{R}}^x_{i+1} \left(\overline{\mathbf{x}}_i + \overline{\mathbf{K}}_3\right) + \Delta\xi_i\, \overline{\mathbf{x}}^s_{i+1} \,. \tag{2.251}$$

These values are inserted into Equation (2.247) and lead to a formula for the matrizant

$$\mathcal{M}_{\xi_i}^{\xi_{i+1}}\left\{\overline{\mathbf{R}}^x\right\} = \overline{\mathbf{E}} + \overline{\mathbf{R}}^{(1)} \Delta\xi_i + \overline{\mathbf{R}}^{(2)} \frac{\Delta\xi_i^2}{2} + \overline{\mathbf{R}}^{(3)} \frac{\Delta\xi_i^3}{6} + \overline{\mathbf{R}}^{(4)} \frac{\Delta\xi_i^4}{24} \tag{2.252}$$

with the auxiliary variables

$$\overline{\mathbf{R}}^{(1)} = \frac{1}{6}\left(\overline{\mathbf{R}}^x_{i+1} + 4\,\overline{\mathbf{R}}^x_{i+\frac{1}{2}} + \overline{\mathbf{R}}^x_i\right) \tag{2.253}$$

$$\overline{\mathbf{R}}^{(2)} = \frac{1}{3}\left[\overline{\mathbf{R}}^x_{i+1}\, \overline{\mathbf{R}}^x_{i+\frac{1}{2}} + \left(\overline{\mathbf{R}}^x_{i+\frac{1}{2}}\right)^2 + \overline{\mathbf{R}}^x_{i+\frac{1}{2}}\, \overline{\mathbf{R}}^x_i\right] \tag{2.254}$$

$$\overline{\mathbf{R}}^{(3)} = \frac{1}{2}\left[\overline{\mathbf{R}}^x_{i+1} \left(\overline{\mathbf{R}}^x_{i+\frac{1}{2}}\right)^2 + \left(\overline{\mathbf{R}}^x_{i+\frac{1}{2}}\right)^2 \overline{\mathbf{R}}^x_i\right] \tag{2.255}$$

$$\overline{\mathbf{R}}^{(4)} = \overline{\mathbf{R}}^x_{i+1} \left(\overline{\mathbf{R}}^x_{i+\frac{1}{2}}\right)^2 \overline{\mathbf{R}}^x_i \,. \tag{2.256}$$

It can easily be seen that in the case of a constant matrix function $\overline{\mathbf{R}}^x(\xi)$ the auxiliary variables $\overline{\mathbf{R}}^{(k)}$ turn to ordinary powers of the matrix function $(\overline{\mathbf{R}}^x)^k$. Then the results in Equation (2.252) agree with the series expansion $\exp \overline{\mathbf{R}}^x\, \Delta\xi_i$ up to the fourth order.

A formula for the source vector at the end of the segments results from Equations (2.245) and (2.246), together with the remaining terms of Equation (2.247)

$$\begin{aligned} \overline{\mathbf{x}}^s_{\mathrm{eq},i} = {} & \frac{\Delta\xi_i}{6}\left(\overline{\mathbf{x}}^s_i + 4\,\overline{\mathbf{x}}^s_{i+\frac{1}{2}} + \overline{\mathbf{x}}^s_{i+1}\right) + \\ & \frac{\Delta\xi_i^2}{6}\left(\overline{\mathbf{R}}^x_{i+\frac{1}{2}}\, \overline{\mathbf{x}}^s_i + \overline{\mathbf{R}}^x_{i+\frac{1}{2}}\, \overline{\mathbf{x}}^s_{i+\frac{1}{2}} + \overline{\mathbf{R}}^x_{i+1}\, \overline{\mathbf{x}}^s_{i+\frac{1}{2}}\right) + \\ & \frac{\Delta\xi_i^3}{12}\left[\left(\overline{\mathbf{R}}^x_{i+\frac{1}{2}}\right)^2 \overline{\mathbf{x}}^s_i + \overline{\mathbf{R}}^x_{i+1}\, \overline{\mathbf{R}}^x_{i+\frac{1}{2}}\, \overline{\mathbf{x}}^s_{i+\frac{1}{2}}\right] + \frac{\Delta\xi_i^4}{24}\left[\overline{\mathbf{R}}^x_{i+1} \left(\overline{\mathbf{R}}^x_{i+\frac{1}{2}}\right)^2 \overline{\mathbf{x}}^s_i\right] . \end{aligned} \tag{2.257}$$

If the value of the matrix function and the inhomogeneous term at the position $\xi_{i+\frac{1}{2}}$ is not given it can be obtained by means of a formula for the mean value, $\overline{\mathbf{R}}^x_{i+\frac{1}{2}} = \frac{1}{2}(\overline{\mathbf{R}}^x_i + \overline{\mathbf{R}}^x_{i+1})$, or by higher approximation formulas. The error of the classic four-level Runge–Kutta method is of the order of $O(\Delta\xi_i^5)$.

By means of Equations (2.243) and (2.244) it is also possible to transfer, besides the classical versions, arbitrary Runge–Kutta methods to the matrix calculus. However, this results in more dense matrices $\mathbf{A} = [a_{m,n}]$ and thus leads to an increase in the number of matrix products and, in turn, to an increase in numerical effort.

2.2.6.4 Hermite Integration

For the computation of the matrizant, the previous integration methods only involve the values of the parameter $\overline{\mathbf{R}}^x$ at various discrete points of an interval. The *Hermite integration* also takes advantage of derivatives. This method was first applied by G. Duffing [31]. It is based on a generalized Taylor formula of Hermite[5]

$$\sum_{v=0}^{k} \overline{\mathbf{f}}^{(v)}(b)\frac{\Delta\xi^{v}}{v!}\frac{\binom{k}{v}}{\binom{k+m}{v}} = \sum_{v=0}^{m} \overline{\mathbf{f}}^{(v)}(a)\frac{\Delta\xi^{v}}{v!}\frac{\binom{m}{v}}{\binom{k+m}{v}} + \overline{\mathbf{R}}_{k,m} \tag{2.258}$$

and for $k = 1$ and $m = 2$ one finds from Equation (2.258) with $a = \xi_i$ and $b = \xi_{i+1}$

$$\overline{\mathbf{x}}(\xi_{i+1}) - \frac{\Delta\xi}{3}\overline{\mathbf{x}}'(\xi_{i+1}) = \overline{\mathbf{x}}(\xi_i) + \frac{2\,\Delta\xi}{3}\overline{\mathbf{x}}'(\xi_i) + \frac{\Delta\xi^2}{6}\overline{\mathbf{x}}''(\xi_i) + \overline{\mathbf{R}}_{1,2}\,. \tag{2.259}$$

Combining Equation (2.259) with the differential Equation (2.130) yields the solution

$$\overline{\mathbf{x}}_{i+1} = \overline{\mathbf{x}}_i + \frac{\Delta\xi_i}{6}\left(4\,\overline{\mathbf{R}}_i^x\overline{\mathbf{x}}_i + 2\,\overline{\mathbf{R}}_{i+1}^x\overline{\mathbf{x}}_{i+1} + \Delta\xi_i\left[\frac{\mathrm{d}}{\mathrm{d}\xi}\overline{\mathbf{R}}^x(\xi)\overline{\mathbf{x}}(\xi)\right]_{\xi=\xi_i}\right) + \frac{\Delta\xi_i}{3}\left(2\,\overline{\mathbf{x}}_i^{\mathrm{s}} + \overline{\mathbf{x}}_{i+1}^{\mathrm{s}}\right) + \frac{\Delta\xi_i^2}{6}\overline{\mathbf{x}}_i^{\mathrm{s}\prime} \tag{2.260}$$

$$\overline{\mathbf{x}}_{i+1} = \underbrace{\left(\overline{\mathbf{E}} - \frac{\Delta\xi_i}{3}\overline{\mathbf{R}}_{i+1}^x\right)^{-1}\left(\overline{\mathbf{E}} + \frac{2\Delta\xi_i}{3}\overline{\mathbf{R}}_i^x + \frac{\Delta\xi_i^{\,2}}{6}\left[\overline{\mathbf{R}}_i^{x\prime} + \left(\overline{\mathbf{R}}_i^x\right)^2\right]\right)}_{\mathcal{M}_{\xi_i}^{\xi_{i+1}}\{\overline{\mathbf{R}}^x\}}\overline{\mathbf{x}}_i + \underbrace{\left(\overline{\mathbf{E}} - \frac{\Delta\xi_i}{3}\overline{\mathbf{R}}_{i+1}^x\right)^{-1}\left[\frac{\Delta\xi_i}{3}\left(2\,\overline{\mathbf{x}}_i^{\mathrm{s}} + \overline{\mathbf{x}}_{i+1}^{\mathrm{s}}\right) + \frac{\Delta\xi_i^2}{6}\left(\overline{\mathbf{R}}_i^x\overline{\mathbf{x}}_i^{\mathrm{s}} + \overline{\mathbf{x}}_i^{\mathrm{s}\prime}\right)\right]}_{\overline{\mathbf{x}}_{eq,i}^{\mathrm{s}}}\,. \tag{2.261}$$

This equation is of the order $O(\Delta\xi_i^4)$.

In a similar way to Section 2.2.6.2 it is possible to avoid matrix inversions if an estimate for the value $\overline{\mathbf{x}}_{i+1}$ on the right-hand side of Equation (2.260) is used. For this estimate it is not possible to use the Euler method since it requires a linear relation between the estimate and the derivative. This would lead to a reduction of the order of the method. To obtain a formula without reduction of the order it is possible to apply the modified trapezoidal rule. It then follows

$$\overline{\mathbf{x}}_{i+1} = \underbrace{\left(\overline{\mathbf{E}} + \frac{\Delta\xi_i}{3}\left[2\overline{\mathbf{R}}_i^x + \overline{\mathbf{R}}_{i+1}^x\right] + \frac{\Delta\xi_i^2}{6}\left[\overline{\mathbf{R}}_i^{x\prime} + \left(\overline{\mathbf{R}}_i^x\right)^2 + \overline{\mathbf{R}}_{i+1}^x\left(\overline{\mathbf{R}}_i^x + \overline{\mathbf{R}}_{i+1}^x\right) + \frac{\Delta\xi_i^3}{6}\left(\overline{\mathbf{R}}_{i+1}^x\right)^2\overline{\mathbf{R}}_i^x\right]\right)}_{\mathcal{M}_{\xi_i}^{\xi_{i+1}}\{\overline{\mathbf{R}}^x\}}\overline{\mathbf{x}}_i + \underbrace{\frac{\Delta\xi_i}{3}\left(2\,\overline{\mathbf{x}}_i^{\mathrm{s}} + \overline{\mathbf{x}}_{i+1}^{\mathrm{s}}\right) + \frac{\Delta\xi_i^2}{6}\left(\overline{\mathbf{x}}_i^{\mathrm{s}\prime} + \overline{\mathbf{R}}_i^x\overline{\mathbf{x}}_i^{\mathrm{s}} + \overline{\mathbf{R}}_{i+1}^x\overline{\mathbf{x}}_{i+1}^{\mathrm{s}}\right) + \frac{\Delta\xi_i^3}{6}\left(\overline{\mathbf{R}}_{i+1}^x\right)^2\overline{\mathbf{x}}_i^{\mathrm{s}}}_{\overline{\mathbf{x}}_{eq,i}^{\mathrm{s}}}\,. \tag{2.262}$$

---

[5] The generalization of the Taylor formula relates the derivatives of a function $\overline{\mathbf{f}}(\xi)$ at the points $\xi = a$ and $\xi = b$, where the function must be $(k+m)$-times continuously differentiable. The last term is $\overline{\mathbf{R}}_{k,m} = \frac{(-1)^{k+m}}{(k+m)!}\int_a^b (\xi - a)^k\,(\xi - b)^m\overline{\mathbf{f}}^{(k+m+1)}(\xi)\,\mathrm{d}\xi$. For $k = 0$ the Taylor series is obtained as a special case.

Assuming that both the matrix function $\overline{\mathbf{R}}^x(\xi)$ and the inhomogeneous term $\overline{\mathbf{x}}^s(\xi)$ are linear within the considered segment Equation (2.262) simplifies to

$$\overline{\mathbf{x}}_{i+1} = \underbrace{\left(\overline{\mathbf{E}} + \frac{\Delta\xi_i}{2}\left[\overline{\mathbf{R}}_i^x + \overline{\mathbf{R}}_{i+1}^x\right] + \frac{\Delta\xi_i^2}{6}\left[\left(\overline{\mathbf{R}}_i^x\right)^2 + \overline{\mathbf{R}}_{i+1}^x\overline{\mathbf{R}}_i^x + \left(\overline{\mathbf{R}}_{i+1}^x\right)^2\right] + \frac{\Delta\xi_i^3}{6}\left(\overline{\mathbf{R}}_{i+1}^x\right)^2\overline{\mathbf{R}}_i^x\right)}_{\mathcal{M}_{\xi_i}^{\xi_{i+1}}\{\overline{\mathbf{R}}^x\}}\overline{\mathbf{x}}_i +$$

$$\underbrace{\frac{\Delta\xi_i}{2}\left(\overline{\mathbf{x}}_i^s + \overline{\mathbf{x}}_{i+1}^s\right) + \frac{\Delta\xi_i^2}{6}\left(\overline{\mathbf{R}}_i^x\overline{\mathbf{x}}_i^s + \overline{\mathbf{R}}_{i+1}^x\overline{\mathbf{x}}_{i+1}^s\right) + \frac{\Delta\xi_i^3}{6}\left(\overline{\mathbf{R}}_{i+1}^x\right)^2\overline{\mathbf{x}}_i^s}_{\overline{\mathbf{x}}_{eq,i}^s} . \tag{2.263}$$

2.2.6.5 Improving Accuracy by the Romberg Method

The Romberg method is based on the assumption that the order of the error of a certain method is known. Then, for a general integration scheme of order $(p-1)$,

$$\mathcal{M}_{\xi_0}^{\xi}\left\{\overline{\mathbf{R}}^x\right\} = {}^{(n)}\mathcal{M}_{\xi_0}^{\xi}\left\{\overline{\mathbf{R}}^x\right\} + \overline{\mathbf{C}}\,\Delta\xi_i^p + O(\Delta\xi_i^{p+1}) . \tag{2.264}$$

Here, the term ${}^{(n)}\mathcal{M}_{\xi_0}^{\xi}\{\overline{\mathbf{R}}^x\}$ denotes an estimate for the matrizant $\mathcal{M}_{\xi_0}^{\xi}\{\overline{\mathbf{R}}^x\}$ which is calculated on the basis of N points within the interval (ξ_0, ξ). Doubling the number of points yields

$$\mathcal{M}_{\xi_0}^{\xi}\left\{\overline{\mathbf{R}}^x\right\} = {}^{(2n)}\mathcal{M}_{\xi_0}^{\xi}\left\{\overline{\mathbf{R}}^x\right\} + \overline{\mathbf{C}}\,\frac{\Delta\xi_i^p}{2^p} + O(\Delta\xi_i^{p+1}) . \tag{2.265}$$

Combining Equations (2.264) and (2.265) makes it possible to eliminate the error term $\overline{\mathbf{C}}\,\Delta\xi_i^p$. This results in a more exact approximation with

$$\mathcal{M}_{\xi_0}^{\xi}\left\{\overline{\mathbf{R}}^x\right\} = \frac{2^p}{2^p-1}\,{}^{(2n)}\mathcal{M}_{\xi_0}^{\xi}\left\{\overline{\mathbf{R}}^x\right\} - \frac{1}{2^p-1}\,{}^{(n)}\mathcal{M}_{\xi_0}^{\xi}\left\{\overline{\mathbf{R}}^x\right\} + O(\Delta\xi_i^{p+1}) . \tag{2.266}$$

For example, for the calculation of the matrizant by means of the four-level Runge–Kutta method the error is of the order $O(\Delta\xi_i^5)$ and the correction

$$\mathcal{M}_{\xi_0}^{\xi}\left\{\overline{\mathbf{R}}^x\right\} = \frac{32}{31}\,{}^{(2n)}\mathcal{M}_{\xi_0}^{\xi}\left\{\overline{\mathbf{R}}^x\right\} - \frac{1}{31}\,{}^{(n)}\mathcal{M}_{\xi_0}^{\xi}\left\{\overline{\mathbf{R}}^x\right\} + O(\Delta\xi_i^{p+1}) \tag{2.267}$$

is obtained. In principle, the method can repeatedly be used to eliminate terms of higher order successively. To this end, however, the order of the errors must be known precisely. It is also necessary to perform more and more computations with a doubling of the number of steps in order to determine the higher correction terms (see Table 2.3).

The Romberg method can also be used to improve integration formulas. If the Romberg method is applied once to the Euler procedure within the segment $[\xi_i, \xi_{i+1}]$ it follows a second-order integration method. For a double number of steps,the intermediate step in the

Table 2.3 Representation of the Romberg method

${}^{(n)}\mathcal{M}_{\xi_0}^{\xi}\left\{\overline{\mathbf{R}}^x\right\} + O(\Delta\xi_i^p)$	${}^{(2n)}\mathcal{M}_{\xi_0}^{\xi}\left\{\overline{\mathbf{R}}^x\right\} + O(\Delta\xi_i^p)$	${}^{(4n)}\mathcal{M}_{\xi_0}^{\xi}\left\{\overline{\mathbf{R}}^x\right\} + O(\Delta\xi_i^p)$	...
$\downarrow$ $\swarrow$		$\downarrow$ $\swarrow$	
$\dot{\mathcal{M}}_{\xi_0}^{\xi}\left\{\overline{\mathbf{R}}^x\right\} + O(\Delta\xi_i^{p+1})$	$\dot{\mathcal{M}}_{\xi_0}^{\xi}\left\{\overline{\mathbf{R}}^x\right\} + O(\Delta\xi_i^{p+1})$		...
$\downarrow$ $\swarrow$			
$\ddot{\mathcal{M}}_{\xi_0}^{\xi}\left\{\overline{\mathbf{R}}^x\right\} + O(\Delta\xi_i^{p+2})$	...		
...			

middle of a segment is given by

$$\overline{\mathbf{x}}_{i+\frac{1}{2}} = \overline{\mathbf{x}}_i + \frac{\Delta\xi_i}{2}\overline{\mathbf{R}}_i^x\overline{\mathbf{x}}_i \tag{2.268}$$

and at the end of the segment one finds

$$\overline{\mathbf{x}}_{i+1} = \overline{\mathbf{x}}_{i+\frac{1}{2}} + \frac{\Delta\xi_i}{2}\overline{\mathbf{R}}_{i+\frac{1}{2}}^x\overline{\mathbf{x}}_{i+\frac{1}{2}} . \tag{2.269}$$

Combining both results yields the matrizant within the segment as

$$\mathcal{M}_{\xi_i}^{\xi_{i+1}}\left\{\overline{\mathbf{R}}^x\right\} = \overline{\mathbf{E}} + \frac{\Delta\xi_i}{2}\left(\overline{\mathbf{R}}_i^x + \overline{\mathbf{R}}_{i+\frac{1}{2}}^x\right) + \frac{\Delta\xi_i^2}{4}\overline{\mathbf{R}}_{i+\frac{1}{2}}^x\overline{\mathbf{R}}_i^x . \tag{2.270}$$

Of course, this method to construct new integration schemes can also be used repeatedly. Thus, arbitrarily exact (and arbitrarily complicated) integration methods can be generated.

2.2.6.6 Controlling Step Size and Error

The efficiency of a numerical integration method very much depends on the size of the necessary steps and their control. In regions where the matrizant varies widely it is necessary to choose small step sizes $\Delta\xi$ to keep the integration error small. In the case of a slow variation it is possible to choose larger steps. It is best to adapt the step size accordingly. An automatic step size control often relies on an estimate of the integration error. If this estimate is smaller or larger than the predefined limiting values then the step size becomes modified.

Error Estimate According to Richardson

In Section 2.2.6.5 the Romberg method used a known error to improve accuracy or to obtain a more exact integration method. It is also possible to use a known error to estimate the integration error and to implement an automatic step size control, as proposed by Richardson [32]. The idea is to perform two calculations for the segment considered:

- one with step size $\Delta\xi_i$ which is defined by the previous or initial step size,
- a second one with twice the step size, that is with step size $2\Delta\xi_i$.

The second calculation with twice the step size involves an error which is approximately 2^p-times the error of the step size $\Delta\xi_i$.[6] It follows for the Runge–Kutta method of Section 2.2.6.3, for example, that the error grows by a factor of 32. Since for the distance $2\,\Delta\xi_i$ two steps of size $\Delta\xi_i$ are necessary, which approximately doubles the error, it follows a resulting factor of 2^{p-1}, that is a factor of 16 for the Runge–Kutta method. The error, approximately, is

$$\overline{\sigma}(\mathcal{M}_{\xi_i}^{\xi_{i+1}}\left\{\overline{\mathbf{R}}^x\right\}) = \frac{1}{2^{p-1}}\left({}^{(\Delta\xi_i)}\mathcal{M}_{\xi_i}^{\xi_{i+1}}\left\{\overline{\mathbf{R}}^x\right\} - {}^{(2\Delta\xi_i)}\mathcal{M}_{\xi_i}^{\xi_{i+1}}\left\{\overline{\mathbf{R}}^x\right\}\right), \tag{2.271}$$

and this result can be used for the correction of the calculation

$$\mathcal{M}_{\xi_i}^{\xi_{i+1}}\left\{\overline{\mathbf{R}}^x\right\} = \frac{2^{p-1}+1}{2^{p-1}}\,{}^{(\Delta\xi_i)}\mathcal{M}_{\xi_i}^{\xi_{i+1}}\left\{\overline{\mathbf{R}}^x\right\} - \frac{1}{2^{p-1}}\,{}^{(2\Delta\xi_i)}\mathcal{M}_{\xi_i}^{\xi_{i+1}}\left\{\overline{\mathbf{R}}^x\right\}. \tag{2.272}$$

On the basis of the calculated error and a given error tolerance it is then possible to control the step size. For a chosen tolerance ϵ the following rules can be applied:

- $0.2\,\epsilon < \|\overline{\sigma}(\mathcal{M}_{\xi_i}^{\xi_{i+1}}\{\overline{\mathbf{R}}^x\})\| < 10\,\epsilon$ – step size $\Delta\xi_i$ is kept and the current result, corrected according to Equation (2.272), is accepted,
- $\|\overline{\sigma}(\mathcal{M}_{\xi_i}^{\xi_{i+1}}\{\overline{\mathbf{R}}^x\})\| \geq 10\,\epsilon$ – step size $\Delta\xi_i$ is divided by two and results are recalculated with this finer step size,
- $\|\overline{\sigma}(\mathcal{M}_{\xi_i}^{\xi_{i+1}}\{\overline{\mathbf{R}}^x\})\| \leq 0.2\,\epsilon$ – current and, according to Equation (2.272), corrected result is accepted and the step size $\Delta\xi_i$ will be doubled for the next step.

This rather simple scheme can be further optimized. In general, an absolute (ϵ) and a relative (κ) error tolerance should be prescribed. If the error is given by Equation (2.271), the norm can be approximated by

$$\|\overline{\sigma}\left(\mathcal{M}_{\xi_i}^{\xi_{i+1}}\left\{\overline{\mathbf{R}}^x\right\}\right)\| \approx \alpha_p\,\Delta\xi_i{}^p \tag{2.273}$$

such that the value α_p can be determined.

If $\|\overline{\sigma}(\mathcal{M}_{\xi_i}^{\xi_{i+1}}\{\overline{\mathbf{R}}^x\})\| < \epsilon$ and $\|\overline{\sigma}(\mathcal{M}_{\xi_i}^{\xi_{i+1}}\{\overline{\mathbf{R}}^x\})\| < \kappa\,\|\mathcal{M}_{\xi_i}^{\xi_{i+1}}\{\overline{\mathbf{R}}^x\}\,\|$, then the result of the current step is accepted and the step size of the next step is determined from

$$\begin{aligned}\Delta\xi_{\text{new}} &= \Delta\xi_{i+1}\\ &= \Delta\xi_i\,\min\left(\sqrt[p]{\frac{\epsilon}{\|\overline{\sigma}\left(\mathcal{M}_{\xi_i}^{\xi_{i+1}}\left\{\overline{\mathbf{R}}^x\right\}\right)\|}},\ \sqrt[p]{\frac{\kappa}{\|\overline{\sigma}\left(\mathcal{M}_{\xi_i}^{\xi_{i+1}}\left\{\overline{\mathbf{R}}^x\right\}\right)\|}\|\mathcal{M}_{\xi_i}^{\xi_{i+1}}\left\{\overline{\mathbf{R}}^x\right\}\|}\right).\end{aligned} \tag{2.274}$$

In practice, the value $\Delta\xi_{\text{new}}$ will be multiplied with a further tolerance (e.g. 0.9) and the nearest dual number will be chosen to avoid unnecessary rounding errors. In the previous method this was guaranteed by the prescription of a dual number and the doubling or division by two of the step size. The use of a specific matrix norm is not prescribed. Rather, it can be chosen freely and involves the weighting of the error matrix $\overline{\sigma}(\mathcal{M}_{\xi_i}^{\xi_{i+1}}\{\overline{\mathbf{R}}^x\})$. If the column-sum norm

[6] The step size $2\,\Delta\xi_i$ must be small enough to neglect error terms of higher orders than the leading error term $O(\Delta\xi_i^p)$

$\|\overline{\mathbf{A}}\|_{\text{row}} = \max_i(\sum_j |a_{i,j}|)$ is chosen then the error is judged by the maximum deviation of one element of the solution vector. Using the two-norm $\|\overline{\mathbf{A}}\|_2 = \sqrt{\sum_i \sum_j |a_{i,j}|^2}$, however, the error will be evaluated of the quadratic mean. For the meaning and calculation of various norms refer to [33–35].

Embedded Methods

In contrast to the error estimate according to Richardson extrapolation it is also possible to estimate the error by means of an embedded method with a pair of s-level Runge–Kutta methods, as proposed by Fehlberg [36, 37] for the classical Runge–Kutta method and Frey [38] for the corresponding Hermite method. For a detailed description of embedded methods one may consult [29].

Global Error

If the step size is controlled by one of the previous methods it is assured that the norm of the local error $\|\overline{\sigma}(\mathcal{M}_{\xi_i}^{\xi_{i+1}}\{\overline{\mathbf{R}}^x\})\|$ within a segment i does not exceed a given tolerance. It is then possible to estimate the global error, that is the error for the matrizant $\mathcal{M}_{\xi_0}^{\xi_N}\{\overline{\mathbf{R}}^x\})$ at the end of a transmission line. The global or resulting error of the matrizant $\mathcal{M}_{\xi_0}^{\xi_k}\{\overline{\mathbf{R}}^x\})$ turns out to be

$$\overline{\sigma}\left(\mathcal{M}_{\xi_0}^{\xi_k}\left\{\overline{\mathbf{R}}^x\right\}\right) = \sum_{i=1}^{k} \mathcal{M}_{\xi_i}^{\xi_k}\left\{\overline{\mathbf{R}}^x\right\} \overline{\sigma}\left(\mathcal{M}_{\xi_{i-1}}^{\xi_i}\left\{\overline{\mathbf{R}}^x\right\}\right). \tag{2.275}$$

An estimate of the norm of the global error yields

$$\|\overline{\sigma}\left(\mathcal{M}_{\xi_0}^{\xi_k}\left\{\overline{\mathbf{R}}^x\right\}\right)\| \leq \sum_{i=1}^{k} \|\mathcal{M}_{\xi_i}^{\xi_k}\left\{\overline{\mathbf{R}}^x\right\}\| \max\left(\epsilon, \kappa\|\mathcal{M}_{\xi_{i-1}}^{\xi_i}\left\{\overline{\mathbf{R}}^x\right\}\|\right) \tag{2.276}$$

$$< \sum_{i=1}^{k} \exp L\,(\xi_k - \xi_i) \max\left(\epsilon, \kappa\|\mathcal{M}_{\xi_{i-1}}^{\xi_i}\left\{\overline{\mathbf{R}}^x\right\}\|\right), \tag{2.277}$$

with the Lipschitz constant L. Recursively, the global error is given by the values of the previous segments,

$$\overline{\sigma}\left(\mathcal{M}_{\xi_0}^{\xi_k}\left\{\overline{\mathbf{R}}^x\right\}\right) = \mathcal{M}_{\xi_{k-1}}^{\xi_k}\left\{\overline{\mathbf{R}}^x\right\} \overline{\sigma}\left(\mathcal{M}_{\xi_0}^{\xi_{k-1}}\left\{\overline{\mathbf{R}}^x\right\}\right) + \overline{\sigma}\left(\mathcal{M}_{\xi_{k-1}}^{\xi_k}\left\{\overline{\mathbf{R}}^x\right\}\right) \tag{2.278}$$

and for the norm one finds

$$\|\overline{\sigma}\left(\mathcal{M}_{\xi_0}^{\xi_k}\left\{\overline{\mathbf{R}}^x\right\}\right)\| \leq \|\mathcal{M}_{\xi_{k-1}}^{\xi_k}\left\{\overline{\mathbf{R}}^x\right\}\| \|\overline{\sigma}\left(\mathcal{M}_{\xi_0}^{\xi_{k-1}}\left\{\overline{\mathbf{R}}^x\right\}\right)\| + \max\left(\epsilon, \kappa\|\mathcal{M}_{\xi_{k-1}}^{\xi_k}\left\{\overline{\mathbf{R}}^x\right\}\|\right). \tag{2.279}$$

2.2.7 Remarks on Efficiency and the Choice of an Appropriate Method

As is often the case it is difficult to recommend one method for all types of conceivable problems. However, to summarize, this section provides a few recommendations.

The Picard iteration is rather involved and, in general, not suitable for a numerical implementation.

By means of the piecewise constant approximation according to Volterra it is possible to determine the matrizant quickly if an efficient implementation of the matrix exponential function is available. Due to the piecewise constant approximation of the parameter function the accuracy is rather limited, though. A disadvanatge is the artificial concentration at discrete points of the continuous scattering along the line. The constant step functions cannot precisely approximate the matrix function $\overline{\mathbf{R}}^{x}(\xi)$. Problems occur for strong variations of the parameter matrix. Therefore, this method is only suitable for slowly varying transmission-line structures.

Because of its relatively large error and the high number of required segments the Euler method should only be used for electrically short transmission lines. A satisfying accuracy usually requires several thousand segments per wavelength. The method is nevertheless of interest for the investigation of the above mentioned electrically short transmission lines, for statistical analysis, and for an estimate of limiting values since the numerical effort and the related computation time per segment is small.

The trapezoidal method and other second-order methods are suitable for problems with a medium demand for accuracy, where methods that require matrix inversion should be avoided, since the related computation time for a method with matrix inversion is of the order of the Runge–Kutta method of the fourth order.

In view of accuracy and computation time, very good results can be obtained by means of the fourth-order Runge–Kutta method and the recursion formula with linear or power series interpolation.

For arbitrary, nonuniform and electrically large transmission lines the recursion formula of Equation (2.153) or, in case of a higher approximation order of the matrix function, the power series of Equation (2.181) probably constitute the best choice. The series converges as long as the segments are kept smaller than twice the wavelengths. A further advantage is that the errors are known and no step size control is necessary since the series can be truncated once the desired accuracy is reached. However, the actual numerical effort cannot be determined since it is not clear from the beginning how many series terms need to be kept. In all investigated cases the Runge–Kutta method turned out to be slower and numerically more involved.

2.3 Semianalytic and Numerical Solutions for Selected Transmission Lines in the TLST

2.3.1 The Straight, Finite Length Wire Above Ground

At first glance the setup shown in Figure 2.6 might appear to be a simple uniform transmission line. For low frequencies ($h \ll \lambda$) this is indeed true. Then the classical transmission-line theory can be applied successfully. However, at higher frequencies, additional effects occur

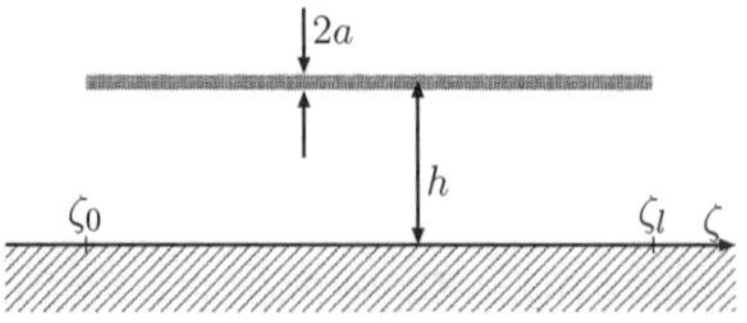

Figure 2.6 A transmission line consisting of a straight, finite length wire parallel to a ground plane.

due to the open ends at the terminals. These influence the per-unit-length parameters and make them position dependent. It is possible to derive analytical formulas for these parameters with the aid of the TLST.

The exact geometry of the line is given in Figure 2.6 and the line starts at ζ_0 and ends at ζ_l. It has a radius of $a = 1$ mm and is at a height $h = 50$ mm above the ground plane. With the aid of the image theory and the thin-wire approximation the Green's function and the integral kernels, which are scalars in this case, are easily found:

$$G\left(\mathbf{C}(\zeta), \mathbf{C}(\zeta')\right) = \frac{1}{4\pi}\left(\frac{\mathrm{e}^{-jk\sqrt{(\zeta-\zeta')^2+a^2}}}{\sqrt{(\zeta-\zeta')^2+a^2}} - \frac{\mathrm{e}^{-jk\sqrt{(\zeta-\zeta')^2+4h^2}}}{\sqrt{(\zeta-\zeta')^2+4h^2}}\right) \tag{2.280}$$

$$k_l\left(\zeta, \zeta'\right) = \mu\, G\left(\mathbf{C}(\zeta), \mathbf{C}(\zeta')\right) \tag{2.281}$$

$$k_c\left(\zeta, \zeta'\right) = \frac{1}{\epsilon}\, G\left(\mathbf{C}(\zeta), \mathbf{C}(\zeta')\right). \tag{2.282}$$

Equation (2.60) then becomes:

$$\overline{\mathbf{I}}^{(0)} = \int_{\zeta_0}^{\zeta_l} \left.\begin{bmatrix} 0 & k_l\left(\zeta, \zeta'\right) \\ k_c\left(\zeta, \zeta'\right) & 0 \end{bmatrix}\right|_{\omega=0} \mathrm{d}\zeta' \tag{2.283}$$

$$= \begin{bmatrix} 0 & \mu \\ \frac{1}{\epsilon} & 0 \end{bmatrix} F_1(\zeta) \tag{2.284}$$

where the closed-form solution of the integral is

$$F_1 = \ln \frac{\sqrt{(\zeta_l-\zeta)^2+a^2}+(\zeta_l-\zeta)}{\sqrt{(\zeta_l-\zeta)^2+4h^2}+(\zeta_l-\zeta)} - \ln \frac{\sqrt{(\zeta_0-\zeta)^2+a^2}+(\zeta_0-\zeta)}{\sqrt{(\zeta_0-\zeta)^2+4h^2}+(\zeta_0-\zeta)}. \tag{2.285}$$

With Equation (2.51) the zeroth-order parameters are eventually found for the finite length transmission-line:

$$\overline{\mathbf{P}}^{(0)} = \begin{bmatrix} \frac{1}{j\omega F_1}\frac{\partial F_1}{\partial \zeta} & \frac{1}{c_0^2} \\ 1 & 0 \end{bmatrix}. \tag{2.286}$$

The new quantity $c_0 = 1/\sqrt{\mu\epsilon}$ indicates the propagation speed of the electromagnetic wave. In the following step the product integral of Equation (2.286) needs to be evaluated. Unfortunately this is not possible in closed form. Because only the starting values for the iteration are involved it is acceptable to simplify these parameters. If the distances from the two termination regions are large enough, the function F_1 becomes constant and its derivative vanishes. Then these parameters can be used everywhere on the line. They can be calculated by the following limit:

$$\overline{\mathbf{I}}^{(0)} = \lim_{\substack{\zeta_0 \to -\infty \\ \zeta_l \to \infty}} \int_{\zeta_0}^{\zeta_l} \begin{bmatrix} 0 & k_l\left(\zeta, \zeta'\right) \\ k_c\left(\zeta, \zeta'\right) & 0 \end{bmatrix}_{\omega=0} \mathrm{d}\zeta' \tag{2.287}$$

$$= \begin{bmatrix} 0 & \mu \\ \frac{1}{\epsilon} & 0 \end{bmatrix} \lim_{\substack{\zeta_0 \to -\infty \\ \zeta_l \to \infty}} F_1(\zeta) \tag{2.288}$$

with the solution

$$\lim_{\substack{\zeta_0\to-\infty\\ \zeta_l\to\infty}} F_1 = F_1^{\infty} = 2\ln\frac{2h}{a}. \tag{2.289}$$

Now the zeroth-order parameters are considerably simpler:

$$\overline{\mathbf{P}}^{(0)} = \begin{bmatrix} 0 & 1/c_0^2 \\ 1 & 0 \end{bmatrix}. \tag{2.290}$$

They are equivalent to the parameters from the classical transmission-line theory. The computation of the product integral can be carried out analytically:

$$\mathcal{M}_{\zeta}^{\zeta'}\left\{-j\omega\overline{\mathbf{P}}^{(0)}\right\} = \begin{bmatrix} \cos k\left(\zeta'-\zeta\right) & -j\frac{1}{c_0}\sin k\left(\zeta'-\zeta\right) \\ -jc_0\sin k\left(\zeta'-\zeta\right) & \cos k\left(\zeta'-\zeta\right) \end{bmatrix}. \tag{2.291}$$

This expression is now inserted into Equation (2.54) and the elements of the $\overline{\mathbf{I}}^{(1)}$ matrix become:

$$I_{11}^{(1)} = I_{22}^{(1)} = -\frac{j}{4\pi}\sqrt{\frac{\mu}{\epsilon}}G_1 \tag{2.292}$$

$$I_{12}^{(1)} = \frac{\mu}{4\pi}G_2 \tag{2.293}$$

$$I_{21}^{(1)} = \frac{1}{4\pi\epsilon}G_2 \tag{2.294}$$

where

$$\begin{aligned}\begin{Bmatrix} G_1 \\ G_2 \end{Bmatrix} &= \int_{\zeta_0}^{\zeta_l} G\left(\mathbf{C}(\zeta),\mathbf{C}\left(\zeta'\right)\right)\begin{Bmatrix} \sin k\left(\zeta'-\zeta\right) \\ \cos k\left(\zeta'-\zeta\right) \end{Bmatrix} \mathrm{d}\zeta' \\ &= \begin{Bmatrix} -j/2 \\ 1/2 \end{Bmatrix} (\mathrm{E}_1(jkR_{1l-}) - \mathrm{E}_1(jkR_{10-}) - \mathrm{E}_1(jkR_{2l-}) + \mathrm{E}_1(jkR_{20-}) \\ &\quad \pm(\mathrm{E}_1(jkR_{1l+}) - \mathrm{E}_1(jkR_{10+}) - \mathrm{E}_1(jkR_{2l+}) + \mathrm{E}_1(jkR_{20+})))\end{aligned} \tag{2.295}$$

and

$$R_{1\left\{\substack{l\\0}\right\}\pm} = \sqrt{\left(\zeta_{\left\{\substack{l\\0}\right\}}-\zeta\right)^2 + a^2} \pm \left(\zeta_{\left\{\substack{l\\0}\right\}}-\zeta\right) \tag{2.296}$$

$$R_{2\left\{\substack{l\\0}\right\}\pm} = \sqrt{\left(\zeta_{\left\{\substack{l\\0}\right\}}-\zeta\right)^2 + 4h^2} \pm \left(\zeta_{\left\{\substack{l\\0}\right\}}-\zeta\right). \tag{2.297}$$

The function E_1 is the exponential integral (see [39])

$$\mathrm{E}_1(z) = \int_z^{\infty}\frac{\mathrm{e}^{-t}}{t}\mathrm{d}t. \tag{2.298}$$

With these results a closed-form solution for the parameter matrix $\overline{\mathbf{P}}$ can be given after one iteration:

$$\overline{\mathbf{P}}^{(1)} = \begin{bmatrix} \frac{1}{j\omega}\frac{1}{G_2}\frac{\partial G_2}{\partial \zeta}\frac{1}{c_0^2} & -\frac{1}{\omega c_0}\frac{1}{G_2}\frac{\partial G_1}{\partial \zeta} \\ 1 & 0 \end{bmatrix}. \tag{2.299}$$

In this setup, the current only has a component in the direction of the conductor, which is straight and in parallel to the ground plane. Then the vector potential is also in this direction and the voltage can be defined in a plane perpendicular to the conductor as

$$v(\zeta) := \varphi(\zeta), \tag{2.300}$$

where φ is the scalar potential at this conductor. Hence, the telegrapher equations can be given in the voltage–current representation, as familiar from the classical theory:

$$\frac{\partial}{\partial \zeta}\begin{bmatrix} v \\ i \end{bmatrix} + j\omega\overline{\mathbf{P}}^{*(1)}\begin{bmatrix} v \\ i \end{bmatrix} = \mathbf{0}, \tag{2.301}$$

with

$$\overline{\mathbf{P}}^{*(1)} = \begin{bmatrix} -j\frac{1}{c_0}\frac{G_1}{G_2} & \frac{\mu}{4\pi}\left(G_2 + \frac{G_1^2}{G_2}\right) \\ 4\pi\epsilon\frac{1}{G_2} & j\frac{1}{c_0}\frac{G_1}{G_2} \end{bmatrix}. \tag{2.302}$$

Solving the above equations with appropriate boundary conditions makes it possible to compute currents and voltages. However, it is almost impossible to compare these results with data from measurements or from computations with other methods, for example the method of moments (MoM). Real setups as well as MoM models require fixed boundary conditions at the terminals. That is, it is necessary to connect a source and a load to the two ends of the transmission line. In this setup, there is no connection from the wire to the ground plane. This would need to be taken into account in the parameter computation. As a consequence, the MoM computation and measurements would refer to a different system than the one computed with the TLST. The results will be different and a comparison does not make sense. It is possible, however, to compare the parameters from the classical transmission-line theory with the new parameters.

For this a line is investigated where the conductor is 50 mm above the ground plane, has a radius of 1 mm and is 1m long. Figure 2.7 shows the per-unit-length inductance for different frequencies related to the per-unit-length inductance calculated with the classical transmission-line theory ($l_0 = \frac{\mu}{2\pi}\ln\frac{2h}{a}$). For low frequencies ($\lambda \gg h$) it is seen that in the central part of the line, this parameter is equal to the parameter from the classical theory. Compared to a point at the central part of the line where the wire is present at both sides, at the terminals the wire is only present at one side. Only this wire contributes to the inductance and, consequently, the value must decrease to one half of the value at the center.

As the frequency is increased and reaches regions where the wavelength and the height are in the same order of magnitude, the real part of the parameters become smaller and an imaginary part occurs which, multiplied by $j\omega$, becomes real and is a part of the radiation resistance. This effect is especially large at the terminals, but exists along the whole conductor. A similar behavior can be observed for the per-unit-length capacitance with the difference that it increases when becoming closer to the terminals. Also the diagonal parameters P_{11}^*

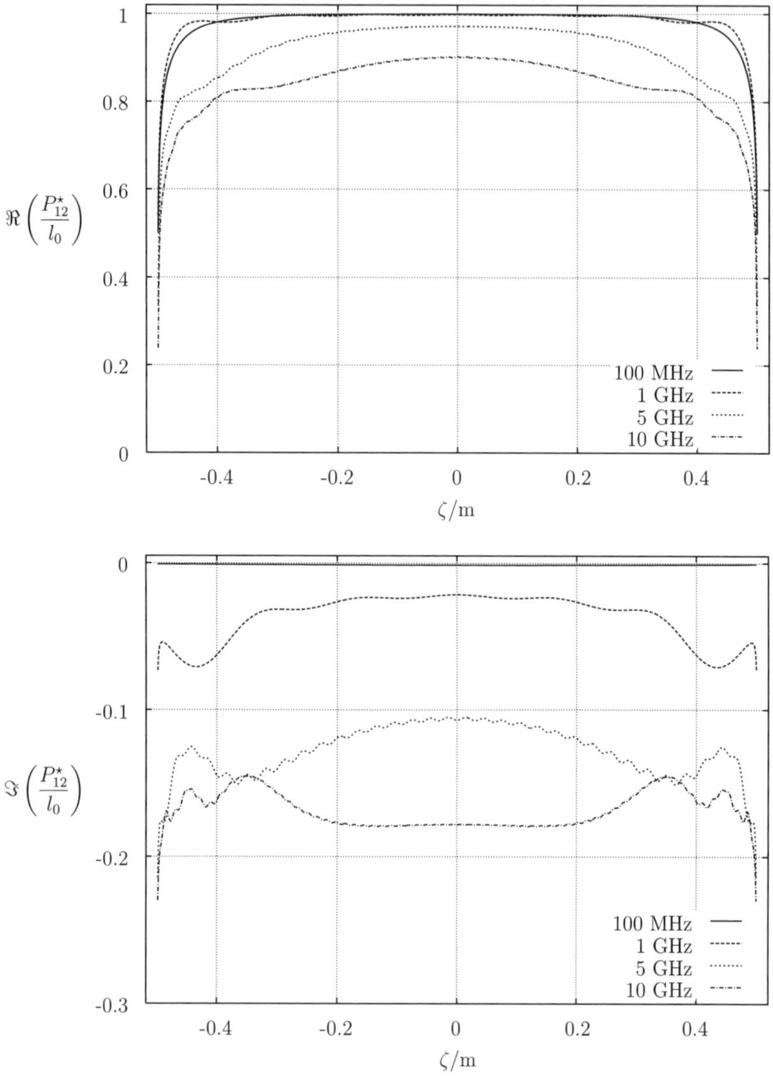

Figure 2.7 Per-unit-length inductance of the finite length transmission line related to the corresponding parameter from the classical transmission-line theory (upper plot real part, lower plot imaginary part).

(see Figure 2.8) and P_{22}^{*} reflect this frequency-dependent behavior. The parameters almost vanish for low frequencies and become larger with higher frequencies. They show the largest values at the terminals. The 'antisymmetric' progression with respect to ζ is caused by the antisymmetric definition of the current which always flows towards a larger ζ.

To further concentrate on the open end, one open end is moved to infinity. Then this end does not influence the other end and the boundary condition can be fixed there. This leads to the model of a semi-infinite line.

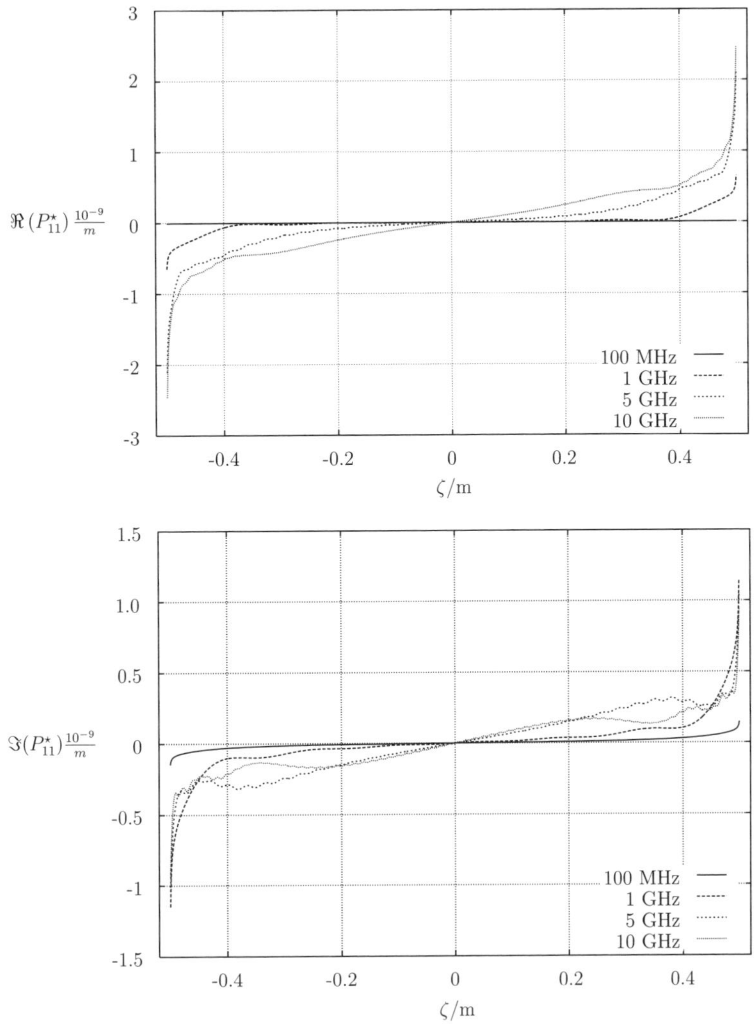

Figure 2.8 Per-unit-length parameter P_{11}^{*} of the finite length transmission line (upper plot real part, lower plot imaginary part).

2.3.2 The Semi-Infinite Line

As noted before, to pass from the finite line to the semi-infinite line one lets ζ_0 approach $-\infty$. Then G_1 and G_2 become

$$\begin{Bmatrix} G_1 \\ G_2 \end{Bmatrix} = \begin{Bmatrix} -j/2 \\ 1/2 \end{Bmatrix} \left(\mathrm{E}_1(jkR_{1l-}) - \mathrm{E}_1(jkR_{2l-}) \pm (\mathrm{E}_1(jkR_{1l+}) - \mathrm{E}_1(jkR_{2l+})) + 2\ln\frac{2h}{a} \right). \quad (2.303)$$

The parameters are calculated with the aid of Equation (2.299). They are very similar to the parameters of the finite line. At the open end they show the same behavior. If one moves away

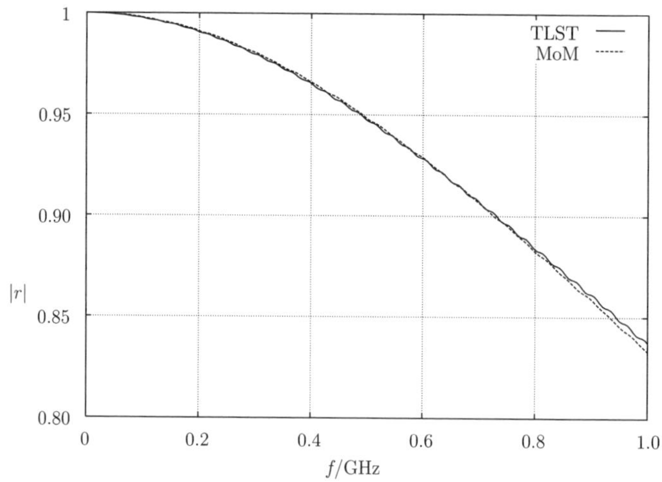

Figure 2.9 The reflection factor at the open end of the semi-infinite transmission line.

from the open end the parameters approach the parameters of the classical transmission-line theory.

Going one step further, one calculates currents and voltages on the line by solving the extended telegrapher equations numerically. A wave can be generated by connecting a voltage source ($v = 1\text{V}$). The source should be far enough away from the open end. This is the case when the per-unit-length parameters are approximately the same as those of the classical transmission-line theory. Furthermore, the internal resistance of the source should match the characteristic impedance of the line to avoid reflections.

When a wave traveling along the line comes into the region where the parameters start to change due to the influence of the open end, at every position a small fraction of the wave will be reflected. Additionally, a small part will be radiated and the remaining part travels further where again a small fraction will be reflected and radiated. When the wave eventually reaches the end, again a part of the energy travels further and, therefore, is radiated and the rest is reflected. This effect is small for low frequencies. Here almost all the energy will be reflected and hence, the reflection factor is one. In this frequency region the classical transmission-line theory is still valid. For higher frequencies the radiation increases and less wave energy is reflected. This behavior can be observed in the reflection factor r vs. frequency. The magnitude of this quantity is presented in Figure 2.9. It was calculated from the actual voltages and currents from the TLST. Moreover, the plot shows the magnitude of the reflection factor calculated with the method of moments (MoM), which is a direct numerical solution of Maxwell's equations. For this computation the program CONCEPT was used. The line was modeled with a thin 5 m long wire. The discretization was chosen such that at least 20 segments per wavelength were used. At one end the line was connected to the ground plane and excited with a voltage source. The other end was left open. Then at the central part of the line, far away from the terminations, one can assume a TEM mode with forward and backward traveling waves. These waves were determined from the current distribution. Their quotient then gives the desired reflection factor. Practically, there is no difference between the two curves. Thus it is concluded that the TLST correctly models the physical conditions at the open end. This already happens after one iteration.

2.3.3 Field Coupling to an Infinite Line

Next the remaining end of the semi-infinite line is moved to infinity. This yields an infinitely long line. In the absence of an external excitation only a TEM mode can propagate. This statement is supported by the per-unit-length parameters, which can be calculated by setting ζ_l in Equation (2.303) to infinity. The result is

$$G_1 = 0 \quad \text{and} \quad G_2 = 2\ln\frac{2h}{a}, \tag{2.304}$$

which yields the parameters

$$\overline{\mathbf{P}}^{(1)} = \begin{bmatrix} 0 & \frac{1}{c_0^2} \\ 1 & 0 \end{bmatrix} \tag{2.305}$$

or, for the v–i representation

$$\overline{\mathbf{P}}^{*(1)} = \begin{bmatrix} 0 & \dfrac{\mu}{2\pi}\ln\dfrac{2h}{a} \\ 2\pi\epsilon\dfrac{1}{\ln\frac{2h}{a}} & 0 \end{bmatrix}. \tag{2.306}$$

These are exactly the same as the zeroth-order parameters which means that the iteration already converged to the exact result. The parameters are frequency and position independent. The diagonal terms in the v–i representation as well as in the q–i representation vanish and the remaining parameters have real values. This indicates that there is no radiation caused by the transmission line itself. However, when illuminated with an external field (see Figure 2.10) a non-TEM mode can be excited and radiation occurs (see also [40]). This is not reflected in the per-unit-length parameters but in the source terms.

The electric field strength along the wire for an incident plane-wave field as in Figure 2.10 is given by

$$v^{(i)'}(\zeta) = E_0 e^{-jk\zeta\cos\theta}, \tag{2.307}$$

where E_0 is the magnitude that also takes into account the reflections from the ground plane. Normally one would use the iterative procedure of Equation (2.62) to determine q_s'. In this special case, however, a closed-form solution is possible. The dependence of the electric field on the spatial coordinate ζ implies the same dependence of the source term:

$$q_s'(\zeta) = q_0' e^{-jk\zeta\cos\theta}. \tag{2.308}$$

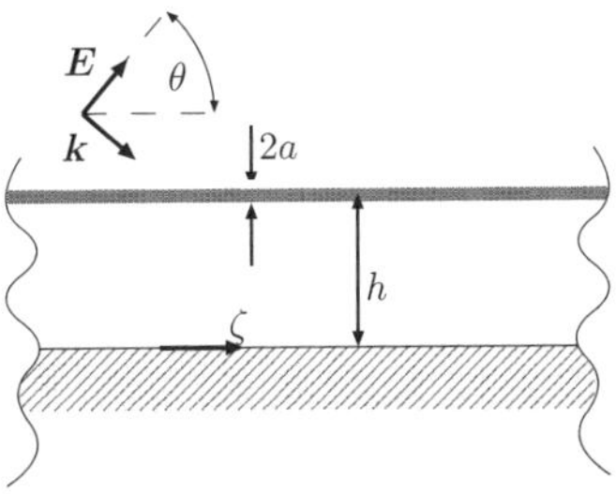

Figure 2.10 The infinitely long, uniform transmission line with field coupling.

Then with Equation (2.49),

$$I_{10} = -jc_0 q_0' \int_{-\infty}^{\infty} k_l\left(\zeta, \zeta'\right) \int_{\zeta}^{\zeta'} \sin\left(k\left(\zeta' - \xi\right)\right) \mathrm{e}^{-jk\xi\cos\theta} \mathrm{d}\xi \mathrm{d}\zeta' \tag{2.309}$$

and

$$I_{20} = q_0' \int_{-\infty}^{\infty} k_c\left(\zeta, \zeta'\right) \int_{\zeta}^{\zeta'} \cos\left(k\left(\zeta' - \xi\right)\right) \mathrm{e}^{-jk\xi\cos\theta} \mathrm{d}\xi \mathrm{d}\zeta'. \tag{2.310}$$

The inner integration is carried out first and yields

$$\begin{aligned}
&\int_{\zeta}^{\zeta'} \sin\left(k\left(\zeta' - \xi\right)\right) \mathrm{e}^{-jk\xi\cos\theta} \mathrm{d}\xi \\
&= \frac{\mathrm{e}^{-jk\zeta\cos\theta}}{k\sin^2\theta}\left[\mathrm{e}^{-jk\left(\zeta'-\zeta\right)\cos\theta} - \cos\left(k\left(\zeta' - \zeta\right)\right) + j\sin\left(k\left(\zeta' - \zeta\right)\right)\cos\theta\right]
\end{aligned} \tag{2.311}$$

and

$$\begin{aligned}
&\int_{\zeta}^{\zeta'} \cos\left(k\left(\zeta' - \xi\right)\right) \mathrm{e}^{-jk\xi\cos\theta} \mathrm{d}\xi \\
&= -j\cos\left(\theta\right)\frac{\mathrm{e}^{-jk\zeta\cos\theta}}{k\sin^2\theta}\left[\mathrm{e}^{-jk\left(\zeta'-\zeta\right)\cos\theta} - \cos\left(k\left(\zeta' - \zeta\right)\right) + j\sin\left(k\left(\zeta' - \zeta\right)\right)\frac{1}{\cos\theta}\right].
\end{aligned} \tag{2.312}$$

Because k_l and k_c are even functions, and the sine is an odd function the sine components in Equations (2.311) and (2.312) do not contribute to the integrals in Equations (2.309) and (2.310). Hence,

$$I_{10} = jc_0 q_0' \frac{\mathrm{e}^{-jk\zeta\cos\theta}}{k\sin^2\theta}\int_{-\infty}^{\infty} k_l\left(\zeta, \zeta'\right)\left(\cos\left(k\left(\zeta' - \zeta\right)\right) - \cos\left(k\left(\zeta' - \zeta\right)\cos\theta\right)\right) \mathrm{d}\zeta' \tag{2.313}$$

and

$$I_{20} = j\cos\theta\, q_0' \frac{\mathrm{e}^{-jk\zeta\cos\theta}}{k\sin^2\theta}\int_{-\infty}^{\infty} k_c\left(\zeta, \zeta'\right)\left(\cos\left(k\left(\zeta'-\zeta\right)\right) - \cos\left(k\left(\zeta'-\zeta\right)\cos\theta\right)\right) \mathrm{d}\zeta'. \tag{2.314}$$

These integrals can be solved (see Appendix F.3) yielding

$$I_{10} = j\frac{\mu}{4\pi} c_0 q_0' \frac{\mathrm{e}^{-jk\zeta\cos\theta}}{k\sin^2\theta}\left(G_2 - G_3\right) \tag{2.315}$$

and

$$I_{20} = j\frac{1}{4\pi\epsilon}\cos\theta\, q_0' \frac{\mathrm{e}^{-jk\zeta\cos\theta}}{k\sin^2\theta}\left(G_2 - G_3\right) \tag{2.316}$$

where

$$G_2 = 4\pi \int_{-\infty}^{\infty} g\left(\mathbf{C}\left(\zeta\right), \mathbf{C}\left(\zeta'\right)\right)\cos\left(k\left(\zeta' - \zeta\right)\right) \mathrm{d}\zeta' \tag{2.317}$$

$$= 2\ln\frac{2h}{a}, \tag{2.318}$$

$$G_3 = 4\pi \int_{-\infty}^{\infty} g\left(\mathbf{C}(\zeta), \mathbf{C}(\zeta')\right) \cos\left(k\left(\zeta' - \zeta\right)\cos\theta\right) \mathrm{d}\zeta' \tag{2.319}$$

$$= -j\pi\left(H_0^{(2)}(ak\sin\theta) - H_0^{(2)}(2hk\sin\theta)\right). \tag{2.320}$$

With the aid of Equations (2.307) and (2.308) it is now possible to solve Equation (2.52) for q_0, yielding the rather simple answer

$$q_0 = \frac{4\pi\epsilon}{G_3} E_0. \tag{2.321}$$

Then the exact extended telegrapher equations for a uniform, infinitly long, ideally conducting transmission line with plane-wave field coupling become

$$\frac{\partial}{\partial\zeta}\begin{bmatrix} q \\ i \end{bmatrix} + j\omega\begin{bmatrix} 0 & \frac{1}{c_0^2} \\ 1 & 0 \end{bmatrix}\begin{bmatrix} q \\ i \end{bmatrix} = \begin{bmatrix} \frac{4\pi\epsilon}{G_3} E_0 \mathrm{e}^{-jk\zeta\cos\theta} \\ 0 \end{bmatrix}. \tag{2.322}$$

This can also be converted to the v–i representation, yielding

$$\frac{\partial}{\partial\zeta}\begin{bmatrix} v \\ i \end{bmatrix} + j\omega\begin{bmatrix} 0 & P_{12}^* \\ P_{21}^* & 0 \end{bmatrix}\begin{bmatrix} v \\ i \end{bmatrix} = \begin{bmatrix} \frac{G_2 - G_3\cos^2\theta}{G_2\sin^2\theta} E_0 \mathrm{e}^{-jk\zeta\cos\theta} \\ -\sqrt{\frac{\epsilon}{\mu}} \frac{4\pi\left(G_2 - G_3\right)\cos\theta}{G_2 G_3 \sin^2\theta} E_0 \mathrm{e}^{-jk\zeta\cos\theta} \end{bmatrix}. \tag{2.323}$$

These equations are identical with the results obtained in [40] with the aid of a spatial Fourier transformation. For small k, if $\lambda \ll h$, the function G_3 converges to G_2 thus $G_3 \approx G_2$. In this case, the source terms can be simplified and the telegrapher equations become

$$\frac{\partial}{\partial\zeta}\begin{bmatrix} v \\ i \end{bmatrix} + j\omega\begin{bmatrix} 0 & P_{12}^* \\ P_{21}^* & 0 \end{bmatrix}\begin{bmatrix} v \\ i \end{bmatrix} = \begin{bmatrix} E_0 \mathrm{e}^{-jk\zeta\cos\theta} \\ 0 \end{bmatrix}. \tag{2.324}$$

This is exactly the same result as obtained from classical transmission-line theory.

2.3.4 *The Skewed Wire Transmission Line*

Now a more complicated configuration is chosen, the skewed wire transmission line. This line is built from a finite length wire which is located above a ground plane. The wire is not parallel to this plane, as depicted in Figure 2.11. Furthermore, at one end the wire is short-circuited to the ground plane. This is the position where the line will be fed and the current is measured.

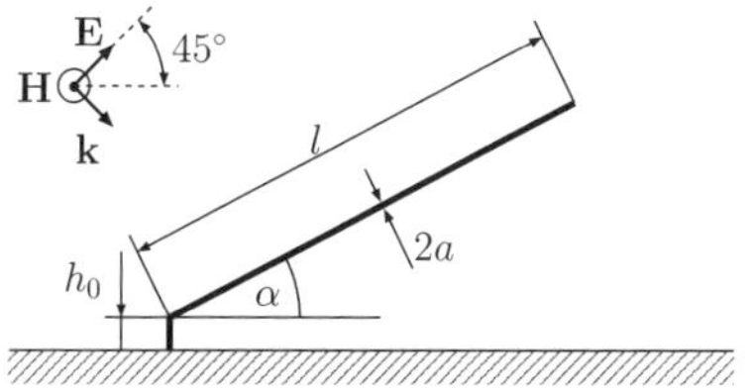

Figure 2.11 The setup of the skewed wire transmission line ($h_0 = 1$ mm, $l = 0.538$ m, $a = 0.2$ mm).

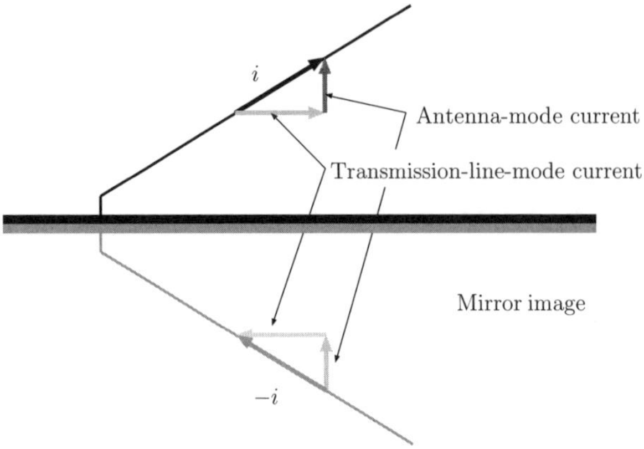

Figure 2.12 Transmission-line-mode current and antenna-mode current in the skewed transmission line.

The current on this line consists of two components which are often referred to as the transmission-line-mode current and the antenna-mode current. In order to determine these two modes one can take into account the mirror image due to the ground plane. Furthermore, it is known that the current is derived from a vector quantity and as such can be decomposed into mutually orthogonal (horizontal and vertical) components (see Figure 2.12). The horizontal parts have opposite directions, such that the fields that are caused by these currents partly compensate each other. This is a property of the transmission-line mode. The vertical components have the same direction and the field contributions add up. This indicates the antenna mode.

Thus, for an angle of $\alpha = 0°$ there are almost no antenna mode currents (except for the riser) and the line behaves like a (classical) transmission line. For an angle of $\alpha = 90°$ there are no transmission-line-mode currents. The wire is an effective antenna. Both cases, and all those in between, will be correctly modeled by the TLST.

Unfortunately, a voltage cannot easily be defined in this setup because the vector potential does not only have components in the direction of the conductor. Therefore the v–i representation is not suitable here and it is better to use the q–i or φ–i representation. As stated before, the quasivoltage has no physical meaning, except for parts of the line that are very close ($\ll \lambda$) to the ground plane, for example at the beginning of the vertical part. Then φ becomes the conventional voltage between the plane and the transmission line.

The per-unit-length parameters are calculated numerically. Figure 2.13 shows the P_{12} parameters of the q–i representation, related to the corresponding parameter of a uniform line. It was calculated for different angles at a frequency of 1GHz. For other frequencies the parameter shows a very similar behavior.

When $\alpha = 0°$ the line behaves almost like a uniform line because its distance to the ground plane is only 1mm. This is much smaller then the wavelength. Only very close to the terminals can a small deviation from the classical TLT parameters be observed. The parameter P_{12} is real and has a value of almost $1/c_0^2$, just like a uniform line. When the angle increases, the value of the parameter starts to change, it becomes position dependent and complex. For a lossless system this indicates radiation. The biggest deviation can be observed for an angle of $\alpha = 90°$, where the wire is an antenna.

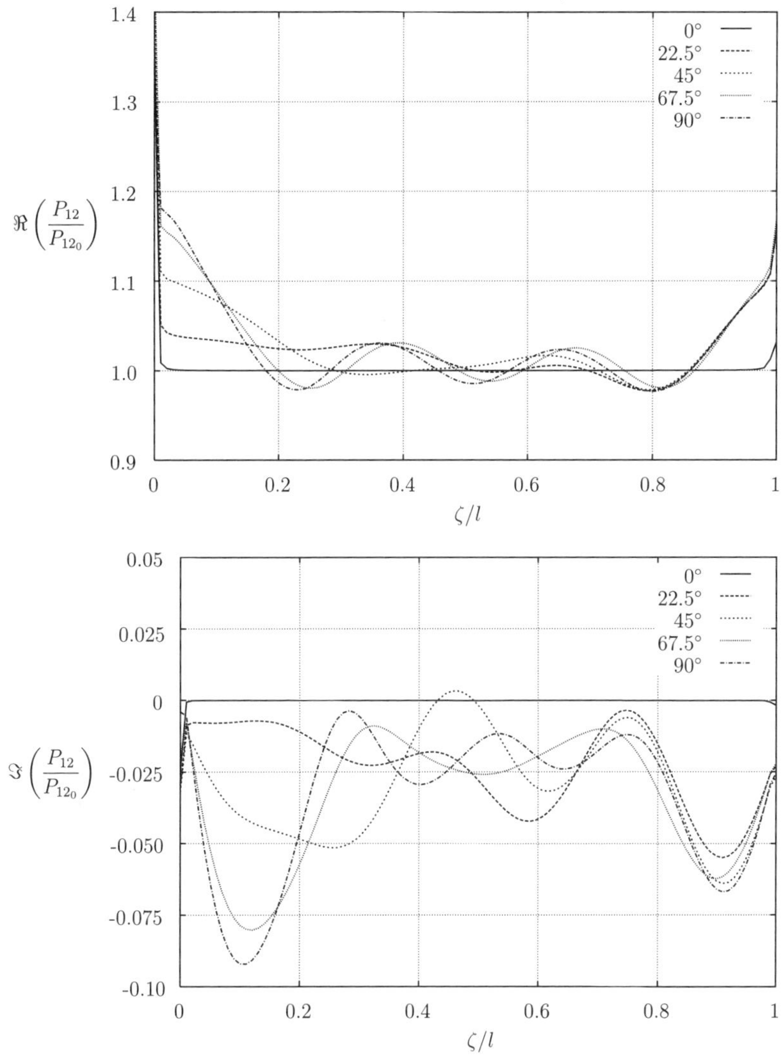

Figure 2.13 The real (upper) and imaginary (lower) parts of the parameter P_{12} for the skewed transmission line, related to the corresponding parameter of a uniform line ($P_{12_0} = 1/c_0^2$).

To validate the TLST it is possible to compare the magnitude of the input impedance $|Z_i|$ of the transmission line with measured data. The measurement was performed on an actual setup of the line with an angle of $\alpha = 21.7°$ with a network analyzer. The results are shown in Figure 2.14. The two currents exhibit a reasonably good agreement. Both curves show the typical behavior of an open-ended transmission line. One can identify the resonances due to standing waves; their frequencies roughly correspond to the length of the wire. For lower frequencies the resonances are very sharp and have very high or very low magnitudes, respectively. With increasing frequency they become smeared out and less sharp which is caused by losses.

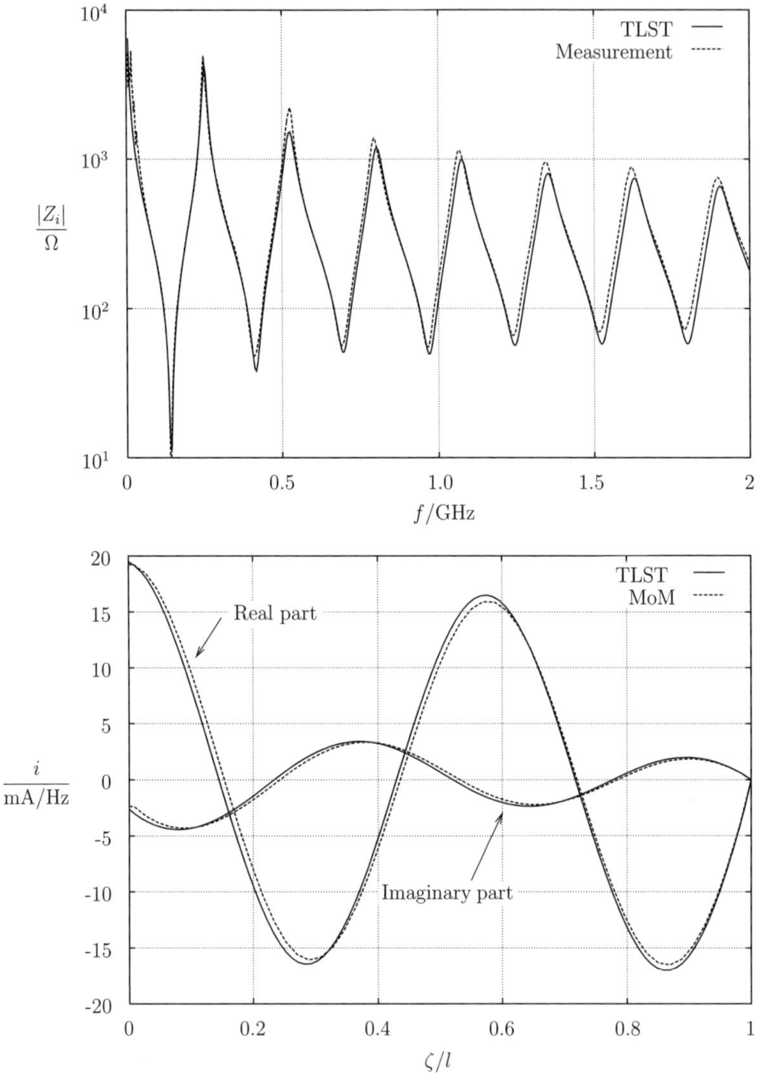

Figure 2.14 The magnitude of the input impedance Z_i vs. frequency (upper) and the current distribution at 0.974GHz (lower) of the skewed transmission line ($\alpha = 21.7°$).

These are radiation losses, because the system is otherwise considered to be lossless. The input impedance looks very similar for other angles, up to 90°.

The validity can also be confirmed with MoM calculations, where the program CONCEPT was used. The line was modeled with a thin wire and the discretization was chosen to have more than 20 segments per wavelength. A 1V voltage source was used as the excitation at the beginning of the line. Then the current distribution along the skewed wire was determined for a frequency of 0.974 GHz, which is very close to a resonance frequency. The results from the TLST are compared to the MoM solution in Figure 2.14. Again, there is a very good agreement between the two results.

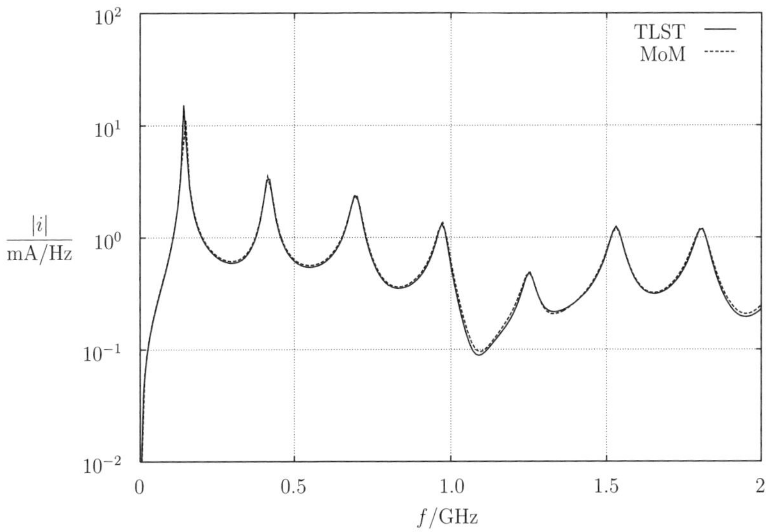

Figure 2.15 The magnitude of the field excited current to the ground plane vs. frequency of the skewed transmission line ($\alpha = 21.7°$).

Eventually, one can also excite the skewed transmission line by an external field, that is use the line as a receiving antenna. The vectors **E** and **k** are directed as shown in Figure 2.11. As can be seen in Figure 2.15, a perfect match between the MoM and the TLST solution is obtained.

2.3.5 The Periodic Transmission Line

Realistic transmission lines often have a periodic structure. Some examples are lines that are tied together at periodic distances, lines that are placed above a periodic structure or helically-shaped lines that are placed above a ground plane. These periodic lines exhibit a very interesting behavior, for example they block certain frequency bands. To demonstrate these effects, as a rather canonical structure a sinusoidally-shaped line is taken, as shown in Figure 2.16. The line is 0.8 m long and has 8.5 periods with an amplitude of 5 cm. The zero-line of the sine is 6 cm above the ground plane and the wire has a radius of 0.2 mm. At both terminals the line is connected to the ground plane via a resistor roughly the size of the low-frequency wave resistance ($R = 304\ \Omega$). Furthermore a voltage source of 1 V is connected at the beginning of the line.

Figure 2.17 shows the input and output current of the transmission line as a function of frequency. Up to 1 GHz the transmission line shows rather typical behavior. The resistor of 304 Ω is not a perfect match, such that some reflections occur at the ports. This causes the ripples on the current. The input and output currents are almost equal and close to $1\ \text{V}/(2 \times 304\ \Omega) = 1.6$ mA, just as it should be for a matched transmission line. There are some irregularities which are caused by the nonuniformity, and they occur above 500 MHz.

Above 1 GHz the behavior dramatically changes. This is the frequency where approximately eight to nine periods of the wave are on the line. The output current decreases rapidly and stays very close to zero for all frequencies. Thus a signal that propagates along the line does

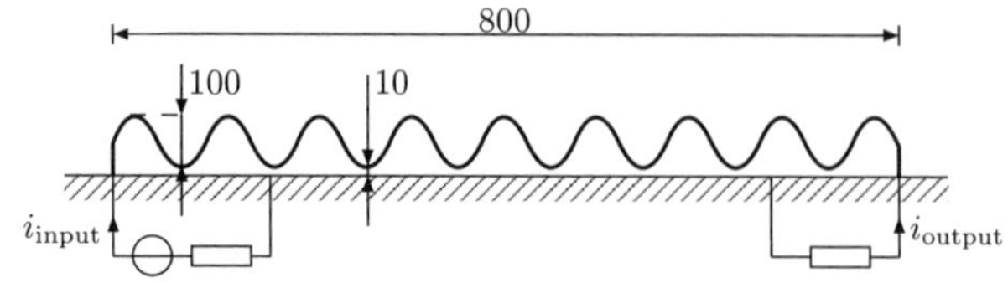

Figure 2.16 The mechanical and electrical setup of the sinusoidal transmission line.

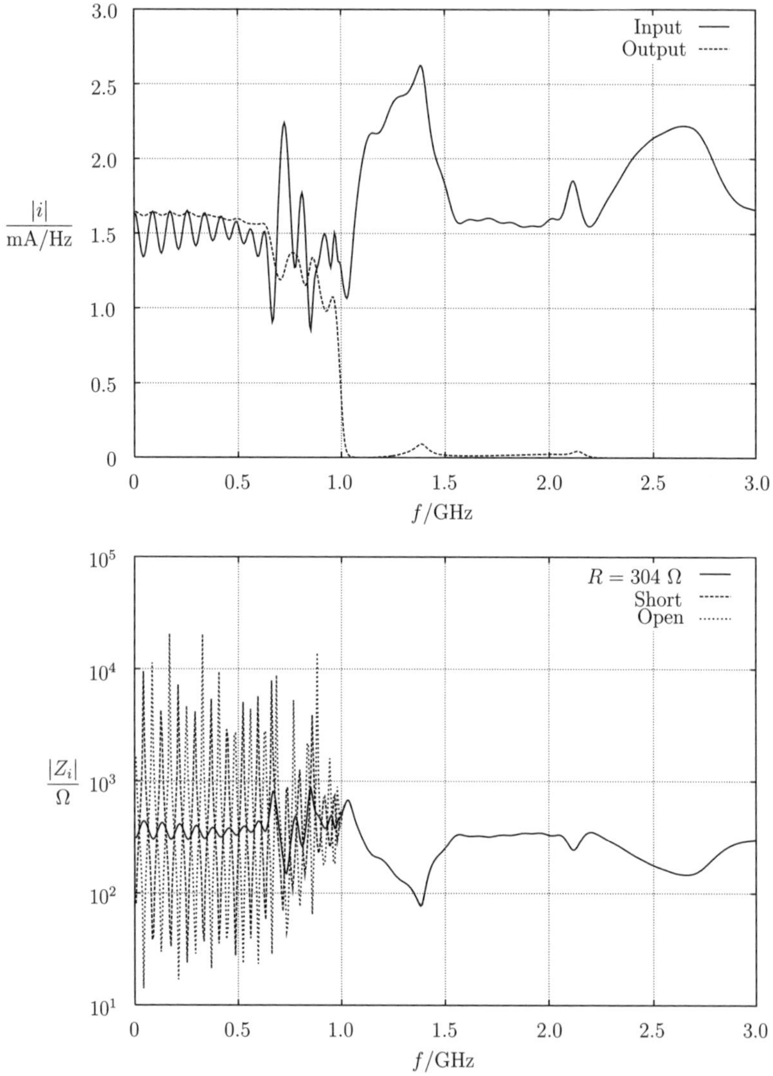

Figure 2.17 The magnitude of the input current vs. frequency (upper) and the input impedance for three different terminations (lower) of the periodic transmission line.

not reach the end. The vanishing current at the end suggests the assumption that the input impedance is independent from the termination resistor above 1 GHz. This is indeed the case as can be seen in Figure 2.17, where the magnitudes of the input impedances for the open, shorted and 'almost matched' far end are shown.

Below 1 GHz all three configurations show the typical transmission-line input impedance. Above 1 GHz, all three input impedances are exactly equal. Hence, the input impedance of the line is independent from the actual termination. Furthermore, above this frequency the input impedance has a large real part (see Figure 2.18) and a rather small imaginary part. The

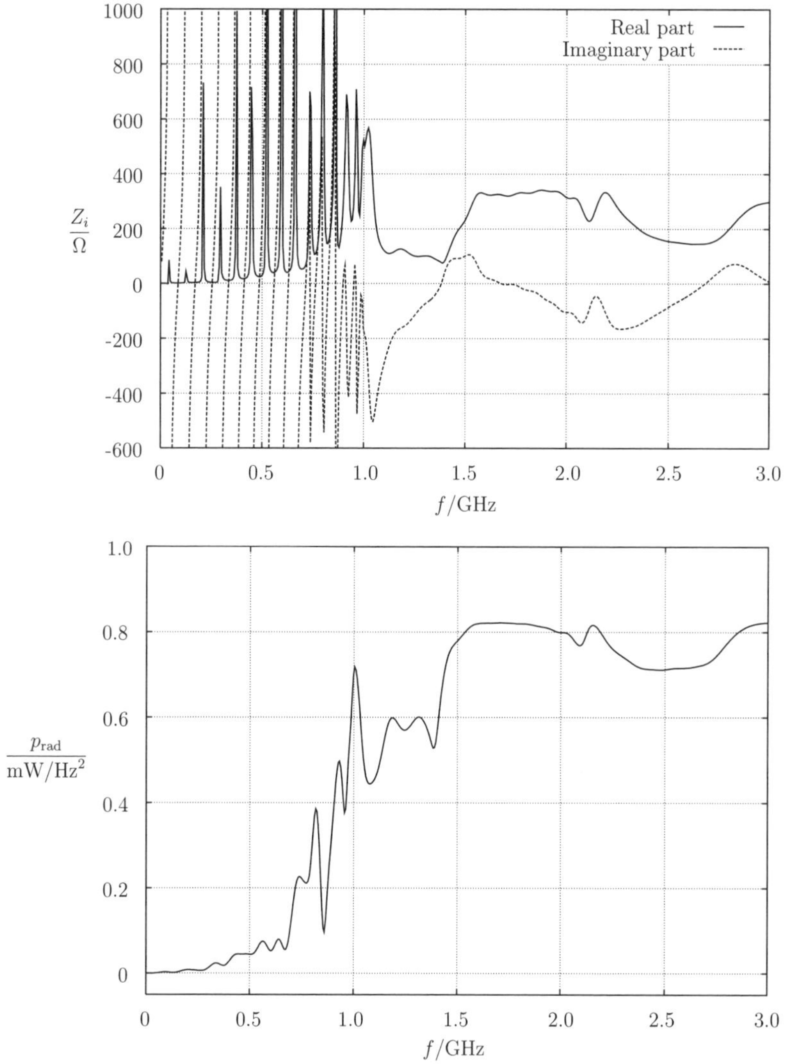

Figure 2.18 The real and imaginary part of the input impedance of the sinusoidal line when short-circuited to the ground (upper) and the radiated power (lower).

line acts almost as a resistor. This is due to the radiation since the transmission line itself is perfectly conducting.

Also the time averaged radiated power can be calculated by subtracting the power dissipated in the termination resistors from the power generated by the source. The result is also shown in Figure 2.18. As it turns out, the line is a quite good wide-band radiator at frequencies above 1.5 GHz.

2.3.6 *Cross-Talk in a Nonuniform Multiconductor Transmission Line*

In this final example the cross-talk in a nonuniform multiconductor transmission line is examined. Figure 2.19 shows the V-shaped wires which are placed above a ground plane. One wire is at a constant height of 20 mm, the height of the other wire changes from 10 mm to 20 mm. Furthermore, the wires are bent at the ends in order to reach the termination resistors which are located below the ground plane.

The wire that changes its height is driven by a voltage source of 1 V with an internal resistance of 50 Ω. All other ends are terminated with 50 Ω resistors.

Figure 2.20 shows the cross-talk current flowing into the termination resistor at the far end of the passive wire. The plot presents the solutions from the TLST with parameters that

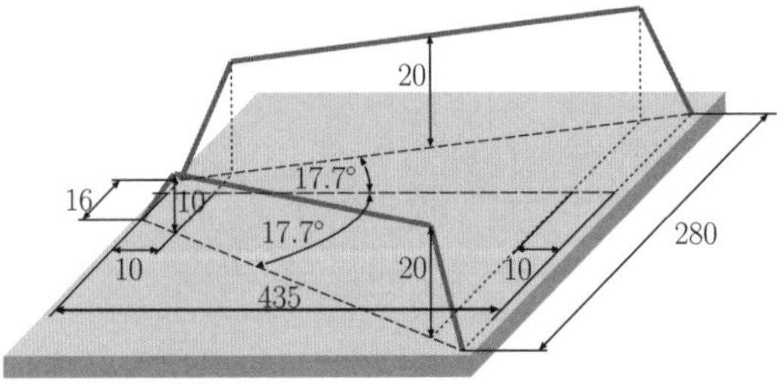

Figure 2.19 A nonuniform multiconductor transmission line.

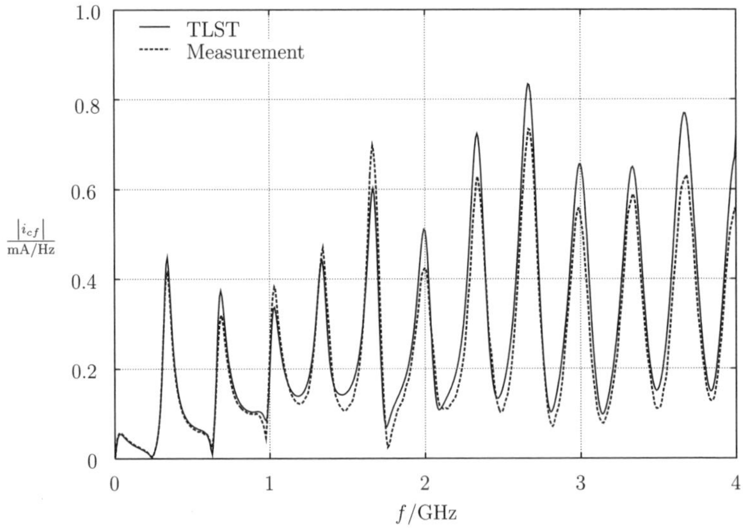

Figure 2.20 The far end cross-talk current of the passive wire of the NMTL.

have been obtained after one iteration. Additionally, the current, measured by means of a network analyzer along a real setup of the wire structure, is shown for validation. Again, good agreement between the measurement and the TLST prediction for the cross-talk current is shown.

2.4 Analytic Approaches

Besides the limited classical transmission-line theory, which is suitable for uniform transmission lines, and the full-wave TLST, which fully encompasses radiation effects, there are generalized transmission-line theories which lie between these two extremes. These theories are formulated in a way such that they can handle, at least to a certain extent, nonuniform transmission line configurations. What is particularly interesting is that analytic or semianalytic solution procedures for the corresponding telegrapher equations exist. The discussion of some of these solution procedures is the topic of this subsection.[7]

2.4.1 Classical Telegrapher Equations for Nonuniform Transmission Lines

2.4.1.1 Basic Definitions and Equations

For a system of N transmission lines and one reference conductor, generalized telegrapher equations in the frequency domain can be written as an ansatz in the combined form

$$\frac{\partial}{\partial z}\begin{pmatrix}\boldsymbol{v}(z,s)\\ \boldsymbol{i}(z,s)\end{pmatrix}=\begin{pmatrix}\mathbf{0} & \boldsymbol{Z}'(z,s)\\ \boldsymbol{Y}'(z,s) & \mathbf{0}\end{pmatrix}\begin{pmatrix}\boldsymbol{v}(z,s)\\ \boldsymbol{i}(z,s)\end{pmatrix}+\begin{pmatrix}\boldsymbol{v}'^{(\mathrm{s})}(z,s)\\ \boldsymbol{i}'^{(\mathrm{s})}(z,s)\end{pmatrix} \tag{2.325}$$

where $s=\Omega+j\omega$ denotes the Laplace-transform variable or complex frequency and z denotes the position along the multiconductor transmission line. The voltage vector $\boldsymbol{v}$, current vector $\boldsymbol{i}$, per-unit-length series voltage source vector $\boldsymbol{v}'^{(\mathrm{s})}$ and per-unit-length shunt current source vector $\boldsymbol{i}'^{(\mathrm{s})}$ have N components while the per-unit-length series impedance matrix $\boldsymbol{Z}'$ and per-unit-length shunt admittance matrix $\boldsymbol{Y}'$ are $N\times N$. The latter are defined (as indicated by the symbol :=) by

$$\boldsymbol{Z}'(z,\mathrm{s}) := \boldsymbol{R}'(z,\mathrm{s})+s\boldsymbol{L}'(z,\mathrm{s}), \tag{2.326}$$

$$\boldsymbol{Y}'(z,\mathrm{s}) := \boldsymbol{G}'(z,\mathrm{s})+s\boldsymbol{C}'(z,\mathrm{s}), \tag{2.327}$$

with the $N\times N$ per-unit-length matrices $\boldsymbol{R}'$ (resistance) $\boldsymbol{L}'$ (inductance), $\boldsymbol{G}'$ (conductance) and $\boldsymbol{C}'$ (capacitance). The propagation matrix is defined as

$$\boldsymbol{\gamma}_{\mathrm{c}}(z,\mathrm{s}) := \sqrt{\boldsymbol{Z}'(z,\mathrm{s})\boldsymbol{Y}'(z,\mathrm{s})} \tag{2.328}$$

and the characteristic impedance matrix can be obtained according to

$$\begin{aligned}\boldsymbol{Z}_{\mathrm{c}}(z,\mathrm{s}) &:= \boldsymbol{\gamma}_{\mathrm{c}}(z,\mathrm{s})\boldsymbol{Y}'^{-1}(z,\mathrm{s})\\ &= \boldsymbol{\gamma}_{\mathrm{c}}^{-1}(z,\mathrm{s})\boldsymbol{Z}'(z,\mathrm{s}).\end{aligned} \tag{2.329}$$

It is noted that the parameter matrix of the telegrapher Equations (2.325) has vanishing diagonal elements.

---

[7] This subsection is adapted from Ref. [18] with the permission of the Publisher, *IEEE Transactions on EMC*.

The telegrapher Equation (2.325) has the general structure

$$\frac{\partial \overline{\boldsymbol{v}}(z,\mathrm{s})}{\partial z} = \overline{\boldsymbol{P}}(z,\mathrm{s})\overline{\boldsymbol{v}}(z,\mathrm{s}) + \overline{\boldsymbol{v}}'^{(\mathrm{s})}(z,\mathrm{s}). \tag{2.330}$$

Here, and in the following text, symbols with an overbar indicate $2N$-vectors or $2N \times 2N$-matrices and are named by a prefix 'super-', in analogy to a similar nomenclature which is used in the context of supersymmetric field theories.

The standard solution of Equation (2.330) with initial value $\overline{\boldsymbol{v}}(z_0,\mathrm{s})$ is given by [15]

$$\overline{\boldsymbol{v}}(z,\mathrm{s}) = \overline{\boldsymbol{M}}^{z}_{z_0}(\overline{\boldsymbol{P}})\overline{\boldsymbol{v}}(z_0,\mathrm{s}) + \int_{z_0}^{z} \overline{\boldsymbol{M}}^{z}_{\zeta}(\overline{\boldsymbol{P}})\,\overline{\boldsymbol{v}}'^{(\mathrm{s})}(\zeta,s)\,\mathrm{d}\zeta, \tag{2.331}$$

where the supermatrizant $\overline{\boldsymbol{M}}^{z}_{z_0}(\overline{\boldsymbol{P}})$ is defined by the expansion

$$\overline{\boldsymbol{M}}^{z}_{z_0}(\overline{\boldsymbol{P}}) := \overline{\mathbf{1}} + \int_{z_0}^{z} \overline{\boldsymbol{P}}(\zeta,\mathrm{s})\mathrm{d}\zeta + \int_{z_0}^{z} \overline{\boldsymbol{P}}(\zeta,\mathrm{s})\int_{z_0}^{\zeta} \overline{\boldsymbol{P}}(\eta,\mathrm{s})\,\mathrm{d}\eta\,\mathrm{d}\zeta + \ldots. \tag{2.332}$$

Therefore the general matrix Equation (2.330) and, in turn, the telegrapher Equation (2.325) are formally solved. The remaining problem is the nontrivial computation of the supermatrizant $\overline{\boldsymbol{M}}^{z}_{z_0}(\overline{\boldsymbol{P}})$.[8]

While passing from Equation (2.325) to Equation (2.330) one should take care of a unique physical dimension of $\overline{\boldsymbol{v}}(z)$. Thus a direct identification $(\boldsymbol{v}(z), \boldsymbol{i}(z)) \longrightarrow \overline{\boldsymbol{v}}(z)$ is not meaningful. Two possibilities to do better are rather immediate: first, the characteristic impedance matrix $\boldsymbol{Z}_\mathrm{c}$ is used to identify the supervectors

$$\overline{\boldsymbol{v}}(z) \equiv \begin{pmatrix} \boldsymbol{v}(z) \\ \boldsymbol{Z}_\mathrm{c}(z)\boldsymbol{i}(z) \end{pmatrix}, \quad \overline{\boldsymbol{v}}'^{(\mathrm{s})}(z) \equiv \begin{pmatrix} \boldsymbol{v}'^{(\mathrm{s})}(z) \\ \boldsymbol{Z}_\mathrm{c}(z)\boldsymbol{i}'^{(\mathrm{s})}(z) \end{pmatrix}. \tag{2.333}$$

(For notational convenience here and in the following the explicit s-dependence of the physical quantities is skipped, except if noted otherwise.) The matrix $\overline{\boldsymbol{P}}$ assumes the form

$$\begin{aligned} \overline{\boldsymbol{P}}(z) &= \begin{pmatrix} \mathbf{0} & \mathbf{0} \\ \mathbf{0} & \boldsymbol{g}(z) \end{pmatrix} + \begin{pmatrix} \mathbf{0} & -\boldsymbol{\gamma}_\mathrm{c}(z) \\ -\boldsymbol{\gamma}_\mathrm{c}(z) & \mathbf{0} \end{pmatrix} \\ &=: \overline{\boldsymbol{G}}(z) + \overline{\boldsymbol{\Gamma}}_\mathrm{c}(z) \end{aligned} \tag{2.334}$$

where

$$\boldsymbol{g}(z) := \boldsymbol{D}_z\Big(\boldsymbol{Z}_\mathrm{c}(z)\Big) := \Big(\frac{\partial}{\partial z}\boldsymbol{Z}_\mathrm{c}(z)\Big)\boldsymbol{Z}_\mathrm{c}^{-1}(z) \tag{2.335}$$

introduces the product derivative $\boldsymbol{D}_z$ [41] of $\boldsymbol{Z}_\mathrm{c}(z)$ with respect to z.

[8] See also Section 2.2. Here, slightly different methods are applied that lead to more explicit results.

Alternatively, in case of a uniform medium with permittivity ε and permeability μ, it is sometimes convenient to employ the constant scalar wave impedance $Z := \sqrt{\mu/\varepsilon}$ and to identify

$$\overline{\boldsymbol{v}}(z) \equiv \begin{pmatrix} \boldsymbol{v}(z) \\ Z\boldsymbol{i}(z) \end{pmatrix}, \quad \overline{\boldsymbol{v}}'^{(\mathrm{s})}(z) \equiv \begin{pmatrix} \boldsymbol{v}'^{(\mathrm{s})}(z) \\ Z\boldsymbol{i}'^{(\mathrm{s})}(z) \end{pmatrix}. \tag{2.336}$$

This yields for $\overline{\boldsymbol{P}}$ the expression

$$\overline{\boldsymbol{P}}(z) = \begin{pmatrix} \mathbf{0} & -\boldsymbol{Z}'(z)Z^{-1} \\ -Z\boldsymbol{Y}'(z) & \mathbf{0} \end{pmatrix}. \tag{2.337}$$

A third useful representation is that in terms of wave variables $\boldsymbol{v}_{(+)}$, $\boldsymbol{v}_{(-)}$ with $(+)$ indicating the direction of increasing z (right-moving wave) and $(-)$ indicating the direction of decreasing z (left-moving wave). This wave representation can be obtained from Equation (2.333) by means of the relation [41]

$$\begin{pmatrix} \boldsymbol{v}_{(+)}(z) \\ \boldsymbol{v}_{(-)}(z) \end{pmatrix} = [\overline{\boldsymbol{R}} + \overline{\boldsymbol{P}}_{\mathrm{er}}] \begin{pmatrix} \boldsymbol{v}(z) \\ \boldsymbol{Z}_{\mathrm{c}}(z)\boldsymbol{i}(z) \end{pmatrix} \tag{2.338}$$

and

$$\begin{pmatrix} \boldsymbol{v}'^{(\mathrm{s})}_{(+)}(z) \\ \boldsymbol{v}'^{(\mathrm{s})}_{(-)}(z) \end{pmatrix} = [\overline{\boldsymbol{R}} + \overline{\boldsymbol{P}}_{\mathrm{er}}] \begin{pmatrix} \boldsymbol{v}'^{(\mathrm{s})}(z) \\ \boldsymbol{Z}_{\mathrm{c}}(z)\boldsymbol{i}'^{(\mathrm{s})}(z) \end{pmatrix} \tag{2.339}$$

where the supermatrices $\overline{\boldsymbol{R}}$ and $\overline{\boldsymbol{P}}_{\mathrm{er}}$ describe reflections

$$\overline{\boldsymbol{R}} := \begin{pmatrix} \mathbf{1} & \mathbf{0} \\ \mathbf{0} & -\mathbf{1} \end{pmatrix} = \overline{\boldsymbol{R}}^{-1} \tag{2.340}$$

and permutations

$$\overline{\boldsymbol{P}}_{\mathrm{er}} := \begin{pmatrix} \mathbf{0} & \mathbf{1} \\ \mathbf{1} & \mathbf{0} \end{pmatrix} = \overline{\boldsymbol{P}}_{\mathrm{er}}^{-1}, \tag{2.341}$$

respectively. This yields the identification

$$\overline{\boldsymbol{v}}(z) \equiv \begin{pmatrix} \boldsymbol{v}_{(+)}(z) \\ \boldsymbol{v}_{(-)}(z) \end{pmatrix}, \quad \overline{\boldsymbol{v}}'^{(\mathrm{s})}(z) \equiv \begin{pmatrix} \boldsymbol{v}'^{(\mathrm{s})}_{(+)}(z) \\ \boldsymbol{v}'^{(\mathrm{s})}_{(-)}(z) \end{pmatrix} \tag{2.342}$$

and the matrix $\overline{\boldsymbol{P}}$ assumes the form

$$\overline{\boldsymbol{P}}(z) = \begin{pmatrix} -\boldsymbol{\gamma}_{\mathrm{c}}(z) + \frac{1}{2}\boldsymbol{g}(z) & -\frac{1}{2}\boldsymbol{g}(z) \\ -\frac{1}{2}\boldsymbol{g}(z) & \boldsymbol{\gamma}_{\mathrm{c}}(z) + \frac{1}{2}\boldsymbol{g}(z) \end{pmatrix}. \tag{2.343}$$

In the following, advantage will be taken of all three types of representation.

For the calculation of the supermatrizant $\overline{M}_{z_0}^{z}(\overline{P})$ there are two general rules [15] which will turn out to be quite useful. These are the *sum rule* (which is valid for any two matrices but here is already applied to $\overline{G}$ and $\overline{\Gamma}_{\mathrm{c}}$)

$$\overline{M}_{z_0}^{z}(\overline{G}+\overline{\Gamma}_{\mathrm{c}}) = \overline{M}_{z_0}^{z}(\overline{G})\overline{M}_{z_0}^{z}(\overline{S}), \tag{2.344}$$

with

$$\overline{S} = \left(\overline{M}_{z_0}^{z}(\overline{G})\right)^{-1}\overline{\Gamma}_{\mathrm{c}}\,\overline{M}_{z_0}^{z}(\overline{G}), \tag{2.345}$$

and the *product rule*

$$\overline{M}_{z_0}^{z} = \overline{M}_{z_{n-1}}^{z}\overline{M}_{z_{n-2}}^{z_{n-1}}\cdots\overline{M}_{z_1}^{z_2}\overline{M}_{z_0}^{z_1}. \tag{2.346}$$

If the representation of Equations (2.333) and (2.334) is chosen the computation of the supermatrizant $\overline{M}_{z_0}^{z}(\overline{P})$ reduces, due to the sum rule of Equation (2.344), to a separate computation of $\overline{M}_{z_0}^{z}(\overline{G})$ and $\overline{M}_{z_0}^{z}(\overline{S})$. It is noted that the computation of $\overline{M}_{z_0}^{z}(\overline{G})$ is rather simple, since

$$\overline{G} = \begin{pmatrix} \mathbf{0} & \mathbf{0} \\ \mathbf{0} & D_z\left(\mathbf{Z}_{\mathrm{c}}(z)\right) \end{pmatrix} = D_z\begin{pmatrix} \mathbf{1} & \mathbf{0} \\ \mathbf{0} & \mathbf{Z}_{\mathrm{c}}(z) \end{pmatrix} \tag{2.347}$$

it follows from the definition of the matrizant and product derivative that

$$\overline{M}_{z_0}^{z}(\overline{G}) = \begin{pmatrix} \mathbf{1} & \mathbf{0} \\ \mathbf{0} & \mathbf{Z}_{\mathrm{c}}(z)\mathbf{Z}_{\mathrm{c}}^{-1}(z_0) \end{pmatrix}. \tag{2.348}$$

Use of this result and application of the sum rule of Equation (2.344) yields for $\overline{S}$ the expression

$$\begin{aligned}\overline{S} &= \begin{pmatrix} \mathbf{0} & -\boldsymbol{\gamma}_{\mathrm{c}}\mathbf{Z}_{\mathrm{c}}(z)\mathbf{Z}_{\mathrm{c}}^{-1}(z_0) \\ -\mathbf{Z}_{\mathrm{c}}(z_0)\mathbf{Z}_{\mathrm{c}}^{-1}(z)\boldsymbol{\gamma}_{\mathrm{c}} & \mathbf{0} \end{pmatrix} \\ &= \begin{pmatrix} \mathbf{0} & -\mathbf{Z}'(z)\mathbf{Z}_{\mathrm{c}}^{-1}(z_0) \\ -\mathbf{Z}_{\mathrm{c}}(z_0)\mathbf{Y}'(z) & \mathbf{0} \end{pmatrix},\end{aligned} \tag{2.349}$$

which is of the same structure as $\overline{P}$ in Equation (2.337) and, in general, cannot be further simplified. In this general case an explicit computation of $\overline{M}_{z_0}^{z}(\overline{S})$ is not possible. The same applies to the computation of $\overline{M}_{z_0}^{z}(\overline{P})$ if the representation of Equations (2.336) and (2.337) is used. Therefore, approximative and simplifying computation schemes are discussed next.

2.4.2 Calculation Methods for the General Solution

2.4.2.1 The Piecewise-Constant Approximation of the Characteristic Impedance Matrix

Suppose an NMTL is approximated by a set of cascaded segments, each representing a uniform MTL in a section $z_l < z < z_{l+1}$, with a total of n sections, that is $0 \le l < n$. For such an approximation it is plausible to assume within each section a constant per-unit-length impedance matrix $\mathbf{Z}'^{(l)}$ and per-unit-length admittance matrix $\mathbf{Y}'^{(l)}$. This, in turn, implies within each interval a constant characteristic impedance matrix $\mathbf{Z}_{\mathrm{c}}^{(l)}$ and a constant propagation matrix $\boldsymbol{\gamma}_{\mathrm{c}}^{(l)}$. Thus, in this case it is found that $\overline{\boldsymbol{S}}^{(l)} = \overline{\boldsymbol{\Gamma}}_{\mathrm{c}}^{(l)}$ is constant along z. This, together with $\overline{\boldsymbol{G}}^{(l)} = \overline{\mathbf{0}}$ and the implication $\overline{\boldsymbol{M}}_{z_l}^{z}(\overline{\boldsymbol{G}}^{(l)}) = \overline{\mathbf{1}}$, can be used immediately to obtain the result

$$\overline{\boldsymbol{M}}_{z_l}^{z}\left(\overline{\boldsymbol{G}}^{(l)} + \overline{\boldsymbol{\Gamma}}_{\mathrm{c}}^{(l)}\right) = \overline{\boldsymbol{M}}_{z_l}^{z}\left(\overline{\boldsymbol{\Gamma}}_{\mathrm{c}}^{(l)}\right) = \begin{pmatrix} \cosh[(z - z_l)\boldsymbol{\gamma}_{\mathrm{c}}^{(l)}] & \sinh[(z - z_l)\boldsymbol{\gamma}_{\mathrm{c}}^{(l)}] \\ -\sinh[(z - z_l)\boldsymbol{\gamma}_{\mathrm{c}}^{(l)}] & \cosh[(z - z_l)\boldsymbol{\gamma}_{\mathrm{c}}^{(l)}] \end{pmatrix} \tag{2.350}$$

which is valid for $z_l < z < z_{l+1}$. Propagation of voltage and current along this section is then expressed as

$$\begin{pmatrix} \boldsymbol{v}(z) \\ \mathbf{Z}_{\mathrm{c}}^{(l)}\boldsymbol{i}(z) \end{pmatrix} = \overline{\boldsymbol{M}}_{z_l}^{z}\left(\overline{\boldsymbol{\Gamma}}_{\mathrm{c}}^{(l)}\right)\begin{pmatrix} \boldsymbol{v}(z_l) \\ \mathbf{Z}_{\mathrm{c}}^{(l)}\boldsymbol{i}(z_l) \end{pmatrix} \tag{2.351}$$

with $\overline{\boldsymbol{M}}_{z_l}^{z}(\overline{\boldsymbol{\Gamma}}_{\mathrm{c}}^{(l)})$ given by Equation (2.350). In order to construct a solution which is valid along an arbitrary number of segments one may write

$$\begin{aligned} &\begin{pmatrix} \boldsymbol{v}(z_{l+1}) \\ \mathbf{Z}_{\mathrm{c}}^{(l+1)}\boldsymbol{i}(z_{l+1}) \end{pmatrix} \\ &\equiv \begin{pmatrix} \mathbf{1} & \mathbf{0} \\ \mathbf{0} & \mathbf{Z}_{\mathrm{c}}^{(l+1)}\mathbf{Z}_{\mathrm{c}}^{(l)-1} \end{pmatrix}\begin{pmatrix} \boldsymbol{v}(z_{l+1}) \\ \mathbf{Z}_{\mathrm{c}}^{(l)}\boldsymbol{i}(z_{l+1}) \end{pmatrix} \\ &= \begin{pmatrix} \mathbf{1} & \mathbf{0} \\ \mathbf{0} & \mathbf{Z}_{\mathrm{c}}^{(l+1)}\mathbf{Z}_{\mathrm{c}}^{(l)-1} \end{pmatrix}\overline{\boldsymbol{M}}_{z_l}^{z_{l+1}}(\overline{\boldsymbol{\Gamma}}_{\mathrm{c}}^{(l)})\begin{pmatrix} \boldsymbol{v}(z_l) \\ \mathbf{Z}_{\mathrm{c}}^{(l)}\boldsymbol{i}(z_l) \end{pmatrix}. \end{aligned} \tag{2.352}$$

Here it is interesting to note that the structure of the matrix

$$\overline{\boldsymbol{\mathcal{M}}}_{z_l}^{z_{l+1}}(\overline{\boldsymbol{G}}^{(l)}) := \begin{pmatrix} \mathbf{1} & \mathbf{0} \\ \mathbf{0} & \mathbf{Z}_{\mathrm{c}}^{(l+1)}\mathbf{Z}_{\mathrm{c}}^{(l)-1} \end{pmatrix}, \tag{2.353}$$

which describes the change of $\mathbf{Z}_{\mathrm{c}}$ along z, is the same as the exact result in Equation (2.348) that was derived under the assumption of a differentiable $\mathbf{Z}_{\mathrm{c}}$.

Let us now define the modified matrizant

$$\overline{\boldsymbol{\mathcal{M}}}_{z_l}^{z_{l+1}}\left(\overline{\boldsymbol{G}}^{(l)} + \overline{\boldsymbol{\Gamma}}_{\mathrm{c}}^{(l)}\right) := \overline{\boldsymbol{\mathcal{M}}}_{z_l}^{z_{l+1}}\left(\overline{\boldsymbol{G}}^{(l)}\right)\overline{\boldsymbol{M}}_{z_l}^{z_{l+1}}\left(\overline{\boldsymbol{\Gamma}}_{\mathrm{c}}^{(l)}\right) \tag{2.354}$$

and, for $l > 0$, the product

$$\overline{\boldsymbol{\mathcal{M}}}_{z_0}^{z_l}(\overline{\boldsymbol{G}} + \overline{\boldsymbol{\Gamma}}_{\mathrm{c}}) := \prod_{k=1}^{l}\overline{\boldsymbol{\mathcal{M}}}_{z_{l-k}}^{z_{l+1-k}}\left(\overline{\boldsymbol{G}}^{(l-k)} + \overline{\boldsymbol{\Gamma}}_{\mathrm{c}}^{(l-k)}\right). \tag{2.355}$$

Then the approximative solution of voltage and current propagation along an arbitrary number of sections is given by

$$\begin{pmatrix} \boldsymbol{v}(z_l) \\ \mathbf{Z}_{\mathrm{c}}^{(l)}\boldsymbol{i}(z_l) \end{pmatrix} = \overline{\boldsymbol{\mathcal{M}}}_{z_0}^{z_l}(\overline{\boldsymbol{G}} + \overline{\boldsymbol{\Gamma}}_{\mathrm{c}})\begin{pmatrix} \boldsymbol{v}(z_0) \\ \mathbf{Z}_{\mathrm{c}}^{(0)}\boldsymbol{i}(z_0) \end{pmatrix}. \tag{2.356}$$

On first sight it may seem that Equation (2.356) constitutes an analytically and physically exact relationship between the propagating voltage v and current i at z_0 and z_l. However, in writing down Equation (2.356) one must keep in mind that v and i, which are assumed to actually propagate along a piecewise-constant MTL, are generally not continuous at points z_l that connect adjacent intervals. This is due to the, in general, discontinuous behavior of Z_c at z_l. In Equation (2.352) $(\boldsymbol{v}(z_{l+1}), \boldsymbol{Z}_c^{(l)}\boldsymbol{i}(z_{l+1}))$ is connected to $(\boldsymbol{v}(z_l), \boldsymbol{Z}_c^{(l)}\boldsymbol{i}(z_l))$, both of which are defined within the interval $[z_l, z_{l+1}]$, by means of an exact solution for a single interval. However, a smooth transition from $(\boldsymbol{v}(z_{l+1}), \boldsymbol{Z}_c^{(l)}\boldsymbol{i}(z_{l+1}))$ in $[z_l, z_{l+1}]$ to $(\boldsymbol{v}(z_{l+1}), \boldsymbol{Z}_c^{(l+1)}\boldsymbol{i}(z_{l+1}))$ in $[z_{l+1}, z_{l+2}]$ was simply adopted, compare the transition from the second to the third row in Equation (2.352). Therefore the sectional solution in Equation (2.356) is only an approximation in order to match together the solutions of adjacent sections. However, in this approach the reflections of the propagating waves at wire discontinuous are omitted. In the following subsection, an improved computation scheme will be given which circumvents this disadvantage.

2.4.2.2 The Continuous Approximation of the Characteristic Impedance Matrix

In order to avoid discontinuities and related reflections at the boundary of adjacent intervals it is sufficient to connect continuously the corresponding characteristic impedance matrices $Z_c^{(l)}(z_l)$ and $Z_c^{(l+1)}(z_{l+1})$. The following method to do so is based on [42] and was considerably refined in [43]. This method will reduce, in a first step, the N-dimensional NMTL problem to that of N one-dimensional NMTL problems.

The main assumption which is employed is the ansatz that the propagation matrix is of the form

$$\boldsymbol{\gamma}_c(z) = \sqrt{\boldsymbol{Z}'(z)\boldsymbol{Y}'(z)} = \gamma(z)\mathbf{1}. \tag{2.357}$$

Physically, this ansatz implies that all modes propagate with the same speed. This may happen, for example, in each cross section of perfectly conducting wires or in a, uniform and isotropic dielectric medium where the wire diameters and relative positions are allowed to be a function of z. It follows that, apart from possible forefactors, $Z'(z)$ and $Y'(z)$ are mutually inverse to each other. As usual it is convenient to introduce a dimensionless geometrical-impedance-factor matrix $f_g = \boldsymbol{f}_g(z)$ and write, with the s-dependence explicitly stated in the rest of this paragraph,

$$\boldsymbol{Z}'(z, s) =: z'(z, s)\boldsymbol{f}_g(z), \quad \boldsymbol{Y}'(z, s) =: y'(z, s)\boldsymbol{f}_g^{-1}(z). \tag{2.358}$$

The natural assumption of reciprocity is used, that is, the matrix f_g is assumed to be symmetric. With $z'(z, s)$, $y'(z, s)$ of the form

$$z'(z, s) = s\mu(z), \qquad y'(z, s) = \sigma(z) + s\varepsilon(z), \tag{2.359}$$

the matrix f_g can also be taken as real and, as describing a passive NMTL, positive definite. The parameters μ, ε and σ characterize the medium which surrounds the transmission line. Let us furthermore restrict ourselves to the case where μ and ε are constant with respect to z. Also, a possible conductivity σ is absorbed in ε by allowing the permittivity to be complex and frequency-dependent, thus formally setting $\sigma = 0$. Then one obtains for the characteristic impedance matrix and its inverse

$$\mathbf{Z}_{\mathrm{c}}(z) = \sqrt{\frac{\mu}{\varepsilon}}\,\boldsymbol{f}_{\mathrm{g}}(z) = Z\boldsymbol{f}_{\mathrm{g}}(z), \tag{2.360}$$

$$\mathbf{Y}_{\mathrm{c}}(z) = \mathbf{Z}_{\mathrm{c}}^{-1}(z) = \sqrt{\frac{\varepsilon}{\mu}}\,\boldsymbol{f}_{\mathrm{g}}^{-1}(z) = Z^{-1}\boldsymbol{f}_{\mathrm{g}}^{-1}(z). \tag{2.361}$$

The propagation matrix simply becomes

$$\boldsymbol{\gamma}_{\mathrm{c}} = s\sqrt{\varepsilon\mu}\,\mathbf{1} = \frac{s}{v}\mathbf{1} = \gamma\mathbf{1} \tag{2.362}$$

with $v = 1/\sqrt{\varepsilon\mu}$ the wave speed.

Analogously to the previous subsection, the interval along z is subdivided into n sections, $z_0 < z_1 < \ldots < z_l < z_{l+1} < \ldots < z_n$, and the $(l+1)$th section $z_l < z < z_{l+1}$ is considered. As dynamical variable within this section one may choose $\overline{\boldsymbol{v}}^{(l)}(z) = (\boldsymbol{v}_1^{(l)}(z), \boldsymbol{v}_2^{(l)}(z))$ as

$$\boldsymbol{v}_1^{(l)}(z) := \boldsymbol{f}_{\mathrm{g}}^{-1/2}(z_l)\boldsymbol{v}(z), \tag{2.363}$$

$$\boldsymbol{v}_2^{(l)}(z) := Z\boldsymbol{f}_{\mathrm{g}}^{-1/2}(z_l)\boldsymbol{i}(z), \tag{2.364}$$

and also put the source term $\overline{\boldsymbol{v}}'^{(l,\mathrm{s})}(z) = (\boldsymbol{v}'^{(l,\mathrm{s})}_1(z), \boldsymbol{v}'^{(l,\mathrm{s})}_2(z))$ into the same form,

$$\boldsymbol{v}'^{(l,\mathrm{s})}_1(z) := \boldsymbol{f}_{\mathrm{g}}^{-1/2}(z_l)\boldsymbol{v}'^{(\mathrm{s})}(z), \tag{2.365}$$

$$\boldsymbol{v}'^{(l,\mathrm{s})}_2(z) := Z\boldsymbol{f}_{\mathrm{g}}^{-1/2}(z_l)\boldsymbol{i}'^{(\mathrm{s})}(z). \tag{2.366}$$

It is noted that the square root of $\boldsymbol{f}_{\mathrm{g}}$ is well defined since $\boldsymbol{f}_{\mathrm{g}}$ is real, symmetric and positive definite. Therefore it can be diagonalized with positive, real eigenvalues and real eigenvectors and the square root of $\boldsymbol{f}_{\mathrm{g}}$ is defined by the positive square roots of the eigenvalues.

The telegrapher equation for the above variables becomes

$$\begin{aligned}
&\frac{\partial}{\partial z}\begin{pmatrix} \boldsymbol{v}_1^{(l)}(z) \\ \boldsymbol{v}_2^{(l)}(z) \end{pmatrix} \\
&\qquad = -\gamma \begin{pmatrix} \mathbf{0} & \boldsymbol{X}^{(l)}(z) \\ \boldsymbol{X}^{(l)^{-1}}(z) & \mathbf{0} \end{pmatrix} \begin{pmatrix} \boldsymbol{v}_1^{(l)}(z) \\ \boldsymbol{v}_2^{(l)}(z) \end{pmatrix} + \begin{pmatrix} \boldsymbol{v}'^{(l,\mathrm{s})}_1(z) \\ \boldsymbol{v}'^{(l,\mathrm{s})}_2(z) \end{pmatrix},
\end{aligned} \tag{2.367}$$

with

$$\boldsymbol{X}^{(l)}(z) := \boldsymbol{f}_{\mathrm{g}}^{-1/2}(z_l)\boldsymbol{f}_{\mathrm{g}}(z)\boldsymbol{f}_{\mathrm{g}}^{-1/2}(z_l) \tag{2.368}$$

It is aimed to continuously connect $\boldsymbol{X}^{(l)}$ between adjacent intervals since this implies, due to its definition in terms of $\boldsymbol{f}_{\mathrm{g}}$ and the relation $\mathbf{Z}_{\mathrm{c}} = Z\boldsymbol{f}_{\mathrm{g}}$, a continuous $\mathbf{Z}_{\mathrm{c}}$ along z.

Due to the properties of the matrix $\boldsymbol{f}_{\mathrm{g}}$ the matrix $\boldsymbol{X}^{(l)}$ is also symmetric, real valued and positive definite. Therefore it can pointwise be expanded in terms of a complete set of real and normalized eigenvectors with real eigenvalues. This is done at the point $z = z_{l+1}$, that is

$$\boldsymbol{X}^{(l)}(z_{l+1}) = \sum_{\beta=1}^{N} x_\beta^{(l)}(z_{l+1})\boldsymbol{x}_\beta^{(l)}\boldsymbol{x}_\beta^{(l)} \tag{2.369}$$

with $\boldsymbol{x}_\beta^{(l)}$ representing the eigenvectors at z_l and $\boldsymbol{x}_\beta^{(l)}(z_{l+1})$ the corresponding eigenvalues. In particular, it follows from the definition in Equation (2.368) and the completeness of $\boldsymbol{x}_\beta^{(l)}$

$$\boldsymbol{X}^{(l)}(z_l) = \mathbf{1} = \sum_{\beta=1}^{N} \boldsymbol{x}_\beta^{(l)}\boldsymbol{x}_\beta^{(l)}. \tag{2.370}$$

In order to connect $\boldsymbol{X}^{(l)}(z_{l+1})$ to $\boldsymbol{X}^{(l)}(z_l)$ continuously, $\boldsymbol{X}^{(l)}(z)$ is approximated, for $z_l < z < z_{l+1}$, by

$$\boldsymbol{X}^{(l,a)}(z) = \sum_{\beta=1}^{N} x_\beta^{(l,a)}(z)\boldsymbol{x}_\beta^{(l)}\boldsymbol{x}_\beta^{(l)} \tag{2.371}$$

with freely choosable functions $x_\beta^{(l,a)}(z)$ that have to fulfill the boundary conditions

$$x_\beta^{(l,a)}(z_l) = 1, \qquad x_\beta^{(l,a)}(z_{l+1}) = x_\beta^{(l)}(z_{l+1}). \tag{2.372}$$

Simple examples for the $x_\beta^{(l,a)}(z)$ are provided by a linear z-dependence,

$$x_\beta^{(l,a)}(z) = 1 + \frac{z - z_l}{z_{l+1} - z_l}\left[x_\beta^{(l,a)}(z_{l+1}) - 1\right] \tag{2.373}$$

or an exponential interpolation which can be taken of the form

$$x_\beta^{(l,a)}(z) = \exp(2\alpha_\beta^{(l)}(z - z_l)), \tag{2.374}$$

where the constant $\alpha_\beta^{(l)}$ is given by

$$\alpha_\beta^{(l)} := \frac{1}{2}\frac{\ln\left(x_\beta^{(l,a)}(z_{l+1})\right)}{z_{l+1} - z_l}. \tag{2.375}$$

The approximation in Equation (2.371) for $\boldsymbol{X}^{(l)}(z)$ also implies an approximation for the geometric-factor matrix $\boldsymbol{f}_\mathrm{g}$ which is of the form

$$\begin{aligned}\boldsymbol{f}_\mathrm{g}^{(l,a)}(z) &= \boldsymbol{f}_\mathrm{g}^{1/2}(z_l)\boldsymbol{X}^{(l,a)}(z)\boldsymbol{f}_\mathrm{g}^{1/2}(z_l)\\ &= \sum_{\beta=1}^{N} x_\beta^{(l,a)}(z)\left(\boldsymbol{f}_\mathrm{g}^{1/2}(z_l)\boldsymbol{x}_\beta^{(l)}\right)\left(\boldsymbol{f}_\mathrm{g}^{1/2}(z_l)\boldsymbol{x}_\beta^{(l)}\right)\end{aligned} \tag{2.376}$$

with

$$\boldsymbol{f}_\mathrm{g}^{(l,a)}(z_l) = \boldsymbol{f}_\mathrm{g}(z_l), \qquad \boldsymbol{f}_\mathrm{g}^{(l,a)}(z_{l+1}) = \boldsymbol{f}_\mathrm{g}(z_{l+1}). \tag{2.377}$$

Analogous formulas are valid for the characteristic impedance matrix $\mathbf{Z}_\mathrm{c} = Z\boldsymbol{f}_\mathrm{g}$.

The matrices $\boldsymbol{X}^{(l,a)}(z)$ define an approximated dynamics of the NMTL by the telegrapher equation

$$\frac{\partial}{\partial z}\begin{pmatrix}\boldsymbol{v}_1^{(l,a)}(z)\\ \boldsymbol{v}_2^{(l,a)}(z)\end{pmatrix} = -\gamma\begin{pmatrix}\mathbf{0} & \boldsymbol{X}^{(l,a)}(z)\\ \boldsymbol{X}^{(l,a)^{-1}}(z) & \mathbf{0}\end{pmatrix}\begin{pmatrix}\boldsymbol{v}_1^{(l,a)}(z)\\ \boldsymbol{v}_2^{(l,a)}(z)\end{pmatrix} + \begin{pmatrix}\boldsymbol{v}_1'^{(l,s)}(z)\\ \boldsymbol{v}_2'^{(l,s)}(z)\end{pmatrix}, \tag{2.378}$$

or, in supermatrix notation,

$$\frac{\partial\overline{\boldsymbol{v}}^{(l,a)}(z)}{\partial z} = \overline{\boldsymbol{P}}^{(l,a)}(z)\overline{\boldsymbol{v}}^{(l,a)}(z) + \overline{\boldsymbol{v}}'^{(l,s)}(z). \tag{2.379}$$

It is still necessary to solve this equation via the computation of the supermatrizant $\overline{\boldsymbol{M}}_{z_l}^{z_{l+1}}(\overline{\boldsymbol{P}}^{(l,a)}(z))$. Now the key observation is that the telegrapher Equations (2.378) and (2.379) can be transformed in a way such that the block matrices $\boldsymbol{X}^{(l,a)}(z)$, $\boldsymbol{X}^{(l,a)-1}(z)$ of $\overline{\boldsymbol{P}}^{(l,a)}(z)$ become diagonal. The corresponding transformation matrix $\chi^{(l)}$ has the constant and normalized eigenvectors $\boldsymbol{x}_\beta^{(l)}$ as its columns, that is formally

$$\chi^{(l)} = \left(\boldsymbol{x}_1^{(l)}\middle|\boldsymbol{x}_2^{(l)}\middle|\dots\middle|\boldsymbol{x}_N^{(l)}\right). \tag{2.380}$$

From the definition

$$\overline{\chi} = \begin{pmatrix}\chi & \mathbf{0}\\ \mathbf{0} & \chi\end{pmatrix}, \quad \overline{\chi}^{-1} = \begin{pmatrix}\chi^{-1} & \mathbf{0}\\ \mathbf{0} & \chi^{-1}\end{pmatrix} \tag{2.381}$$

the desired similarity transformation reads

$$\overline{\chi}^{-1}\frac{\partial\overline{\boldsymbol{v}}^{(l,a)}(z)}{\partial z} = \overline{\chi}^{-1}\overline{\boldsymbol{P}}^{(l,a)}(z)\overline{\chi}\,\overline{\chi}^{-1}\overline{\boldsymbol{v}}^{(l,a)}(z) + \overline{\chi}^{-1}\overline{\boldsymbol{v}}'^{(l,s)}(z). \tag{2.382}$$

To compute the supermatrizant $\overline{\boldsymbol{M}}_{z_l}^{z_{l+1}}(\overline{\chi}^{-1}\overline{\boldsymbol{P}}^{(l,a)}(z)\overline{\chi})$ it is sufficient to concentrate on the homogeneous part of the telegrapher equation. The homogeneous part of the transformed Equation (2.382) is given by

$$\frac{\partial}{\partial z}\begin{pmatrix}\chi^{-1}\boldsymbol{v}_1^{(l,a)}(z)\\ \chi^{-1}\boldsymbol{v}_2^{(l,a)}(z)\end{pmatrix} = -\gamma\begin{pmatrix}\mathbf{0} & \chi^{-1}\boldsymbol{X}^{(l,a)}(z)\chi\\ \chi^{-1}\boldsymbol{X}^{(l,a)-1}(z)\chi & \mathbf{0}\end{pmatrix}\begin{pmatrix}\chi^{-1}\boldsymbol{v}_1^{(l,a)}(z)\\ \chi^{-1}\boldsymbol{v}_2^{(l,a)}(z)\end{pmatrix}, \tag{2.383}$$

with diagonal block matrices that explicitly read

$$\chi^{-1}\boldsymbol{X}^{(l,a)}(z)\chi = \begin{pmatrix} x_1^{(l,a)}(z) & & \mathbf{0} \\ & \ddots & \\ \mathbf{0} & & x_N^{(l,a)}(z) \end{pmatrix} \tag{2.384}$$

$$\chi^{-1}\boldsymbol{X}^{(l,a)-1}(z)\chi = \begin{pmatrix} x_1^{(l,a)-1}(z) & & \mathbf{0} \\ & \ddots & \\ \mathbf{0} & & x_N^{(l,a)-1}(z) \end{pmatrix}. \tag{2.385}$$

Therefore the telegrapher Equation (2.383) for an N-dimensional NMTL decouples into N telegrapher equations for a one-dimensional NTL. These are of the two-component form

$$\begin{aligned}\frac{\partial}{\partial z}\begin{pmatrix} v_{1\beta}^{(l,a)}(z) \\ v_{2\beta}^{(l,a)}(z) \end{pmatrix} \\ = -\gamma \begin{pmatrix} 0 & x_\beta^{(l,a)}(z) \\ x_\beta^{(l,a)-1}(z) & 0 \end{pmatrix}\begin{pmatrix} v_{1\beta}^{(l,a)}(z) \\ v_{2\beta}^{(l,a)}(z) \end{pmatrix},\end{aligned} \tag{2.386}$$

with

$$v_{i\beta}^{(l,a)}(z) := (\chi^{-1}\boldsymbol{v}_i^{(l,a)}(z))_\beta \tag{2.387}$$

the βth component of the transformed voltage $\boldsymbol{v}_i^{(l,a)}$ for $\beta = 1, \ldots, N$. How to deal with these telegrapher equations of a one-dimensional NTL analytically and arrive at corresponding exact solutions has already been demonstrated in [42, 43]. An alternative solution procedure is provided in Section 2.4.3.

Once the solutions for the N Equations (2.386) are found the supermatrizant $\overline{\boldsymbol{M}}_{z_l}^{z_{l+1}}(\overline{\chi}^{-1}\overline{\boldsymbol{P}}^{(l,a)}(z)\overline{\chi})$ can be constructed from Equation (2.383). Since, due to the constant $\overline{\chi}$,

$$\overline{\boldsymbol{M}}_{z_l}^{z_{l+1}}\left(\overline{\chi}^{-1}\overline{\boldsymbol{P}}^{(l,a)}(z)\overline{\chi}\right) = \overline{\chi}^{-1}\overline{\boldsymbol{M}}_{z_l}^{z_{l+1}}\left(\overline{\boldsymbol{P}}^{(l,a)}(z)\right)\overline{\chi} \tag{2.388}$$

one then also obtains the supermatrizant $\overline{\boldsymbol{M}}_{z_l}^{z_{l+1}}(\overline{\boldsymbol{P}}^{(l,a)}(z))$ which provides an approximate solution for the telegrapher Equation (2.367). Reversing the transformations of Equations (2.363) and (2.364) allows this solution to be expressed for the $(l+1)$th section in terms of the original variables. Finally, by construction, the solutions of the single sections are continuously fitted together over the whole interval.

2.4.2.3 Circulant Nonuniform MTLs

It has just been seen how the solution of a telegrapher equation for an N-dimensional NMTL was reduced to the analytically known solution of N telegrapher equations for one-dimensional NTLs. The key ingredient for this procedure was the diagonalization of the block matrices which occur in the supermatrix $\overline{\boldsymbol{P}}$, the supermatrizant of which describes the complete solution.

Now there is an interesting class of matrices which form a commutative group and thus can be simultaneously diagonalized; this is the set of all $N \times N$ nonsingular *circulant* matrices. All elements of this group can be simultaneously diagonalized by means of a *single* unitary matrix [44].

A general $N \times N$ circulant matrix has the structure

$$\begin{pmatrix} C_1 & C_2 & C_3 & \dots & C_N \\ C_N & C_1 & C_2 & \dots & C_{N-1} \\ \vdots & \vdots & \vdots & \ddots & \vdots \\ C_2 & C_3 & C_4 & \dots & C_1 \end{pmatrix} \tag{2.389}$$

and it is tempting to ask what type of NMTL is characterized by matrices $\boldsymbol{Z}'$ and $\boldsymbol{Y}'$ that are circulant. In this case, also $\boldsymbol{Z}_c$ and $\boldsymbol{\gamma}_c$ are circulant. As it turns out, the corresponding type of NMTL exhibit a cross section with a discrete rotational symmetry. Then it is also possible to reduce the N-dimensional NMTL to the solution of one-dimensional NTLs. This can be shown along the same lines as for the continuous approximation of $\boldsymbol{Z}_c$ in the last subsection.

It is possible to start, for example, with the representation of Equations (2.336) and (2.337) of $\overline{\boldsymbol{v}}(z)$ and $\overline{\boldsymbol{P}}(z)$. If $\boldsymbol{Z}'$ and $\boldsymbol{Y}'$ are also circulant it trivially follows that the block matrices of $\overline{\boldsymbol{P}}(z)$ are circulant also. Then a similarity transformation of the corresponding telegrapher equation can be transformed which is analogous to Equation (2.382),

$$\overline{\boldsymbol{U}}^{-1}\frac{\partial \overline{\boldsymbol{v}}}{\partial z} = \overline{\boldsymbol{U}}^{-1}\overline{\boldsymbol{P}\boldsymbol{U}}\,\overline{\boldsymbol{U}}^{-1}\overline{\boldsymbol{v}} + \overline{\boldsymbol{U}}^{-1}\overline{\boldsymbol{v}}'^{(s)}. \tag{2.390}$$

The unitary matrix $\overline{\boldsymbol{U}}$ is defined by

$$\overline{\boldsymbol{U}} = \begin{pmatrix} \boldsymbol{U} & \boldsymbol{0} \\ \boldsymbol{0} & \boldsymbol{U} \end{pmatrix} \tag{2.391}$$

with

$$\boldsymbol{U} = \left(\boldsymbol{u}_1 \middle| \boldsymbol{u}_2 \middle| \dots \middle| \boldsymbol{u}_N\right). \tag{2.392}$$

Here, the vectors $\boldsymbol{u}_\beta$ are the constant and orthonormal eigenvectors that correspond to *all* circulant matrices. They explicitly read, in a complex representation,

$$\boldsymbol{u}_\beta = \frac{1}{\sqrt{N}} \begin{pmatrix} \exp(j2\pi\frac{\beta}{N}) \\ \exp(j2\pi\frac{2\beta}{N}) \\ \exp(j2\pi\frac{3\beta}{N}) \\ \vdots \\ \exp(j2\pi\beta) \end{pmatrix}. \tag{2.393}$$

Since they do not depend on z the block matrices of $\overline{\boldsymbol{P}}(z)$ can be diagonalized over the *whole* interval along z. Thus there is no need to subdivide the interval into single sections and to worry about possible discontinuities between adjacent sections. Once the block matrices of

$\overline{\boldsymbol{P}}(z)$ are diagonalized the desired decoupling into two-component telegrapher equations is achieved in an analytically *exact* way. Then the solution of the initial telegrapher equation is constructed, via the supermatrizant $\overline{\boldsymbol{M}}_{z_0}^{z}(\overline{\boldsymbol{P}})$, along the lines of the previous subsection from the solutions of the two-component telegrapher equations.

For more information on circulant matrices and NMTLs the interested reader is referred to [28, 44].

2.4.2.4 General Approach to Calculate the Matrizant

In the previous paragraphs it has been shown how to find exact solutions for the matrizant. This could be achieved under the assumption of certain symmetries of the physical matrices describing the NMTLs. In this section no symmetry properties on the NMTL matrices are imposed and a solution for an almost uniform transmission line is derived. The wave representation of Equations (2.342) and (2.343) is used such that the supermatrizant of

$$\overline{\boldsymbol{P}} = \begin{pmatrix} -\boldsymbol{\gamma}_{\mathrm{c}} & \boldsymbol{0} \\ \boldsymbol{0} & \boldsymbol{\gamma}_{\mathrm{c}} \end{pmatrix} + \frac{1}{2}\begin{pmatrix} \boldsymbol{g} & -\boldsymbol{g} \\ -\boldsymbol{g} & \boldsymbol{g} \end{pmatrix} \quad =: \overline{\boldsymbol{\Gamma}}_{c\sim} + \overline{\boldsymbol{G}}_{\sim} \tag{2.394}$$

has to be found.

For a host of lines it is an excellent approximation to assume only a small z-dependent deviation for $\overline{\boldsymbol{\Gamma}}_{c\sim}$ from the unit matrix times a z-dependent scalar function γ. At least this is certainly true if the wires do not diverge too much. In this case also for the $\overline{\boldsymbol{G}}_{\sim}$ matrix it is justified to treat its z-dependence as a kind of perturbation. Thus one proceeds by splitting up $\overline{\boldsymbol{G}}_{\sim}$ into a block-diagonal and block-off-diagonal supermatrix, respectively,

$$\overline{\boldsymbol{G}}_{\sim} = \frac{1}{2}\begin{pmatrix} \boldsymbol{g} & \boldsymbol{0} \\ \boldsymbol{0} & \boldsymbol{g} \end{pmatrix} + \frac{1}{2}\begin{pmatrix} \boldsymbol{0} & -\boldsymbol{g} \\ -\boldsymbol{g} & \boldsymbol{0} \end{pmatrix} \equiv \overline{\boldsymbol{G}}_{1\sim} + \overline{\boldsymbol{G}}_{2\sim} \tag{2.395}$$

and treating both summands as perturbations up to the first order. Then the matrizant $\overline{\boldsymbol{M}}_{z_0}^{z}(\overline{\boldsymbol{\Gamma}}_{c\sim} + \overline{\boldsymbol{G}}_{\sim})$ is calculated with

$$\begin{aligned} \overline{\boldsymbol{\Gamma}}_{c\sim} &= \gamma\begin{pmatrix} -\boldsymbol{1} & \boldsymbol{0} \\ \boldsymbol{0} & \boldsymbol{1} \end{pmatrix} + \begin{pmatrix} -\Delta\boldsymbol{\gamma}_{\mathrm{c}} & \boldsymbol{0} \\ \boldsymbol{0} & \Delta\boldsymbol{\gamma}_{\mathrm{c}} \end{pmatrix} \\ &=: \overline{\boldsymbol{\Gamma}}_{c0\sim} + \Delta\overline{\boldsymbol{\Gamma}}_{c\sim} \end{aligned} \tag{2.396}$$

where $\Delta\gamma_{\mathrm{c}}$ describes the deviation from an ideal line and is also taken into account up to the first order.

First, the expression $\overline{\boldsymbol{M}}_{z_0}^{z}(\overline{\boldsymbol{G}}_{1\sim})$ is estimated. For this purpose, the characteristic impedance matrix is written as

$$\mathbf{Z}_{\mathrm{c}}(z) = \mathbf{Z}_{\mathrm{c}}(z_0) + \Delta\mathbf{Z}_{\mathrm{c}}(z), \quad \text{with} \quad \Delta\mathbf{Z}_{\mathrm{c}}(z_0) = \mathbf{0} \tag{2.397}$$

and for the product derivative one obtains, up to the first order,

$$\boldsymbol{D}_z\mathbf{Z}_{\mathrm{c}}(z) \cong \frac{\partial}{\partial z}\Big(\Delta\mathbf{Z}_{\mathrm{c}}(z)\Big)\mathbf{Z}_{\mathrm{c}}^{-1}(z_0). \tag{2.398}$$

With this ansatz the matrizant $\overline{\boldsymbol{M}}_{z_0}^{z}(\overline{\boldsymbol{G}}_{1\sim})$, truncated after the second term, turns out to be

$$\overline{\boldsymbol{M}}_{z_0}^{z}(\overline{\boldsymbol{G}}_{1\sim}) \cong \overline{\mathbf{1}} + \begin{pmatrix} \frac{1}{2}\Delta \mathbf{Z}_{\mathrm{c}}(z)\mathbf{Z}_{\mathrm{c}}^{-1}(z_0) & \mathbf{0} \\ \mathbf{0} & \frac{1}{2}\Delta \mathbf{Z}_{\mathrm{c}}(z)\mathbf{Z}_{\mathrm{c}}^{-1}(z_0) \end{pmatrix}. \tag{2.399}$$

For $\overline{\boldsymbol{M}}_{z_0}^{z}(\overline{\boldsymbol{\Gamma}}_{c0\sim})$ one obtains the exact result

$$\overline{\boldsymbol{M}}_{z_0}^{z}(\overline{\boldsymbol{\Gamma}}_{c0\sim}) = \begin{pmatrix} \exp(-\gamma(z-z_0))\mathbf{1} & \mathbf{0} \\ \mathbf{0} & \exp(\gamma(z-z_0))\mathbf{1} \end{pmatrix}. \tag{2.400}$$

Both terms can be collected under the observation of the sum rule of Equation (2.344) to yield approximately

$$\begin{aligned} &\overline{\boldsymbol{M}}_{z_0}^{z}(\overline{\boldsymbol{\Gamma}}_{c0\sim} + \overline{\boldsymbol{G}}_{1\sim}) \\ &\cong \begin{pmatrix} \exp(-\gamma(z-z_0))\Big(\mathbf{1} - \frac{1}{2}\Delta \mathbf{Z}_{\mathrm{c}}(z)\mathbf{Z}_{\mathrm{c}}^{-1}(z_0)\Big) & \mathbf{0} \\ \mathbf{0} & \exp(-\gamma(z-z_0))\Big(\mathbf{1} + \frac{1}{2}\Delta \mathbf{Z}_{\mathrm{c}}(z)\mathbf{Z}_{\mathrm{c}}^{-1}(z_0)\Big) \end{pmatrix}. \end{aligned} \tag{2.401}$$

Next, the expressions $\overline{\boldsymbol{M}}_{z_0}^{z}(\Delta\overline{\boldsymbol{\Gamma}}_{c\sim})$ and $\overline{\boldsymbol{M}}_{z_0}^{z}(\overline{\boldsymbol{G}}_{2\sim})$ are evaluated to the first order and the sum rule is again applied twice. This eventually leads to the final result

$$\begin{aligned} &\overline{\boldsymbol{M}}_{z_0}^{z}(\overline{\boldsymbol{\Gamma}}_{c\sim} + \overline{\boldsymbol{G}}_{\sim}) \\ &\cong \begin{pmatrix} \boldsymbol{M}_{z_0}^{z}(\overline{\boldsymbol{\Gamma}}_{c\sim} + \overline{\boldsymbol{G}}_{\sim})_{11} & \boldsymbol{M}_{z_0}^{z}(\overline{\boldsymbol{\Gamma}}_{c\sim} + \overline{\boldsymbol{G}}_{\sim})_{12} \\ \boldsymbol{M}_{z_0}^{z}(\overline{\boldsymbol{\Gamma}}_{c\sim} + \overline{\boldsymbol{G}}_{\sim})_{21} & \boldsymbol{M}_{z_0}^{z}(\overline{\boldsymbol{\Gamma}}_{c\sim} + \overline{\boldsymbol{G}}_{\sim})_{22} \end{pmatrix} \end{aligned} \tag{2.402}$$

where

$$\begin{aligned} &\boldsymbol{M}_{z_0}^{z}(\overline{\boldsymbol{\Gamma}}_{c\sim} + \overline{\boldsymbol{G}}_{\sim})_{11} \\ &= \exp(-\gamma(z-z_0))\Big[\mathbf{1} - \int_{z_0}^{z} \Delta\boldsymbol{\gamma}_{\mathrm{c}}(\zeta)\,\mathrm{d}\zeta + \frac{1}{2}\Delta \mathbf{Z}_{\mathrm{c}}(z)\mathbf{Z}_{\mathrm{c}}^{-1}(z_0)\Big], \end{aligned} \tag{2.403}$$

$$\begin{aligned} &\boldsymbol{M}_{z_0}^{z}(\overline{\boldsymbol{\Gamma}}_{c\sim} + \overline{\boldsymbol{G}}_{\sim})_{12} \\ &= -\frac{1}{2}\exp(\gamma(z-z_0))\Big[\Delta \mathbf{Z}_{\mathrm{c}}(z)\mathbf{Z}_{\mathrm{c}}^{-1}(z_0) - 2\gamma\exp(-2\gamma(z-z_0)) \\ &\quad \times \int_{z_0}^{z} \exp(2\gamma(\zeta-z_0))\Delta \mathbf{Z}_{\mathrm{c}}(\zeta)\,\mathrm{d}\zeta\,\Delta \mathbf{Z}_{\mathrm{c}}^{-1}(z_0)\Big], \end{aligned} \tag{2.404}$$

$$\begin{aligned} &\boldsymbol{M}_{z_0}^{z}(\overline{\boldsymbol{\Gamma}}_{c\sim} + \overline{\boldsymbol{G}}_{\sim})_{21} \\ &= -\frac{1}{2}\exp(-\gamma(z-z_0))\Big[\Delta \mathbf{Z}_{\mathrm{c}}(z)\mathbf{Z}_{\mathrm{c}}^{-1}(z_0) + 2\gamma\exp(2\gamma(z-z_0)) \\ &\quad \times \int_{z_0}^{z} \exp(-2\gamma(\zeta-z_0))\Delta \mathbf{Z}_{\mathrm{c}}(\zeta)\mathrm{d}\zeta\,\Delta \mathbf{Z}_{\mathrm{c}}^{-1}(z_0)\Big], \end{aligned} \tag{2.405}$$

$$\begin{aligned} &\boldsymbol{M}_{z_0}^{z}(\overline{\boldsymbol{\Gamma}}_{c\sim} + \overline{\boldsymbol{G}}_{\sim})_{22} \\ &= \exp(\gamma(z-z_0))\Big[\mathbf{1} + \int_{z_0}^{z} \Delta\boldsymbol{\gamma}_{\mathrm{c}}(\zeta)\,\mathrm{d}\zeta + \frac{1}{2}\Delta \mathbf{Z}_{\mathrm{c}}(z)\mathbf{Z}_{\mathrm{c}}^{-1}(z_0)\Big]. \end{aligned} \tag{2.406}$$

As expected, it is observed that the coupling between forward and backward running waves is due, solely, to the dependence of the characteristic impedance matrix on the local coordinate z. If the matrizant for the entire interval $[z_0, z]$ is decomposed into a product of matrizants for 'smaller' intervals, say

$$\overline{\boldsymbol{M}}_{z_0}^{z}(\overline{\boldsymbol{\Gamma}}_{c\sim} + \overline{\boldsymbol{G}}_{\sim}) = \overline{\boldsymbol{M}}_{z_{N-1}}^{z}(\overline{\boldsymbol{\Gamma}}_{c\sim} + \overline{\boldsymbol{G}}_{\sim}) \cdot \ldots \cdot \overline{\boldsymbol{M}}_{z_1}^{z_2}(\overline{\boldsymbol{\Gamma}}_{c\sim} + \overline{\boldsymbol{G}}_{\sim})\overline{\boldsymbol{M}}_{z_0}^{z_1}(\overline{\boldsymbol{\Gamma}}_{c\sim} + \overline{\boldsymbol{G}}_{\sim}) \tag{2.407}$$

(with $z_0 < z_1 < \ldots < z_{l-1} < z_l \ldots < z_{N-1} < z_N$) then the solution in Equation (2.402) might become a very good approximation and in this case $\Delta \mathbf{Z}_c(z)$ can be thought of as the first term of a Taylor expansion for $\mathbf{Z}_c(z)$ at some initial point z_{l-1} of an interval $[z_{l-1}, z_l]$,

$$\Delta \mathbf{Z}_c(z_{l-1}) = \left.\frac{\partial \mathbf{Z}_c}{\partial z}\right|_{z=z_{l-1}} (z - z_{l-1}). \tag{2.408}$$

If the perturbation of the propagation matrix $\Delta \boldsymbol{\gamma}_c(z)$ is analogously approximated, the integrals in the above solution can easily be calculated, for example as

$$\int_{z_0}^{z} \Delta \boldsymbol{\gamma}_c(\zeta)\, d\zeta = \frac{1}{2}\left.\frac{\partial \boldsymbol{\gamma}_c}{\partial z}\right|_{z_0} (z - z_0)^2 \tag{2.409}$$

and

$$\int_{z_0}^{z} \exp(2\gamma(\zeta - z_0))\Delta \mathbf{Z}_c(\zeta)\, d\zeta = \frac{1}{4\gamma^2}\Big[1 + \exp(2\gamma(z - z_0))(2\gamma(z - z_0) - 1)\Big]. \tag{2.410}$$

Thus, an exact solution is derived as a product of matrizants over sufficiently small interval sections.

2.4.3 Matrizant Reduction

In order to construct the supermatrizant $\overline{\boldsymbol{M}}_{z_0}^{z}(\overline{\boldsymbol{P}})$ an alternative solution procedure is applied to the telegrapher Equation (2.330). The basic idea [45] is to shift the solution of the telegrapher equations as a first-order linear $2N \times 2N$ supermatrix differential equation to the solutions of a nonlinear $N \times N$ matrix Riccati equation which is supposed to be solved with more ease.

To begin with, it is noted again that for the determination of the supermatrizant it is sufficient to solve the homogeneous part of the telegrapher equations which is written in the sourceless form

$$\frac{\partial \overline{\boldsymbol{V}}(z)}{\partial z} = \overline{\boldsymbol{P}}\,\overline{\boldsymbol{V}}(z) \tag{2.411}$$

with $\overline{\boldsymbol{V}} = (\overline{v}_{ik})$ a $2N \times 2N$ integral supermatrix, the columns of which are composed from $2N$ linear independent (and so far unknown) solutions of Equation (2.330) without source terms. Alternatively, Equation (2.411) can be put in the form

$$D_z \overline{\boldsymbol{V}}(z) = \overline{\boldsymbol{P}}(z). \tag{2.412}$$

This equation is transformed by means of the relative supermatrix [46]

$$\overline{T} = \begin{pmatrix} \mathbf{1} & \boldsymbol{T}_{12} \\ \mathbf{0} & \mathbf{1} \end{pmatrix}, \text{ with } \overline{T}^{-1} = \begin{pmatrix} \mathbf{1} & -\boldsymbol{T}_{12} \\ \mathbf{0} & \mathbf{1} \end{pmatrix} \tag{2.413}$$

and a so far undetermined matrix $\boldsymbol{T}_{12}$, into an equivalent equation for $\overline{\boldsymbol{W}} := \overline{\boldsymbol{T}}^{-1}\overline{\boldsymbol{V}}$. This results in

$$D_z\overline{\boldsymbol{W}} = \overline{\boldsymbol{T}}^{-1}\overline{\boldsymbol{P}}\overline{\boldsymbol{T}} - \overline{\boldsymbol{T}}^{-1}\frac{\partial \overline{\boldsymbol{T}}}{\partial z} =: \overline{\boldsymbol{Q}}. \tag{2.414}$$

Computation of the block elements of $\overline{\boldsymbol{Q}}$ yields

$$\begin{aligned} \boldsymbol{Q}_{11} &= \boldsymbol{P}_{11} - \boldsymbol{T}_{12}\boldsymbol{P}_{21}, \\ \boldsymbol{Q}_{12} &= \boldsymbol{P}_{11}\boldsymbol{T}_{12} - \boldsymbol{T}_{12}\boldsymbol{P}_{21}\boldsymbol{T}_{12} - \boldsymbol{T}_{12}\boldsymbol{P}_{22} + \boldsymbol{P}_{12} - \frac{\partial \boldsymbol{T}_{12}}{\partial z}, \\ \boldsymbol{Q}_{21} &= \boldsymbol{P}_{21}, \\ \boldsymbol{Q}_{22} &= \boldsymbol{P}_{21}\boldsymbol{T}_{12} + \boldsymbol{P}_{22}. \end{aligned} \tag{2.415}$$

Note that in the present representation one has chosen $\boldsymbol{P}_{11} = \mathbf{0}$. As will become clear in a few lines (see Equation (2.423) and below), the vanishing of the block matrix $\boldsymbol{Q}_{12}$ considerably simplifies the construction of the solution $\overline{\boldsymbol{M}}^{z}_{z_0}(\overline{\boldsymbol{P}})$. The vanishing of $\boldsymbol{Q}_{12}$ is equivalent to the validity of the *matrix Riccati equation*

$$\frac{\partial \boldsymbol{T}_{12}}{\partial z} = \boldsymbol{P}_{11}\boldsymbol{T}_{12} - \boldsymbol{T}_{12}\boldsymbol{P}_{21}\boldsymbol{T}_{12} - \boldsymbol{T}_{12}\boldsymbol{P}_{22} + \boldsymbol{P}_{12}. \tag{2.416}$$

This equation is a nontrivial condition which must be solved for $\boldsymbol{T}_{12}$. In the following it is assumed that Equation (2.416) is solved for $\boldsymbol{T}_{12}$ and one can continue by decomposing $\overline{\boldsymbol{Q}}$ into two summands according to

$$\overline{\boldsymbol{Q}} = \overline{\boldsymbol{Q}}_1 + \overline{\boldsymbol{Q}}_2 := \begin{pmatrix} \boldsymbol{Q}_{11} & \mathbf{0} \\ \mathbf{0} & \boldsymbol{Q}_{22} \end{pmatrix} + \begin{pmatrix} \mathbf{0} & \mathbf{0} \\ \boldsymbol{Q}_{21} & \mathbf{0} \end{pmatrix}. \tag{2.417}$$

With this decomposition the sum rule is applied to $\overline{\boldsymbol{M}}^{z}_{z_0}(\overline{\boldsymbol{Q}})$, yielding

$$\overline{\boldsymbol{M}}^{z}_{z_0}(\overline{\boldsymbol{Q}}) = \overline{\boldsymbol{M}}^{z}_{z_0}(\overline{\boldsymbol{Q}}_1)\overline{\boldsymbol{M}}^{z}_{z_0}(\overline{\boldsymbol{\mathcal{S}}}) \tag{2.418}$$

with

$$\overline{\boldsymbol{\mathcal{S}}} = \left(\overline{\boldsymbol{M}}^{z}_{z_0}(\overline{\boldsymbol{Q}}_1)\right)^{-1}\overline{\boldsymbol{Q}}_2\overline{\boldsymbol{M}}^{z}_{z_0}(\overline{\boldsymbol{Q}}_1). \tag{2.419}$$

From the definition of the supermatrizant in Equation (2.332) one derives

$$\overline{\boldsymbol{M}}^{z}_{z_0}(\overline{\boldsymbol{Q}}_1) = \begin{pmatrix} \boldsymbol{M}^{z}_{z_0}(\boldsymbol{Q}_{11}) & \mathbf{0} \\ \mathbf{0} & \boldsymbol{M}^{z}_{z_0}(\boldsymbol{Q}_{22}) \end{pmatrix}. \tag{2.420}$$

The inverse of this expression is

$$\left(\overline{\boldsymbol{M}}_{z_0}^{z}(\overline{\boldsymbol{Q}}_1)\right)^{-1} = \begin{pmatrix} \left(\boldsymbol{M}_{z_0}^{z}(\boldsymbol{Q}_{11})\right)^{-1} & \boldsymbol{0} \\ \boldsymbol{0} & \left(\boldsymbol{M}_{z_0}^{z}(\boldsymbol{Q}_{22})\right)^{-1} \end{pmatrix}$$
$$= \begin{pmatrix} \boldsymbol{M}_{z}^{z_0}(\boldsymbol{Q}_{11}) & \boldsymbol{0} \\ \boldsymbol{0} & \boldsymbol{M}_{z}^{z_0}(\boldsymbol{Q}_{22}) \end{pmatrix} \tag{2.421}$$

and for $\overline{\boldsymbol{\mathcal{S}}}$ one finds

$$\overline{\boldsymbol{\mathcal{S}}} = \begin{pmatrix} \boldsymbol{0} & \boldsymbol{0} \\ \boldsymbol{M}_{z}^{z_0}(\boldsymbol{Q}_{22})\boldsymbol{Q}_{21}(z)\boldsymbol{M}_{z_0}^{z}(\boldsymbol{Q}_{11}) & \boldsymbol{0} \end{pmatrix}. \tag{2.422}$$

Due to the three trivial entries of $\overline{\boldsymbol{\mathcal{S}}}$ the matrizant $\overline{\boldsymbol{M}}_{z_0}^{z}(\overline{\boldsymbol{\mathcal{S}}})$ only consists of two terms,

$$\overline{\boldsymbol{M}}_{z_0}^{z}(\overline{\boldsymbol{\mathcal{S}}}) = \overline{\boldsymbol{1}} + \int_{z_0}^{z} \begin{pmatrix} \boldsymbol{0} & \boldsymbol{0} \\ \boldsymbol{M}_{\zeta}^{z_0}(\boldsymbol{Q}_{22})\boldsymbol{Q}_{21}(\zeta)\boldsymbol{M}_{z_0}^{\zeta}(\boldsymbol{Q}_{11}) & \boldsymbol{0} \end{pmatrix} \mathrm{d}\zeta$$
$$= \begin{pmatrix} \boldsymbol{1} & \boldsymbol{0} \\ \int_{z_0}^{z} \boldsymbol{M}_{\zeta}^{z_0}(\boldsymbol{Q}_{22})\boldsymbol{Q}_{21}(\zeta)\boldsymbol{M}_{z_0}^{\zeta}(\boldsymbol{Q}_{11})\mathrm{d}\zeta & \boldsymbol{1} \end{pmatrix} \tag{2.423}$$

The expressions above are collected to obtain the preliminary result

$$\overline{\boldsymbol{M}}_{z_0}^{z}(\overline{\boldsymbol{Q}}) = \begin{pmatrix} \boldsymbol{M}_{z_0}^{z}(\boldsymbol{Q}_{11}) & \boldsymbol{0} \\ \boldsymbol{M}_{z_0}^{z}(\boldsymbol{Q}_{22}) \int_{z_0}^{z} \boldsymbol{M}_{\zeta}^{z_0}(\boldsymbol{Q}_{22})\boldsymbol{Q}_{21}(\zeta)\boldsymbol{M}_{z_0}^{\zeta}(\boldsymbol{Q}_{11})\,\mathrm{d}\zeta & \boldsymbol{M}_{z_0}^{z}(\boldsymbol{Q}_{22}) \end{pmatrix} \tag{2.424}$$

which in a final step must be transformed back to $\overline{\boldsymbol{M}}_{z_0}^{z}(\overline{\boldsymbol{P}})$. In order to find this transformation it is observed that

$$\overline{\boldsymbol{V}}(z) = \overline{\boldsymbol{M}}_{z_0}^{z}(\overline{\boldsymbol{P}})\overline{\boldsymbol{V}}(z_0) = \overline{\boldsymbol{T}}(z)\overline{\boldsymbol{W}}(z)$$
$$= \overline{\boldsymbol{T}}(z)\overline{\boldsymbol{M}}_{z_0}^{z}(\overline{\boldsymbol{Q}})\overline{\boldsymbol{W}}(z_0)$$
$$= \overline{\boldsymbol{T}}(z)\overline{\boldsymbol{M}}_{z_0}^{z}(\overline{\boldsymbol{Q}})\overline{\boldsymbol{T}}^{-1}(z_0)\overline{\boldsymbol{V}}(z_0). \tag{2.425}$$

Therefore, one finally obtains

$$\overline{\boldsymbol{M}}_{z_0}^{z}(\overline{\boldsymbol{P}}) = \overline{\boldsymbol{T}}(z)\overline{\boldsymbol{M}}_{z_0}^{z}(\overline{\boldsymbol{Q}})\overline{\boldsymbol{T}}^{-1}(z_0), \tag{2.426}$$

and thus the computation of $\overline{\boldsymbol{M}}_{z_0}^{z}(\overline{\boldsymbol{P}})$ is reduced to that of $\overline{\boldsymbol{M}}_{z_0}^{z}(\overline{\boldsymbol{Q}})$ in the form of Equation (2.424).

For $N = 1$, that is a one-dimensional nonuniform transmission line, the matrices $\boldsymbol{Q}_{ii}(z)$ reduce to complex scalar functions $Q_{ii}(z)$ that naturally commute. In this case, the matrizant $M_{z_0}^{z}(Q_{ii}(z))$ can be written directly in exponential form [15] such that Equation (2.424) reduces to

$$\overline{\boldsymbol{M}}_{z_0}^{z}(\overline{\boldsymbol{Q}}) = \begin{pmatrix} f_1(z, z_0) & 0 \\ f_{21}(z, z_0) & f_2(z, z_0) \end{pmatrix} \tag{2.427}$$

with

$$f_1(z, z_0) := \exp\left(\int_{z_0}^{z} Q_{11}(z')\mathrm{d}z'\right), \tag{2.428}$$

$$f_2(z, z_0) := \exp\left(\int_{z_0}^{z} Q_{22}(z')\mathrm{d}z'\right), \tag{2.429}$$

$$f_{21}(z, z_0) := \exp\left(\int_{z_0}^{z} Q_{22}(z')\mathrm{d}z'\right) \times \int_{z_0}^{z} Q_{21}(\zeta)\exp\left(\int_{z_0}^{\zeta} (Q_{11}(z') - Q_{22}(z'))\mathrm{d}z'\right) \mathrm{d}\zeta. \tag{2.430}$$

In a last step, the transformation of Equation (2.426) is performed and one obtains

$$\overline{\boldsymbol{M}}_{z_0}^{z}(\overline{\boldsymbol{P}}) = \begin{pmatrix} g_{11}(z, z_0) & g_{12}(z, z_0) \\ g_{21}(z, z_0) & g_{22}(z, z_0) \end{pmatrix} \tag{2.431}$$

with

$$g_{11}(z, z_0)) := f_1(z, z_0) + T_{12}(z)f_{21}(z, z_0), \tag{2.432}$$

$$g_{12}(z, z_0)) := -f_1(z, z_0)T_{12}(z_0) - T_{12}(z)f_{21}(z, z_0)T_{12}(z_0) + T_{12}(z)f_2(z, z_0), \tag{2.433}$$

$$g_{21}(z, z_0)) := f_{21}(z, z_0), \tag{2.434}$$

$$g_{11}(z, z_0)) := -f_{21}(z, z_0)T_{12}(z_0) + f_2(z, z_0). \tag{2.435}$$

This solution is general and applies to *all* one-dimensional NTLs. It also provides the missing piece in the solution procedures discussed in Sections 2.4.2.2 and 2.4.2.3 where, due to diagonalization, the N-dimensional NMTL problem could be reduced to N decoupled one-dimensional NTL problems. Note that at any rate the nonlinear Riccati equation still needs to be solved and this,in general, has to be carried out numerically.

References

[1] King, R.W.P. '*Transmission Line Theory*', McGraw-Hill, New York, USA, 1955.

[2] Paul, C.R. *Analysis of Multiconductor Transmission Lines*, John Wiley & Sons, Inc., New York, USA, 1994.

[3] Unger, H.G. *Elektromagnetische Wellen auf Leitungen*, 4th edn, Hüthig Buch Verlag, Heidelberg, Germany, 1996.

[4] Tesche, F.M., Ianoz, M.V. and Karlsson, T. *EMC Analysis Methods and Computational Methods*, John Wiley & Sons, Inc., New York, USA, 1997.

[5] Taylor, C.D., Satterwhite, R.S. and Harrison C.H. 'The Response of a Terminated Two-Wire Transmission Line Excited by a Nonuniform Electromagnetic Field', *IEEE Trans. Antennas Propag.*, **13**, 1965, 987–989.

[6] Agrawal, A.K., Price, H.J. and Gurbaxani, S.H. 'Transient response of multiconductor transmission-lines excited by a nonuniform electromagnetic field', *IEEE Trans. EMC*, **22**, 1980, 119–129.

[7] Rachidi, F. 'Formulation of the Field-to-Transmission Line Coupling Equations in Terms of Magnetic Excitation Field', *IEEE Trans. EMC*, **35**, 1993, 404–407.

[8] Haase, H. 'Full-Wave Filed Interactions of Nonuniform Transmission Lines', *Res Electricae Magdeburgenses* (eds J. Nitsch and Z. Styczynski) Magdeburg, Germany, 2005.

[9] Steinmetz, T. 'Ungleichförmige und zufällig geführte Mehrfachleitungen in komplexen, technischen Systemen', *Res Electricae Magdeburgenses* (eds J. Nitsch and Z. Styczynski) Magdeburg, Germany, 2006.

[10] O'Neill, B. *Elementary Differential Geometry*, Academic Press, New York, USA, 1966.

[11] Mie, G. 'Elektrische Wellen an zwei parallelen Drähten (Electrical waves along two parallel wires)', *Annalen der Physik*, **2**, 1900, 201.

[12] Sommerfeld, A. *Electrodynamics*, Academic Press, New York, USA, 1964.

[13] Kuznetsov, P.I. and Stratonovich, R.L. 'The Propagation of Electromagnetic Waves in Multiconductor Transmission Lines', (translated from the Russian original). Pergamon Press, Oxford, UK, 1964.

[14] Nitsch, J. and Tkachenko, S. 'Newest Developments in Transmission-Line Theory and Applications', *Interaction Note 592*, 2004.

[15] Gantmacher, F.R. *The Theory of Matrices*, vol. 2, Chelsea Publishing Company, New York, USA, 1984.

[16] Dollard, J.D. and Friedman, C.N. *'Product Integration with Application to Differential Equations'*, Addison-Wesley, Reading, Massachusetts, USA, 1979.

[17] Baum, C.E., Nitsch, J. and Sturm, R. 'Analytical solution for uniform and nonuniform multiconductor transmission lines with sources', in *Review of Radio Science*, (ed. W.R., Stone) Oxford University Press, Oxford, UK, 1996, pp. 433–464.

[18] Nitsch, J. and Gronwald, F. 'Analytical Solutions in Multiconductor Transmission Line Theory', *IEEE Trans. EMC*. **41**, 1999, 469–479.

[19] Zurmühl, R. *Praktische Mathematik für Ingenieure und Physiker*, Springer, Berlin, Germany, 1984.

[20] Nitsch, J.B. and Tkachenko, S.V. 'Global and Modal Parameters in the Generalized Transmission Line Theory and their Physical Meaning', *Radio Sci. Bull.*, **312**, 2005, 21–31.

[21] Jackson, J.D. *Classcial Electrodynamics*, 3rd edn, John Wiley & Sons, Inc., New York, USA, 1998.

[22] Balanis, C.A. *Advanced Engineering Electromagnetics*, John Wiley & Sons, Inc., New York, USA, 1989.

[23] Lindell, I.V., *Methods for Electromagnetic Field Analysis*, IEEE Press, New Jersey, USA, 1995.

[24] Harrington, R.F. 'Matrix methods for field problems', *Proc. IEEE*, **55**, 1967, 136–149.

[25] Gantmacher, F.R. *Matrizentheorie*, VEB Deutscher Verlag der Wissenschaften, Berlin, Germany, 1986.

[26] Volterra, V. and Hostinsky, B. *Operations infinitesimales lineaires*, Gauthier villars, Paris, France, 1938.

[27] Baum, C.E. 'Approximation of Nonuniform Multiconductor Transmission Lines by Analytically Solvable Sections', *Interaction Note 490*, 1992.

[28] Baum, C.E., Nitsch, J.B. and Sturm, R.J. 'Analytical Solution for Uniform and Nonuniform Multiconductor Transmission Lines with Sources', in *Review of Radio Sience* 1993–1996, chapter 18, Oxford University Press, Oxford, UK, 1996.

[29] Strehmel, K. and Weiner, R. *Numerik gewöhnlicher Differentialgleichungen*, Springer, Berlin, Germany, 1994.

[30] Shampine, L.F. *Numerical Solution of Ordinary Differential Equations*, Champman & Hall, New York, USA, 1993.

[31] Duffing, G. 'Zur numerischen Integration gewöhnlicher Differentialgleichungen 1. und 2. Ordnung', *Forsch.-Arb. Ing.-Wes.*, 1920, 29–50.

[32] Richardson, L.F. 'The Deferred Approach to the Limit, I - Single Lattice', *Trans. R. Soc.*, London, **226**, 1927, 299–349.

[33] Baum, C.E. 'Norms and Eigenvector Norms', *Mathematics Notes 63*, 1979.

[34] Baum, C.E. 'Energy Norms and 2-Norms', *Mathematics Notes 89*, 1988.

[35] Lüttkepohl, H. *Handbook of Matrices*, John Wiley & Sons, Inc., New York, USA, 1996.

[36] Fehlberg, E. 'New high-order Runge–Kutta formulas with step size control for systems of first- and second order differential equations', *ZAMM*, **44**, 1964, T17–T29.

[37] Fehlberg, E. 'Klassische Runge–Kutta-Formeln vierter und niedrigerer Ordnung mit Schrittweiten-Kontrolle und ihre Anwendung auf Wärmeleitprobleme', *Computing*, **6**, 1970, 61–71.

[38] Frey, E. 'Explizite Hermite Runge-Kutta-Formeln mit automatischer Schrittweitenkontrolle zur Lösung von Anfangswertaufgaben bei gewöhnlichen Differentialgleichungen', PhD Thesis, University of Gie βen, Germany, 1980.

[39] Abramowitz, M. and Stegun, I.A. *Handbook of Mathematical Functions with Formulas, Graphs and Mathematical Tables*, 9th printing, Dover, 1972.

[40] Nitsch, J.B. and Tkachenko, S.V. 'Complex-valued transmission-line parameters and their relation to the radiation resistance', *IEEE Trans. EMC*, **46**, 2004, 477–487.

[41] Baum, C.E., Nitsch, J. and Sturm, R. 'Nonuniform Multiconductor Transmission Lines', *Interaction Note 516*, 1996.

[42] Baum, C.E. 'Approximation of Non-Uniform Multiconductor Transmission Lines by Analytically Solvable Sections', *Interaction Note 490*, 1992.

[43] Baum, C.E. 'Symmetric Renormalization of the Nonuniform Multiconductor-Transmission-Line equations with a Single Modal Speed for Analytically Solvable Sections', *Interaction Note 537*, 1998.

[44] Nitsch, J., Baum, C.E. and Sturm, R. 'Analytical Treatment of Circulant Nonuniform Multiconductor Transmission Lines', *IEEE Trans. EMC,* **34** 1992, 28–38.

[45] Nitsch, J.: 'Exact analytical solution for nonuniform multiconductor transmission lines with the aid of the solution of a corresponding matrix Riccati equation', *Electrical Engineering*, **81** 1998 117–120.

[46] Rasch, G. 'Zur Theorie und Anwendung des Produktintegrals', *Journal für Mathematik*, **171/2** 1933, 65–119.

3

Complex Systems and Electromagnetic Topology

The starting point of this chapter[1] is given by the results of Baum, Liu and Tesche [1, 2][2], Lee [3], Karlsson [4] and Parmantier [5], but also by the results on the theory of uniform multiconductor transmission lines [6–13]. It is an outline of current research on the modeling of complex systems involving transmission-line networks and serves as a basis for further developments. Within this outline, which will be presented in the following section, the term 'transmission lines' will be restricted to uniform multiconductor transmission lines.

3.1 The Concept of Electromagnetic Topology

During the investigation of the EMC of complex systems one quickly reaches the limits of analytical considerations, numerical computations as well as experimental investigations. The reasons for this circumstance are rooted in the complexity itself, which makes analytic considerations only possible if considerable simplifications are assumed in the modeling process. The number of signals that need to be taken into account is hard to fix. The electrical size of the systems to be analyzed often makes a numerical modeling very difficult or even impossible. Normally, the number of relevant signals and quantities is very high such that experimental measurements quickly become out of hand. Since the time for measurements is limited the number of results obtained, like the number of measured transfer functions, is also limited, leading to limited insight into a system and preventing a complete EMC analysis. This conceptual problem cannot be resolved by means of *scaled* experiments. Even though scaled experiments allow a simplified handling the large number of quantities that need to be taken into account remains. Moreover, results a priori are only relevant for a specific setup. Only in some cases do they lead to insights in the case of modified systems. Also the financial expense that is required to perform experimental measurements constitutes an important factor.

[1] With the permission of T. Steinmetz portions of his dissertation are adapted to this chapter.

[2] In [1] the governing equation of a network of multiconductor transmission lines is named BLT the equation, thus resembling the initial letters of the authors' last names.

Radiating Nonuniform Transmission-Line Systems and the Partial Element Equivalent Circuit Method Jürgen Nitsch, Frank Gronwald, and Günter Wollenberg

Another problem is that in order to perform measurements the actual system or prototype must be available. Nowadays, this is not always guaranteed within the processes of design and development since many of these processes are carried out virtually with no need for an actual prototype. Nevertheless delayed consideration of the EMC often leads to an increase in cost due to redesign and hardening.

This motivates the search for new methods to analyze and design complex systems. These new methods need to provide tools to determine the significant variables of a system. The main reason for the limited applicability of conventional numerical and experimental methods is the fact that with these methods one often attempts to solve the whole problem with a single method. In contrast to this, the concept of electromagnetic topology (EMT) is based on a division of the complete system into coupled subsystems. This division, in turn, is based on hierarchical and/or electromagnetic assumptions, such as the shielding by means of an enclosure. The interactions between the subsystems are graphically represented by means of an interaction graph and a topological diagram. This purely formal representation of the complete problem is eventually transformed into a network formulation. The characterization of each single subsystem can be done by a particular method which is most suitable for this subsystem. The coupling between the single subsystems is realized by the network formulation. An advantage of EMT is the possibility of combining numerical simulations and experimental measurements. Thus, results of different methods can be combined, allowing for the most efficient method to tackle each subsystem.

In summary, the aim of EMT is maximum modularization. This yields three new possibilities for the design process and for system analysis:

- *Hybridization* – application of different solution methods to single subsystems, including measurements,
- *Data Bases* – establishing topological data bases for commonly used components,
- *Parametrization* – analysis of modifications of a system due to the change of a parameter within a subsystem.

The rapid development and improvement of computer hardware and three-dimensional simulation packages for the solution of Maxwell equations nowadays allows the so-called *exterior problem*, that is the excitation of a complete system by external sources, to be conveniently solved. As a consequence, the surface currents and the scattered electromagnetic fields that are due to external sources can be caluclated with high accuracy. In contrast to this the solution of the *interior problem*, that is the propagation of electromagnetic interference fields into a system and the related effects, unfortunately requires a highly detailed description and discretization of the system's geometry. This poses a problem to all common three-dimensional simulation packages.

A treatment of the interior problem by means of EMT consists of the following three steps:

(1) decomposition of the complete system into subsystems,
(2) characterization of each subsystem by an appropriate method,
(3) solution of the network problem which is based on the coupled subsystems.

The division of a complete system into subsystems is usually a quite intuitive process since, a priori, it contains a certain degree of arbitrariness. This is in contrast to the classical three-dimensional methods of field computation. While these normally yield acceptable results as long as thought has been given to common discretization rules and other rules of thumb, it is is possible that an incorrect decomposition of the entire system into subsystems yields no meaningful results at all.

It is often reasonable to have physical surfaces, as provided by metallic walls, screens, cables, for example as boundaries of a subsystem. Due to physical relations, such as given by the notion of shielding effectiveness, the topological network subsequently established often simplifies since certain branches might no longer be relevant or parts of networks might even decouple. In this context it is common to distinguish between two types of topological surfaces:

(1) *Surfaces that act as ideal screens*: they strongly damp electromagnetic signals, thus truly separate electromagnetic environments, and are called *proper surfaces*. Examples of proper surfaces are provided by the outer shell of an aircraft, a cable shield or a screening box for electronic equipment. These surfaces allow a separate treatment of coupling and propagation phenomena on each side and serve to define the *shielding level*.
(2) *Surfaces that do not act as ideal screens*: they serve to divide subsystems according to functional or physical-geometric aspects. This implies that electromagnetic signals on both sides are coupled. These surfaces are called *elementary surfaces*. Their natural boundaries give some evidence for the decomposition of the complete system and the interaction between subsystems. An example of an elementary surface is a screen with a hole, such as a cable feedthrough. Here, the screen itself provides an ideal boundary for the decomposition into subsystems. The hole corresponds to a propagation path, that is to an interaction between subsystems.

Domains that are completely enclosed by topological surfaces are called topological volumes. In analogy to the topological surfaces the topological volumes are divided into *proper volumes* and *elementary volumes*. Here, a volume is termed a proper volume if its boundaries are exclusively given by proper surfaces. All other volumes are automatically defined as elementary volumes. The division into subvolumes is graphically represented in the topological diagram. Due to the *good shielding approximation* a proper volume only contains interactions that do not couple to its exterior.

On the basis of a topological diagram, which resembles representations of set theory, it is possible to establish interaction graphs. These describe the interaction between single topological volumes. It is possible that an interaction graph represents a specific mechanism, such as the coupling and propagation of electromagnetic disturbances along transmission lines. Therefore, a topological decomposition of a complex system requires many interaction graphs (see also [14]).

The following example clarifies the procedure of the topological decomposition. Figure 3.1 displays a passenger aircraft as an example of a complex system. Four subsystems of the aircraft are considered:

- a communication module with attached antenna, close to the cockpit (R),
- a central control unit (C) within the aircraft body,
- an actuator (A) within the right wing, and
- a sensor (B) within the left wing.

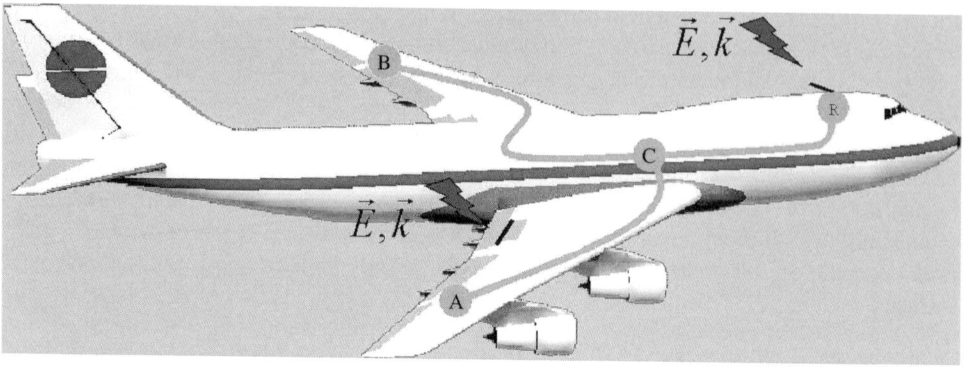

Figure 3.1 Passenger aircraft and parts of its wiring as an example of a complex system.

Cables connect the subsystems to each other, as indicated in Figure 3.1. To keep things relatively simple, only two coupling paths will be considered in the following discussion: one path represents the coupling via an aperture at the pitch elevator of the right wing; the second path represents the coupling of the wanted signal (and the unwanted interference signal) via the receiving antenna to the communication module close to the cockpit. Furthermore, the interaction via cables and transmission lines is taken into account.

In the application of the topological concept, the first step is to divide the complete volume 'Aircraft' into subvolumes. Ideally, metallic boundaries provide major guidelines for this step. The topological diagram is drawn in Figure 3.2. In this diagram, the aircraft body $S_{0.1;1.1}$ separates the interior $V_{1.1}$ of the aircraft from its surrounding volume $V_{0.1}$, where the first index k of the volumes $V_{k,i}$ indicated the screening level, the second index i is used for a consecutive numbering of volumes of a certain screening level. The actual geometric form of the boundaries is not modeled, rather it is symbolically represented as an ellipse. The decomposition further continues towards the interior. The surface of the enclosure of the central control unit $S_{1.1;2.1}$ separates the aircraft interior $V_{1.1}$ from the device volume $V_{2.1}$ of

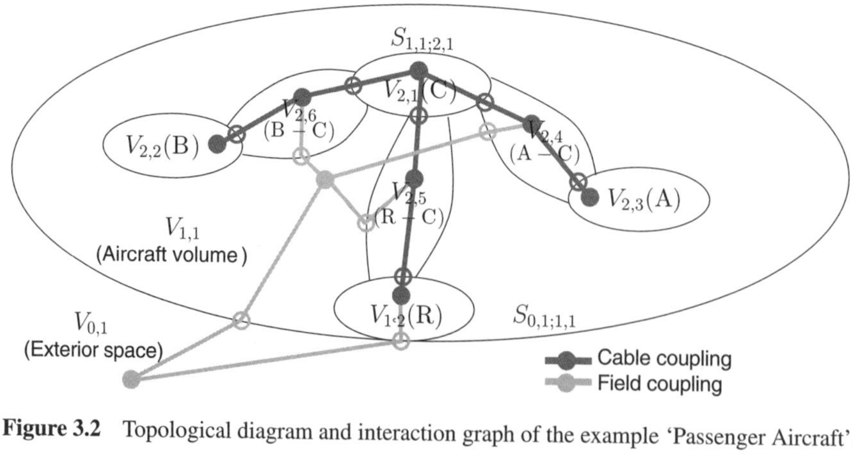

Figure 3.2 Topological diagram and interaction graph of the example 'Passenger Aircraft'.

the control unit. This process continues until the complete system is divided into sufficiently small subvolumes.

In the second step the interactions need to be described by means of interaction graphs. It is possible to use different graphs for different interaction mechanisms. Frequency-dependent graphs, where edges are added or eliminated above or below a specific limiting frequency, are also conceivable.

Figure 3.2 contains two interaction graphs where one graph represents the conductive cable coupling and the other graph represents the electromagnetic field coupling. For example, an electromagnetic field coupling, given by a branch of the corresponding graph, leads from the exterior volume $V_{0.1}$ to the interior volume $V_{1.1}$. This branch corresponds to the coupling through the aperture. A further electromagnetic field coupling is given by the coupling to the communication antenna above the cockpit. Within the interaction graph this is represented as a lumped source in volume $V_{1.2}$. Looking at the interaction graph of the cable coupling it is recognized that the control unit C (volume $V_{2.1}$) interacts via the cable volume $V_{2.4}$ with the volume $V_{2.3}$ of actuator A inside the right wing.

In this way, all interactions can be collected and accounted for by means of an interaction graph. It is important to know as much as possible about the system in order to either apply simplifications to keep the interaction graph as simple as possible or to avoid ignoring significant interactions. If in doubt it is, obviously, better to include an interaction rather than neglecting it.

3.2 Topological Networks and BLT Equations

For a quantitative description of interaction graphs it is necessary to transform these into topological networks. These consist of the following two building blocks:

- *tubes* – describe the propagation of electromagnetic quantities along a certain path,
- *junctions* – describe the scattering at previously defined nodes.

The topological network represents the electromagnetic properties of the complete system as coupling between subvolumes. To this end, volumes are modeled as *junctions* and coupled via *tubes*. The excitation of the network is provided on the basis of equivalent sources and happens either along the tubes (BLT1) or is concentrated at the ports of junctions (BLT2).

If pure transmission line networks are considered the corresponding network topology emerges in a natural way: cables and transmission lines are represented by tubes and attached devices, ports or nodes become junctions. It then turns out to be advantageous that attached devices are often enclosed by metallic cavities. Then, according to the *good shielding approximation*, a decoupling of junction volume and exterior volume can be assumed and electromagnetic disturbances or signals only significantly interact via cables and transmission-lines with junctions. One possibility to describe propagation along tubes (cables and transmission lines) is given by classical multiconductor transmission-line theory. Then, external field coupling is incorporated into the interaction graph by means of the usual coupling models of classical transmission-line theory. However, other methods can also be used to model propagation along tubes. The scattering properties of attached devices can follow from analytical considerations or numerical simulations. For example, the scattering matrix of an ideal junction, like the direct connection of two transmission-line ports, immediately follows from the characteristic impedance of the attached transmission lines [15]. To determine the scattering matrix of a junction within a metallic enclosure numerically it is obvious to take advanatge of volume-discretizing numerical methods, such as the FDTD method or the TLM method. These methods easily and effectively incorporate the boundary conditions of closed domains. However, it is also conceivable to use other numerical methods or to determine scattering parameters experimentally.

The topological concept is not restricted to cables or transmission lines. A tube can also model propagation which is not based on transmission-line theory. One may think of the propagation and transmission of signals between two antennas. The electromagnetic field between two antennas is usually of a rather complex form. Nevertheless, for the integration of the propagation between two antennas into a topological network it is sufficient to consider the effect on the voltages and currents at the antenna input ports.

3.2.1 Wave Quantities

The concept of electromagnetic topology is not restricted to a certain choice of physical quantities. For the description of waves, various quantities are conceivable, such as TE- or TM-waves in waveguides, TEM-waves or voltages and currents along transmission lines. What is important is the possibility of describing the effect of quantities that are defined along the tubes on quantities that are defined at the junctions.

In the following, wave quantities will be derived, that describe the propagation along cables and transmission lines, from currents and voltages with

$$\mathbf{w}_{+}(\xi) = \mathbf{v}(\xi) + \mathbf{Z}_{\mathrm{c}}\,\mathbf{i}(\xi) \tag{3.1}$$

$$\mathbf{w}_{-}(\xi) = \mathbf{v}(\xi) - \mathbf{Z}_{\mathrm{c}}\,\mathbf{i}(\xi). \tag{3.2}$$

In the context of electromagnetic topology, this has been introduced for transmission line networks in [1]. Depending on the value of the normalization impedance $\mathbf{Z}_{\mathrm{c}}$ one obtains different values for the parameters of the wave representation. For example, it is common to distinguish between $S50$-parameters and topological S-parameters. For the $S50$-parameters a diagonal matrix with 50 Ω entries is chosen for normalization, that is the S-parameters are determined for ports that are terminated by 50 Ω resistances. For the topological S-parameters the ports are terminated by the wave impedance of the attached tubes. Clearly, in the latter case the S-parameters depend on the network topology which involve the information on specific tubes.

In the literature [7] one often finds the notion of forward and backward running voltage waves

$$\mathbf{v}_{+}(\xi) = \frac{1}{2}\left[\mathbf{v}(\xi) + \mathbf{Z}_{\mathrm{c}}\,\mathbf{i}(\xi)\right] = \frac{\mathbf{w}_{+}}{2} \tag{3.3}$$

$$\mathbf{v}_{-}(\xi) = \frac{1}{2}\left[\mathbf{v}(\xi) - \mathbf{Z}_{\mathrm{c}}\,\mathbf{i}(\xi)\right] = \frac{\mathbf{w}_{-}}{2}\,. \tag{3.4}$$

The relations between the wave quantities are such that a shift to these variables leaves the values of scattering and propagation matrices invariant.

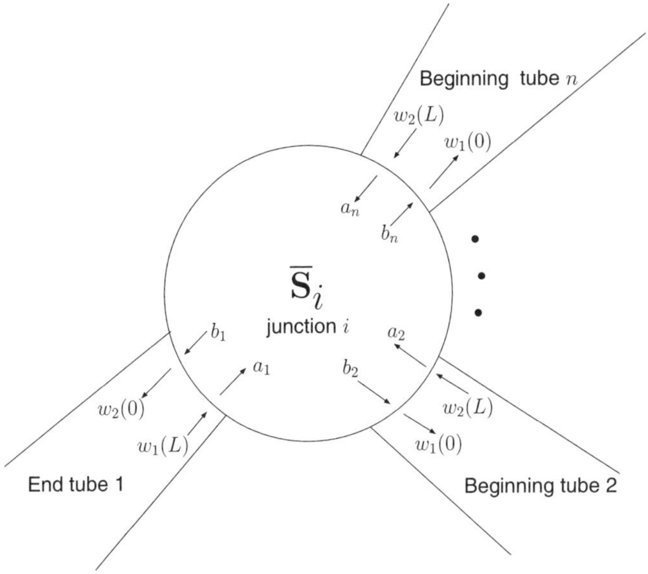

Figure 3.3 Forming blocks of the scattering matrix $\overline{S}_i$ – a junction with n tubes attached.

3.2.2 BLT 1 Equation

A *junction i* is described by the scattering equation

$$\begin{bmatrix} \mathbf{b}_1 \\ \vdots \\ \mathbf{b}_n \end{bmatrix}_i = \overline{\mathbf{S}}_i \begin{bmatrix} \mathbf{a}_1 \\ \vdots \\ \mathbf{a}_n \end{bmatrix}_i . \tag{3.5}$$

Here, blocks are formed according to the attached tubes, that is the vectors $\mathbf{a}_k$ and $\mathbf{b}_k$ comprise those waves that are related to the kth tube (compare Figure 3.3).

In the context of the BLT 1 equation one usually works with topological scattering parameters. The term 'topological' indicates, as mentioned before, that the waves are calculated under the assumption that the ports are terminated by the wave impedance of the attached tubes.

The propagation parameters of a tube $\overline{\mathbf{\Gamma}}_i$, for example, can be obtained by means of multiconductor transmission-line theory (see Section 3.3). A tube i is described by the equation

$$\begin{bmatrix} \mathbf{w}_1(L) \\ \mathbf{w}_2(L) \end{bmatrix}_i = \overline{\mathbf{\Gamma}}_i \begin{bmatrix} \mathbf{w}_1(0) \\ \mathbf{w}_2(0) \end{bmatrix}_i + \begin{bmatrix} \mathbf{w}_1^{(s)}(L) \\ \mathbf{w}_2^{(s)}(L) \end{bmatrix}_i \tag{3.6}$$

where the notation is evident from Figure 3.4. It should be noted that the propagation direction of the backwards running wave $\mathbf{w}_2$ is defined to be positive such that the starting point 0 of the wave vector is located at $z = z_L$ and the end point L is located at $z = z_0$.

In order to describe a complete network of tubes and junctions and to establish the corresponding network equations it turns out to be convenient to establish from the scattering Equations (3.5) of the single junctions and the propagation Equations (3.6) of the single tubes

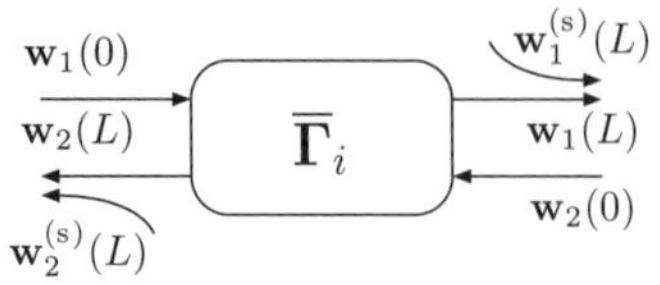

Figure 3.4 Definition of the propagation matrix $\overline{\Gamma}_i$ for the tube of a topological network.

a super matrix equation. This yields a block-diagonal scattering equation

$$\begin{bmatrix} \begin{bmatrix} \mathbf{b}_1 \\ \vdots \\ \mathbf{b}_{n_1} \end{bmatrix}_1 \\ \vdots \\ \begin{bmatrix} \mathbf{b}_1 \\ \vdots \\ \mathbf{b}_{n_k} \end{bmatrix}_k \end{bmatrix} = \begin{bmatrix} \overline{S}_1 & \ldots & \overline{0} \\ \vdots & \ddots & \vdots \\ \overline{0} & \ldots & \overline{S}_k \end{bmatrix} \begin{bmatrix} \begin{bmatrix} \mathbf{a}_1 \\ \vdots \\ \mathbf{a}_{n_1} \end{bmatrix}_1 \\ \vdots \\ \begin{bmatrix} \mathbf{a}_1 \\ \vdots \\ \mathbf{a}_{n_k} \end{bmatrix}_k \end{bmatrix} \tag{3.7}$$

$$\overline{\overline{\mathbf{b}}} = \overline{\overline{\mathbf{S}}}\,\overline{\overline{\mathbf{a}}} \tag{3.8}$$

and a propagation equation for the complete network

$$\begin{bmatrix} \begin{bmatrix} \mathbf{w}_1(L) \\ \mathbf{w}_2(L) \end{bmatrix}_1 \\ \vdots \\ \begin{bmatrix} \mathbf{w}_1(L) \\ \mathbf{w}_2(L) \end{bmatrix}_l \end{bmatrix} = \begin{bmatrix} \overline{\mathbf{\Gamma}}_1 & \ldots & \overline{\mathbf{0}} \\ \vdots & \ddots & \vdots \\ \overline{\mathbf{0}} & \ldots & \overline{\mathbf{\Gamma}}_l \end{bmatrix} \begin{bmatrix} \begin{bmatrix} \mathbf{w}_1(0) \\ \mathbf{w}_2(0) \end{bmatrix}_1 \\ \vdots \\ \begin{bmatrix} \mathbf{w}_1(0) \\ \mathbf{w}_2(0) \end{bmatrix}_l \end{bmatrix} + \begin{bmatrix} \begin{bmatrix} \mathbf{w}_1^{(s)}(L) \\ \mathbf{w}_2^{(s)}(L) \end{bmatrix}_1 \\ \vdots \\ \begin{bmatrix} \mathbf{w}_1^{(s)}(L) \\ \mathbf{w}_2^{(s)}(L) \end{bmatrix}_l \end{bmatrix} \tag{3.9}$$

$$\overline{\overline{\mathbf{w}}}(L) = \overline{\overline{\mathbf{\Gamma}}}\,\overline{\overline{\mathbf{w}}}(0) + \overline{\overline{\mathbf{w}}}^{(s)}(L)\,. \tag{3.10}$$

The dimension of the matrices in both equations is the same and given by $d = \sum_{i=1}^{k} n_{p_i}$ with k the number of junctions and n_{p_i} the number of ports of the ith junction.

To combine Equations (3.7) and (3.9) it is necessary to express the quantities $\overline{\overline{\mathbf{a}}}$ and $\overline{\overline{\mathbf{b}}}$ by $\overline{\overline{\mathbf{w}}}(0)$ and $\overline{\overline{\mathbf{w}}}(L)$. From Figure 3.3 it is evident that at each junction the outgoing wave vector of a port $\mathbf{b}_i$ can be associated to an incoming wave vector of an attached tube $\mathbf{w}_n(0)$. The same is true for $\mathbf{a}_i$ and $\mathbf{w}_n(L)$. Then, according to the topology of the network, the supervectors $\overline{\overline{\mathbf{a}}}$ and $\overline{\overline{\mathbf{b}}}$ can be expressed as

$$\overline{\overline{\mathbf{a}}} = \overline{\overline{\mathbf{JT}}}_a\,\overline{\overline{\mathbf{w}}}(L) \text{ and } \overline{\overline{\mathbf{b}}} = \overline{\overline{\mathbf{JT}}}_b\,\overline{\overline{\mathbf{w}}}(0)\,. \tag{3.11}$$

The supermatrices $\overline{\overline{\mathbf{JT}}}_a$ and $\overline{\overline{\mathbf{JT}}}_b$ are permutation matrices which incorporate the information on which junction is connected to which tube. The entries of these super matrices are either zero or one, the sum of each column or row is one. The inverse of each matrix is equal to its transpose. This yields the super scattering equation

$$\overline{\overline{\mathbf{w}}}(L) = \overline{\overline{\mathbf{JT}}}_a^{\mathsf{T}} \, \overline{\overline{\mathbf{S}}}^{-1} \, \overline{\overline{\mathbf{JT}}}_b \, \overline{\overline{\mathbf{w}}}(0) \, , \tag{3.12}$$

which can be inserted into the super propagation equation. As a result, the BLT 1 equation is obtained,

$$\left(\overline{\overline{\mathbf{E}}} - \overline{\overline{\mathbf{JT}}}_b^{\mathsf{T}} \, \overline{\overline{\mathbf{S}}} \, \overline{\overline{\mathbf{JT}}}_a \, \overline{\overline{\mathbf{\Gamma}}}\right) \overline{\overline{\mathbf{w}}}(0) = \overline{\overline{\mathbf{JT}}}_b^{\mathsf{T}} \, \overline{\overline{\mathbf{S}}} \, \overline{\overline{\mathbf{JT}}}_a \, \overline{\overline{\mathbf{w}}}^{(s)}(L) \, . \tag{3.13}$$

The expression $\overline{\overline{\mathbf{JT}}}_b^{\mathsf{T}} \, \overline{\overline{\mathbf{S}}} \, \overline{\overline{\mathbf{JT}}}_a$ is a reordered super scattering matrix. The number of nonzero elements is unchanged, just their position has been changed due to the reordering of columns and rows.

Solving the BLT 1 equation for $\overline{\overline{\mathbf{w}}}(0)$ and inserting the result in Equation (3.12) yields all propagating waves of the network.

3.2.3 BLT 2 Equation

A second possibility to construct a topological network is to use only junctions that are directly connected to each other by their ports. This formulation of a network does not incorporate tubes. Therefore, it is necessary to add all types of excitations, such as concentrated or distributed sources, as additional terms to the outgoing waves. The scattering parameters of two connected ports must refer to the same characteristic wave impedance. For simplicity, and in contrast to the BLT 1 equation, it is common to use scattering parameters that refer to a unique characteristic wave impedance at all ports. An example is given by the S50-parameters which have a 50 Ω coax-system as reference ($\overline{\mathbf{Z}}_0 = 50\Omega \, \overline{\mathbf{E}}$).

The extended scattering equation of a junction i reads

$$\begin{bmatrix} \mathbf{b}_1 \\ \vdots \\ \mathbf{b}_n \end{bmatrix}_i = \overline{\mathbf{S}}_i \begin{bmatrix} \mathbf{a}_1 \\ \vdots \\ \mathbf{a}_n \end{bmatrix}_i + \begin{bmatrix} \mathbf{b}_1^{(s)} \\ \vdots \\ \mathbf{b}_n^{(s)} \end{bmatrix}_i \, . \tag{3.14}$$

As displayed in Figure 3.5, ports of two junctions are directly connected to each other. Branchings are not allowed as connections, these are represented by additional junctions. The structure of the network is described by a connecting super matrix $\overline{\overline{\mathbf{C}}}$. It indicates which outgoing wave b_i of one junction corresponds to which incoming wave a_j of another junction

$$\overline{\overline{\mathbf{a}}} = \overline{\overline{\mathbf{C}}} \, \overline{\overline{\mathbf{b}}} \, . \tag{3.15}$$

This allows outgoing waves of the network to substituted by incoming waves. The connecting supermatrix is a permutation matrix which, in this case, is symmetric.

Combining the scattering Equation (3.14) of all junctions to a supermatrix equation,

$$\overline{\overline{\mathbf{b}}} = \overline{\overline{\mathbf{S}}} \, \overline{\overline{\mathbf{a}}} + \overline{\overline{\mathbf{b}}}^{(s)} \, , \tag{3.16}$$

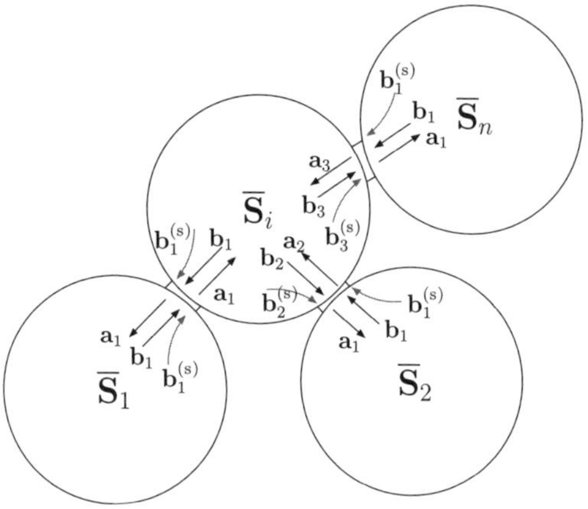

Figure 3.5 A topological network of junctions; wave vectors are not labelled by an additional index to indicate that they relate to a specific junction and all vectors within a circle are related to a specific junction and differ from vectors with identical notation of other junctions.

where $\overline{\overline{\mathbf{S}}}$ is block diagonal, allows the BLT 2 equation to be obtained by means of Equation (3.15)

$$\left(\overline{\overline{\mathbf{E}}} - \overline{\overline{\mathbf{S}}}\,\overline{\overline{\mathbf{C}}}\right)\overline{\overline{\mathbf{b}}} = \overline{\overline{\mathbf{b}}}^{(s)} . \tag{3.17}$$

It should be mentioned that the matrix product $\overline{\overline{\mathbf{S}}}\,\overline{\overline{\mathbf{C}}}$, in general, will no longer be block diagonal. It is a reordered scattering super matrix which, however, follows an order scheme which is different from that of the BLT 1 equation.

Even though the representation of a topological network by means of BLT 2 equations does not contain any tubes it is possible to model a tube as an equivalent junction within the network. To this end, the tube is transformed into an equivalent junction, as will be described in Section 3.3.2, and merged into the network according to Figure 3.6. If within a network which reflects the BLT 1 equations all tubes are replaced by equivalent junctions one obtains a BLT 1 network of double dimension. The advantage of the BLT 2 equation is its rather straightforward implementation. Multiplication by the permutation matrix $\overline{\overline{\mathbf{C}}}$ from the right corresponds to a multiple interchange of columns. This interchange, as well as the addition of 1 along the diagonal and a sign change, is easy to implement. Apart from the inversion of the system matrix there are no more matrix operations necessary.

Another relation between the representation according to the BLT 1 and BLT 2 equations appears during the transition from general tubes to zero length tubes. In this case, the propagation supermatrix $\overline{\overline{\mathbf{\Gamma}}}$ becomes unity and all junctions are coupled according to the BLT 2 representation. Then the BLT 1 equation becomes

$$\left(\overline{\overline{\mathbf{E}}} - \overline{\overline{\mathbf{JT}}}_b^{\mathsf{T}}\,\overline{\overline{\mathbf{S}}}\,\overline{\overline{\mathbf{JT}}}_a\right)\overline{\overline{\mathbf{w}}}(0) = \overline{\overline{\mathbf{JT}}}_b^{\mathsf{T}}\,\overline{\overline{\mathbf{S}}}\,\overline{\overline{\mathbf{JT}}}_a\,\overline{\overline{\mathbf{w}}}^{(s)}(L) . \tag{3.18}$$

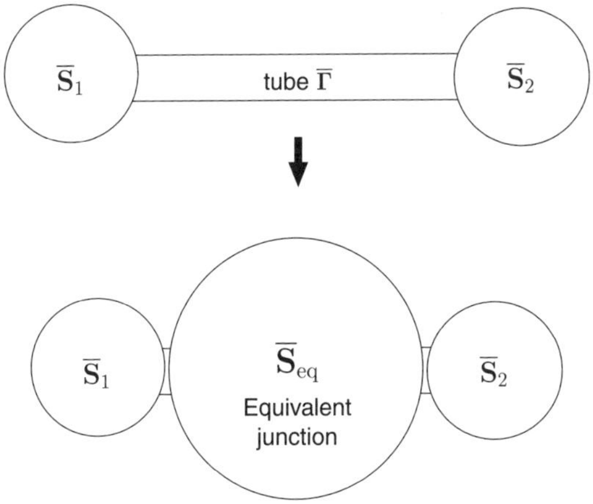

Figure 3.6 Transformation of a tube into an equivalent junction.

Together with Equation (3.11) this yields

$$\left(\overline{\overline{\mathbf{E}}} - \overline{\overline{\mathbf{S}}}\,\overline{\overline{\mathbf{JT}}}_a \overline{\overline{\mathbf{JT}}}_b^{\mathsf{T}}\right) \overline{\overline{\mathbf{b}}} = \overline{\overline{\mathbf{b}}}^{(s)}, \tag{3.19}$$

and it further follows that

$$\overline{\overline{\mathbf{JT}}}_a \overline{\overline{\mathbf{JT}}}_b^{\mathsf{T}} = \overline{\overline{\mathbf{C}}}\,. \tag{3.20}$$

This relates the ordering schemes of the BLT 1 and BLT 2 representations, where the supervectors of the BLT 1 equation are ordered according to tubes and the supervectors of the BLT 2 equation are ordered according to junctions.

3.2.4 Admittance Representation

The two representations of the topological network that have been discussed so far are based on wave quantities. In this section a representation is added which is based on voltages and currents. It takes advantage of the characterization of junctions and tubes by admittance matrices, the so-called *Y* parameters. To this end tube equations are written as

$$\underbrace{\begin{bmatrix} \mathbf{i}_1 \\ \mathbf{i}_2 \end{bmatrix}}_{\bar{\mathbf{i}}_t} = \overline{\mathbf{Y}}_t \underbrace{\begin{bmatrix} \mathbf{v}_1 \\ \mathbf{v}_2 \end{bmatrix}}_{\bar{\mathbf{v}}_t} + \underbrace{\begin{bmatrix} \mathbf{i}_1^{(s)} \\ \mathbf{i}_2^{(s)} \end{bmatrix}}_{\bar{\mathbf{i}}_t^{(s)}}, \tag{3.21}$$

with symmetric input (index 1) and output (index 2) ports (see Figure 3.7).

Ports at junctions can be grouped arbitrarily (see Figure 3.8), such that a junction is described by the equation

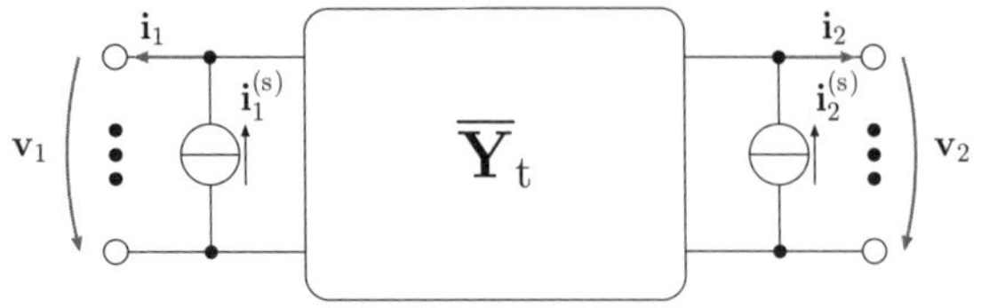

Figure 3.7 Representation of tubes by means of admittance parameters – input and output ports are of the same dimension.

$$\underbrace{\begin{bmatrix} \mathbf{i}_1 \\ \vdots \\ \mathbf{i}_2 \end{bmatrix}}_{\mathbf{i}_j} = \overline{\mathbf{Y}}_j \underbrace{\begin{bmatrix} \mathbf{v}_1 \\ \vdots \\ \mathbf{v}_2 \end{bmatrix}}_{\overline{\mathbf{v}}_j} + \underbrace{\begin{bmatrix} \mathbf{i}_1^{(s)} \\ \vdots \\ \mathbf{i}_2^{(s)} \end{bmatrix}}_{\overline{\mathbf{j}}_j^{(s)}} . \tag{3.22}$$

For an ideal junction[3] an equation of the type shown in Equation (3.22) cannot be established since in this case no admittance matrix exists. However, it is possible to obtain from Kirchhoff's laws a relation between currents and voltages,

$$\begin{bmatrix} \mathbf{K} \\ \mathbf{0} \end{bmatrix} \mathbf{i}_{ij} = \begin{bmatrix} \mathbf{0} \\ \mathbf{M} \end{bmatrix} \mathbf{v}_{ij} . \tag{3.23}$$

Here, $\mathbf{K}$ contains $k-1$ equations corresponding to Kirchhoff's point rule at k nodes of the ideal junction and $\mathbf{M}$ contains $n-k+1$ equations corresponding to Kirchhoff's loop rule. It is seen that the matrices $\begin{bmatrix} \mathbf{K} \\ \mathbf{0} \end{bmatrix}$ and $\begin{bmatrix} \mathbf{0} \\ \mathbf{M} \end{bmatrix}$ are singular. Therefore, Equation (3.23) cannot be solved for currents or voltages.

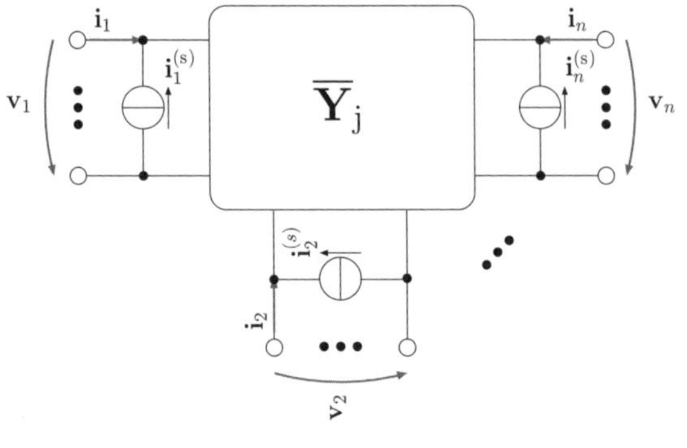

Figure 3.8 Representation of junctions by means of admittance parameters; by definition, the direction of current is reversed if compared to the representation of tubes.

[3] An ideal junction contains no network elements but only nodes. It models an ideal electric connection between ports.

Again, the network equations can be expressed by supermatrix equations. For the n_t tubes one obtains

$$\begin{bmatrix} [\bar{\mathbf{i}}_\mathrm{t}]_1 \\ \vdots \\ [\bar{\mathbf{i}}_\mathrm{t}]_{n_\mathrm{t}} \end{bmatrix} = \begin{bmatrix} [\bar{\mathbf{Y}}_\mathrm{t}]_1 & \cdots & \bar{\mathbf{0}} \\ \vdots & \ddots & \vdots \\ \bar{\mathbf{0}} & \cdots & [\bar{\mathbf{Y}}_\mathrm{t}]_{n_\mathrm{t}} \end{bmatrix} \begin{bmatrix} [\bar{\mathbf{v}}_\mathrm{t}]_1 \\ \vdots \\ [\bar{\mathbf{v}}_\mathrm{t}]_{n_\mathrm{t}} \end{bmatrix} + \begin{bmatrix} [\bar{\mathbf{i}}_\mathrm{t}^{(\mathrm{s})}]_1 \\ \vdots \\ [\bar{\mathbf{i}}_\mathrm{t}^{(\mathrm{s})}]_{n_\mathrm{t}} \end{bmatrix} \tag{3.24}$$

$$\bar{\bar{\mathbf{i}}}_\mathrm{t} = \bar{\bar{\bar{\mathbf{Y}}}}_\mathrm{t}\, \bar{\bar{\mathbf{v}}}_\mathrm{t} + \bar{\bar{\mathbf{i}}}_\mathrm{t}^{(\mathrm{s})} \tag{3.25}$$

and all junctions, including ideal junctions, yield

$$\begin{bmatrix} \bar{\mathbf{E}} & \cdots & \bar{\mathbf{0}} & \bar{\mathbf{0}} & \cdots & \bar{\mathbf{0}} \\ \vdots & \ddots & \vdots & \vdots & & \vdots \\ \bar{\mathbf{0}} & \cdots & \bar{\mathbf{E}} & \bar{\mathbf{0}} & \cdots & \bar{\mathbf{0}} \\ \bar{\mathbf{0}} & \cdots & \bar{\mathbf{0}} & \bar{\mathbf{K}}_1 & \cdots & \bar{\mathbf{0}} \\ \vdots & & \vdots & \vdots & \ddots & \vdots \\ \bar{\mathbf{0}} & \cdots & \bar{\mathbf{0}} & \bar{\mathbf{0}} & \cdots & \bar{\mathbf{K}}_{n_{ij}} \end{bmatrix} \begin{bmatrix} [\bar{\mathbf{i}}_\mathrm{j}]_1 \\ \vdots \\ [\bar{\mathbf{i}}_\mathrm{j}]_{n_j} \\ [\bar{\mathbf{i}}_{ij}]_1 \\ \vdots \\ [\bar{\mathbf{i}}_{ij}]_{n_{ij}} \end{bmatrix} =$$

$$\begin{bmatrix} [\bar{\mathbf{Y}}_\mathrm{j}]_1 & \cdots & \bar{\mathbf{0}} & \bar{\mathbf{0}} & \cdots & \bar{\mathbf{0}} \\ \vdots & \ddots & \vdots & \vdots & & \vdots \\ \bar{\mathbf{0}} & \cdots & [\bar{\mathbf{Y}}_\mathrm{j}]_{n_\mathrm{j}} & \bar{\mathbf{0}} & \cdots & \bar{\mathbf{0}} \\ \bar{\mathbf{0}} & \cdots & \bar{\mathbf{0}} & \bar{\mathbf{M}}_1 & \cdots & \bar{\mathbf{0}} \\ \vdots & & \vdots & \vdots & \ddots & \vdots \\ \bar{\mathbf{0}} & \cdots & \bar{\mathbf{0}} & \bar{\mathbf{0}} & \cdots & \bar{\mathbf{M}}_{n_{ij}} \end{bmatrix} \begin{bmatrix} [\bar{\mathbf{v}}_\mathrm{j}]_1 \\ \vdots \\ [\bar{\mathbf{v}}_\mathrm{j}]_{n_\mathrm{j}} \\ [\bar{\mathbf{v}}_{ij}]_1 \\ \vdots \\ [\bar{\mathbf{v}}_{ij}]_{n_{ij}} \end{bmatrix} + \begin{bmatrix} [\bar{\mathbf{i}}_\mathrm{j}^{(\mathrm{s})}]_1 \\ \vdots \\ [\bar{\mathbf{i}}_\mathrm{j}^{(\mathrm{s})}]_{n_\mathrm{j}} \\ \bar{\mathbf{0}} \\ \vdots \\ \bar{\mathbf{0}} \end{bmatrix} \tag{3.26}$$

$$\bar{\bar{\bar{\mathbf{K}}}}\, \bar{\bar{\mathbf{i}}}_\mathrm{j} = \overline{\overline{\overline{\mathbf{Y}_\mathrm{j}||\boldsymbol{M}}}}\, \bar{\bar{\mathbf{v}}}_\mathrm{j} + \bar{\bar{\mathbf{i}}}_\mathrm{j}^{(\mathrm{s})}. \tag{3.27}$$

The topological structure of the network is expressed by a connection supermatrix $\overline{\overline{\overline{\mathbf{J}\boldsymbol{T}}}}_Y$

$$\bar{\bar{\mathbf{i}}}_\mathrm{t} = \overline{\overline{\overline{\mathbf{J}\boldsymbol{T}}}}_Y\, \bar{\bar{\mathbf{i}}}_\mathrm{j} = \bar{\bar{\mathbf{i}}} \quad \text{and} \quad \bar{\bar{\mathbf{v}}}_\mathrm{t} = \overline{\overline{\overline{\mathbf{J}\boldsymbol{T}}}}_Y\, \bar{\bar{\mathbf{v}}}_\mathrm{j} = \bar{\bar{\mathbf{v}}}\,. \tag{3.28}$$

It describes which ports of the tubes and junctions are connected to each other. It is a symmetric permutation matrix.

Combining Equations (3.25), (3.27) and (3.28) yields a network representation which is based on admittance matrices,

$$\left(\bar{\bar{\bar{\mathbf{K}}}}\, \overline{\overline{\overline{\mathbf{J}\boldsymbol{T}}}}_Y\, \bar{\bar{\bar{\mathbf{Y}}}}_\mathrm{t} - \overline{\overline{\overline{\mathbf{Y}_\mathrm{j}||\boldsymbol{M}}}}\, \overline{\overline{\overline{\mathbf{J}\boldsymbol{T}}}}_Y\right) \bar{\bar{\mathbf{v}}} = \bar{\bar{\mathbf{i}}}_\mathrm{j}^{(\mathrm{s})} - \bar{\bar{\bar{\mathbf{K}}}}\, \overline{\overline{\overline{\mathbf{J}\boldsymbol{T}}}}_Y\, \bar{\bar{\mathbf{i}}}_\mathrm{t}^{(\mathrm{s})}\,. \tag{3.29}$$

Once the voltages are known it is possible to obtain the currents from Equation (3.25). The dimension of this system equation equals that of the BLT 1 equation. No scattering matrices of junctions need to be calculated. This implies that a change in the characteristic wave impedances of the attached tubes does not require the scattering parameters to be recalculated, in contrast to what would be needed in the BLT 1 formalism [15].

The calculation of admittance matrices of a given circuit follows the usual rules of circuit theory. In the case of a junction where all nodes represent accessible ports the admittance matrix is straightforwardly calculated from nodal analysis. The resulting matrix is a symmetric square matrix and its dimension corresponds to the number of nodes/ports (not counting reference nodes). Its diagonal elements contain the sum of all admittances that are connected to a specific node. Off-diagonal elements contain the negative admittance between two specific nodes.

Nodes can also be within junctions where they are not accessible and do not represent ports. Then the Y parameters of the reduced circuit are derived from the admittance matrix of the complete circuit. To this end, nodes are divided in two groups of ports:

- port group 1 – exterior and accessible nodes,
- port group 2 – interior nodes that are not accessible.

The complete circuit of a junction, including interior nodes, is then characterized by

$$\hat{\mathbf{i}}_k = \overline{\mathbf{Y}}_k \, \overline{\mathbf{v}}_k + \mathbf{i}_k^{(s)} , \tag{3.30}$$

where $\overline{\mathbf{Y}}_k$ is obtained from circuit theory. The vector $\mathbf{i}_k^{(s)}$ contains the source currents at a specific node.

A division of this equation with respect to the two port groups yields

$$\begin{bmatrix} \mathbf{i}_1 \\ \mathbf{0} \end{bmatrix} = \begin{bmatrix} \mathbf{Y}_{k11} & \mathbf{Y}_{k12} \\ \mathbf{Y}_{k21} & \mathbf{Y}_{k22} \end{bmatrix} \begin{bmatrix} \mathbf{v}_1 \\ \mathbf{v}_2 \end{bmatrix} + \begin{bmatrix} \mathbf{i}_{k\;1}^{(s)} \\ \mathbf{i}_{k\;2}^{(s)} \end{bmatrix} \tag{3.31}$$

where the fact that the current at an interior node vanishes has been used. Expanding this equation and elimination of $\mathbf{v}_2$, leads to an equation for the unknown quantities at the junction

$$\underbrace{\mathbf{i}_1}_{\mathbf{i}_j} = \underbrace{\left(\mathbf{Y}_{k11} - \mathbf{Y}_{k12}\,\mathbf{Y}_{k22}^{-1}\,\mathbf{Y}_{k21}\right)}_{\overline{\mathbf{Y}}_j} \underbrace{\mathbf{v}_1}_{\overline{\mathbf{v}}_j} + \underbrace{\mathbf{i}_{k\;1}^{(s)} - \mathbf{Y}_{k12}\,\mathbf{Y}_{k22}^{-1}\,\mathbf{i}_{k\;2}^{(s)}}_{\mathbf{i}_j^{(s)}} . \tag{3.32}$$

The solution of this equation can be inserted into a corresponding block of the super matrix Equation (3.29).

3.3 Transmission Lines and Topological Networks

The representations of topological networks that have been discussed in the previous section describe the elements of the network by means of propagation, scattering and admittance matrices. This embeds the structure of a system into a network but does not contain dynamical properties. These are taken into account by the actual values of the parameter matrices $[\mathbf{\Gamma}]_i$, $[\mathbf{S}]_i$ and $[\mathbf{Y}]_i$. They model the behavior of subsystems with respect to their ports by frequency dependent and complex parameter matrices. The interactions and couplings between the various subsystems is represented by the structure of the topological network. For numerical analysis the parameter matrices must be known quantitatively. Depending on the type of subsystems various methods for the determination of parameter matrices are conceivable.

For complex interconnection structures the use of purely numerical field computation methods is often necessary and possible. This involves placing ports at the boundaries of the subsystem, that is at the topological surface which limits the topological volume, which can also be used as ports within the topological network. It is often that quasi-TEM ports are used that allow a unique definition of voltages and currents. Then the wave quantities can be derived according to Section 3.2.1. In the numerical field computation it might be necessary to assign integration paths to determine the parameter matrices from uniquely defined voltages and currents.

For interconnection structures such as cables and transmission lines it is often meaningful to use transmission-line theory for the calculation of parameter matrices. The standard procedure is to calculate in a first step static transmission-line parameters and to solve in a second step the resulting transmission-line equations. The solution, once it is obtained, allows required parameter matrices to be determined from linear transformations and comparison of coefficients. This circumvents time-consuming three-dimensional field computations.

To become more precise, a uniform multiconductor transmission line is considered [6, 7, 12]. The relevant classical transmission-line equations can be written in supermatrix notation according to

$$\frac{\mathrm{d}}{\mathrm{d}\xi}\begin{bmatrix}\mathbf{v}(\xi)\\ \mathbf{i}(\xi)\end{bmatrix} = \begin{bmatrix}\mathbf{0} & -\mathbf{Z}'\\ -\mathbf{Y}' & \mathbf{0}\end{bmatrix}\begin{bmatrix}\mathbf{v}(\xi)\\ \mathbf{i}(\xi)\end{bmatrix} + \begin{bmatrix}\mathbf{v}^{(\mathrm{s})\prime}(\xi)\\ \mathbf{i}^{(\mathrm{s})\prime}(\xi)\end{bmatrix}, \tag{3.33}$$

where $\mathbf{v}(z)$ and $\mathbf{i}(z)$ denote the position dependent vectors of voltages and currents along the transmission line. The per-unit-length impedance matrix $\mathbf{Z}' = \mathbf{R}' + j\omega\mathbf{L}'$ contains the per-unit-length resistance matrix $\mathbf{R}'$ and the matrix of the per-unit-length inductance $\mathbf{L}'$. Equivalently, the matrix of the per-unit-length admittance $\mathbf{Y}' = \mathbf{G}' + j\omega\mathbf{C}'$ contains the per-unit-length conductance $\mathbf{G}'$ and the per-unit-length capacitance $\mathbf{C}'$. Methods and formulas to calculate these per-unit-length parmeters and the distributed sources due to exterior field coupling, together with extensive remarks on their meaning and interrelations, can be found in [6, 7, 12, 13].

The solution of Equation (3.33) is given by

$$\begin{bmatrix}\mathbf{v}(\xi)\\ \mathbf{i}(\xi)\end{bmatrix} = \underbrace{\exp\left(\begin{bmatrix}\mathbf{0} & -\mathbf{Z}'\\ -\mathbf{Y}' & \mathbf{0}\end{bmatrix}(\xi-\xi_0)\right)}_{\mathcal{M}_{\xi_0}^{\xi}\left\{\begin{bmatrix}\mathbf{0} & -\mathbf{Z}'\\ -\mathbf{Y}' & \mathbf{0}\end{bmatrix}\right\}}\begin{bmatrix}\mathbf{v}(\xi_0)\\ \mathbf{i}(\xi_0)\end{bmatrix} + \begin{bmatrix}\mathbf{v}_{\mathrm{eq}}^{(\mathrm{s})}(\xi)\\ \mathbf{i}_{\mathrm{eq}}^{(\mathrm{s})}(\xi)\end{bmatrix} \tag{3.34}$$

with the equivalent sources

$$\begin{bmatrix}\mathbf{v}_{\mathrm{eq}}^{(\mathrm{s})}(\xi)\\ \mathbf{i}_{\mathrm{eq}}^{(\mathrm{s})}(\xi)\end{bmatrix} = \int_{\xi_0}^{\xi_L}\exp\left(\begin{bmatrix}\mathbf{0} & -\mathbf{Z}'\\ -\mathbf{Y}' & \mathbf{0}\end{bmatrix}(\zeta-\xi_0)\right)\begin{bmatrix}\mathbf{v}^{(\mathrm{s})\prime}(\zeta)\\ \mathbf{i}^{(\mathrm{s})\prime}(\zeta)\end{bmatrix}\mathrm{d}\zeta\,. \tag{3.35}$$

The expression $\mathcal{M}_{\xi_0}^{\xi}\{\overline{\mathbf{A}}\}$ is the matrizant which is familiar from Chapter 2. Here it is calculated from the constant matrix $\overline{\mathbf{A}}$ by means of the matrix exponential.

The electromagnetic field along the transmission line can, in general, be rather complex. However, only its effect on the beginning and the end of the transmission line is of importance. This corresponds to the methodology of the topological concept. Hence, to calculate the parameter matrices of the topological network it is sufficient to know the values of the relevant quantities at the beginning and the end of the line. They are related by an equation of the form

$$\begin{bmatrix}\mathbf{v}(L)\\ \mathbf{i}(L)\end{bmatrix} = \underbrace{\exp\left(\begin{bmatrix}\mathbf{0} & -\mathbf{Z}'\\ -\mathbf{Y}' & \mathbf{0}\end{bmatrix} L\right)}_{\overline{\mathbf{M}}^{vi}} \begin{bmatrix}\mathbf{v}(0)\\ \mathbf{i}(0)\end{bmatrix} + \begin{bmatrix}\mathbf{v}_{\mathrm{eq}}^{(\mathrm{s})}(L)\\ \mathbf{i}_{\mathrm{eq}}^{(\mathrm{s})}(L)\end{bmatrix} . \tag{3.36}$$

Here the vectors $\mathbf{v}(L)$ and $\mathbf{i}(L)$ denote the voltages and currents at the end of the line, $\mathbf{v}(0)$ and $\mathbf{i}(0)$ are the quantities at the beginning of the line. The symbol $\overline{\mathbf{M}}^{vi}$ represents the matrizant along the whole line.

It is thus evident that the multiconductor transmission line is viewed as a multiple port with two groups of ports. The first group is formed by the beginning of the line, the second group by the end of the line. Both groups contain the same number of ports.

From the representation of the solution of Equation (3.36) the admittance and impedance matrix of the multiple port can be immediately calculated. If scattering or propagation parameters are required it is necessary to transform the solution into wave quantities. As explained in Section 3.2.1, for complex transmission-line networks wave quantities have been defined in the framework of electromagnetic topology [1] according to

$$\begin{bmatrix}\mathbf{w}_{+}(\xi)\\ \mathbf{w}_{-}(\xi)\end{bmatrix} = \begin{bmatrix}\mathbf{E} & \mathbf{Z}_{\mathrm{c}}\\ \mathbf{E} & -\mathbf{Z}_{\mathrm{c}}\end{bmatrix} \begin{bmatrix}\mathbf{v}(z)\\ \mathbf{i}(z)\end{bmatrix} . \tag{3.37}$$

This transformation is also valid for equivalent sources that are expressed in terms of waves. Application of this linear transformation to the transmission-line equations yields

$$\frac{\mathrm{d}}{\mathrm{d}\xi}\begin{bmatrix}\mathbf{w}_{+}(\xi)\\ \mathbf{w}_{-}(\xi)\end{bmatrix} = \underbrace{\begin{bmatrix}\mathbf{E} & \mathbf{Z}_{\mathrm{c}}\\ \mathbf{E} & -\mathbf{Z}_{\mathrm{c}}\end{bmatrix}\begin{bmatrix}\mathbf{0} & -\mathbf{Z}'\\ -\mathbf{Y}' & \mathbf{0}\end{bmatrix}\begin{bmatrix}\mathbf{E} & \mathbf{Z}_{\mathrm{c}}\\ \mathbf{E} & -\mathbf{Z}_{\mathrm{c}}\end{bmatrix}^{-1}}_{\begin{bmatrix}-\mathbf{P} & \mathbf{0}\\ \mathbf{0} & +\mathbf{P}\end{bmatrix}} \begin{bmatrix}\mathbf{w}_{+}(\xi)\\ \mathbf{w}_{-}(\xi)\end{bmatrix} + \begin{bmatrix}\mathbf{w}_{+}^{(\mathrm{s})\prime}(\xi)\\ \mathbf{w}_{-}^{(\mathrm{s})\prime}(\xi)\end{bmatrix} \tag{3.38}$$

where $\mathbf{P}^2 = \mathbf{Z}'\,\mathbf{Y}' = \mathbf{Y}'\,\mathbf{Z}'$ and $\mathbf{Z}_{\mathrm{c}}^2 = \mathbf{Z}'\,\mathbf{Y}'^{-1} = \mathbf{Y}'^{-1}\,\mathbf{Z}'$. Thus, a decoupling of the super matrix blocks is achieved and the solution is easily determined. The solution for the wave quantities at the beginning and the end of the line turns out to be

$$\begin{bmatrix}\mathbf{w}_{+}(L)\\ \mathbf{w}_{-}(L)\end{bmatrix} = \underbrace{\begin{bmatrix}\exp -\mathbf{P}L & \mathbf{0}\\ \mathbf{0} & \exp +\mathbf{P}L\end{bmatrix}}_{\overline{\mathbf{M}}^{w}} \begin{bmatrix}\mathbf{w}_{+}(0)\\ \mathbf{w}_{-}(0)\end{bmatrix} + \begin{bmatrix}\mathbf{w}_{+}^{(\mathrm{s})}(L)\\ \mathbf{w}_{-}^{(\mathrm{s})}(L)\end{bmatrix} . \tag{3.39}$$

If the solution is already known in terms of the v–i representations it can be turned into the wave representation according to Equation (E.18)

$$\overline{\mathbf{M}}^{w} = \begin{bmatrix}\mathbf{E} & \mathbf{Z}_{\mathrm{c}}\\ \mathbf{E} & -\mathbf{Z}_{\mathrm{c}}\end{bmatrix} \overline{\mathbf{M}}^{vi} \begin{bmatrix}\mathbf{E} & \mathbf{Z}_{\mathrm{c}}\\ \mathbf{E} & -\mathbf{Z}_{\mathrm{c}}\end{bmatrix}^{-1} \tag{3.40}$$

and vice versa. With these two representations of the solution of the transmission-line equations it is possible to obtain the parameter matrices and source terms that are necessary to characterize the dynamics of the topological network. In the following, the relevant transformations will explicitly be given.

3.3.1 Transformation into a Propagation Matrix

In order to incorporate a multiconductor transmission line into a topological network along the lines of the BLT 1 formalism it is necessary to transform the solution of the transmission-line equations into a propagation matrix $\overline{\mathbf{\Gamma}}_i$ and its corresponding source wave vector $\overline{\mathbf{w}}_i^{(s)}(L)$. The propagation matrix $\overline{\mathbf{\Gamma}}_i$, which corresponds to a specific tube, is then inserted into the super matrix $\overline{\overline{\mathbf{\Gamma}}}$.

The propagation matrix is defined on the basis of wave quantities. If the wave matrizant has not been calculated within the wave representation it must be transformed accordingly. This yields the result

$$\begin{bmatrix} \mathbf{w}_+(L) \\ \mathbf{w}_-(L) \end{bmatrix} = \underbrace{\begin{bmatrix} \mathbf{M}_{11}^w & \mathbf{M}_{12}^w \\ \mathbf{M}_{21}^w & \mathbf{M}_{22}^w \end{bmatrix}}_{\overline{\mathbf{M}}^w} \begin{bmatrix} \mathbf{w}_+(0) \\ \mathbf{w}_-(0) \end{bmatrix} + \begin{bmatrix} \mathbf{w}_+^{(s)}(L) \\ \mathbf{w}_-^{(s)}(L) \end{bmatrix}. \tag{3.41}$$

The desired propagation matrix $\overline{\mathbf{\Gamma}}_i$ is obtained from the comparison of the propagation equation of the tube

$$\begin{bmatrix} \mathbf{w}_1(L) \\ \mathbf{w}_2(L) \end{bmatrix} = \begin{bmatrix} \mathbf{\Gamma}_{11} & \mathbf{\Gamma}_{12} \\ \mathbf{\Gamma}_{21} & \mathbf{\Gamma}_{22} \end{bmatrix} \begin{bmatrix} \mathbf{w}_1(0) \\ \mathbf{w}_2(0) \end{bmatrix} + \begin{bmatrix} \mathbf{w}_1^{(s)}(L) \\ \mathbf{w}_2^{(s)}(L) \end{bmatrix} \tag{3.42}$$

with Equation (3.41). This yields

$$\overline{\mathbf{\Gamma}}_i = \begin{bmatrix} \mathbf{M}_{11}^w - \mathbf{M}_{12}^w \mathbf{M}_{22}^{w\ -1} \mathbf{M}_{21}^w & \mathbf{M}_{12}^w \mathbf{M}_{22}^{w\ -1} \\ -\mathbf{M}_{22}^{w\ -1} \mathbf{M}_{21}^w & \mathbf{M}_{22}^{w\ -1} \end{bmatrix} \tag{3.43}$$

and the source terms are given by

$$\begin{bmatrix} \mathbf{w}_1^{(s)}(L) \\ \mathbf{w}_2^{(s)}(L) \end{bmatrix} = \begin{bmatrix} \mathbf{w}_+^{(s)}(L) - \mathbf{M}_{12}^w \mathbf{M}_{22}^{w\ -1} \mathbf{w}_-^{(s)}(L) \\ -\mathbf{M}_{22}^w \mathbf{w}_-^{(s)}(L) \end{bmatrix}. \tag{3.44}$$

These quantities are illustrated by Figure 3.9. The reversed direction of the backwards running wave vector $\mathbf{w}_2$ from the end of the beginning of the line shoud be kept in mind. Its starting position (Index 0) is located at the geometric end of the line, that is $\mathbf{w}_2(0) = \mathbf{w}_-(L)$ and $\mathbf{w}_2(L) = \mathbf{w}_-(0)$. This circumstance is taken care of within the transformation equations.

The resulting quantities $\overline{\mathbf{\Gamma}}_i$ and $\overline{\mathbf{w}}_i^{(s)}(L)$ can be inserted directly into the BLT equation of the nonuniform transmission line such that the network problem can be solved. This formalism allows a unified treatment of uniform and nonuniform transmission lines in the framework of

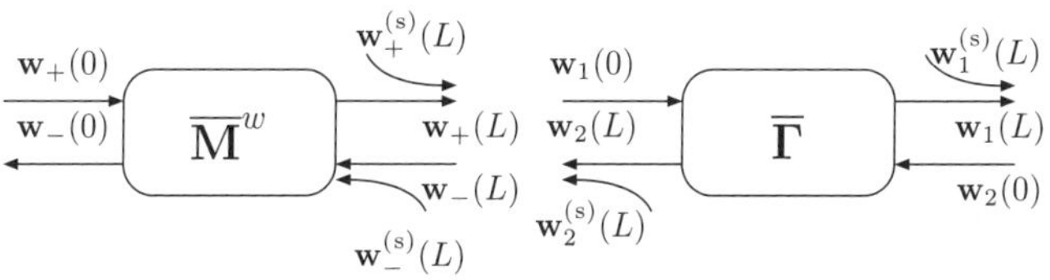

Figure 3.9 Matrizant and propagation matrix.

the BLT 1 equation. The ansatz of the so-called NBLT equation that has been proposed in [14] appears to be more complicated in this respect.

3.3.2 Equivalent Scattering Matrices

The representation of a multiconductor transmission line in terms of an equivalent scattering matrix is motivated by the implementation within a network solver. Often the propagation matrices $\overline{\mathbf{\Gamma}}_i$ of single tubes are required to be of diagonal form and the use of nondiagonal propagation matrices considerably complicates the implementation. In these cases it is desirable to include transmission lines as equivalent junctions into the topological network. A transformation into equivalent scattering matrices allows the transmission line to be represented as a junction with two identically-sized port groups and to use the BLT 1 or BLT 2 formalism.

If the transmission line is supposed to be merged into a BLT 2 network it is necessary to calculate the S50-parameters (Sections 3.2.3 and 3.2.1). To this end, the characteristic impedance matrix $\mathbf{Z}_c(\xi) = 50\Omega\,\mathbf{E}$ of Equation (3.37) is used for the transformation.

A further possibility is the inclusion of the transmission line as an equivalent junction by means of topological scattering parameters, where the wave quantities are formed according to Equation (3.37) from the wave impedances at the beginning and the end of the line. The topological scattering matrix is connected to the network with the aid of *zero length tubes* that exhibit these wave impedances.

The required equivalent scattering matrix of a tube follows, similar to the propagation matrix of the previous section, by means of comparing Equation (3.41) and the definition of the scattering parameters,

$$\overline{\mathbf{S}}_{\text{eq}} = \begin{bmatrix} -\mathbf{M}_{22}^{w\ -1}\mathbf{M}_{21}^{w} & \mathbf{M}_{22}^{w\ -1} \\ \mathbf{M}_{11}^{w} - \mathbf{M}_{12}^{w}\mathbf{M}_{22}^{w\ -1}\mathbf{M}_{21}^{w} & \mathbf{M}_{12}^{w}\mathbf{M}_{22}^{w\ -1} \end{bmatrix}. \tag{3.45}$$

The representation of a nonuniform transmission line by a junction is

$$\begin{bmatrix} \mathbf{b}_1 \\ \mathbf{b}_2 \end{bmatrix} = \begin{bmatrix} \mathbf{S}_{\text{eq}_{11}} & \mathbf{S}_{\text{eq}_{12}} \\ \mathbf{S}_{\text{eq}_{21}} & \mathbf{S}_{\text{eq}_{22}} \end{bmatrix} \begin{bmatrix} \mathbf{a}_1 \\ \mathbf{a}_2 \end{bmatrix} + \begin{bmatrix} \mathbf{b}_1^{(\text{s})} \\ \mathbf{b}_2^{(\text{s})} \end{bmatrix} \tag{3.46}$$

with

$$\begin{bmatrix} \mathbf{b}_1^{(\text{s})} \\ \mathbf{b}_2^{(\text{s})} \end{bmatrix} = \begin{bmatrix} -\mathbf{M}_{22}^{w}\mathbf{w}_-^{(\text{s})}(L) \\ \mathbf{w}_+^{(\text{s})}(L) - \mathbf{M}_{12}^{w}\mathbf{M}_{22}^{w\ -1}\mathbf{w}_-^{(\text{s})}(L) \end{bmatrix}. \tag{3.47}$$

The quantities of Equations (3.41) and (3.46) are illustrated in Figure 3.10

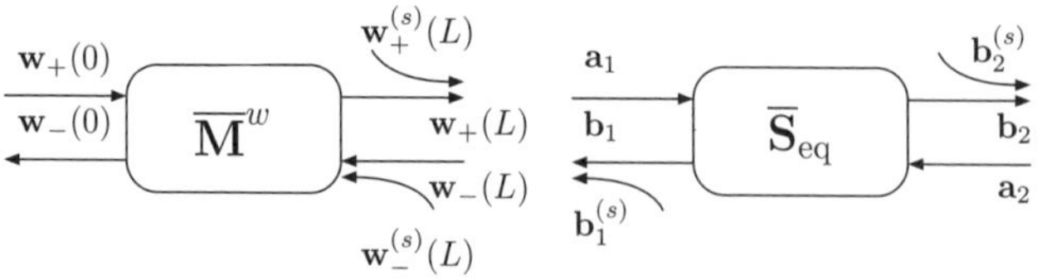

Figure 3.10 Matrizant and equivalent scattering matrices.

A further possibility leads via the admittance representation of a tube as an intermediate step to Section 3.3.3. The use of the Y-matrix according to Equation (3.50) yields the scattering matrix as

$$\overline{\mathbf{S}} = \left(\overline{\mathbf{E}} + \overline{\mathbf{Z}}_c\,\overline{\mathbf{Y}}_t\right)\left(\overline{\mathbf{E}} - \overline{\mathbf{Z}}_c\,\overline{\mathbf{Y}}_t\right)^{-1}. \tag{3.48}$$

It should be noted that the definition of the admittance representation implies that the currents flow out of the tubes. The supermatrix $\overline{\mathbf{Z}}_c$ contains the impedances of the ports

$$\overline{\mathbf{Z}}_c = \begin{bmatrix} \mathbf{Z}_c(0) & \mathbf{0} \\ \mathbf{0} & \mathbf{Z}_c(L) \end{bmatrix}, \tag{3.49}$$

the explicit values of which depend on the choice of scattering parameters, such as S50-parameters or topological S-parameters.

3.3.3 Admittance Matrix

To use the admittance representation of the network it is necessary to express the dynamical properties of the tubes as admittance matrices. These are, eventually, inserted into the system's Equation (3.29).

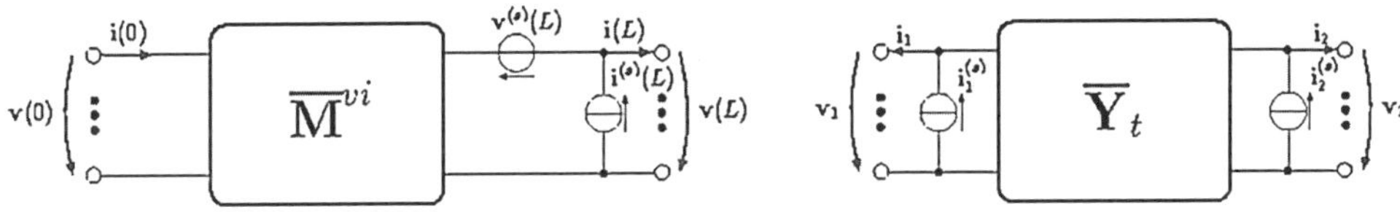

Figure 3.11 Matrizant and admittance matrix.

According to Figure 3.11 it is sufficient to express the matrizant in terms of its current–voltage representation. Then a comparison yields the Y matrix

$$\overline{\mathbf{Y}}_t = \begin{bmatrix} \mathbf{M}_{12}^{vi\;-1}\,\mathbf{M}_{11}^{vi} & -\mathbf{M}_{12}^{vi\;-1} \\ \mathbf{M}_{21}^{vi} - \mathbf{M}_{22}^{vi}\,\mathbf{M}_{12}^{vi\;-1}\,\mathbf{M}_{11}^{vi} & \mathbf{M}_{22}^{vi}\,\mathbf{M}_{12}^{vi\;-1} \end{bmatrix} \tag{3.50}$$

and the vector of the source currents becomes

$$\begin{bmatrix} \mathbf{i}_1^{(s)} \\ \mathbf{i}_2^{(s)} \end{bmatrix} = \begin{bmatrix} \mathbf{M}_{12}^{vi\;-1}\,\mathbf{v}^{(s)}(L) \\ \mathbf{i}^{(s)}(L) - \mathbf{M}_{22}^{vi}\,\mathbf{M}_{12}^{vi\;-1}\,\mathbf{v}^{(s)}(L) \end{bmatrix}. \tag{3.51}$$

3.4 Shielding

The following two subsections present simple examples for elementary and proper surfaces and corresponding interactions, respectively (transfer impedance and shielding).[4]

[4] The content of Section 3.4.1 was published by Haase and Nitsch in the proceedings of the IEEE International Symposium on EMC in 2001, 13-17 August 2001, pp. 1241–1247. The permission to reuse it for this book was provided by the Publisher, *IEEE Transactions on EMC*.

3.4.1 Model for the Transfer Impedance Based on NMTLT

Cable shields often consist of braided wires. This allows a maximum flexibility of the cable, but an incident electromagnetic field can penetrate this shield through the holes and couple into the inner conductors. This behavior of the cable in an electromagnetic environment is described by the transfer impedance and transfer admittance which have been the subject of intensive research in EMC.

There are several computational models for the transfer impedance [16]. All of them are based on interpretations of the actual physical behavior. The transfer impedance is composed of several terms each representing a different effect. The first term gives the diffusion of the electromagnetic energy through the shield and is modeled by the diffusion through a tubular shield [17]. Another term, the aperture inductance, represents the coupling through the rhombic apertures and is modeled by a tubular thin-walled shield with elliptical holes [18]. In [19] an additional term is introduced, which takes into account the magnetic coupling between the inner and the outer layers of the braided shield.

There are also several methods to measure the transfer parameters; however, due to resonances on the cables these methods are restricted to lower frequencies.

Moreover, the models as well as the data from measurements give the integral behavior of the cable and therefore do not account for the nonuniformities occurring due to the nonhomogeneous shields. Usually this is not important since at normal operation frequencies the cables behave like uniform transmission lines.

A first step to another model for the transfer impedance of cables with braided shields is presented in this subsection. Different to the known models, here the individual conductors of the shield, the center conductor and the current return path, which is an infinite good conducting ground plane furnish a nonuniform multiconductor transmission line. With the assumption that either the voltages or the currents are equal in all shield conductors, this transmission line can be reduced to a two-conductor transmission line. The transfer impedance can than be given in terms of the per-unit-length parameters of this two-conductor transmission line. Since these transmission lines are nonuniform, the TLST of Section 2.1 must be used, in order to determine the per-unit-length parameters. However, only the quasi-static solution is necessary because radiation plays only a secondary role and is neglected here. Moreover, this theory restricts the computation to arrangements where the distance between wires is large compared to the radius, since for the thin-wire approximation infinite good conducting wires are assumed. This reflects the behavior of the cables at high frequencies, where the resistive parts become small compared with the inductive part of the transfer impedance and therefore are negligible.

3.4.1.1 Transfer Parameters of Cables

Definition of the Transfer Impedance and the Transfer Admittance

If a cable is placed in an electromagnetic environment, the cable shield and a current return path (e.g. a ground plane) form a transmission line (see Figure 3.12). Incident electromagnetic fields can now couple into these transmission lines and also influence the transmission line formed by the shield and the center conductor of the cable. These phenomena are usually described by the transfer impedance and transfer admittance. Both systems are represented by the corresponding telegrapher's equations. Furthermore, the systems are coupled by controlled

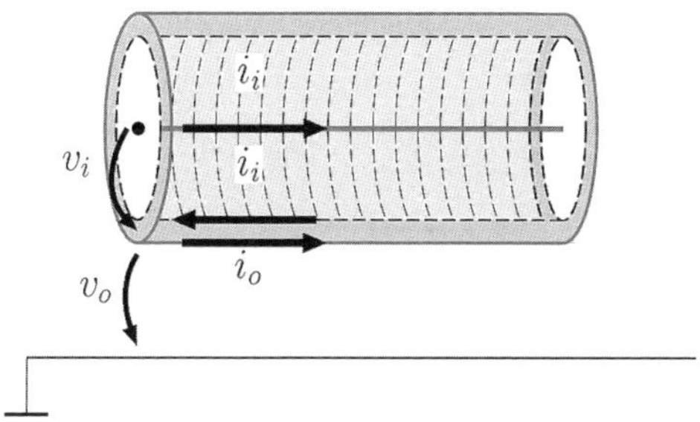

Figure 3.12 Cable in an electromagnetic environment.

current and voltage sources [7], representing the link between both.

$$\frac{\partial}{\partial z}\begin{bmatrix} v_i \\ i_i \end{bmatrix} = -\begin{bmatrix} 0 & Z'_i \\ Y'_i & 0 \end{bmatrix}\begin{bmatrix} v_i \\ i_i \end{bmatrix} + \begin{bmatrix} 0 & Z'_{t_{i,0}} \\ Y'_{t_{i,0}} & 0 \end{bmatrix}\begin{bmatrix} v_0 \\ i_0 \end{bmatrix} \tag{3.52}$$

$$\frac{\partial}{\partial z}\begin{bmatrix} v_0 \\ i_0 \end{bmatrix} = -\begin{bmatrix} 0 & Z'_0 \\ Y'_0 & 0 \end{bmatrix}\begin{bmatrix} v_0 \\ i_0 \end{bmatrix} + \begin{bmatrix} 0 & Z'_{t_{0,i}} \\ Y'_{t_{0,i}} & 0 \end{bmatrix}\begin{bmatrix} v_i \\ i_i \end{bmatrix}. \tag{3.53}$$

From these equations it is easy to find definition for the transfer impedance and transfer admittance:

$$Z'_{t_{0,i}} = Z'_{t_{i,0}} = Z'_t := \frac{1}{i_0}\left.\frac{\partial v_i}{\partial z}\right|_{i_i=0} \tag{3.54}$$

$$Y'_{t_{0,i}} = Y'_{t_{i,0}} = Y'_t := \frac{1}{v_0}\left.\frac{\partial i_i}{\partial z}\right|_{v_i=0}. \tag{3.55}$$

Relations to the Per-unit-length Parameters

The cable in the electromagnetic environment can also be regarded as a multiconductor transmission line which consists of the cable shield, the center conductor and the ground plane as the reference conductor. Then the voltages and currents are defined in accordance with Figure 3.13. Here, the telegrapher's equations have the form

$$\frac{\partial}{\partial z}\begin{bmatrix} v_c \\ v_s \end{bmatrix} = \underbrace{\begin{bmatrix} Z'_{c,c} & Z'_{c,s} \\ Z'_{s,c} & Z'_{s,s} \end{bmatrix}}_{Z'_e}\begin{bmatrix} i_c \\ i_s \end{bmatrix} \tag{3.56}$$

$$\frac{\partial}{\partial z}\begin{bmatrix} i_c \\ i_s \end{bmatrix} = \underbrace{\begin{bmatrix} Y'_{c,c} & Y'_{c,s} \\ Y'_{s,c} & Y'_{s,s} \end{bmatrix}}_{Y'_e}\begin{bmatrix} v_c \\ v_s \end{bmatrix} \tag{3.57}$$

Because both representations constitute the same system, the following relations for the voltages and currents can be elaborated:

$$v_c = v_i + v_0 \qquad i_c = i_i \tag{3.58}$$

$$v_s = v_0 \qquad i_s = i_0 - i_i. \tag{3.59}$$

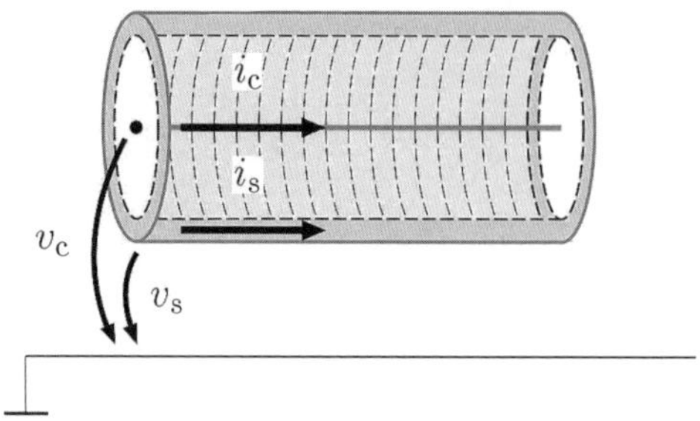

Figure 3.13 Definition of voltage and current for a cable as a multiconductor.

Additionally, the following identities for the parameters can be found:

$$Z'_i = Z'_{c,c} + Z'_{s,s} - Z'_{c,s} - Z'_{s,c} \tag{3.60}$$
$$Z'_0 = Z'_{s,s} \tag{3.61}$$
$$Z'_{t_{i,0}} = Z'_{s,s} - Z'_{c,s} \tag{3.62}$$
$$Z'_{t_{0,i}} = Z'_{s,s} - Z'_{s,c} \tag{3.63}$$
$$Y'_i = Y'_{c,c} \tag{3.64}$$
$$Y'_0 = Y'_{c,c} + Y'_{s,s} + Y'_{s,c} + Y'_{c,s} \tag{3.65}$$
$$Y'_{t_{i,0}} = -Y'_{c,c} - Y'_{c,s} \tag{3.66}$$
$$Y'_{t_{0,i}} = -Y'_{c,c} - Y'_{s,c}. \tag{3.67}$$

Thus the transfer impedance is given by the difference of the self impedance

$$Z'_{s,s} = R'_s + j\omega L'_{s,s} \tag{3.61a}$$

of the cable shield over the ground plane and the mutual impedance

$$Z'_{s,c} = Z'_{c,s} = j\omega L'_{s,c} = j\omega L'_{c,s} \tag{3.63a}$$

between the cable shield and the center conductor.

3.4.1.2 Determination of the Per-unit-length Parameters

Description of the Cable as a Nonuniform Multiconductor Transmission Line

The generalized telegrapher's equations for a nonuniform multiconductor transmission line are:

$$\frac{\partial}{\partial z}\mathbf{v}(z,\omega) = -\mathbf{Z}'(z,\omega)\mathbf{i}(z,\omega) \tag{3.69}$$
$$\frac{\partial}{\partial z}\mathbf{i}(z,\omega) = -\mathbf{Y}'(z,\omega)\mathbf{v}(z,\omega). \tag{3.70}$$

The matrix $\mathbf{Z}'(z,\omega) = \mathbf{R}'(z,\omega) + j\omega\mathbf{L}'(z,\omega)$ denotes the per-unit-length impedance matrix and $\mathbf{Y}'(z,\omega) = \mathbf{G}'(z,\omega) + j\omega\mathbf{C}'(z,\omega)$ the per-unit-length admittance matrix. $\mathbf{v}(z,\omega) = [v_c v_{s_1} \ldots v_{s_n}]^T$ and $\mathbf{i}(z,\omega) = [i_c i_{s_1} \ldots v_{s_n}]^T$ are the corresponding voltage and current column vectors, respectively. Since ideal conductors in a lossless media are considered, $\mathbf{R}'(z\omega)$ and $\mathbf{G}'(z\omega)$ vanish. All other values, in general, are functions of the frequency ω as well as of the position z along the wire. To simplify matters these dependencies will be omitted in subsequent equations.

The per-unit-length inductance and the per-unit-length capacitance matrices are, for the quasi-static case, given by the following formulas

$$\mathbf{L}' = \begin{bmatrix} L'_{1,1} & \cdots & L'_{1,n} \\ \vdots & \ddots & \vdots \\ L'_{n,1} & \cdots & L'_{n,n} \end{bmatrix} \tag{3.71}$$

$$\mathbf{C}' = P^{-1} \tag{3.72}$$

$$\mathbf{P}' = \begin{bmatrix} P_{1,1} & \cdots & P_{1,n} \\ \vdots & \ddots & \vdots \\ P_{n,1} & \cdots & P_{n,} \end{bmatrix} \tag{3.73}$$

where the matrix elements of L and P are obtained by the evaluation of the integrals:

$$L'_{i,j}(z) = \mu \int_{z_0}^{z_1} k_{l_{i,j}}(z, \xi)\, \mathrm{d}\xi \tag{3.74}$$

$$P_{i,j}(z) = \frac{1}{\varepsilon} \int_{z_0}^{z_1} k_{p_{i,j}}(z, \xi)\, \mathrm{d}\xi. \tag{3.75}$$

For an infinite good conducting ground plane the integral kernels k_l and k_p are:

$$k_{l_{i,j}}(z, \xi) = \frac{1}{4\pi\,|\mathbf{r}_i(z) - \mathbf{r}_j(\xi)|} \frac{\mathrm{d}\mathbf{r}_i(z)}{\mathrm{d}z} \frac{\mathrm{d}\mathbf{r}_j(\xi)}{\mathrm{d}\xi} - \frac{1}{4\pi\,|\mathbf{r}_i(z) - \mathbf{r}^*_j[\xi]|} \frac{\mathrm{d}\mathbf{r}_i(z)}{\mathrm{d}z} \frac{\mathrm{d}\mathbf{r}^*_j(\xi)}{\mathrm{d}\xi} \tag{3.76}$$

$$k_{p_{i,j}}(z, \xi) = \frac{1}{4\pi\,|\mathbf{r}_i(z) - \mathbf{r}_j(\xi)|} - \frac{1}{4\pi\,|\mathbf{r}_i(z) - \mathbf{r}^*_j(\xi)|}. \tag{3.77}$$

The variables are defined according to Figure 3.14. The superscript * indicates the spatial vector to the image of the wire (i.e. $\mathbf{r}^*_x = -\mathbf{r}_x, \mathbf{r}^*_y = \mathbf{r}_y, \mathbf{r}^*_z = \mathbf{r}_z$).

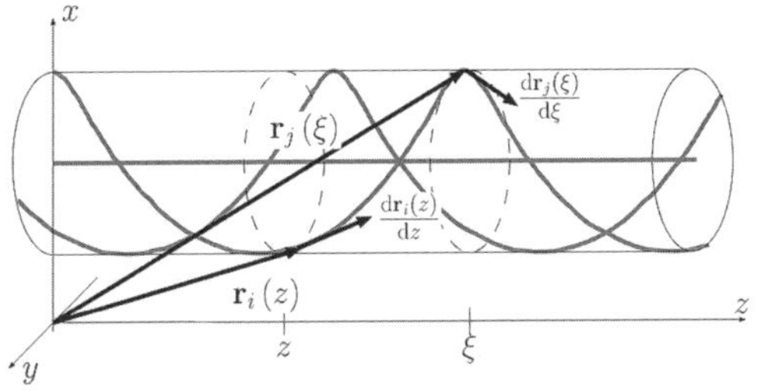

Figure 3.14 Nonuniform conductors of a cable shield.

These equations take into account the nonuniformity of the conductors. However, current displacement effects (skin effect, proximity effect) are not considered due to the thin-wire approximation. It is assumed that the current as well as the charge is equally distributed over the surface of each conductor.

Equivalent Per-unit-length Parameters

The telegrapher's Equations (3.69) and (3.70) relate the voltages and currents in the center conductor ($v_{S_1} \dots v_{S_n}, i_{S_1} \dots i_{S_n}$) of the cable. Because shields usually consist of several hundred conductors, the involved matrices are quite large. To reduce their size, it is possible to merge all shield conductors into one equivalent conductor with equivalent per-unit-length parameters. For this, certain assumptions about either the voltages or currents of the shield conductors must be made:

(1) It can be assumed that the current is equally distributed in all shield conductors. This, in general, yields a different voltage for every conductor, unless the setup is completely cylinder symmetrical. An equivalent voltage for the shield can be defined by averaging all the individual voltages.
(2) It can be assumed that the voltages in all conductors at each position z is the same. The overall current is then given by the sum of all shield conductor currents. This particular setup is equivalent to a parallel connection of all conductors and would automatically lead to a current distribution, such that the magnetic fluxes between the shield conductors vanish.

(i) Equally Distributed Currents

A shield conductor current is given by

$$i_{S_k} := \frac{1}{n} i_S \quad \text{for} \quad k = 1 \dots n \tag{3.78}$$

where i_S is the overall shield current. The equivalent shield voltage is determined by:

$$v_S := \frac{1}{n} \sum_{k=1}^{n} v_{S_k} \tag{3.79}$$

Inserting these conditions into Equations (3.69) and (3.70) yields the telegrapher's Equations (3.56) and (3.57) for the cable, with the per-unit-length parameters

$$\boldsymbol{Z}'_e := \boldsymbol{T}\boldsymbol{Z}'\boldsymbol{T}^T \tag{3.80}$$

$$\boldsymbol{Y}'_e := (\boldsymbol{T}\boldsymbol{Y}'^{-1}\boldsymbol{T}^T)^{-1}. \tag{3.81}$$

The $2 \times (n+1)$ matrix $\boldsymbol{T}$ is given by:

$$\boldsymbol{T} = \begin{bmatrix} 1 & 0 \\ 0 & \frac{1}{n} \end{bmatrix} \begin{bmatrix} 1 & 0 & 0 & \dots & 0 \\ 0 & 1 & 1 & \dots & 1 \end{bmatrix}. \tag{3.82}$$

(ii) Equal Voltages

Analogously, the per-unit-length parameters can be defined for the case where the voltages are assumed to be equal. Here the overall shield current is given by the sum of the currents through each of the shield conductors:

$$i_s := \sum_{k=1}^{n} i_{s_k} \tag{3.83}$$

and the shield voltages are:

$$v_{s_k} := v_s \quad \text{for} \quad k = 1 \ldots n. \tag{3.84}$$

Then the equivalent per-unit-length parameters become:

$$\boldsymbol{Z}'_e := (\boldsymbol{T}\boldsymbol{Z}'^{-1}\boldsymbol{T}^T)^{-1} \tag{3.85}$$

$$\boldsymbol{Y}'_e := \boldsymbol{T}\boldsymbol{Y}'\boldsymbol{T}^T, \tag{3.86}$$

where the $2 \times (n+1)$ matrix $\mathbf{T}$ is

$$\mathbf{T} = \begin{bmatrix} 1 & 0 & 0 & \ldots & 0 \\ 0 & 1 & 1 & \ldots & 1 \end{bmatrix}. \tag{3.87}$$

3.4.1.3 Computation of the Transfer Impedance

With the above shown set of equations it is now possible to compute the transfer impedance of cable shields. In a first step the per-unit-length parameters need to be determined, which is done by applying Equations (3.71)–(3.77) to the mathematical description of the cable shield geometry. Unfortunately there are no analytical solutions to the involved integrals for braided shields available, so that numerical integration techniques were used. Then equivalent per-unit-length parameters are computed from these results by using either Equations (3.80) and (3.81) or Equations (3.87) and (3.86). For simplicity only the first variant will be used, since no inversion of the matrices is necessary. From these equivalent per-unit-length parameters the transfer impedance can now be computed by utilizing Equation (3.62).

Because ideal conductors are considered, the transfer impedance only has a frequency independent inductive part. This parameter will be analyzed in the upcoming examples.

Cylindrical Cage

The first cable shield that is investigated is a cylindrical cage as shown in Figure 3.15. This particular structure has already been investigated by Kaden [24], who derived an analytical

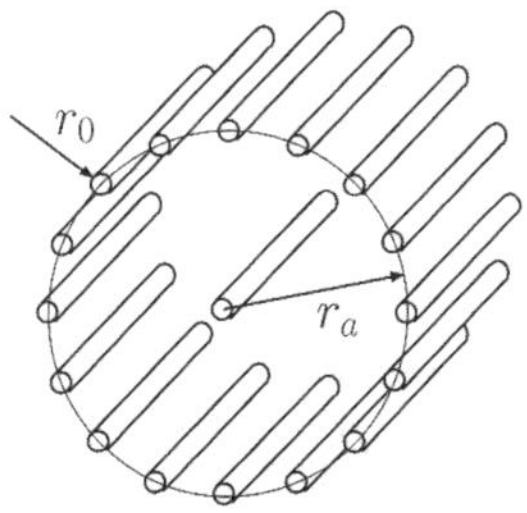

Figure 3.15 Cylindrical cage as a cable shield of radius r_a with n wires having a radius of r_0.

equation which even approximatively includes the proximity effect. For infinite good conductors Kaden's equation becomes:

$$Z'_t = j\omega L'_t = \frac{j\omega\mu_0}{2\pi n}\left[\ln\frac{r_a}{nr_0} + \frac{(n^2-1)r_0^2}{4r_a^2}\right], \tag{3.88}$$

where r_a is the radius of the cage, r_0 the radius of the wires and n the number of wires. The second term in this equation takes into account the proximity effect and is valid for $r_0 \le \pi r_a/2n$. This analytical formula allows the verification of the proposed new method. For this two different configurations were studied.

(i) Constant Number of Wires

In this configuration, the number of wires is kept constant at 50, while the radius of the shield wires is changed. The results of the computations are depicted in Figure 3.16.

As expected, for small wire radii the proximity effect is negligible; only when the radius of the wires becomes larger than one tenth of the distance between the wires does there occur a significant difference between the analytic values with and without the consideration of the proximity effect. Furthermore, the proposed method reproduces the results obtained with the analytic formula without the proximity effect.

(ii) Constant Radius

Different to the previous section the radius r_0 can also be held constant and the number of wires can be changed. The results are shown in Figure 3.17.

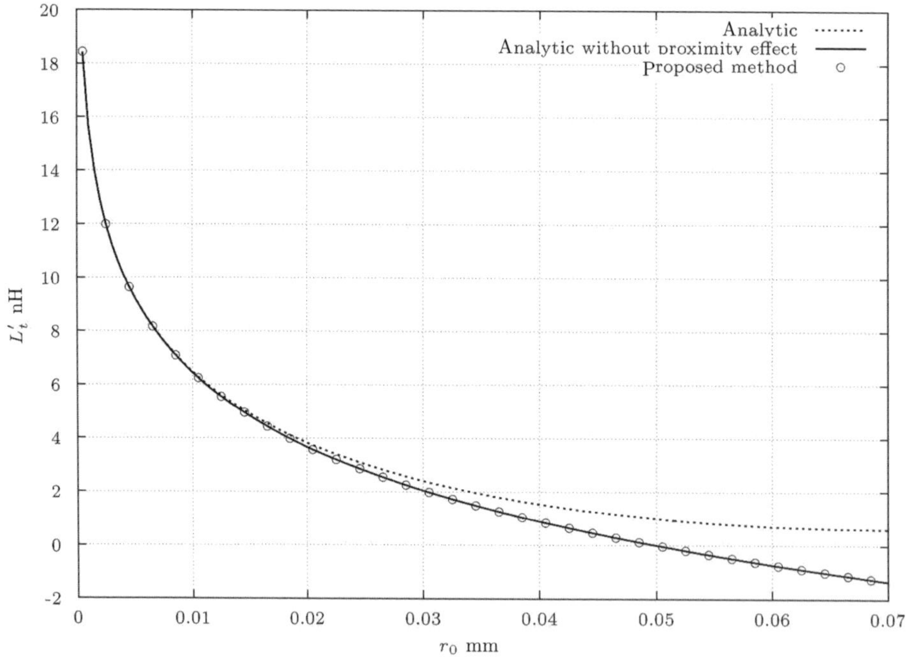

Figure 3.16 Inductive part of the transfer impedance for the cylindrical cage ($n = 50$, $r_a = 2.5$ mm, $r_0 = 0.0005 \ldots 0.07$ mm).

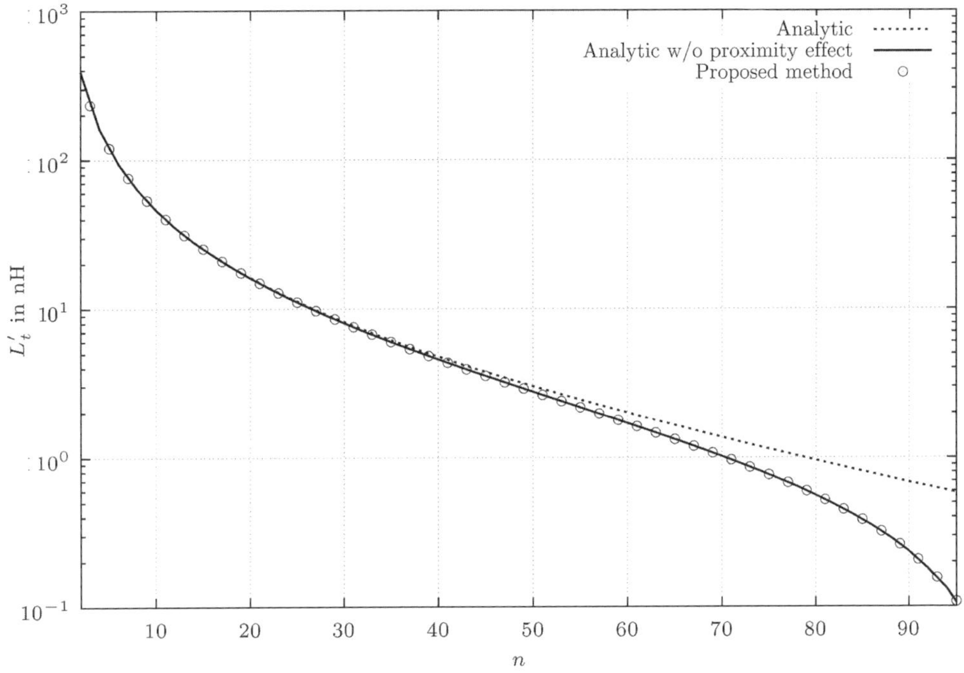

Figure 3.17 Inductive part of the transfer impedance for the cylindrical cage ($n = 1 \ldots 95$, $r_a = 2.5\,\text{mm}$, $r_0 = r_a/100$).

Also here a very good agreement between the proposed method and the analytical solution without the proximity effect can be observed. Again, if the distance between the wires becomes smaller than $10r_a$, the proximity effect becomes important.

Braided Shield

In this section a simple braided shield will be studied. The proposed method permits the determination of the transfer impedance as a function of the coordinate z along the line. Thus different values of the transfer impedance at different positions of the braided shield can be expected (e.g. if the position is near an aperture the transfer impedance should become larger). However, in order to perform the calculations a mathematical description of the cable shield conductors is necessary, which will be given first.

(i) Mathematical Description of the Shield

Every wire is formed like a helix and is described by

$$\mathbf{r}(z) = \begin{bmatrix} h_0 + r_a(z)\cos\left(\dfrac{2\pi z}{l_w} + \varphi_w\right) \\ \pm r_a(z)\sin\left(\dfrac{2\pi z}{l_w} + \varphi_w\right) \\ z \end{bmatrix}. \qquad (3.89)$$

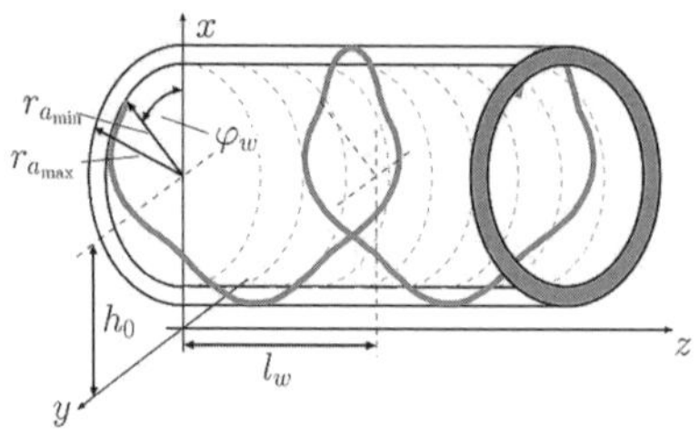

Figure 3.18 Helix structure of a braided shield conductor.

The parameters l_ω and φ_ω characterize the length of one period of the helix and the position of the conductor at the circumference of the shield, respectively (see Figure 3.18). The $\pm$ in front of the y coordinate determines whether the helix turns clockwise or counter clockwise. The porpoising of the wire is achieved by modulating the helix radius ($r_a(z)$) with a cosine function:

$$r_a(z) = r_{a_0} + m \cos\left(\frac{2\pi}{l_\mathrm{p}} z + \varphi_\mathrm{p}\right) \tag{3.90}$$

where

$$m = \frac{1}{2}(r_{a_{\max}} - r_{a_{\min}}) \tag{3.91}$$

$$r_{a_0} = \frac{1}{2}(r_{a_{\max}} + r_{a_{\min}}). \tag{3.92}$$

The minimum and maximum radii of the shield are $r_{a_{\max}}$ and $r_{a_{\min}}$, respectively. The variables l_p and φ_p determine where and how often the wire is on the outside or the inside of the shield.

Usually a cable shield is constructed of several carriers, where each carrier is shifted by an angle of $2{\cdot}360/n_{\mathrm{carrier}}$ with n_{carrier} being the total number of carriers. Within each carrier there

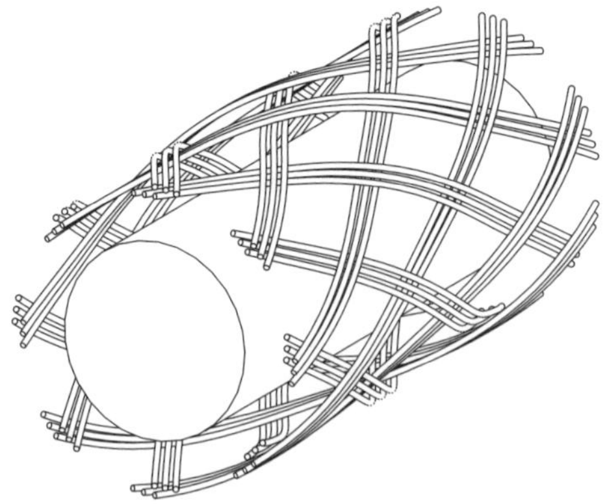

Figure 3.19 Example for a braided shield ($n_{\mathrm{carrier}} = 16$, $n_{\mathrm{wire}} = 3$, $r_{a_{\min}} = 1.45\,\mathrm{mm}$, $r_{a_{\max}} = 1.55\,\mathrm{mm}$, $r_0 = 0.01\,\mathrm{mm}$, center conductor radius $r_i = 0.02\,\mathrm{mm}$, $l_w = 2\,\mathrm{cm}$, $l_\mathrm{p} = 0.5\,\mathrm{cm}$, $h_0 = 30\,\mathrm{mm}$, $\Delta\varphi_w = 4.5°$).

are n_{wire} wires. The wires are phase shifted by $\Delta\varphi_\omega$. Figure 3.19 depicts such a cable shield calculated with the aid of the above equations.

(ii) Computational Results

For the cable structure shown in Figure 3.19 the equivalent mutual and self inductances ($L'_{\text{c,c}}, L'_{\text{s,c}}, L'_{\text{c,s}}, L'_{\text{s,s}}$) for different numbers of wires per carrier were computed along the shield. The results are in accordance with the inductances that can be computed with the formulas from the classical transmission-line theory. For the outer system (solid cylinder over ground plane) and the inner system (coaxial line) one obtains:

$$L'_0 = \frac{\mu}{2\pi} \ln \frac{2h_0}{r_a} = 737.8 \text{ nH} \tag{3.93}$$

$$L'_i = \frac{\mu}{2\pi} \ln \frac{r_a}{r_i} = 863.5 \text{ nH} \tag{3.94}$$

Figure 3.20 depicts the corresponding values obtained from the equivalent per-unit-length parameters by using Equations (3.60) and (3.61). The values do not completely match, since Equations (3.93) and (3.94) are only valid for ideal configurations, but there is a reasonable coincidence. Figure 3.21 shows the self inductance of the cable shield and the mutual inductance between the shield and the center conductor. Additonally, at the bottom of this picture the cable shield itself is shown. The mutual inductance is the same for all configurations, while

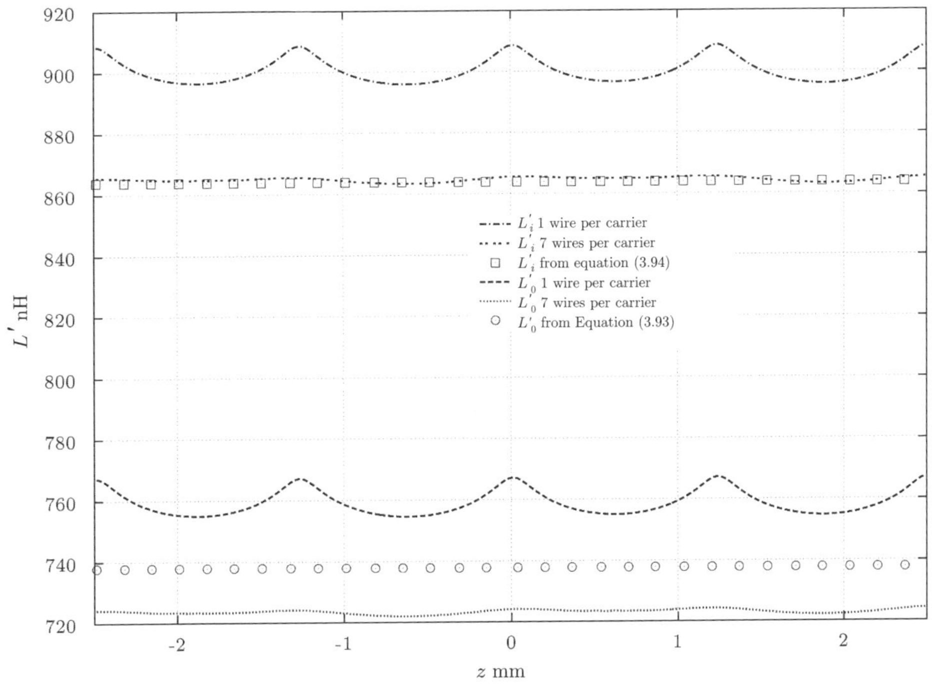

Figure 3.20 Comparison between the inductances of the outer and the inner system, obtained with the proposed method and with the formulas from the classical transmission–line theory.

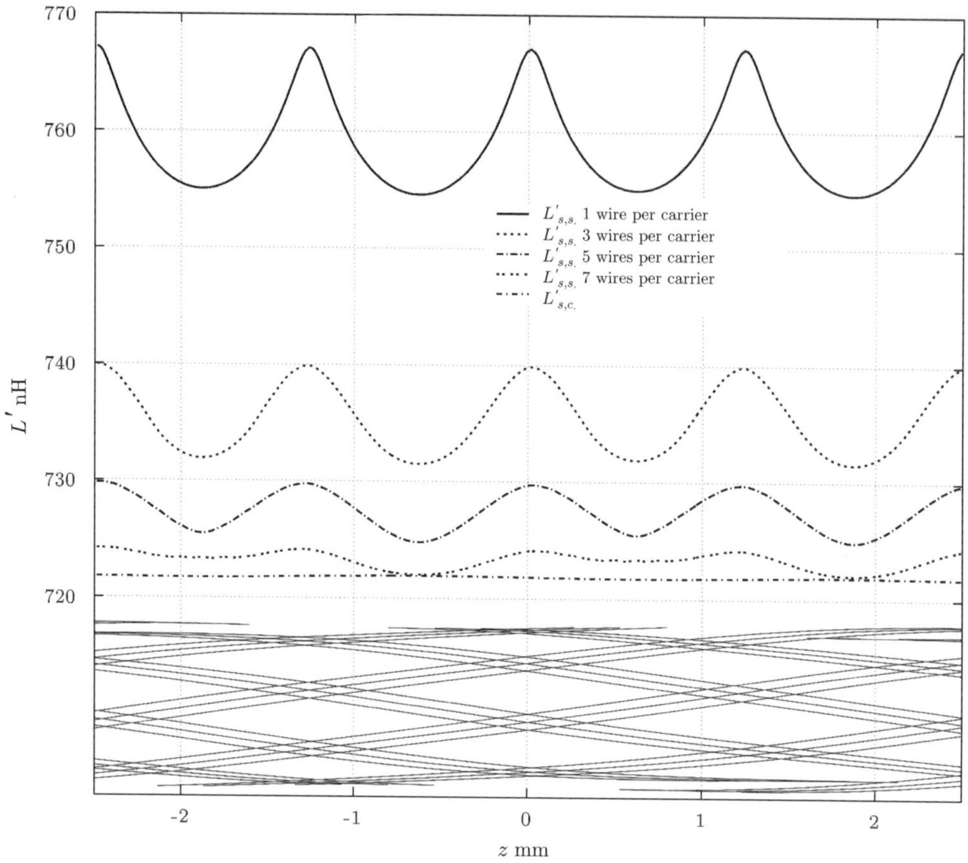

Figure 3.21 Equivalent self and mutual inductances ($L'_{s,s}$, $L'_{s,c}$) along a braided shield cable for different number of wires per carrier (other parameters as in Figure 3.19).

the self inductance clearly depends on the number of wires per carrier. For more wires, the self inductance becomes smaller.

The inductive part of the transfer impedance is given as the difference of these two values (see Equation (3.62)). The picture reveals some interesting aspects of the transfer impedance. It has a maximum when it is evaluated at the center of an aperture and a minimum in between. If more wires per carrier are used the apertures become smaller and so does the difference between the maximum and the minimum. Simultaneously the self inductance becomes smaller. Moreover, every second minimum is slightly larger than the other minima. This is caused by the changeover of the carriers from the outside to the inside and vice versa.

3.4.2 *Shielding of an Anisotropic Spherical Shell*

Carbon fiber reinforced plastic (CFRP) is a composite material that is frequently used for building the outer structure of modern aircraft. As a multilayer system it exhibits a strong

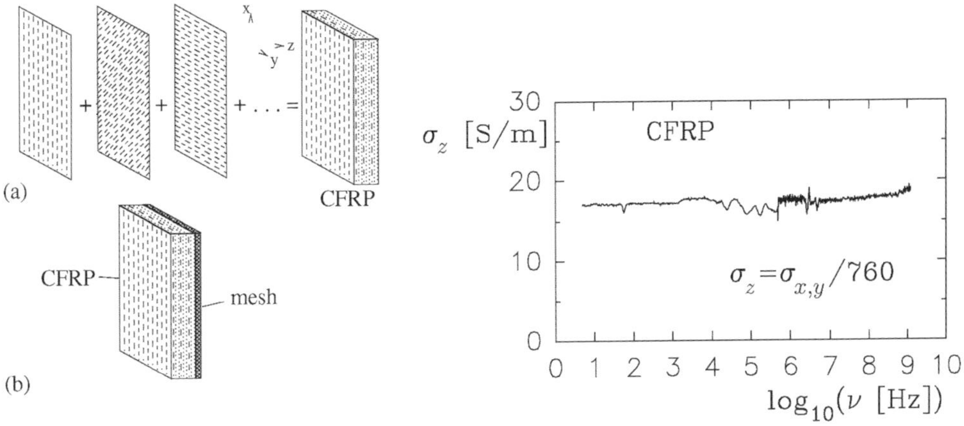

Figure 3.22 Left side : (a) Carbon fiber reinforced plastic as a multilayer system with different layer orientations. For lightning protection of aircraft the surface may be covered with a metallic mesh (b). Right side: Conductivity of CFRP measured perpendicular to the layers vs. frequency in a semilog plot. The conductivity in the direction of the layers is larger by a factor of 760. For the measurement technique see the literature [30].

anisotropic conductivity. These properties are measured in a broad range of frequencies from d.c. to 2 GHz. The experimental data are used to evaluate electric, magnetic and electromagnetic shielding of a spherical shell. At low frequencies up to 1 MHz the effect of anisotropy can be taken into account analytically. It is shown that for thin shells (thickness $\ll$ radius) only the effective conductivity parallel to the surface determines the shielding. Using this conductivity the equations for isotropic shells apply. An exact analytical solution is used to obtain simple expressions for the shielding at low frequencies as well as in the high-frequency range, where resonances become important.

CFRP is a heterogeneous material consisting of layers with different fiber orientations (see Figure 3.22). Due to its small weight and high mechanical stability it is used, for example, in the aircraft industry forming the outer structure of aircraft. In this case, the surface or parts of it may be covered with an additional metallic mesh in order to guarantee lightning protection. The structural anisotropy induces an electric one. In numerical studies the anisotropic microstructure is generally modeled in a simplified way as a perfect periodic fiber-matrix composite and then either shielding or reflection are calculated [21, 22, 25, 33]. However, the anisotropic conductivity of the material depends markedly on details of the microstructure, especially on the quality of the contacts between single fibers. Therefore, these simplified models cannot yield realistic values for the shielding properties. Here a different approach is used, that is the multilayer system is treated as a homogeneous effective medium, the anisotropic conductivity of which is determined by experiment. This is correct as long as the wavelength is large compared to the length scale of the inhomogenities (fiber thickness and distances between the fibers), that is up to about 1 GHz. Then the shielding is calculated in an analytical way.

The effective conductivity of CFRP samples parallel to the surface, $\sigma_{x,y}$ was measured, as well as perpendicular to it, σ_z. Contact resistances between electrodes and samples were reduced applying a silver paste. Apart from d.c. measurements a transmission method in the frequency range from 5 Hz to 2 GHz was also used (see Figure 3.22). As a result one obtains

$$\begin{aligned} \sigma_{x,y} &= 1.3 \cdot 10^4 \text{ S/m} \qquad \text{(parallel to the layers)} \\ \sigma_z &= 17.1 \qquad \text{S/m} \qquad \text{(perpendicular to the layers).} \end{aligned} \tag{3.95}$$

Note that these are effective conductivities of the hetereogeneous material: $\sigma_{x,y}$ is an average value for all fiber orientations (see Figure 3.22(a)). The conductivity perpendicular to the fiber layers, σ_z, is much smaller due to the contact resistances between the fibers. These high conductivities dominate the interaction with electromagnetic waves up to microwave frequencies, that is the complex permittivity is governed by energy dissipation due to free charge carriers:

$$\varepsilon_a = -i\frac{\sigma_a}{\varepsilon_0 \omega} \qquad \text{with} \quad a = (x, y) \text{ or } z, \tag{3.96}$$

where $\omega = 2\pi\nu$ denotes the angular frequency and $\varepsilon_0 = 8854 \times 10^{-12}$ F/m is the permittivity of free space. The anisotropy of the material can be described by the parameter

$$A = \frac{\varepsilon_z}{\varepsilon_{x,y}} = \frac{\sigma_z}{\sigma_{x,y}} = \frac{1}{760}. \tag{3.97}$$

Here, interest focusses on the influence of electric anisotropy on the shielding properties of CFRP. Since it is already a difficult task to calculate the coupling of electromagnetic waves into complex geometric structures made of isotropic materials, focus is on simple geometries without any openings. In addition, it is assumed that the shielding walls are manufactured as a single piece (real structures are assembled from CFRP sheets so that contact resistances at the joints weaken the shielding). At first, a short review on the shielding properties of a plain (anisotropic) wall is given, then focus is on a spherical shell (see Figure 3.23).

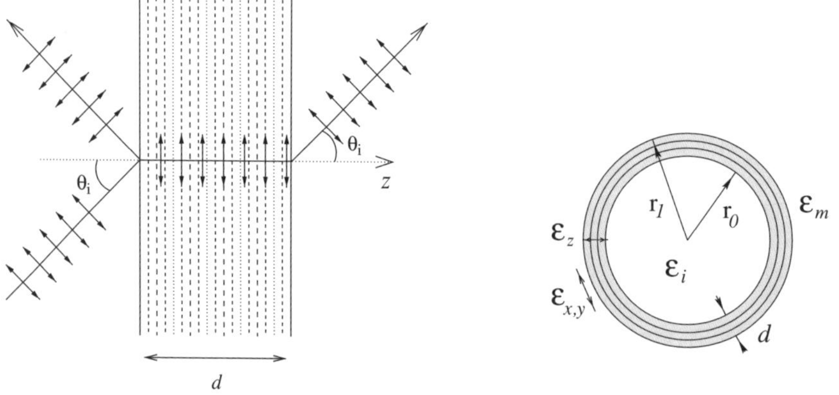

Figure 3.23 Left side: plain wall made of CFRP exposed to a TEM-mode. Right side: spherical shell made of CFRP.

3.4.2.1 Shielding of a Plane Wall

In the following, a plane wall is considered exposed to an incident TEM mode (excluding the case of grazing incidence; for details see [31]). The high conductivity values of CFRP imply a high refraction index, $n = \mathrm{Re}\sqrt{\varepsilon_{x,y}}$, and thus the angle of refraction is close to 0 (see Figure 3.23). As a consequence, inside the material electric and magnetic fields oscillate in the direction of the layers so that the above anisotropy does not affect the shielding. Using the experimental value for $\sigma_{x,y}$ the transmission coefficients for both the electric and the magnetic field are easily calculated, see [29, 32, 34]. For normal incidence

$$S_{21} = \frac{E_{\mathrm{trans}}}{E_{\mathrm{inc}}} = \frac{H_{\mathrm{trans}}}{H_{\mathrm{inc}}} = \frac{(1-r^2)a}{1-r^2a^2}, \qquad \text{where} \qquad r = \frac{1-\sqrt{\varepsilon_{x,y}}}{1+\sqrt{\varepsilon_{x,y}}} \tag{3.98}$$

$$\text{and} \qquad a = \exp(-ik_1 \times d). \tag{3.99}$$

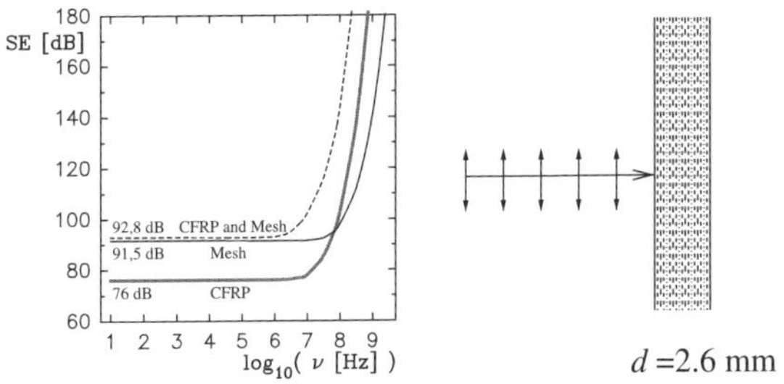

Figure 3.24 Shielding effectiveness of CFRP vs. frequency for a plain electromagnetic wave of normal incidence (from 10 Hz to 1 GHz in a semilog plot). Also the shielding of a metallic mesh and the shielding of CFRP with a mesh on its surface (see Figure 3.22) is shown.

Here, $k_1 = \omega/c \times \sqrt{\varepsilon_{x,y}}$ denotes the wave number and d is the thickness of the wall (in this case $d = 2.6$ mm). The shielding effectiveness in the far field, $SE = -20 \times \log_{10}(|S_{21}|)$ is displayed in Figure 3.24. A metallic mesh at the surface of the CFRP (see Figure 3.22) enhances the shielding effectiveness (please keep in mind that contact resistances between different sheets of CFRP are neglected). For details of the calculation see [31].

3.4.2.2 Spherical Shell

Effect of Anisotropy at Quasi-static Frequencies

Consider a spherical shell made of anisotropic material as sketched in Figure 3.23. The quantity ε_i denotes the permittivity inside the shell, ε_m the permittivity outside, r_0 and r_1 the inner and outer radius, respectively (wall thickness $d = r_1 - r_0$). The electric shielding at quasi-static frequencies ($\lambda_0 \gg 2\pi r_1$) has been calculated in [23]: when the shell is exposed to an arbitrary electric field (excitation), it is polarized. The resulting total field has to fulfill the equation div $\boldsymbol{\varepsilon}(\omega)\mathbf{E} = 0$ with permittivity tensor $\varepsilon(\omega)$. It can be written as $\mathbf{E} = -\nabla V(\mathbf{r})$, whereas the potential takes the following form in spherical coordinates:

$$V_i(\mathbf{r}) = \sum_l P_l(\cos\Theta) \quad \times A_l \times r^l \qquad \text{inside } (0 < r < r_0) \tag{3.100}$$

$$V_s(\mathbf{r}) = \sum_l P_l(\cos\Theta) \quad \times (C_l \times r^{u_+} + D_l \times r^{u_-}) \qquad \text{shell } (r_0 < r < r_1) \tag{3.101}$$

$$V_a(\mathbf{r}) = \sum_l P_l(\cos\Theta) \quad \times (\underbrace{F_l \times r^l}_{\text{excitation}} + \underbrace{G_l \times r^{-(l+1)}}_{\text{reflection}}) \qquad \text{outside } (r > r_1) \tag{3.102}$$

P_l are the Legendre polynomials and the exponent

$$u_\pm = 0.5 \times \left(-1 \pm \sqrt{1 + 4l(l+1)/A}\right) \tag{3.103}$$

depends on the anisotropy parameter A defined above (Equation (3.97)). For $A = 1$ one has $u_+ = l$ and $u_- = -(l+1)$, that is the above equations reduce to the well-known result describing spherical shells made of isotropic material. The field inside is connected to the field outside via the usual boundary conditions, that is the continuity of the normal component for $\boldsymbol{\varepsilon}(\omega)\mathbf{E}$ (here this is the radial term $\varepsilon_z \partial V/\partial r$). The same holds for the tangential component of the electric field, that is for $1/r \times \partial V/\partial\Theta$. Solving the linear set of equations leads to expressions for the ratio of the polynomial coefficients A_l/F_l, C_l/F_l, D_l/F_l and G_l/F_l. The ratio of the potential inside to the potential of the excitation is given by [23]

$$\frac{A_l}{F_l} = \frac{-r_0^{-1-l-u_-} r_1^{l-u_+} \varepsilon_z \varepsilon_m (2l+1) w_l}{(l\varepsilon_i - \varepsilon_z u_+)(\varepsilon_z u_- + \varepsilon_m(l+1)) \times (r_0/r_1)^{w_l} - (l\varepsilon_i - \varepsilon_z u_-)(\varepsilon_z u_+ + \varepsilon_m(l+1))} \tag{3.104}$$

where $w_l = \sqrt{1 + 4l(l+1)/A}$. A shell exposed to a homogeneous field corresponds to a dipolar excitation ($l = 1$), that is $E_a = -(F_1 r\cos\Theta) = -F_1$, whereas for the field inside $E_i = -A_1$ holds. So the ratio of the electric fields is given by $E_i/E_a = A_1/F_1$. For this very case the above equation can be written in a modified form:

$$\frac{E_i}{E_a} = \frac{(r_0/r_1)^{-3/2+w/2} \cdot 3\varepsilon_z\varepsilon_m \cdot w}{[\varepsilon_i + \frac{w+1}{2}\varepsilon_z] \cdot [2\varepsilon_m + \frac{w-1}{2}\varepsilon_z] - (r_0/r_1)^w \cdot [\varepsilon_i - \frac{w-1}{2}\varepsilon_z] \cdot [2\varepsilon_m - \frac{w+1}{2}\varepsilon_z]} \tag{3.105}$$

where the effect of anisotropy is described by the parameter

$$w = \sqrt{1 + 8/A} = \sqrt{1 + 8\varepsilon_{x,y}/\varepsilon_z}. \tag{3.106}$$

In the isotropic case ($A = 1$ and thus $\varepsilon = \varepsilon_z = \varepsilon_{x,y}$) one has $w = 3$ and the above result simplifies to

$$\left.\frac{E_i}{E_a}\right|_{A=1} = \frac{9\varepsilon_m\varepsilon}{(2\varepsilon + \varepsilon_i) \times (\varepsilon + 2\varepsilon_m) - 2(r_0/r_1)^3 \times (\varepsilon - \varepsilon_i) \cdot (\varepsilon - \varepsilon_m)}. \tag{3.107}$$

3.4.2.3 Application to Thin Conductive Shells

The above result shows that in general the coupling of electric fields into a spherical cell depends on the anisotropy of the material. However, it will be shown in the following that this is not necessarily the case for cells with thin conductive walls [$\varepsilon_{x,y} = -i\sigma_{x,y}/(\varepsilon_0\omega)$ and $\varepsilon_z = -i\sigma_z/(\varepsilon_0\omega)$ with $\sigma_{x,y}, \sigma_z > 0$]. The surrounding medium is supposed to be insulating, that is ε_i and ε_m are real quantities (see Figure 3.23). For sufficiently high conductivities (or equivalently, for sufficiently low frequencies) $|(w \pm 1)\varepsilon_z| = |\sqrt{\varepsilon_z^2 + 8\varepsilon_{x,y}\varepsilon_z} \pm \varepsilon_z| \gg \varepsilon_i, \varepsilon_m$ holds and Equation (3.105) becomes

$$\frac{E_i}{E_a} \simeq i \times \varepsilon_m \times \frac{3\varepsilon_0\omega}{2\sigma_{x,y}} \cdot \underbrace{\left(\frac{w \cdot \left(\frac{r_0}{r_1}\right)^{-3/2}}{\left(\frac{r_0}{r_1}\right)^{-w/2} - \left(\frac{r_0}{r_1}\right)^{w/2}} \right)}_{\to r_0/d \text{ for } |3 \pm w|/2 \times d \ll r_0} . \tag{3.108}$$

As indicated, a Taylor expansion of the latter term yields for a thin shell

$$\frac{E_i}{E_a} \simeq i \times \varepsilon_m \times \frac{3\varepsilon_0\omega}{2\sigma_{x,y}} \times \frac{r_0}{d} = i \times \varepsilon_m \cdot \frac{3R_s}{2Z_0} \times k_0 r_0 \qquad \text{for} \quad k_0 r_0 \ll 1, \tag{3.109}$$
$$|(w \pm 1)\varepsilon_z| \gg \varepsilon_i, \varepsilon_m \quad \text{and} \quad |3 \pm w|/2 \times d \ll r_0$$

$R_s = 1/(\sigma_{x,y} d)$ denotes the film resistance of the shell and $c = 1/\sqrt{\varepsilon_0\mu_0}$ is the speed of light. $Z_0 = \sqrt{\mu_0/\varepsilon_0} = 1/(\varepsilon_0 c) = 376.7\ \Omega$ and $k_0 = \omega/c$ denote the wave impedance and wave number of free space, respectively.

Note that the above result (Equation (3.109)), does not depend on the anisotropy A. Therefore, it describes isotropic as well as anisotropic materials (but the smaller the anisotropy parameter A, that is the more the conductivity parallel to the surface dominates, the smaller the thickness of the shell has to be). Similar to the case of large plain shielding walls, only the conductivity parallel the the surface, $\sigma_{x,y}$, determines the shielding. Summarizing, the electric shielding of thin conductive shells is given by

$$a_{\text{el}} = -20 \times \log_{10}\left(\frac{|E_i|}{|E_a|}\right) = -20 \times \log_{10}\left(\varepsilon_m \times \frac{3R_s}{2Z_0} \times k_0 r_0\right) . \tag{3.110}$$

For a shell made of CFRP (wall thickness $d = 2.6$ mm and conductivity as in Equation (3.95)) this yields

$$a_{\text{el}}^{\text{CFRP}} = 78.45 - 20 \times \log_{10}(k_0 r_0)\,[\text{dB}] \quad \text{for} \quad k_0 r_0 \leq 0.2 \quad \text{and} \quad r_0 \geq 0.25 \text{ m}. \tag{3.111}$$

The result is displayed in Figure 3.25. The CFRP shell behaves as a Faraday screen, that is the smaller the frequency, the film resistance or the radius, the better the electric shielding.

Electric and Magnetic Shielding of a Thin Conductive Shell in the Entire Frequency Range

It was shown above that for thin conductive shells ($d \ll r_0$) the conductivity parallel to the surface solely determines the electric shielding properties in the quasi-static frequency range ($k_0 r_0 \ll 1$). This means, that the equations derived for isotropic spherical shields can be used for an anisotropic material setting $\sigma = \sigma_{x,y}$. In the following, it is assumed that this remains a good approximation even at higher frequencies up to 1 GHz. Furthermore, one can expect without any further proof that the situation is similar for the magnetic shielding, at least for materials that do not show magnetic anisotropy (CFRP is nonmagnetic, that is for the permeability $\mu = 1$ holds). The incident electromagnetic wave excites eddy currents inside

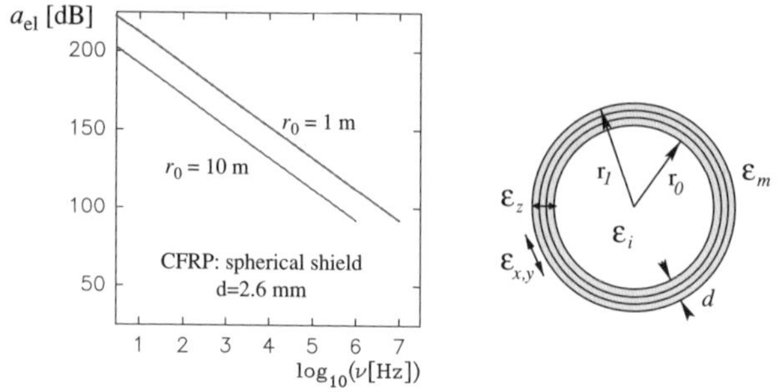

Figure 3.25 Electric shielding of a CFRP shell in air ($\varepsilon_i = \varepsilon_m = 1$). In the frequency range displayed $\lambda_{vac}\pi \gg 2r_0 > 0.5$ m holds and thus Equations (3.110) and (3.111) apply.

the thin shell, so that for the weakening of magnetic fields only the conductivity parallel to the surface is of importance.

In the literature there are several approximating equations describing the shielding of isotropic shields (e.g., [24]). Since these are obtained neglecting a term in the underlying differential equation, their range of validity is not obvious. For this reason, the exact analytical solution from [26–28] is used. There one obtains for the fields inside the cell at $r = 0$ (note that there is a spatial variation in the high frequency range, i.e., for $k_0 r_0 \gg 1$):

$$\left.\frac{E_i}{E_a}\right|_{r=0} = -\frac{1}{(k_1 r_0 k_0 r_1)^2} \times \frac{1}{F_1} \quad \text{and} \quad \left.\frac{H_i}{H_a}\right|_{r=0} = -\frac{1}{(k_1 r_0 k_0 r_1)^2} \times \frac{1}{G_1} \tag{3.112}$$

where

$$\begin{aligned} F_1 = {} & \left[\alpha(x_{0,0})h_1^{(2)}(x_{1,0}) - \sqrt{\frac{\mu}{\varepsilon}}\, j_1(x_{0,0})\beta(x_{1,0})\right] \times \left[\alpha(x_{1,1})h_1^{(2)}(x_{0,1}) - \sqrt{\frac{\varepsilon}{\mu}}\, j_1(x_{1,1})\beta(x_{0,1})\right] \\ & - \left[\alpha(x_{0,0})j_1(x_{1,0}) - \sqrt{\frac{\mu}{\varepsilon}}\, j_1(x_{0,0})\alpha(x_{1,0})\right] \times \left[\beta(x_{1,1})h_1^{(2)}(x_{0,1}) - \sqrt{\frac{\varepsilon}{\mu}}\, h_1^{(2)}(x_{1,1})\beta(x_{0,1})\right] \end{aligned} \tag{3.113}$$

$$\begin{aligned} G_1 = {} & \left[\alpha(x_{0,0})h_1^{(2)}(x_{1,0}) - \sqrt{\frac{\varepsilon}{\mu}}\, j_1(x_{0,0})\beta(x_{1,0})\right] \times \left[\alpha(x_{1,1})h_1^{(2)}(x_{0,1}) - \sqrt{\frac{\mu}{\varepsilon}}\, j_1(x_{1,1})\beta(x_{0,1})\right] \\ & - \left[\alpha(x_{0,0})j_1(x_{1,0}) - \sqrt{\frac{\varepsilon}{\mu}}\, j_1(x_{0,0})\alpha(x_{1,0})\right] \times \left[\beta(x_{1,1})h_1^{(2)}(x_{0,1}) - \sqrt{\frac{\mu}{\varepsilon}}\, h_1^{(2)}(x_{1,1})\beta(x_{0,1})\right]. \end{aligned} \tag{3.114}$$

Here $x_{m,n} = k_m r_n$ ($m, n = 0$ or 1), $k_0 = \omega/c$ and $k_1 = \sqrt{\varepsilon\mu} \times \omega/c$ is the wave number in the material. The spherical Bessel function j_1, the spherical Hankel function of the second kind $h_1^{(2)}$ and the functions α and β are defined as:

$$\begin{aligned} j_1(x) &= \frac{\sin x}{x^2} - \frac{\cos x}{x}, & h_1^{(2)}(x) &= i \times \frac{e^{-ix}}{x^2} - \frac{e^{-ix}}{x}, \\ \alpha(x) &= \frac{1}{x} \times \frac{d}{dx}[x j_1(x)], & \beta(x) &= \frac{1}{x} \times \frac{d}{dx}[x h_1^{(2)}(x)]. \end{aligned} \tag{3.115}$$

In the following this exact analytic solution is used in order to derive simple expressions that allow the features that determine the shielding properties of a spherical shell to be understood. This is done by inserting Taylor expansions or asymptotic forms of the functions in Equation (3.115) into Equations (3.113) and (3.114). In doing so it is always assumed that $|\varepsilon| = \sigma/(\varepsilon_0\omega) \gg |\mu|$. At low frequencies one obtains for the electric and the magnetic shielding, $a_{\mathrm{el}} = -20 \times \log_{10}(|E_i/E_a|)$ and $a_{\mathrm{mag}} = -20 \times \log_{10}(|H_i/H_a|)$, respectively:

$$a_{\mathrm{el}} = -20 \times \log_{10}\left(\varepsilon_m \times \frac{3R_s}{2Z_0} \times k_0 r_0\right) \quad \text{for} \quad k_0 r_0 \ll 1, \text{ and } |k_1 d| \ll 1 \times R_s/Z_0 \ll 1 \tag{3.116}$$

$$a_{\mathrm{mag}} \simeq 20 \times \log_{10}\left|1 + \frac{2d}{3r_0} \times \frac{(\mu-1)^2}{\mu} + i \times \frac{k_0 r_0}{3} \times \frac{Z_0}{R_s}\right| \quad \text{for} \quad k_0 r_0 \ll 1 \text{ and } |k_1 d| \ll 1. \tag{3.117}$$

Equation (3.116) corresponds to the result being obtained above in Equation (3.110) for thin isotropic or anisotropic shells. Due to the high conductivity a good electric shielding is obtained (Figure 3.26 up to 10 MHz). However, the magnetic shielding at low frequencies is very weak. CFRP is nonmagnetic, that is $\mu = 1$ holds and thus $a_{\mathrm{mag}}(\nu \to 0) = 0\,\mathrm{dB}$. Only with increasing frequency the excitation of eddy currents yields an enhanced magnetic shielding (Figure 3.26 between 10 kHz and 10 MHz).

At high frequencies, where $k_0 r_0 \gg 1$ and $|k_1 d| \ll 1$ hold, the shielding increases due to the skin effect (skin depth $\delta = 1/\sqrt{\pi\mu_o\mu\sigma\nu}$). One obtains

$$a_{\mathrm{el}}|_{r=0} \simeq \underbrace{20 \times \log_{10}(e)}_{\simeq 8,69} \cdot \frac{d}{\delta} + 10 \times \log_{10}\left(\left[\sqrt{\frac{\sigma}{4\varepsilon_0\mu\omega}} \times \sin(k_0 r_0) + \frac{\cos(k_0 r_0)}{\sqrt{8}}\right]^2 + \frac{\cos^2(k_0 r_0)}{8}\right) \tag{3.118}$$

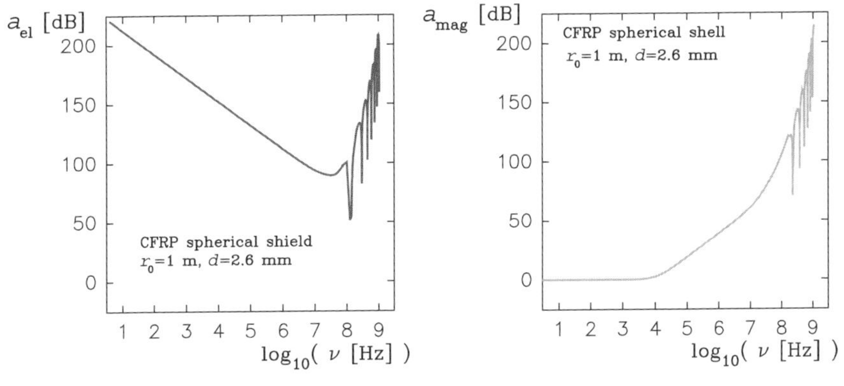

Figure 3.26 Electric and magnetic shielding of a CFRP shell in air.

and

$$a_{\mathrm{mag}}|_{r=0} \simeq \underbrace{20 \times \log_{10}(e)}_{\simeq 8{,}69} \cdot \frac{d}{\delta} +$$

$$+ 10 \times \log_{10}\left(\left[\sqrt{\frac{\sigma}{4\varepsilon_0\mu\omega}} \times \cos(k_0 r_0) - \frac{\sin(k_0 r_0)}{\sqrt{8}}\right]^2 + \frac{\sin^2(k_0 r_0)}{8}\right) . \quad (3.119)$$

Due to multiple reflections inside the cell the shielding breaks down at so-called geometric resonances, that is at $2r_0 \simeq n \times \lambda$ (electric field) and at $2r_0 \simeq (2n+1) \times \lambda/2$ (magnetic field). This behavior can be seen in Figure 3.26 above 100 MHz. Starting with an electric resonance ($n = 1$), electric and magnetic resonances alternate with increasing frequency.

Electromagnetic Shielding

As one could see above, electric and magnetic shielding differ considerably. Since half of the power of an incident TEM mode is due to its electric field while the other 50 % is due to its magnetic field, the so-called electromagnetic shielding is defined via the average value of electric and magnetic coupling, that is

$$a_{\mathrm{el-mag}}[\mathrm{dB}] = -10 \times \log_{10}\left(\frac{1}{2} \times \left|\frac{E_i}{E_a}\right|^2 + \frac{1}{2} \times \left|\frac{H_i}{H_a}\right|^2\right) . \quad (3.120)$$

The values for a spherical shell made of CFRP are displayed in Figure 3.27. At low frequencies, the electric shielding is very good, that is $|E_i/E_a| \to 0$, while magnetic fields are not shielded at all, $H_i/H_a \simeq 1$. Therefore, half of the energy is coupled into the cell and thus $a_{\mathrm{el-mag}} \simeq 3$ dB. The coupling of magnetic fields dominates the entire low-frequency range, so that a good electromagnetic shielding is only obtained above some MHz. In the high-frequency range $a_{\mathrm{el-mag}}$ shows both the electric and the magnetic geometrical resonances, that is the shielding breaks down for $2r_0 \simeq n \times \lambda/2$.

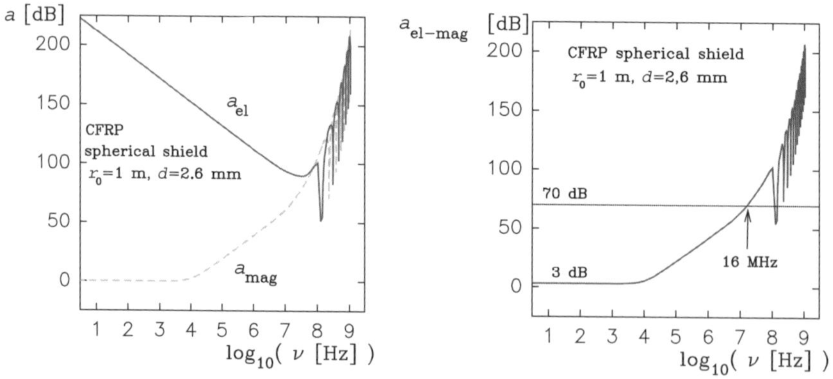

Figure 3.27 Left side: electric and magnetic shielding of a CFRP shell vs. frequency in a semilog plot. Right side: electromagnetic shielding vs. frequency in a semilog plot. At low frequencies the magnetic coupling determines the shielding, so that $a_{\mathrm{el-mag}} < 70$ dB for $\nu < 16$ MHz.

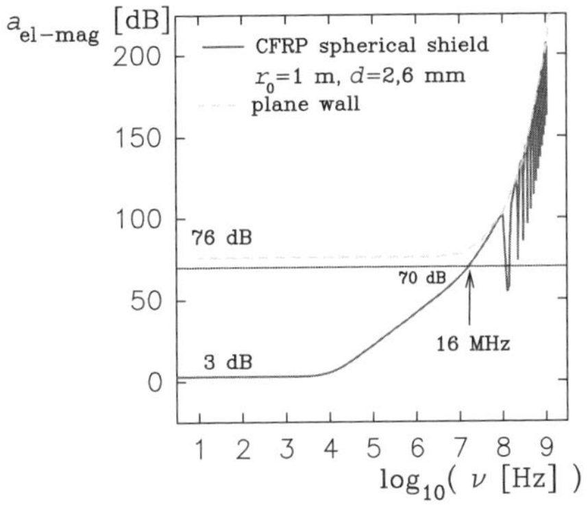

Figure 3.28 Electromagnetic shielding of a spherical shell made of CFRP compared to that of a plain wall.

3.4.2.4 Conclusions

The anisotropic conductivity of CFRP up to 2 GHz was measured. Using an effective medium approach, the shielding properties of this material were calculated. For a plane wall only the conductivity parallel to the surface, $\sigma_{x,y}$, determines the shielding, that is the anisotropy is not important and $a_{\mathrm{el}} = a_{\mathrm{mag}} = a_{\mathrm{el-mag}}$ holds. In the case of a conductive spherical shell, the effect of anisotropy can only be neglected when the thickness of the shell is small compared to its radius. In this case, once again only $\sigma_{x,y}$ determines the shielding. In general, $a_{\mathrm{el}} \neq a_{\mathrm{mag}} \neq a_{\mathrm{el-mag}}$ holds and magnetic coupling dominates in the low-frequency range (see Figure 3.27). In Figure 3.28 the electromagnetic shielding of a plane wall to that of a thin shell is compared. Note that the values displayed refer to perfect geometries, while real structures are assembled from smaller sheets. Joints with contact resistances will considerably weaken the shielding. The same holds for apertures (see, e.g., [20]).

References

[1] Baum, C.E., Liu, T.K. and Tesche, F.M. 'On the Analysis of General Multiconductor Transmission-Line Networks', *Interaction Note 350*, 1978.

[2] Baum, C.E. 'Electromagnetic Topology : A Formal Approach to the Analysis and Design of Complex Electronic Systems', *Interaction Note 400*, 1980.

[3] Lee, K.S.H. *EMP Interaction : Principles, Techniques and Reference Data*, Hemisphere Publishing Corporation, Washington, USA, 1986.

[4] Karlsson, T. 'The Topological Concept of a Generalized Shield', *Interaction Note 461*, 1988.

[5] Parmantier, J.P. 'Application of EM Topology on Complex Wiring Systems', in Proceedings of the International Symposium on Electromagnetic Compatibility, Magdeburg, Germany, 1999, pp. 1–8.

[6] Paul, C.R. *Analysis of Multiconductor Transmission Lines*, John Wiley & Sons, Inc., New York, USA, 1994.

[7] Tesche, F.M., Ianoz, M.V. and Karlsson, T. *EMC Analysis Methods and Computational Models*, John Wiley & Sons, Inc., New York, USA, 1997.

[8] Granzow, K.D. *Digital Transmission Lines*, Oxford University Press, Oxford, UK, 1998.

[9] Dworsky, L.N. *Modern Transmission Line Theory And Application*, Robert E. Krieger Publishing Company, Malabar, Florida, USA, 1988.

[10] Faria, J.A.B. *Multiconductor Transmission-Line Structures – Modal Analysis Techniques*, John Wiley & Sons, Inc., New York, USA, 1993.

[11] Faché, N., Olyslager, F. and De Zutter, D. *Electromagnetic and Circuit Modelling of Multiconductor Transmission Lines*, Clarendon Press, Oxford, UK, 1993.
[12] King, R.W.P. *Transmission-Line Theory*, McGraw Hill, New York, USA, 1955.
[13] Frankel, S. *Multiconductor Transmission Line Analysis*, Artech House, Norwood, USA, 1977.
[14] Baum, C.E. 'Generalization of the BLT Equation', *Interaction Note 511*, 1995.
[15] Parmantier, J.-P. 'An Efficient Technique to Calculate Ideal Junction Scattering Parameteres in Multiconductor Transmission Line Networks', *Interaction Note 536*, 1998.
[16] Bennani, S. and Rifi M. 'A comparison between improved models for transfer impedance calculations of braided coaxial cable', in Proceedings of the 4th European Symposium on EMC, vol. 1, Brugge, 2000 pp. 89–92.
[17] Schelkunoff, S. 'The electromagnetic theory of coaxial transmission lines and cylindrical shields', *Bell Syst. Tech. J.* **13**, 1934, pp. 532–579.
[18] Vance, E. *Coupling to Shielded Cables*, John Wiley and Sons, Inc., New York, USA, 1978.
[19] Tyni, M. 'The transfer impecance of coaxial cables with braided outer conductor in Proceedings of the, 3rd International Symposium on Electromagnetic Compatibility, Poland, Wroclaw, 1976, pp. 410–419.
[20] Blume, S. and Klinkenbusch, L. 'Scattering of a Plane Wave by a Spherical Shell with an Elliptical Aperture', *IEEE Trans. EMC*, **34**, 1992, 308–314.
[21] Chiu, C.N. and Chen, C.H. 'Scattering from an Advanced Composite Cylindrical Shell', *IEEE Trans. EMC*, **38**, 1996, 62–67.
[22] Chu, H.-C. and Chen, C.H. 'Shielding and Reflection Properties of Periodic Fiber-Matrix Composite Structures', *IEEE Trans. EMC*, **38**, 1996, 1–6.
[23] Henrard, L. Etude théorique des excitations électroniques collectives des hyperfullerènes de carbone. Application à l'analyse du spectre d'absorption ultraviolet des poussières interstellaires, Doctorate thesis, Science Faculty, University of Notre-Dame de la Paix, Namur, Belgique, 1996.
[24] Kaden, H. *Wirbelströme und Schirmung in der Nachrichtentechnik* (ed. W. Meissner). Springer Verlag, Berlin, Germany, 1959.
[25] Kimmel, M. Brüns, H.-D. and Singer, H. 'Erweiterung der Momentenmethode zur Analyse inhomogener anisotroper Schirmmaterialien', Proceedings of EMV'96, Stuttgart, Germany, VDE-Verlag, 1996, pp. 269–276.
[26] Klinkenbusch, L. 'Rigorous field analysis of a spherical shield by using the multipole technique', *Archiv für Elektrotechnik*, **77**, 1994, pp. 315–325.
[27] Klinkenbusch, L. 'Multipole analysis of the electromagnetic field in a spherical anechoic chamber, Part I : Theory', *Electrical Engineering*, **78**, 1994, 9–17.
[28] Klinkenbusch, L. *Theorie der sphärischen Absorberkammer und des mehrschaligen Kugelschirms*, Habilitationsschrift, Ruhr-Universität Bochum, Germany, 1996.
[29] Nicolson, A.M. and Ross, G.F. 'Measurement of the intrinsic properties of materials by time-domain techniques', *IEEE Trans. Instr. & Meas.*, **19**, 1970, 377–382.
[30] Pelster, R. 'A Novel Analytical Method for the Broadband Determination of Electromagnetic Impedances and Material Parameters', *IEEE Trans. MTT*, **43**, 1995, 1494–1501.
[31] Pelster, R. and Nitsch, J. 'Electromagnetic Shielding of Anisotropic Multilayered Materials in Modern Aircraft', Proceedings of the International Symposium on Electromagnetic Compatability, Magdeburg, Germany, 1999, 171–178.
[32] Rost, A. *Messung dielektrischer Stoffeigenschaften*, Vieweg-Verlag, Braunschweig, Germany, 1978.
[33] Sarto, M.S. and Holloway, C.L. 'Effective Boundary Conditions for the Time-Domain Analysis of the EMC Performances of Fiber Composites', Proceedings of the IEEE International Symposium on EMC, Seattle, Washington, 1999, pp. 462–467.
[34] Schelkunoff, S.A. *Electromagnetic Waves*, van Nostrand Company, Toronto, New York, London, 1951.

4

The Method of Partial Element Equivalent Circuits (PEEC Method)

The partial element equivalent circuit (PEEC) method is a numerical method for modeling the electromagnetic behavior of arbitrary three-dimensional electrical interconnection structures. Starting from Maxwell equations in real media a mixed potential electric field integral equation (MPIE) is developed. After discretizing the structure, introducing evaluation functions for current and charge densities and applying the Galerkin method, the PEEC equation system in the frequency domain is derived in a general form. With the introduction of generalized partial elements the circuit interpretation results directly in PEEC models. These models are classified depending on the media and on the relation of the maximum frequency of interest to the discretization length. Different geometrical discretizations, for example orthogonal, nonorthogonal or triangular, only result in different formulas for the partial elements, but do not change the structure of the models. Particular attention is paid to the development of stable time domain PEEC models. Inclusion of the skin effect in frequency and time domain PEEC models and the development of PEEC models based on dyadic Green's functions for layered media substantially extend the application range of the PEEC method. This chapter is completed with a discussion of the relationship between PEEC models and uniform transmission lines as well as power considerations within PEEC models. The latter considerations yield insights into the origin and representation of radiation power at the circuit level.

4.1 Fundamental Equations

4.1.1 Maxwell Equations and Real Media for Interconnections

The PEEC method is derived from the Maxwell equations. The system of the Maxwell equations is explained in depth in Chapter 1. Complex interconnection structures to be modeled are composed of both conducting and dieleclric regions that are located in free or half space. The last one is characterized by the electric and magnetic field constants ϵ_0 and μ_0 respectively.

The finite regions of real media composing interconnection structures are further assumed to be linear, homogeneous, isotropic and nonmagnetic. The constitutive parameters for characterizing electric and magnetic features are the electric conductivity σ, the permittivity ϵ and the permeability $\mu = \mu_0$. In the sense of the intended modeling these parameters span a wide range of values from perfect electric conductors (PEC) to perfect lossless electric isolators (PEI). In many cases, especially for high frequencies and wide-band applications, the frequency dependence of constitutive parameters has to be considered. The electric conductivity σ of good conductors is independent of the frequency for all frequencies of practical interest (free electrons dominate the conducting mechanism).

In general, the permittivity $\underline{\epsilon} = \epsilon_0 \underline{\epsilon}_{\mathrm{r}}$ is a complex value with

$$\underline{\epsilon}_{\mathrm{r}} = \epsilon_r' - j\epsilon_{\mathrm{r}}'' \tag{4.1}$$

as the dispersion equation of the complex relative permittivity, where both ϵ_{r}' and ϵ_{r}'' can be functions of ω. The part ϵ_{r}'' characterizes the dielectric or polarization losses in a real dielectric (e. g. microwave heating).

Equation (1.138) may be written in the form

$$\underline{\mathbf{D}} = \underline{\epsilon}\,\underline{\mathbf{E}} = \epsilon_0 \underline{\epsilon}_{\mathrm{r}} \underline{\mathbf{E}} = \epsilon_0 \underline{\mathbf{E}} + \epsilon_0 (\underline{\epsilon}_{\mathrm{r}} - 1) \underline{\mathbf{E}} = \epsilon_0 \underline{\mathbf{E}} + \underline{\mathbf{P}} \tag{4.2}$$

with

$$\underline{\mathbf{P}} = \epsilon_0 (\underline{\epsilon}_{\mathrm{r}} - 1) \underline{\mathbf{E}} = \epsilon_0 \underline{\chi}_{\mathrm{e}} \underline{\mathbf{E}}. \tag{4.3}$$

$\underline{\mathbf{P}}$ is the vector of polarization of the medium and $\underline{\chi}_{\mathrm{e}}$ the electric susceptibility. Thus, the electric behavior of a dielectric is represented by the polarization of its bound charges in free space. Using Equation (1.161) for a homogeneous region holds

$$\nabla \cdot \underline{\mathbf{D}} = \underline{\epsilon} \nabla \cdot \underline{\mathbf{E}} = \epsilon_0 \nabla \cdot \underline{\mathbf{E}} + \nabla \cdot \underline{\mathbf{P}} = \underline{\varrho}_{\mathrm{f}} \tag{4.4}$$

with

$$\nabla \cdot \underline{\mathbf{P}} = -\underline{\varrho}_{\mathrm{b}}, \tag{4.5}$$

where $\underline{\rho}_{\mathrm{f}}$ is the density of free and $\underline{\varrho}_{\mathrm{b}}$ the density of bound charges. Rewriting Equation (4.4) yields

$$\nabla \cdot \underline{\mathbf{E}} = \frac{\underline{\varrho}_{\mathrm{f}}}{\underline{\epsilon}} = \frac{\underline{\varrho}_{\mathrm{f}} + \underline{\varrho}_{\mathrm{b}}}{\epsilon_0} = \frac{\underline{\varrho}}{\epsilon_0} \tag{4.6}$$

with

$$\underline{\varrho} = \underline{\varrho}_{\mathrm{f}} + \underline{\varrho}_{\mathrm{b}}. \tag{4.7}$$

Equation (4.7) means that both the free and the bound charges are sources of the electric field in free space. Further, the Maxwell Equation (1.162) is used and written in the form:

$$\begin{aligned} \nabla \times \underline{\mathbf{H}} = \underline{\mathbf{J}}_{\mathrm{C}} + \underline{\mathbf{J}}_{\mathrm{D}} &= \sigma \underline{\mathbf{E}} + j\omega \underline{\epsilon}\, \underline{\mathbf{E}} = \\ &= \sigma \underline{\mathbf{E}} + j\omega\epsilon_0 (\underline{\epsilon}_{\mathrm{r}} - 1)\mathbf{E} + j\omega\epsilon_0 \underline{\mathbf{E}}. \end{aligned} \tag{4.8}$$

In Equation (4.8) are:

- $\underline{\mathbf{J}}_{\mathrm{C}} = \sigma \underline{\mathbf{E}}$ conduction current density
- $\underline{\mathbf{J}}_{\mathrm{D}} = j\omega\underline{\epsilon}\underline{\mathbf{E}}$ displacement current density in the medium
- $\underline{\mathbf{J}}_{\mathrm{P}} = j\omega\epsilon_0(\underline{\epsilon}_{\mathrm{r}} - 1)\underline{\mathbf{E}}$ polarization current density
- $\underline{\mathbf{J}}_{\mathrm{Do}} = j\omega\epsilon_0\underline{\mathbf{E}}$ displacement current density in free space
- $\underline{\mathbf{J}} = \underline{\mathbf{J}}_{\mathrm{C}} + \underline{\mathbf{J}}_{\mathrm{P}}$ total current density based on moving charges.

Introducing the dispersion Equation (4.1) in Equation (4.8) yields

$$\nabla \times \underline{\mathbf{H}} = (\sigma + \omega\epsilon_0\epsilon_{\mathrm{r}}'')\underline{\mathbf{E}} + j\omega\epsilon_0(\epsilon_{\mathrm{r}}' - 1)\underline{\mathbf{E}} + j\omega\epsilon_0\underline{\mathbf{E}}. \tag{4.9}$$

Additional frequency-dependent heat losses described by the imaginary part ϵ_{r}'' of the complex permittivity can be included in an equivalent electric conductivity

$$\sigma_{\mathrm{e}} = \sigma + \omega\epsilon_0\epsilon_{\mathrm{r}}''. \tag{4.10}$$

The real part ϵ_{r}' of the complex permittivity describes the part of the polarization current that does not cause losses. From a practical point of view, the media for interconnection structures are categorized as:

- *conductors*, characterized by the frequency independent σ with

$$\sigma \gg |\omega\underline{\epsilon}|, \tag{4.11}$$

  a perfect electric conductor is an idealization for $\sigma \to \infty$;
- *dielectrics*, characterized by

$$|\omega\underline{\epsilon}| \gg \sigma, \tag{4.12}$$

  a lossless dielectric is an idealization for $\epsilon'' = 0$.

Taking the divergence of Equations (4.8) or (4.9) for a homogeneous region yields

$$\nabla \cdot \nabla \times \underline{\mathbf{H}} = (\sigma + j\omega\underline{\epsilon})\nabla \cdot \underline{\mathbf{E}} = 0, \tag{4.13}$$

which implies

$$\nabla \cdot \underline{\mathbf{E}} = 0\,.$$

On the other hand, Equation (4.6) holds. This seeming contradiction can be resolved: free charges cannot exist in a homogeneous medium in the steady state. Then, according to Equation (4.6):

$$\nabla \cdot \underline{\mathbf{P}} = -\underline{\varrho}_{\mathrm{b}} = 0$$

is valid.

In compliance with the boundary conditions between two different media (see Section 1.4.2) charges exist only as surface charges at the boundaries between different homogeneous regions.

This fact has to be taken into account in the discretization process for derivation of the numerical model (see Section 4.2.1):

- Boundaries between different homogeneous media also have to be boundaries of discretization elements (cells).
- One has to differentiate between outer cells (having common surface elements with boundaries of regions) and inner cells (not having such common elements).

In the case that one of the media is a very good conductor, the so-called Leontovich surface impedance $\underline{Z}_s$ [1]

$$\underline{Z}_s = \frac{\underline{E}_{t1}}{\underline{J}_s} = R_s + jX_s = (1 + j)\sqrt{\frac{\omega\mu_0}{2\sigma}} \tag{4.14}$$

can be derived from the boundary conditions (see Fig. 4.1). $\underline{Z}_s$ results from the skin effect calculation for the conducting half space.

Taking the divergence from Ampère's law of Equation (4.8), the continuity equation is obtained in the form

$$\nabla \cdot \left(\underline{\mathbf{J}}_C + \underline{\mathbf{J}}_P + \underline{\mathbf{J}}_{Do}\right) = 0 \tag{4.15}$$

with the conclusion that in free space the total current density consisting of the conducting current density, polarization current density and displacement current density is source-free. The continuity equation can also be written in the form

$$\nabla \cdot \underline{\mathbf{J}} + j\omega\underline{\varrho} = 0. \tag{4.16}$$

Equation (4.16) expresses the link between the sources of the total current density $\underline{\mathbf{J}}$ and the total charge density $\underline{\varrho}$ (see Equation (4.7)) in free space.

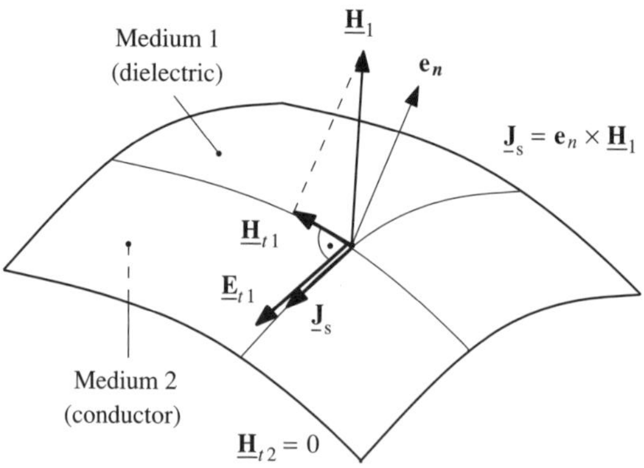

Figure 4.1 Leontovich surface boundary condition.

4.1.2 Mixed Potential Integral Equations (MPIE)

The electric field strength $\underline{\mathbf{E}}$ at a point $\mathbf{r}$ is given by

$$\underline{\mathbf{E}} = \underline{\mathbf{E}}^{\text{inc}} + \underline{\mathbf{E}}^{\text{sc}}. \tag{4.17}$$

$\underline{\mathbf{E}}^{\text{inc}}$ is the incident electric field defined to be the field due to an impressed source in the absence of the scatterer and $\underline{\mathbf{E}}^{\text{sc}}$ represents the scattered electric field that can be expressed by electrodynamic potentials:

$$\underline{\mathbf{E}}^{\text{sc}} = -j\omega\underline{\mathbf{A}} - \text{grad}\underline{\varphi}. \tag{4.18}$$

The magnetic vector potential $\underline{\mathbf{A}}$ is introduced according to Equation (1.148).Using the Lorenz gauge of Equation (1.154) for free space and taking into account conducting and polarization currents as well as free and bound charges one may decouple the wave equations for $\underline{\varphi}$ and $\underline{\mathbf{A}}$ in regions consisting of linear, isotropic and homogeneous media as considered in Section 4.1.1:

$$\Delta\underline{\mathbf{A}} + \omega^2\epsilon_0\mu_0\underline{\mathbf{A}} = -\mu_0\left(\underline{\mathbf{J}}_{\text{C}} + j\omega\epsilon_0(\underline{\epsilon}_{\text{r}} - 1)\underline{\mathbf{E}}\right) = -\mu_0\underline{\mathbf{J}}, \tag{4.19}$$

$$\Delta\underline{\varphi} + \omega^2\epsilon_0\mu_0\underline{\varphi} = -\frac{\underline{\varrho}_{\text{f}} + \underline{\varrho}_{\text{b}}}{\epsilon_0} = -\frac{\underline{\varrho}}{\epsilon_0}. \tag{4.20}$$

The solutions for free space are given by

$$\underline{\varphi}(\mathbf{r}) = \frac{1}{\epsilon_0}\int_{V'} g(\mathbf{r}, \mathbf{r}')\underline{\varrho}(\mathbf{r}')\text{d}V', \tag{4.21}$$

$$\underline{\mathbf{A}}(\mathbf{r}) = \mu_0\int_{V'} g(\mathbf{r}, \mathbf{r}')\underline{\mathbf{J}}(\mathbf{r}')\text{d}V', \tag{4.22}$$

with $\underline{\varrho}$ and $\underline{\mathbf{J}}$ according to Equations (4.7) and (4.9) respectively,

$$g(\mathbf{r}, \mathbf{r}') = \frac{1}{4\pi}\frac{\text{e}^{-jk|\mathbf{r}-\mathbf{r}'|}}{|\mathbf{r} - \mathbf{r}'|} \tag{4.23}$$

as the Green's function for free space,

$$k = \frac{\omega}{c} = \frac{2\pi}{\lambda} \tag{4.24}$$

as the wave number, and

$$c = \frac{1}{\sqrt{\epsilon_0\mu_0}} \tag{4.25}$$

as the propagation velocity in free space.

Table 4.1 Electric field strength in dependence on the location of the field point

Location of field point $\mathbf{r}$	Field Strength $\underline{\mathbf{E}}(\mathbf{r})$	Unknowns $\underline{\mathbf{J}}(\mathbf{r})$	Unknowns $\underline{\varrho}(\mathbf{r})$
Perfect electric conductor (PEC)	0	$\underline{\mathbf{J}}_{\mathrm{C}}$	$\underline{\varrho}_{\mathrm{f}}$
Good conductor $\sigma \gg \lvert\omega\underline{\epsilon}'\rvert$	$\dfrac{\underline{\mathbf{J}}_C}{\sigma}$	$\underline{\mathbf{J}}_{\mathrm{C}}$	$\underline{\varrho}_{\mathrm{f}}$
Lossless dielectric	$\dfrac{\underline{\mathbf{J}}_P}{j\omega\epsilon_0(\epsilon_{\mathrm{r}}' - 1)}$	$\underline{\mathbf{J}}_{\mathrm{P}}$	$\underline{\varrho}_{\mathrm{b}}$
Lossy dielectric $\sigma \ll \lvert\omega\underline{\epsilon}\rvert$	$\dfrac{\underline{\mathbf{J}}_P}{j\omega\epsilon_0[\epsilon_{\mathrm{r}}'' + j(\epsilon_{\mathrm{r}}' - 1)]}$	$\underline{\mathbf{J}}_{\mathrm{P}}$	$\underline{\varrho}_{\mathrm{b}}$

With Equations (4.21) and (4.22) and replacing $\underline{\mathbf{E}}$ by $\underline{\mathbf{J}}$ according to Equation (4.9) Equation (4.18) results in

$$\begin{aligned}\underline{\mathbf{E}}^{\mathrm{inc}} = &\frac{\underline{\mathbf{J}}}{\sigma_{\mathrm{e}} + j\omega\epsilon_0(\epsilon_{\mathrm{r}}' - 1)} + j\omega\mu_0 \int_{V'} g(\mathbf{r}, \mathbf{r}')\underline{\mathbf{J}}(\mathbf{r}')\mathrm{d}V' \\ &+ \frac{\nabla}{\epsilon} \int_{V'} g(\mathbf{r}, \mathbf{r}')\underline{\varrho}(\mathbf{r}')\mathrm{d}V'.\end{aligned} \tag{4.26}$$

Equation (4.26) is an MPIE that will be used in the next section for the derivation of the PEEC method. The unknowns in this Equation are the current density $\underline{\mathbf{J}}$ and the charge density $\underline{\varrho}$. Both kinds of unknowns are linked by the continuity Equation (4.16). The first term on the right-hand side of Equation (4.26) can vary depending on the features of the medium in which the field point $\mathbf{r}$ is located (see Table 4.1).

The method of Green's functions (see Section 1.4.6) is a general method for the calculation of electromagnetic fields (and other physical fields) for given sources and boundary conditions. An MPIE for arbitrarily shaped scatterers in free space has been developed by numerous authors (e.g. [2]). For an observation point $\mathbf{r}$ on a conductor with electric conductivity σ the dyadic Green's function – MPIE (DGF–MPIE) can be written as

$$\underline{\mathbf{E}}^{\mathrm{inc}}(\mathbf{r}) = \frac{\underline{\mathbf{J}}(\mathbf{r})}{\sigma} + j\omega\mu_0 \int_{V'} \overline{\overline{\mathbf{G}}}^A(\mathbf{r}, \mathbf{r}') \cdot \underline{\mathbf{J}}(\mathbf{r}')\mathrm{d}V' + \frac{\nabla}{\epsilon} \int_{V'} G^{\Phi}(\mathbf{r}, \mathbf{r}')\underline{\varrho}(\mathbf{r}')\mathrm{d}V' \tag{4.27}$$

with $\overline{\overline{\mathbf{G}}}^A(\mathbf{r}, \mathbf{r}')$ as the dyadic Green's function of the magnetic vector potential and $G^{\Phi}(\mathbf{r}, \mathbf{r}')$ as the scalar Green's function of the scalar electric potential. In this context, the MPIE according to Equation (4.27) is equivalent to the DGF–MPIE of Equation (4.27) for $\epsilon = \epsilon_0$ and

$$\begin{aligned}\overline{\overline{\mathbf{G}}}^A(\mathbf{r}, \mathbf{r}') &= g(\mathbf{r}, \mathbf{r}')\overline{\overline{\mathbf{I}}} \\ G^{\Phi}(\mathbf{r}, \mathbf{r}') &= g(\mathbf{r}, \mathbf{r}')\end{aligned} \tag{4.28}$$

with $\overline{\overline{\mathbf{I}}}$ as the identity dyadic.

In many practical applications, very good conducting ground planes are used that are relatively large in comparison with the structure under investigation. An usual theoretical

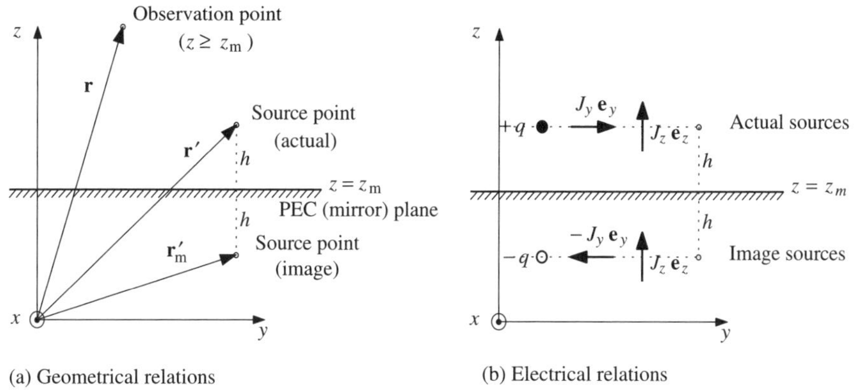

Figure 4.2 Geometrical and electrical relations of a plane half space $z \geq z_m$.

model for such configurations is the plane half space bounded by a PEC (see Figure 4.2). If the boundary is located in the x,y plane at $z = z_m$, the scalar Green's function $g_{\mathrm{H}}(\mathbf{r}, \mathbf{r}')$ of the half space $z \geq z_{\mathrm{m}}$ can easily be found by applying the mirror principle as composition of the Green's functions of the free space:

$$g_{\mathrm{H}}(\mathbf{r}, \mathbf{r}') = g(\mathbf{r}, \mathbf{r}') \pm g(\mathbf{r}, \mathbf{r}'_{\mathrm{m}}) \tag{4.29}$$

with

$$\mathbf{r}' = x'\mathbf{e}_x + y'\mathbf{e}_y + z'\mathbf{e}_z$$

as location of the actual source and

$$\mathbf{r}'_m = \mathbf{r}' - 2(z' - z_m)\mathbf{e}_z$$

as location of the image source. The signs in Equation (4.29) result from boundary conditions according to Figure 4.2.

With it, the Green's functions of the MPIE for the half space are composed as

$$\overline{\overline{\mathbf{G}}}{}^{A}_{\mathrm{H}}(\mathbf{r}, \mathbf{r}') = g(\mathbf{r}, \mathbf{r}')\overline{\overline{\mathbf{I}}} + g(\mathbf{r}, \mathbf{r}' - 2(z' - z_{\mathrm{m}})\mathbf{e}_z) \cdot (\mathbf{e}_z\mathbf{e}_z - \overline{\overline{\mathbf{I}}}_t) \tag{4.30}$$

with $\overline{\overline{\mathbf{I}}}_t = (\mathbf{e}_x\mathbf{e}_x + \mathbf{e}_y\mathbf{e}_y)$ as the identity transverse dyadic and

$$G^{\Phi}_{\mathrm{H}}(\mathbf{r}, \mathbf{r}') = g(\mathbf{r}, \mathbf{r}') - g(\mathbf{r}, \mathbf{r}' - 2(z' - z_{\mathrm{m}})\mathbf{e}_z). \tag{4.31}$$

Thus, one obtains the appropriate MPIE

$$\underline{\mathbf{E}}^{\mathrm{inc}} = \frac{\underline{\mathbf{J}}}{\sigma} + j\omega\mu_0 \int_{V'} \overline{\overline{\mathbf{G}}}{}^{A}_{\mathrm{H}}(\mathbf{r}, \mathbf{r}') \cdot \underline{\mathbf{J}}(\mathbf{r}')\mathrm{d}V' + \frac{\nabla}{\epsilon_0} \int_{V'} G^{\Phi}_{\mathrm{H}}(\mathbf{r}, \mathbf{r}')\underline{\varrho}(\mathbf{r}')\mathrm{d}V'. \tag{4.32}$$

The chosen representation of media by their electric sources (conducting and polarization currents, free and bound charges) in free space (ϵ_0, μ_0) means that the following derivation process of the numerical PEEC method requires the discretization of all source regions, that is both conductors and dielectric regions. Therefore, when it is necessary to take into account dielectrics in the analysis of wiring structures the number of unknowns increases significantly.

To avoid this fact for the practical very important case of conducting scatterers (PCB traces, antennas) embedded in a stratified medium consisting of an arbitrary number of planar layers a DGF–MPIE has been developed ([3],[4]). The geometrical structure of the stratified configuration and the electric features of the planar dielectric layers are included in appropriate and relatively complicated dyadic and scalar Green's functions for the magnetic vector and electric scalar potential respectively. However, the number of unknowns is substantially reduced and represented only by conducting currents and free charges of the conducting structures embedded in stratified dielectrics. For clarity and simplicity, in Section 4.2 in a first step the PEEC method is derived on the basis of Equations (4.27) through (4.32). Later, in Section 4.8, the PEEC method is extented to layered media with a description by the appropriate dyadic and scalar Green's functions ([5]).

4.2 Derivation of the Generalized PEEC Method in the Frequency Domain

4.2.1 The PEEC Equation System in the Frequency Domain

The starting points for the derivation of the PEEC method are the electric field MPIE and the continuity equation explained in Section 4.1. For characterization of the media, a simplified notation is introduced with the replacements σ_e by σ and ϵ' by ϵ. A medium is then generally characterized by the real constitutive parameters σ, ϵ, μ_0. The first step in the development of the PEEC equation system is the geometrical discretization of the interconnection structure to be modeled. The whole structure is dissected into a set of N volume or current cells (see Figure 4.3). The current density $\underline{\mathbf{J}}$ can be developed in the form

$$\underline{\mathbf{J}}(\mathbf{r}) = \sum_n \mathbf{b}_n^c(\mathbf{r})\underline{I}_n,\ n \in N \tag{4.33}$$

with $\mathbf{b}_n^c(\mathbf{r})$ as the vectorial basis functions. First of all, constant basis functions are chosen in the form

$$\mathbf{b}_n^c(\mathbf{r}) = \begin{cases} \dfrac{\mathbf{e_n}}{a_n} & \mathbf{r} \in v_n \\ 0 & \text{otherwise} \end{cases} \tag{4.34}$$

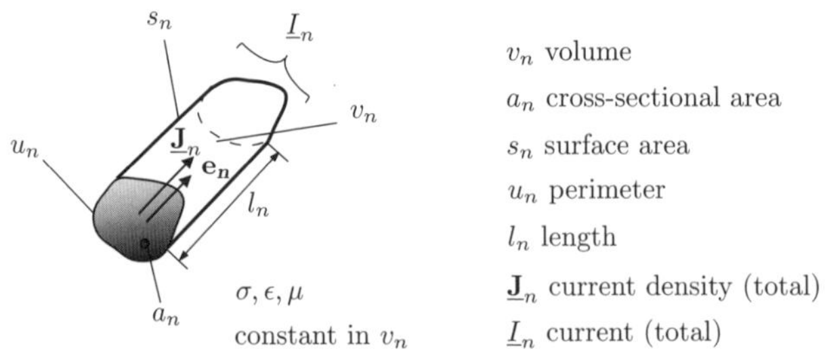

Figure 4.3 Volume cell n ($n \in N$).

with the assumption that $\underline{I}_n$ has the direction $\mathbf{e}_n$ and is uniformly distributed over a_n. Later, the considerations are extended to nonorthogonal and triangular meshing (Section 4.5). For the determination of the unknown charges, a second set M of cells called potential or charge or capacitive cells is introduced. Because of the difference approximation of the gradient operator in the MPIE (e.g. Equation (4.27)), the potential cells are shifted from the current cells at a half of their length. This fact will be explained in more detail later in this section. The charge density distribution is also developed by constant basis functions in the form

$$\underline{\varrho}(\mathbf{r}) = \sum_m b_m^{\mathrm{P}}(\mathbf{r})\underline{q}, m \in M \tag{4.35}$$

with

$$b_m^{\mathrm{P}} = \begin{cases} \dfrac{1}{s_m} & \mathbf{r} \in s_m \\ 0 & \text{otherwise} \end{cases} \tag{4.36}$$

and s_m as the surface of cell m.

With regard to Section 4.1.1, it should be noticed that the charges are distributed as surface charges on the interfaces between different media only. Thus, potential cells which are fully inside a homogeneous region (so-called inner cells) do not have surface charges.

As well as the discontinuities concerning the media, geometrical discontinuities (e.g. changing cross sections, aspect ratio) and the frequency of operation have to be taken into account in the discretization process.

If f_{m} is the maximum frequency of interest, the length l of the cells or, generally speaking, the discretization length Δl, is usually given by

$$\Delta l \leq \frac{\lambda_{\min}}{10 \times \nu} = \frac{c}{10 \times \nu \times f_{\mathrm{m}}} \tag{4.37}$$

with $\nu = 1 \ldots 2$.

For transverse dimensions of the cells, the same condition as Equation (4.37) has to be applied. An additional problem for the segmentation arises for good conductors and high frequencies due to the skin effect. To hold the condition of a homogeneous current distribution over the cross section of a volume cell needs either extremely small cross sections or requires a discretization of the cross section in a high number of filaments with an approximately homogeneous current distribution. The skin effect will be considered in detail in Section 4.7.

Discretizing the structure and introducing the basis function expansions for $\underline{\mathbf{J}}$ and $\underline{\varrho}$ according to Equations (4.33) to (4.36) in the MPI Equation (4.26) result in N unknown currents $\underline{I}_n (n \in N)$ and M unknown charges $\underline{q}_m (m \in M)$. To determine the $N + M$ unknowns, at first, the Galerkin method (weighted residual method) is applied to the MPIE obtaining N equations. For each current cell α ($\alpha \in N$), the inner product with the weighting function $\mathbf{w}_\alpha = \mathbf{b}_\alpha^c$ is built for all terms of the discretized MPI Equation (4.26):

$$\int_{\mathrm{v}_\alpha} \omega_\alpha \ldots \mathrm{d}v_\alpha \quad \forall \alpha \in N. \tag{4.38}$$

On the left-hand side of Equation (4.26) one obtains

$$\int_{v_\alpha} \frac{\mathbf{e}_\alpha}{a_\alpha} \underline{\mathbf{E}}^{\text{inc}} \mathrm{d}v_\alpha = \int_{l_\alpha} \underline{E}_\alpha^{\text{inc}} \mathrm{d}l_\alpha = \underline{U}_\alpha^{\text{inc}}. \tag{4.39}$$

$\underline{E}_\alpha^{\text{inc}}$ is the component of $\underline{\mathbf{E}}^{\text{inc}}$ in the $\mathbf{e}_\alpha$ direction averaged over the cross section a_α. $\underline{U}_\alpha^{\text{inc}}$ represents the voltage caused by $\underline{\mathbf{E}}^{\text{inc}}$ in cell α.

Now, the first term on the right-hand side of Equation (4.26) is considered:

$$\int_{v_\alpha} \frac{\mathbf{e}_\alpha}{a_\alpha} \underline{\mathbf{E}} \mathrm{d}v_\alpha = \int_{l_\alpha} \underline{E}_\alpha \mathrm{d}l_\alpha = \underline{U}_\alpha. \tag{4.40}$$

Introducing $\underline{E}_\alpha = \underline{J}_\alpha/(\sigma + j\omega\varepsilon_0(\varepsilon_r - 1))$ and replacing $\underline{J}_\alpha = \underline{I}_\alpha/a_\alpha$ yields $\underline{U}_\alpha = \underline{Z}_\alpha \underline{I}_\alpha$ with $\underline{Y}_\alpha = 1/\underline{Z}_\alpha = G_\alpha + j\omega C_\alpha^+$ and

$$G_\alpha = \frac{1}{R_\alpha} = \sigma \frac{a_\alpha}{l_\alpha}, \tag{4.41}$$

$$C_\alpha^+ = \epsilon_0(\epsilon_r - 1)\frac{a_\alpha}{l_\alpha}. \tag{4.42}$$

$\underline{U}_\alpha$ is the voltage in the current cell α caused by the current $\underline{I}_\alpha$ because of the total electric field strength and the features of the medium of cell α. Equation (4.41) represents an active resistance and Equation (4.42) is the so-called 'excess capacitance' of cell α introduced in [6] for a medium with $\epsilon_r > 1$.

Now, the second right-hand side term of Equation (4.26), the magnetic vector potential term, is under consideration. In the first step, the current basis functions are introduced:

$$j\omega\mu_0 \sum_{n\in N} \frac{\mathbf{e}_n}{a_n} \int_{v_n} g(\mathbf{r}, \mathbf{r}')\mathrm{d}v_n' \underline{I}_n.$$

Applying the inner product according to Equation (4.38) yields

$$j\omega\mu_0 \sum_{n\in N} \frac{\cos\vartheta_{\alpha n}}{a_\alpha a_n} \int_{v_\alpha} \int_{v_n} g(\mathbf{r}, \mathbf{r}')\mathrm{d}v_n' \mathrm{d}v_\alpha \underline{I}_n \tag{4.43}$$

with $\cos\vartheta_{\alpha n} = \mathbf{e}_\alpha \cdot \mathbf{e}_n$. Equation (4.43) represents a voltage resulting from a linear superposition of n contributions of the magnetic field caused by the cell currents $\underline{I}_n$, $n \in N$. This equation can be written formally as

$$j\omega \sum_{n\in N} \underline{\Lambda}_{\alpha,n} \underline{I}_n, \quad \forall\alpha \in N \tag{4.44}$$

with

$$\underline{\Lambda}_{\alpha,n}(j\omega) = \mu_0 \frac{\cos\vartheta_{\alpha n}}{a_\alpha a_n} \int_{v_\alpha} \int_{v_n} g(\mathbf{r}, \mathbf{r}')\mathrm{d}v_n' \mathrm{d}v_\alpha, \quad \forall(\alpha, n) \in N. \tag{4.45}$$

The complex coefficients $\underline{\Lambda}_{\alpha,n}(\alpha, n \in N)$ have the unit of an inductance and are named generalized partial inductances [7], in particular partial self ($\alpha = n$) and partial mutual ($\alpha \neq n$) inductances. The partial inductances according to Equation (4.45) form the partial inductance matrix $\mathbf{\Lambda}$ and physically describe the action and interaction of and between all current cells due to the magnetic field of the cell currents $\underline{I}_n (n \in N)$.

The last term of the right-hand side of Equation (4.27) refers to the scalar electrical potential and the charges:

$$\nabla \underline{\varphi}(\mathbf{r}) = \nabla\left(\frac{1}{\varepsilon_0}\int_{V'} g(\mathbf{r}, \mathbf{r}')\underline{\varrho}(\mathbf{r}')\mathrm{d}V'\right).$$

Again, the inner product with the weighting function is taken:

$$\int_{v_\alpha} \frac{\mathbf{e}_\alpha}{a_\alpha}\nabla \underline{\varphi}(\mathbf{r})\mathrm{d}v_\alpha = \frac{1}{a_\alpha}\int_{v_\alpha} \nabla_\alpha \underline{\varphi}(\mathbf{r})\mathrm{d}v_\alpha,$$

where ∇_α represents the derivative in direction $\mathbf{e}_\alpha$ and is approximated as a difference quotient:

$$\underbrace{\frac{1}{a_\alpha l_\alpha}\int_{v_\alpha} \underline{\varphi}(\mathbf{r}+\frac{l_\alpha}{2})\mathrm{d}v_\alpha}_{\underline{\varphi}_{\alpha+}} - \underbrace{\frac{1}{a_\alpha l_\alpha}\int_{v_\alpha} \underline{\varphi}(\mathbf{r}-\frac{l_\alpha}{2})\mathrm{d}v_\alpha}_{\underline{\varphi}_{\alpha-}} + \frac{1}{a_\alpha}\int_{v_\alpha} \epsilon_\nabla \mathrm{d}v_\alpha \quad (\mathbf{r} \in v_\alpha). \tag{4.46}$$

ϵ_∇ is the error because of the difference approximation. Further, $\epsilon_\nabla = 0$ is assumed. In the case of strong discontinuities, for example nonuniform discretization, bends, branching and ends of conductors, a multipoint difference quotient can be applied to reduce ϵ_∇ (see [8]).

The left-hand term of Equation (4.46) represents the mean value $\varphi_{\alpha+}$ of the potential of the cell moved at $+\frac{l_\alpha}{2}$ to right on the cell α, analogously, $\varphi_{\alpha-}$ is the mean value of the potential of the cell moved at $-\frac{l_\alpha}{2}$ to left the on the cell α (see Figure 4.4). This is the reason for introducing a second set M of cells, the so-called potential cells, that are moved at a half of the discretization length to the current cells. The potential cells are surface cells since the charge is distributed on surfaces only.

With $\underline{\varphi}_{\alpha-} = \underline{\varphi}_i$ and $\underline{\varphi}_{\alpha+} = \underline{\varphi}_j$ respectively, and considering Equations (4.36) and (4.37) one obtains

$$\begin{aligned} \underline{\varphi}_i &= \sum_{m\in M} \frac{1}{\varepsilon_0 s_i s_m}\int_{s_i}\int_{s_m} g(\mathbf{r}, \mathbf{r}')\mathrm{d}s'_m \mathrm{d}s_i \, \underline{q}_m \quad \forall i \in M \\ &= \sum_{m\in M} \underline{\Pi}_{i,m}\underline{q}_m. \end{aligned} \tag{4.47}$$

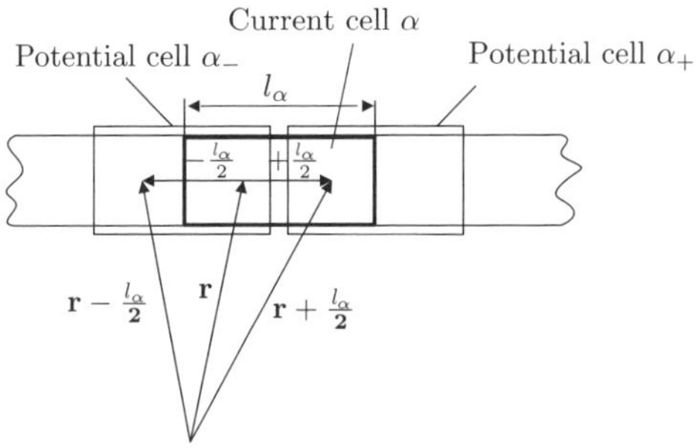

Figure 4.4 Introduction of potential cells.

The complex coefficients

$$\underline{\Pi}_{i,m}(j\omega) = \frac{1}{\varepsilon_0 s_i s_m} \int_{s_i} \int_{s_m} g(\mathbf{r}, \mathbf{r}')\mathrm{d}s'_m \mathrm{d}s_i \quad \forall i, m \in M \tag{4.48}$$

are the generalized partial coefficients of potential, in particular self $(i = m)$ and mutual $(i \neq m)$ potential coefficients. The potential coefficients in Equation (4.48) compose the matrix Π of potential coefficients and physically describe the interdependencies between potentials and charges (see Equation (4.47)).

By applying the Galerkin method to the MPIE one obtains N Equations for the discretized structure consisting of N current cells and M potential cells in the form

$$\underline{U}_{\alpha}^{\mathrm{inc}} = \underline{Z}_{\alpha}\underline{I}_{\alpha} + j\omega \sum_{n \in N} \underline{\Lambda}_{\alpha,n}\, \underline{I}_n + \underline{\varphi}_i - \underline{\varphi}_j, \quad \forall \alpha \in N \tag{4.49}$$

where $\underline{\varphi}_i$ and $\underline{\varphi}_j$ are the potentials of the potential cells adjacent to the current cell α. All other parameters and values of Equation (4.49) are determined by the equations above in this section. The potentials in Equation (4.49) can be expressed by the M unknown charges $\underline{q}_m$ $(m \in M)$ according to Equation (4.47).

The M equations still required for the unique determination of $N + M$ unknowns are given by application of the continuity Equation (4.16) in integral form to each of the potential cells $m \in M$ (see Figure 4.5)

$$\sum_{j \in j_m} \underline{I}_{mj} + j\omega \underline{q}_m = 0, \quad \forall m \in M \tag{4.50}$$

where j_m is the number of current cells adjacent to the potential cell m.

The assignment of the oriented currents in the current cells and the charges or potentials of potential cells is described by a connectivity matrix $\mathbf{A}$ with elements

$$a_{n,m} = \begin{cases} +1 & \text{current } \underline{I}_n \text{ away from cell m} \\ -1 & \text{current } \underline{I}_n \text{ to cell m} \\ 0 & \text{no connection} \end{cases} \quad \forall n \in N, \forall m \in M. \tag{4.51}$$

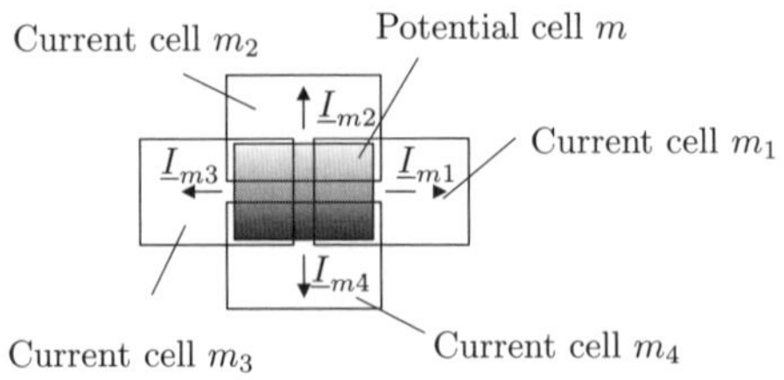

Figure 4.5 Potential cell m with four adjacent current cells (two-dimensional discretizing).

The entire equation system of Equations (4.49) and (4.50) can be written in a compact matrix form

$$\begin{bmatrix} \mathbf{Z} + j\omega\mathbf{\Lambda} & -\mathbf{A}^T\mathbf{\Pi} \\ \mathbf{A} & j\omega\mathbf{1} \end{bmatrix} \begin{bmatrix} \mathbf{I} \\ \mathbf{q} \end{bmatrix} = \begin{bmatrix} \mathbf{U}^{\text{inc}} \\ \mathbf{0} \end{bmatrix}. \tag{4.52}$$

In Equation (4.52) are:

- $\mathbf{Z}$ diagonal matrix (N, N) of local impedances (Equations (4.41) and (4.42))
- $\mathbf{\Lambda}$ matrix (N, N) of generalized partial inductances (Equation (4.44))
- $\mathbf{\Pi}$ matrix (M, M) of generalized potential coefficients (Equation (4.47))
- $\mathbf{A}$ connectivity matrix (M, N) (Equation (4.51))
- $\mathbf{I}$ vector $(N, 1)$ of currents (Equation (4.33))
- $\mathbf{q}$ vector $(N, 1)$ of charges (Equation (4.35))
- $\mathbf{U}^{\text{inc}}$ vector $(N, 1)$ of voltages (Equation (4.39)).

4.2.2 *Generalized Partial Elements and Circuit Interpretation*

The generalized partial elements have been already introduced in Section 4.2.1:

- $\underline{Z}_\alpha (\alpha \in N)$ partial (local) impedances
- $\underline{\Lambda}_{\alpha,n} (\alpha, n \in N)$ partial self $(\alpha = n)$ and mutual $(\alpha \neq n)$ inductances
- $\underline{\Pi}_{i,m} (i, m \in M)$ partial self $(i = m)$ and mutual $(i \neq m)$ potential coefficients.

In the following, the circuit interpretation of the system of Equation (4.52) derived in the previous section and written in a condensed matrix form is developed.

Each of the N Equations (4.49) can be interpreted as a Kirchhoff's voltage law (KVL) pictured in Figure 4.6. Formally, the usual symbols of the circuit theory for resistances, inductances, capacitances, independent and controlled voltage and current sources are used regardless of the complex values of the generalized partial elements. The remaining M equations relate to the potential cells, and to Equations (4.50) and (4.47). Each of the continuity Equations (4.50) can be interpreted as a Kirchhoff's current law (KCL). Let

$$j\omega \underline{q}_m = \underline{I}^{\text{c}}_m \quad , \quad \forall m \in M \tag{4.53}$$

be introduced as additional currents assigned to each node and named 'capacitive currents' in analogy to the 'inductive currents' assigned to the current cells. Using Equation (4.47) one

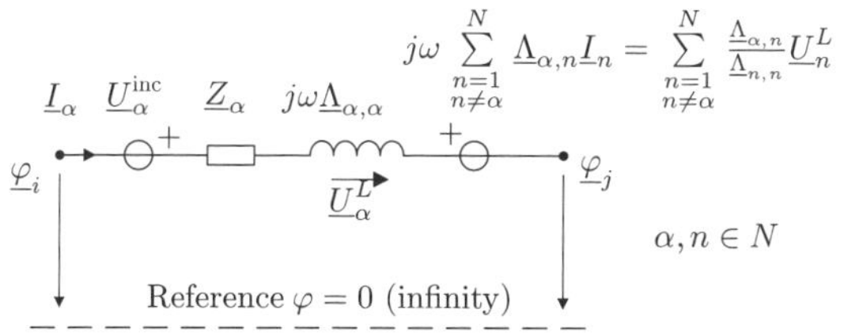

Figure 4.6 Circuit interpretation of Equation (4.49) for the current cell α $(\alpha \in N)$.

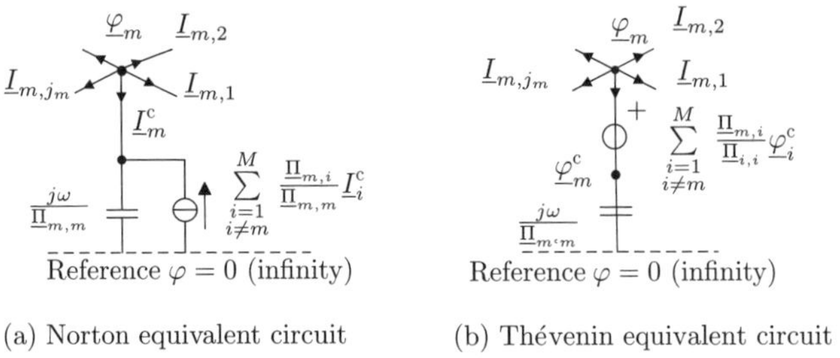

Figure 4.7 Circuit interpretation of Equations (4.50) and (4.47) for the potential cell m ($m \in M$).

obtains equations of the form

$$\underline{I}^c_i = \frac{j\omega}{\underline{\Pi}_{i,i}} \cdot \underline{\varphi}_i - \sum_{\substack{m=1 \\ m\neq i}}^{M} \frac{\underline{\Pi}_{i,m}}{\underline{\Pi}_{i,i}} \underline{I}^c_m \quad \forall i, m \in M \tag{4.54}$$

or of the form

$$\underline{\varphi}_i = \underline{\varphi}^c_i + \sum_{\substack{m=1 \\ m\neq i}}^{M} \frac{\underline{\Pi}_{i,m}}{\underline{\Pi}_{m,m}} \underline{\varphi}^c_m \quad \forall i, m \in M \quad \text{and} \quad \underline{\varphi}^c_i = \underline{\Pi}_{i,i}\, \underline{q}_i. \tag{4.55}$$

Equations (4.54) can be interpreted as Norton and Equations (4.55) as Thévenin equivalent circuits (see Figure 4.7).

For inner cells which have no charges $\underline{q}_{\text{inner}} = 0$ holds and, consequently, $\underline{I}^c_{\text{inner}} = 0$. In this case, the continuity Equations (4.50) relate only to the currents of the current cells linked with the appropriate node representing the potential cell and there is no branch from the appropriate node to the reference $\varphi = 0$. The circuit interpretation of the equation system derived in Section 4.1.1 is the characteristic step and constitutes the notation partial element equivalent circuit (PEEC) method. A PEEC model of a real interconnection structure is a complex electric circuit composed of elementary circuits according to Figure 4.6 and 4.7. By the PEEC method an electromagnetic field problem is mapped on an electric circuit problem. The electromagnetic interactions between different parts (cells) of the structure are respresented by controlled sources whose control parameters are the partial mutual inductances (magnetic interactions) and mutual potential coefficients (electric interactions).

In this way, the retardation of electromagnetic potentials, in the PEEC method according to the Lorenz gange (see Section 4.1.2), is transformed into the electric circuit representation. This means a generalization of 'electric or Kirchhoff circuits' to 'electromagnetic or Maxwell circuits' [11]. Additionally, it should be noted that in the PEEC interpretation an incident electric field $\underline{\mathbf{E}}^{\text{inc}}$ is realized as an independent voltage source $\underline{U}^{\text{inc}}_\alpha$, $\alpha \in N$ (see Equation (4.39) and Figure 4.6). This represents one kind of excitation besides lumped voltage or current sources externally connected to the PEEC model.

Besides providing a good understanding for electrical engineers, a main advantage of the PEEC method consists in using the highly developed methods of the circuit theory. This concerns both the formation of the describing equation system, for example by the modified nodal analysis (MNA) or the modified loop analysis (MLA) and the opportunity of using appropriate SPICE-like solvers.

4.3 Classification of PEEC Models

4.3.1 Classification in Dependence on Media

From a practical point of view, interconnection structures are composed of conductors and dielectrics. Therefore, this aspect will be considered in the following.

4.3.1.1 PEEC Models for Conductors

For conductor cells, Equation (4.12) holds with the electric conductivity σ independent of the frequency for all frequencies of practical interest. Thus, the complex local impedance $\underline{Z}_\alpha$ in Equation (4.41) and Figure 4.6 yields an ohmic resistance

$$\underline{Z}_\alpha = R_\alpha = \frac{1}{\sigma}\frac{l_\alpha}{a_\alpha} \tag{4.56}$$

with the assumption of a uniform current distribution over the cross section a_α. All other elements of the equivalent circuits are to be calculated by the formulas derived in Section 4.2. However, the assumption of a uniform current distribution is only correct up to frequencies where the skin effect has not yet had an essential influence, that is roughly estimated, when the skin depth $\delta = \sqrt{2/\omega\mu_0\sigma}$ is bigger than the half of the cross-sectional dimensions. A PEEC adequate possibility for taking into account a nonuniform current distribution is to choose a finer cross-sectional segmentation, so that in each of the segments the current distribution remains uniform (see Figure 4.8). The mechanism of magnetic couplings through the mutual partial inductances of the segments leads to a nonuniform current distribution over the whole cross section. Unfortunately, because of the big effort for discretization and the high number of unknowns this approach is not practicable in many cases.

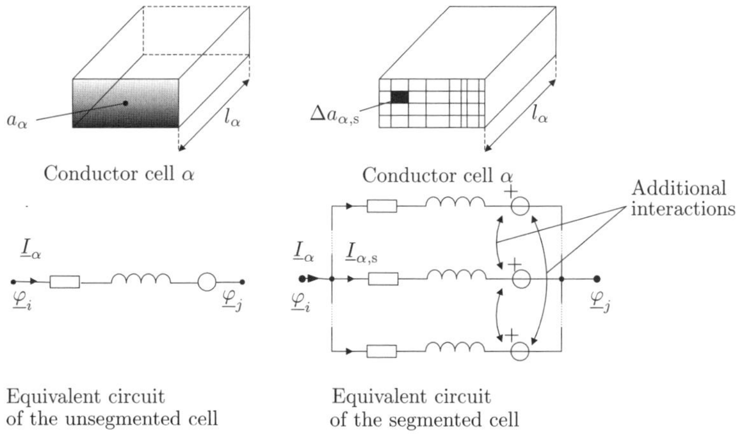

Figure 4.8 Segmentation of a conductor element for considering the skin effect.

In Section 4.7, the skin effect in PEEC models is addressed in more detail. In the special case of a perfect electric conductor with $\sigma \to \infty$, for cell α $\underline{Z}_\alpha = R_\alpha = 0$ holds, that is in the equivalent circuit (see Figure 4.6) the element $\underline{Z}_\alpha$ is replaced by a short. Further, since currents in a PEC are surface currents, the appropriate volume integrals in the formulas for the calculation of self and mutual partial inductances (see Equation (4.45)) have to be taken as surface integrals. In this case, the partial inductances have the physical meaning of only external inductances.

4.3.1.2 PEEC Models for Dielectrics

Considering the explanations about dielectrics in Section 4.2.1, for the local complex impedance $\underline{Z}_\alpha$ and the admittance $\underline{Y}_\alpha$ of cell α with lossy dielectric respectively one obtains

$$\underline{Y}_\alpha = G_\alpha + j\omega C_\alpha^+ = \epsilon_o \epsilon_{\mathrm{r}}'' \omega \frac{a_\alpha}{l_\alpha} + j\omega\epsilon_o(\epsilon_{\mathrm{r}}' - 1)\frac{a_\alpha}{l_\alpha}. \tag{4.57}$$

Then, in the circuit interpretation according to Figure 4.6 $\underline{Z}_\alpha$ is represented by a parallel circuit of a capacitance C_α^+ and an ohmic resistance R_α. Both elements C_α^+ (because of $\varepsilon_{\mathrm{r}}'$) and R_α (because of $\varepsilon_{\mathrm{r}}''$) are generally frequency dependent. For a lossless dielectric $G_\alpha = 0$ holds because $\varepsilon_{\mathrm{r}}'' = 0$. Thus, the circuit representation consists only of the excess capacitance C_α^+.

All other equations and circuit interpretations derived in Section 4.2 remain formally unchanged, notwithstanding the different physical background of conductors and dielectrics.

4.3.2 Classification in Dependence on the Relation of the Maximum Frequency of Interest to the Discretization Length

4.3.2.1 PEEC Models with Generalized Partial Elements

According to the chosen basis functions and the discretization criterion, the PEEC model with generalized partial elements describes the correct physical behavior of the modeled structure for frequencies f in the range $0 \le f \le f_{\mathrm{m}}$. However, a correct modeling of the physical behavior for frequencies $f > f_{\mathrm{m}}$, is still awaited. In practical cases, especially with pulse excitations and in time domain simulations, frequencies higher than f_{m} cannot be avoided. Therefore, the frequency response of the generalized partial elements is also of interest for those frequencies, at least, that the global features such as passivity and stability of the PEEC model (the modeled physical interconnection structure is passive!) can be preserved. The frequency response of the generalized partial elements resulting from the double integral over the complex Green's functions (see Equations (4.45) and (4.48)) is analyzed and represented in a normalized form $|\underline{\Lambda}_{a,\mathrm{n}}(j\omega)/|\underline{\Lambda}_{a,\mathrm{n}}(0)|$ in Figure 4.9. One can see the strong damping of the functions for frequencies beyond f_{m}. This feature will be used for the development of stable time domain PEEC models (see Section 4.6.4). In general, one has to state that the main disadvantage of the PEEC models with generalized partial elements is the high computational effort for the frequency dependent parameters. To reduce this effort, in the following text approximations and simplifications are introduced and their limits are discussed.

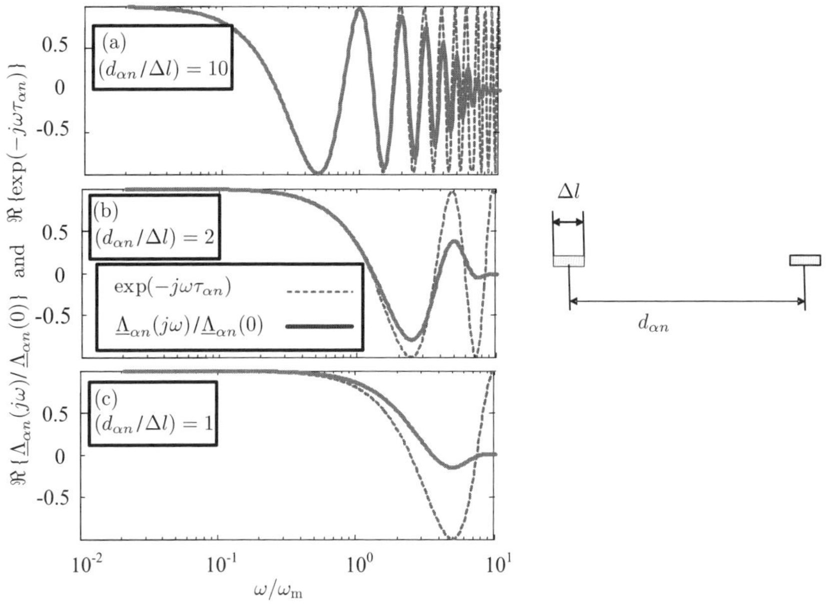

Figure 4.9 Frequency behavior of $\Lambda_{\alpha,n}(j\omega)/\Lambda_{\alpha,n}(0)$ and of $\mathrm{e}^{-jkd_{\alpha n}}$ for different distances between the cells α and n.

4.3.2.2 PEEC Models with Center-to-Center Retardation

If in the Equations (4.45) and (4.48) for the generalized partial elements the 'distributed' retardation of the Green's function is approximated by a lumped retardation between the centers of cells which can be drawn out of the double integrals, one obtains:

$$\begin{aligned}\underline{\Lambda}_{\alpha,n}(j\omega) &\approx \frac{\mu_o \cos\vartheta_{\alpha n}}{a_\alpha a_n}\int_{v_\alpha}\int_{v_n}\frac{1}{4\pi|\mathbf{r}-\mathbf{r}'|}\mathrm{d}v_n'\mathrm{d}v_\alpha \times \mathrm{e}^{-jkd_{\alpha n}} \\ &= \underline{\Lambda}_{\alpha,n}(0)\times \mathrm{e}^{-jkd_{\alpha n}}, \quad \alpha, n \in N\end{aligned} \tag{4.58}$$

$$\begin{aligned}\underline{\Pi}_{i,m}(j\omega) &\approx \frac{1}{\epsilon_o s_i s_m}\int_{s_i}\int_{s_m}\frac{1}{4\pi|\mathbf{r}-\mathbf{r}'|}\mathrm{d}s_m'\mathrm{d}s_i \times \mathrm{e}^{-jkd_{im}} \\ &= \underline{\Pi}_{i,m}(0)\times \mathrm{e}^{-jkd_{im}}, \quad i, m \in M.\end{aligned} \tag{4.59}$$

Thus, one may introduce the frequency independent partial inductances

$$L_{p\alpha,n} = \underline{\Lambda}_{\alpha,n}(0) = \frac{\mu_0 \cos\vartheta_{\alpha n}}{a_\alpha a_n}\int_{v_\alpha}\int_{v_n}\frac{1}{4\pi|\mathbf{r}-\mathbf{r}'|}\mathrm{d}v_n'\mathrm{d}v_\alpha, \quad \alpha, n \in N \tag{4.60}$$

and potential coefficients

$$P_{i,m} = \underline{\Pi}_{i,m}(0) = \frac{1}{\varepsilon_0 s_i s_m}\int_{s_i}\int_{s_m}\frac{1}{4\pi|\mathbf{r}-\mathbf{r}'|}\mathrm{d}s_n'\mathrm{d}s_i, \quad i, m \in M. \tag{4.61}$$

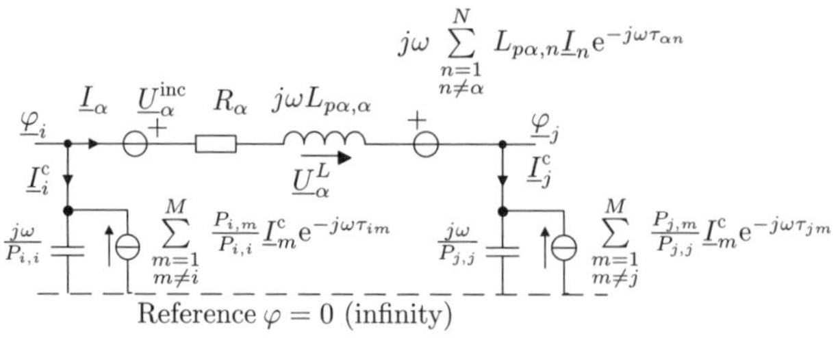

Figure 4.10 PEEC model with center-to-center retardation for a conductor element with Norton equivalent circuit for the potential cells.

The frequency dependence of Equations (4.58) and (4.59) is represented by the retardation factor

$$\exp(-jk(d_{\alpha n} \text{ or } d_{im})) = \exp(-j\omega(\tau_{\alpha n} \text{ or } \tau_{im})), \quad \alpha, n \in N \quad i, m \in M \tag{4.62}$$

with $d_{\alpha n}$, d_{im} as the distances, and $\tau_{\alpha n} = d_{\alpha n}/c$ and $\tau_{im} = d_{im}/c$ as the propagation times between the centers of cells respectively. The main advantage of this model is that the parameters L_{p}, P and τ are independent of frequency. Thus they may be calculated from a knowledge of the material parameters after the geometrical discretization in advance of the circuit calculation. The only frequency-dependent parameters in the circuit interpretation are, besides $\underline{Z}_{\alpha}$ for skin effect or dielectric losses, the complex factors of Equation (4.62) which compose together with the mutual partial elements the coefficients of the controlled voltage and current sources that respresent the electric and magnetic interactions between the cells (see Figure 4.10).

In comparison with the PEEC model based on the generalized partial elements (see Section 4.3.2.1) it can be stated that the approximations according to Equations (4.58) and (4.59) show a very good compliance with the exact behavior up to the maximum frequency of interest ω_{m} used for discretization. However, in contrast to the strongly damped behavior for frequencies beyond ω_{m} the approximation is fully undamped corresponding to $\exp(-j\omega\tau_{\alpha n})$ up to infinite frequencies (see Figure 4.9).

4.3.2.3 Quasi-Static PEEC Models

In contrast to the full-wave PEEC models considered in Section 4.3.2.1 and 4.3.2.2 which take into account the retardation, quasi-static PEEC models neglect the retardation. The condition for quasi-stationarity or, in other words, for 'electrically small' structures is

$$d \ll \lambda, \tag{4.63}$$

where d is the maximum distance between the cells of the model and λ is the wavelength of the highest frequency for investigation. The relation in Equation (4.63) is equivalent to replacing the dynamic free space Green's function by the static one:

$$g_{\mathrm{s}}(\mathbf{r}, \mathbf{r}') = \frac{1}{4\pi |\mathbf{r} - \mathbf{r}'|}. \tag{4.64}$$

Generally, quasi-static PEEC models cover the lower frequency range. However, for special configurations, such as when conductors are running very close to the return conductors or above perfectly conducting planes serving as reference planes ($h \ll \lambda$, with h as the distance between the parallel conductors or the conductors and the reference plane), quasi-static PEEC models can also describe quasi-TEM wave propagation phenomena up to relatively high frequencies. This fact is based on the high selectivity of the static half space Green's function. As a consequence, the $\mathbf{L}_\mathrm{p}$ and $\mathbf{P}$ matrices may essentially be sparsified which opens a broad application to transmission line-like and PCB-like structures. Section 4.9 addresses this problem in more detail.

In the quasi-static case, the matrices $\mathbf{L}_\mathrm{p}$ and $\mathbf{P}$ with the parameters $L_{\mathrm{p}\alpha,n}(\alpha, n \in N)$ and $P_{i,m}(i, m \in M)$ may be inverted:

$$\mathbf{C} = \mathbf{P}^{-1} \quad \text{and} \quad \mathbf{K}_\mathrm{p} = \mathbf{L}_\mathrm{p}^{-1}. \tag{4.65}$$

$\mathbf{C}$ is the matrix of capacitance coefficients with the physical meaning of 'short-circuit capacitances'. The coefficients $C_{i,m}(i, m \in M)$ of the matrix $\mathbf{C}$ can be expressed by partial capacitances $C_{\mathrm{p}i,m}$ $(i, m \in M)$ as coefficients of the matrix $\mathbf{C}_\mathrm{p}$ built between each of the potential cells to all others and to the reference (see textbooks, e.g. [9]):

$$C_{i,m} = \begin{cases} \sum\limits_{m=1}^{M} C_{\mathrm{p}i,m} & i = m \quad \forall i, m \in M \\ -C_{\mathrm{p}i,m} & i \neq m. \end{cases} \tag{4.66}$$

In Equation (4.65), $\mathbf{K}_\mathrm{p}$ is the matrix of partial reluctances. Quasi-static PEEC models in a SPICE compatible form for both current cells and charge cells built by the set of parameters $\underline{Z}$, L_p or K_p, P or C or C_p are presented in Figures 4.11 and 4.12.

In dependence on the features of the interconnect structure to be analyzed, the quasi-static PEEC models can be further simplified. For higher current applications, for example in power electronics structures, the subset $\underline{Z}$, L_p or K_p can already provide a good model (e.g. [12]). On the other hand, for low current applications, for example microelectronic circuits a subset $\underline{Z}$, P or C or C_p can be a reasonable approach (e.g. [13]). It may be pointed out that the advantage of using partial reluctances and capacitance coefficients instead of partial inductances and potential coefficients is based on the dominance of the diagonal elements of the $\mathbf{K}_\mathrm{p}$ and $\mathbf{C}$ matrices, so that off-diagonal elements with very small values can be ignored under retention of the passivity of the interconnect structure [12], [14].

4.4 PEEC Models for the Plane Half Space

The PEEC models derived in Section 4.2 and classified in Section 4.3.1 and 4.3.2 are free space models based on the free space Green's function of Equation (4.23). As already mentioned in Section 4.1.2, the plane half space bounded by a PEC is well suited for modeling interconnect structures that are located above a relatively large ground plane. Using the MPI Equation (4.32) with the Green's functions for the plane half space and applying the derivation process of the PEEC method (see Section 4.3), one obtains the appropriate PEEC model pictured in Figure 4.13.

The generalized partial inductances of the half space $\underline{\Lambda}^\mathrm{H}_{\alpha,n}$ are calculated by

$$\underline{\Lambda}^\mathrm{H}_{\alpha,n} = \frac{\mu_0}{a_\alpha a_n} \int_{v_\alpha} \mathbf{e}_\alpha \cdot \int_{v_n} \overline{\overline{G}}{}^A_\mathrm{H}(\mathbf{r}, \mathbf{r}')\mathbf{e}_n \mathrm{d}v'_n \mathrm{d}v_\alpha = \underline{\Lambda}_{\alpha,n} + \underline{\Lambda}_{\alpha,n_i}, \quad \alpha, n \in N \tag{4.67}$$

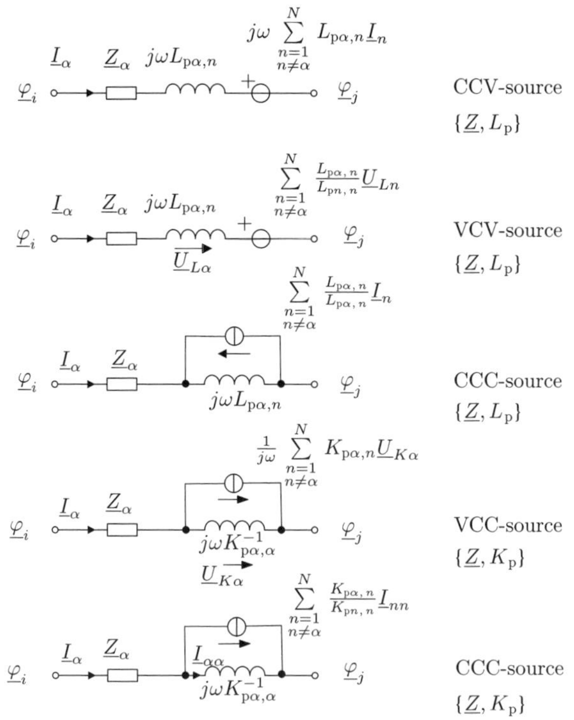

Figure 4.11 SPICE compatible, quasi-static PEEC models of a current cell α ($\alpha \in N$).

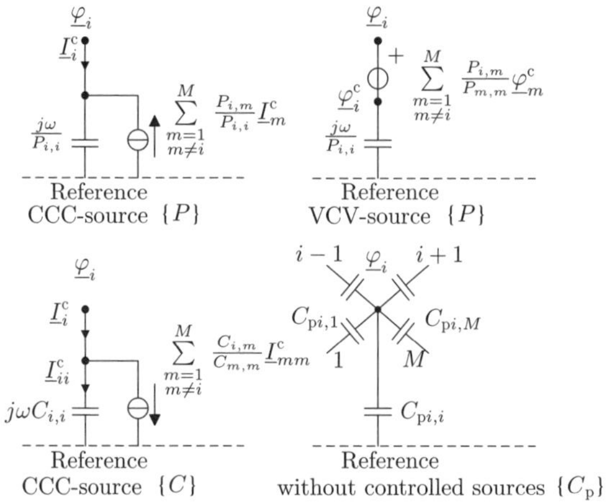

$C_{i,m}$ $(i, m \in M)$ capacitance coefficients

$C_{\mathrm{p}i,m}$ $(i, m \in M)$ partial capacitances

Figure 4.12 SPICE compatible, quasi-static PEEC models of a potential cell i ($i \in M$).

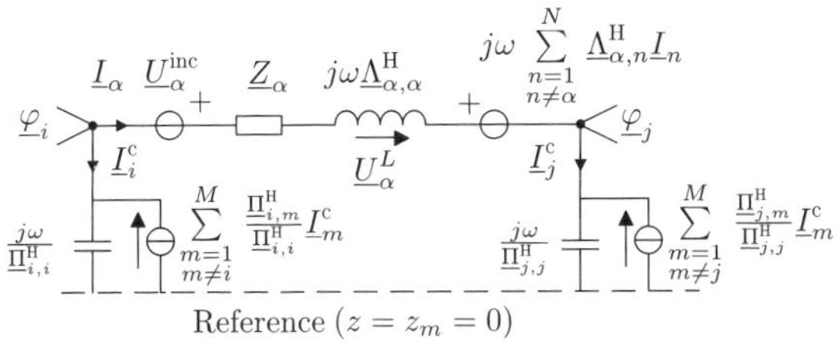

Figure 4.13 Plane half space PEEC model of a volume cell α with adjacent charge cells i and j.

with

$$\underline{\Lambda}_{\alpha,n} = \frac{\mathbf{e}_\alpha \cdot \mathbf{e}_n}{a_\alpha a_n} \mu_0 \int_{v_\alpha} \int_{v_n} g(\mathbf{r}, \mathbf{r}') \mathrm{d}v'_n \mathrm{d}v_\alpha, \tag{4.68}$$

$$\underline{\Lambda}_{\alpha,n_i} = \frac{\mathbf{e}_\alpha \cdot \mathbf{e}_{n_i}}{a_\alpha a_n} \mu_0 \int_{v_\alpha} \int_{v_n} g(\mathbf{r}, \mathbf{r}' - 2z'\mathbf{e}_z) \mathrm{d}v'_n \mathrm{d}v_\alpha. \tag{4.69}$$

Each partial inductance of the half space $\underline{\Lambda}^{\mathrm{H}}_{\alpha,n}$ is composed of two partial inductances of the free space: $\underline{\Lambda}_{\alpha,n}$ (Equation (4.68)) is the partial mutual free space inductance between the cells α and n, whereas $\underline{\Lambda}_{\alpha,n_i}$ (Equation (4.69)) is the partial mutual free space inductance between the cell α and the image cell n_i of the cell n. The unit vector $\mathbf{e}_{n_i}$ is given by (see Figure 4.14)

$$\mathbf{e}_{n_i} = (\mathbf{e}_z \mathbf{e}_z - \bar{\bar{\mathbf{I}}}_t) \cdot \mathbf{e}_n, \qquad n \in N. \tag{4.70}$$

In particular, the partial self inductance of the half space $\underline{\Lambda}^{\mathrm{H}}_{\alpha,\alpha}$ is composed of the partial self inductance $\underline{\Lambda}_{\alpha,\alpha}$ and the partial mutual inductance of cell α with the image cell α_i in free space.

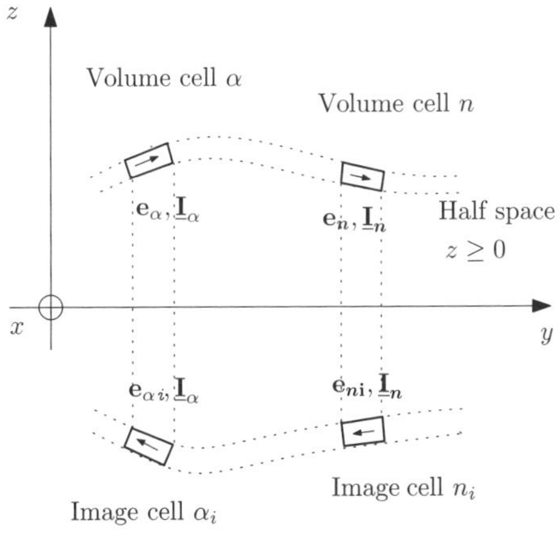

Figure 4.14 Half space and mirror principle.

In a similar way, the potential coefficients of the half space are calculated:

$$\underline{\Pi}^{\mathrm{H}}_{i,m} = \frac{1}{\epsilon_o s_i s_m} \int_{s_i} \int_{s_m} G^{\Phi}_{\mathrm{H}}(\mathbf{r}, \mathbf{r}')\mathrm{d}s'_m \mathrm{d}s_i = \underline{\Pi}_{i,m} - \Pi_{i,m_i}, \quad i, m \in M \tag{4.71}$$

$$\underline{\Pi}_{i,m} = \frac{1}{\epsilon_o s_i s_m} \int_{s_i} \int_{s_m} g(\mathbf{r}, \mathbf{r}')\mathrm{d}s'_m \mathrm{d}s_i, \tag{4.72}$$

$$\underline{\Pi}_{i,m_i} = \frac{1}{\epsilon_o s_i s_m} \int_{s_i} \int_{s_m} g(\mathbf{r}, \mathbf{r}' - 2z'\mathbf{e}_z)\mathrm{d}s'_m \mathrm{d}s_i. \tag{4.73}$$

The potential coefficient of the half space $\underline{\Pi}^{\mathrm{H}}_{i,m}$ results in a difference of the free space coefficient of the potential cell i with the potential cell m and of the potential coefficient of potential cell i with the image cell m_i. The parameters $\underline{Z}_\alpha$ and $\underline{U}^{\mathrm{inc}}_\alpha$ are calculated with the same equations as for the free space (see Equations (4.39) through (4.42)). However, for the calculation of $\underline{\mathbf{E}}^{\mathrm{inc}}$ in the half space one has to take into account the reflections at the reference plane $z_m = 0$. In the following, the practically interesting case of an excitation by an external plane wave is considered. An incoming plane wave $\underline{\mathbf{E}}^{\mathrm{i}}(\mathbf{r})$ causes by reflection at the reference plane $\mathrm{z}_m = 0$ a reflected plane wave $\underline{\mathbf{E}}^{\mathrm{r}}(\mathbf{r})$. The total electric field strength $\underline{\mathbf{E}}^{\mathrm{inc}}(\mathbf{r})$ for calculation of $\underline{U}^{\mathrm{inc}}_\alpha$ according to Equation (4.39) is

$$\underline{\mathbf{E}}^{\mathrm{inc}}(\mathbf{r}) = \underline{\mathbf{E}}^{i}(\mathbf{r}) + \underline{\mathbf{E}}^{r}(\mathbf{r}) = \underline{\mathbf{E}}^{i}_0 \mathrm{e}^{-j\mathbf{k}^{i}\cdot\mathbf{r}} + \underline{\mathbf{E}}^{r}_0 \mathrm{e}^{-j\mathbf{k}^{r}\cdot\mathbf{r}}. \tag{4.74}$$

An incident plane wave is usually described (e.g. [9], [15]) in spherical coordinates by the angles of incidence φ and ϑ, and the angle of polarization γ (see Figure 4.15).

With this, the incident plane wave $\underline{\mathbf{E}}^{\mathrm{i}}$ can be represented in the form

$$\underline{\mathbf{E}}^{\mathrm{i}}_0 = \underline{E}^{\mathrm{i}}_{0x}\mathbf{e}_x + \underline{E}^{\mathrm{i}}_{0y}\mathbf{e}_y + \underline{E}^{\mathrm{i}}_{0z}\mathbf{e}_z \tag{4.75}$$

with

$$\begin{aligned}
\underline{E}^{\mathrm{i}}_{0x} &= \underline{E}_0(-\sin\gamma\cos\vartheta\cos\varphi - \cos\gamma\sin\varphi) \\
\underline{E}^{\mathrm{i}}_{0y} &= \underline{E}_0(-\sin\gamma\cos\vartheta\sin\varphi - \cos\gamma\cos\varphi) \\
\underline{E}^{\mathrm{i}}_{0z} &= \underline{E}_0\sin\gamma\sin\vartheta
\end{aligned}$$

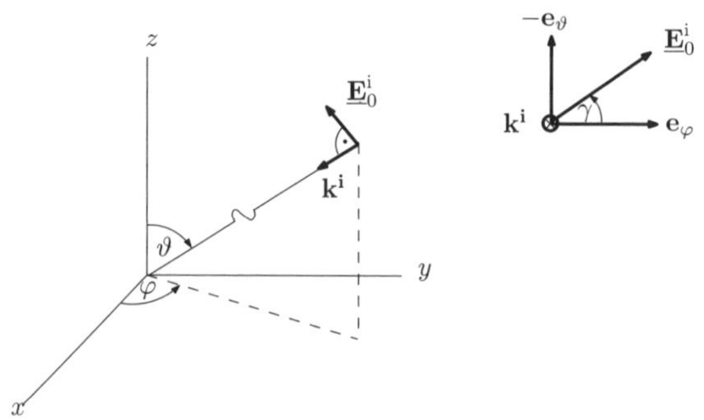

Figure 4.15 Spatial description of a plane wave.

and

$$\mathbf{k}^{\mathrm{i}} = k_x \mathbf{e}_x + k_y \mathbf{e}_y + k_z \mathbf{e}_z \tag{4.76}$$

with

$$\begin{aligned} k_x &= -k \sin \vartheta \cos \varphi \\ k_y &= -k \sin \vartheta \sin \varphi \\ k_z &= -k \cos \vartheta. \end{aligned}$$

Taking into account the boundary conditions at the perfectly conducting reference plane $z_m = 0$, the reflected plane wave $\underline{\mathbf{E}}^{\mathrm{r}}(\mathbf{r})$ can be described by

$$\underline{\mathbf{E}}_0^{\mathrm{r}} = \underline{E}_{0x}^{\mathrm{r}} \mathbf{e}_x + \underline{E}_{0y}^{\mathrm{r}} \mathbf{e}_y + \underline{E}_{0z}^{\mathrm{r}} \mathbf{e}_z \tag{4.77}$$

with

$$\begin{aligned} \underline{E}_{0x}^{\mathrm{r}} &= -\underline{E}_0^{\mathrm{i}} \\ \underline{E}_{0y}^{\mathrm{r}} &= -\underline{E}_{0y}^{\mathrm{i}} \\ E_{0z}^{\mathrm{r}} &= E_{0z}^{\mathrm{i}} \end{aligned}$$

and

$$\mathbf{k}^{\mathrm{r}} = k_x \mathbf{e}_x + k_y \mathbf{e}_y - k_z \mathbf{e}_z. \tag{4.78}$$

With this, $\underline{U}_{\alpha}^{\mathrm{inc}}$ for the half space PEEC model in the case of a plane wave excitation can be calculated by Equation (4.39).

4.5 Geometrical Discretization in PEEC Modeling

4.5.1 Orthogonal Cells

In the volume cell according to Figure 4.3 used for the derivation of the generalized PEEC method (see Section 4.2) the uniformly distributed current density $\underline{\mathbf{J}}$ is directed orthogonally to the cross section a. In general, the cross section may have an arbitrary shape and $\underline{\mathbf{J}}$ an arbitrary direction in space. For this type of cell, constant basis functions for developing $\underline{\mathbf{J}}$ and $\underline{\varrho}$ (see Equations (4.33) through (4.36) may be introduced. The most widespread shapes in the PEEC literature are rectangular cells, that conform with Cartesian coordinates. Reasons for this are:

- There are analytical solutions for calculating the double integrals in the expressions of the partial inductances and potential coefficients for cells oriented parallel to the coordinate axes, which enormously reduces the calculation expense (e.g. [16–19]).
- In practice, many interconnects, for example PCB traces, traces on VLSI and MMIC, but also busbars in power electronics, can be dissected in brick-shaped elements.
- Besides a good suitability for a one-dimensional discretization of wire-like structures, brick-shaped cells or an orthogonal rectilinear meshing are very well suited for a two-dimensional discretization of thin, finite plates and a three-dimensional discretization of finite volumes, for example if dielectrics have to be considered.

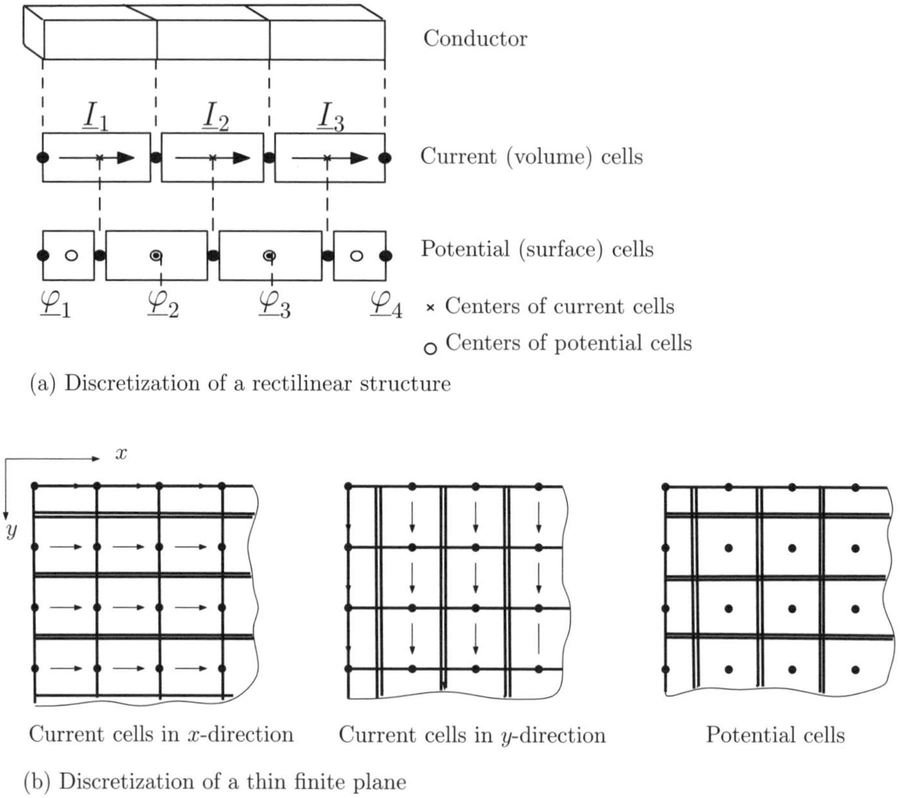

Figure 4.16 One- and two-dimensional discretization with rectangular cells.

Figure 4.16 shows the discretization in current and potential cells for one- and two-dimensional problems. It is evident, that at the ends of linear structures (in the case of branching out, too) and at edges and vertices of planes and volumes half and quarter cells result from the discretization process. This nonuniformity leads to changing the distances between the centers of cells and is the reason for an additional error in the numerical model (see, e.g., Equation (4.46) and [8]).

4.5.2 Nonorthogonal Cells

Many practical electromagnetic problems, such as arbitrary shaped printed circuit antennas or connectors whose EMI radiation has to be modeled, contain nonorthogonal geometries. In [20–22] the PEEC approach is applied in a consistent way to nonorthogonal geometries, especially to quadrilateral (two-dimensional) and hexahedral (three-dimensional) cells. With the higher flexibility in discretization, obtained in this way the number of cells can be reduced and the accuracy of the modeling increased.

An example of a nonorthogonal hexahedral element is shown in Figure 4.17. Each point of the element can be described in both local normalized nonorthogonal coordinates $\mathbf{r}_l(u, v, w)$ with $u, v, w \in [-1, +1]$, and global Cartesian coordinates $\mathbf{r}_g(x, y, z)$. Mapping a point of a

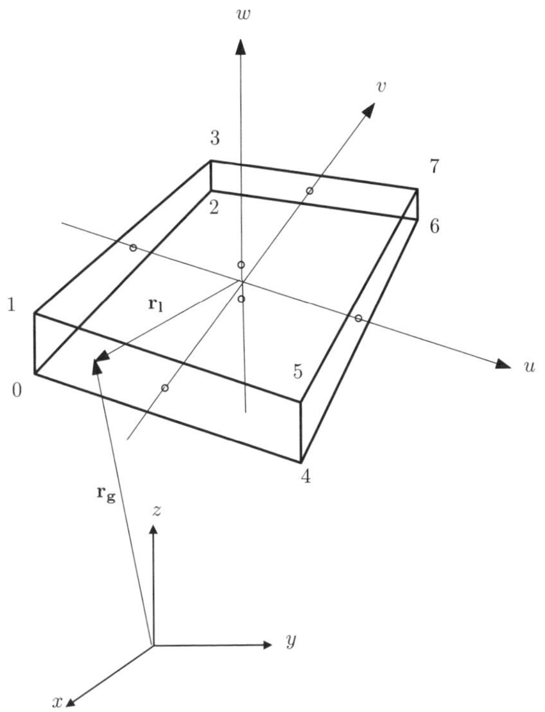

Figure 4.17 Description of a nonorthogonal hexahedral element by local and global coordinates.

hexahedron from local to global coordinates is described by

$$\mathbf{r}_\mathrm{g} = \sum_{k=0}^{7} F_k(u, v, w)\mathbf{r}_{\mathrm{g}k}, \tag{4.79}$$

where $\mathbf{r}_{\mathrm{g}k}$, $k \in [0, 7]$, are the local vectors to the vertices of the hexahedron known from the geometrical discretization. The functions F_k in Equation (4.79) are given by

$$\begin{aligned}
F_0 &= (1/8)(1-u)(1-v)(1-w)\\
F_1 &= (1/8)(1-u)(1-v)(1+w)\\
F_2 &= (1/8)(1-u)(1+v)(1-w)\\
&\;\;\vdots\\
F_7 &= (1/8)(1+u)(1+v)(1+w)
\end{aligned} \tag{4.80}$$

with a permutation of signs according to the binary representation of $k \in [0, 7]$ (binary 0: minus, binary 1: +).

Since in nonorthogonal cells the current density in each of the directions u, v, w is not constant, one has to choose basic functions depending on $\mathbf{r}_\mathrm{g}$ within the cell, for example for the direction $\mathbf{e}_u$ in the form ($\underline{I}_u$ is the cell current in u-direction)

$$\underline{\mathbf{J}}_u(\mathbf{r}_\mathrm{g}) = \mathbf{b}_u^\mathrm{c}(\mathbf{r}_\mathrm{g})\,\underline{I}_u \tag{4.81}$$

with

$$\mathbf{b}_u^\mathrm{c}(\mathbf{r}_\mathrm{g}) = \begin{cases} \dfrac{h_u \mathbf{e}_u}{\left|\dfrac{\partial \mathbf{r}_\mathrm{g}}{\partial u}\left(\dfrac{\partial \mathbf{r}_\mathrm{g}}{\partial v} \times \dfrac{\partial \mathbf{r}_\mathrm{g}}{\partial w}\right)\right|} \cdot \dfrac{1}{a_{vw}} & \mathbf{r}_\mathrm{g} \in \mathrm{cell}u \\ 0 & \text{otherwise.} \end{cases} \tag{4.82}$$

Equation (4.82) is derived from the general definition of the current density by applying generalized curvilinear coordinates [30] under the assumption that the current density related to the normalized v, w coordinates has a constant value. In Equation (4.82), h_u is the metric coefficient, definded by $h_\alpha = |\partial \mathbf{r}_\mathrm{g}/\partial \alpha|(\alpha = u, v, w)$, and $a_{v,w}$ is the cross section for the current $\underline{I}_u$ in normalized coordinates v, w. Further, $[a_{v,w}] = 1$ and $a_{v,w}$ can take the values one for a quarter, two for a half and four for a full cross section (see Figure 4.17). For other directions of the hexahedron, the coordinates u, v, w have to be permuted cyclically.

Applying the Galerkin method one obtains for the inner product

$$\begin{aligned}\int_v \mathbf{w}_u \ldots \mathrm{d}v = \int_u\int_v\int_w \frac{\mathbf{e}_u \cdot h_u}{\left|\dfrac{\partial \mathbf{r}_\mathrm{g}}{\partial u}\left(\dfrac{\partial \mathbf{r}_\mathrm{g}}{\partial v} \times \dfrac{\partial \mathbf{r}_\mathrm{g}}{\partial w}\right)\right|} \cdot \frac{1}{a_{vw}} \cdots \\ \cdots \frac{\partial \mathbf{r}_\mathrm{g}}{\partial u}\left(\frac{\partial \mathbf{r}_\mathrm{g}}{\partial v} \times \frac{\partial \mathbf{r}_\mathrm{g}}{\partial w}\right) \mathrm{d}u\mathrm{d}v\mathrm{d}w\end{aligned} \tag{4.83}$$

with $\mathbf{r}_\mathrm{g} \in$ cell considered. By means of Equation (4.83) the partial elements for the PEEC model in the case of nonorthogonal cell geometries can be calculated.

The voltage source caused by the component of the incident electric field strength $\underline{\mathbf{E}}^\mathrm{inc}$ in the u-direction results in

$$\underline{U}_u^\mathrm{inc} = \int_u\int_v\int_w \frac{1}{a_{vw}} \underline{\mathbf{E}}^\mathrm{inc}(u, v, w) \times \mathbf{e}_u h_u \mathrm{d}u\mathrm{d}v\mathrm{d}w. \tag{4.84}$$

For the local material dependent elements one obtains (e.g., u-direction):

$$R_u = \frac{1}{\sigma}\int_u\int_v\int_w \frac{h_u^2}{a_{vw}^2 \left|\dfrac{\partial \mathbf{r}_\mathrm{g}}{\partial u}\left(\dfrac{\partial \mathbf{r}_\mathrm{g}}{\partial v} \times \dfrac{\partial \mathbf{r}_\mathrm{g}}{\partial w}\right)\right|} \mathrm{d}u\mathrm{d}v\mathrm{d}w \tag{4.85}$$

$$C_u^+ = \varepsilon_0(\varepsilon_\mathrm{r} - 1)\left[\int_u\int_v\int_w \frac{h_u^2}{a_{vw}^2 \left|\dfrac{\partial \mathbf{r}_\mathrm{g}}{\partial u}\left(\dfrac{\partial \mathbf{r}_\mathrm{g}}{\partial v} \times \dfrac{\partial \mathbf{r}_\mathrm{g}}{\partial w}\right)\right|} \mathrm{d}u\mathrm{d}v\mathrm{d}w\right]^{-1}. \tag{4.86}$$

The partial inductances concerning the nonorthogonal cells u and u' with the directions $\mathbf{e}_u$ and $\mathbf{e}_u'$ result in

$$\underline{\Lambda}_{u,u'}(jw) = \frac{\mu_0}{a_{vw} \cdot a'_{vw}} \int_u\int_v\int_w\int_{u'}\int_{v'}\int_{w'} g(\mathbf{r}_\mathrm{g}, \mathbf{r}'_\mathrm{g})\mathbf{e}_u\mathbf{e}_{u'} h_u h_{u'} \mathrm{d}u\mathrm{d}v\mathrm{d}w\mathrm{d}u'\mathrm{d}v'\mathrm{d}w'. \tag{4.87}$$

In a similar way to the basic functions for the current density the adequate basic functions for developing the surface charge density is derived. Writing the equation for the surface charge density in generalized curvilinear coordinates and assuming a constant charge density related to the normalized coordinates the charge density $\underline{\varrho}_m(\mathbf{r}_g)$ on a quadrilateral with the surface s_m yields (e.g. in the local u_m, v_m plane)

$$\underline{\varrho}_m(\mathbf{r}_g) = b_m^{\mathrm{P}}(\mathbf{r}_g) \cdot \underline{q}_m$$

with the basis function

$$b_m^P(\mathbf{r}_g) = \begin{cases} \dfrac{1}{\left| \dfrac{\partial \mathbf{r}_g}{\partial_u} \times \dfrac{\partial \mathbf{r}_g}{\partial v} \right|} \cdot \dfrac{1}{s_{uv}^m} & \mathbf{r}_g \in s_m \\ 0 & \text{otherwise} \end{cases} \tag{4.88}$$

and $\underline{q}_m$ as the total charge on s_m. In Equation (4.88), s_{uv}^m is the area of s_m in normalized u, v coordinates. It can take the values one for a quarter, two for a half and four for a full quadrilateral (e.g. the quadrilateral 1-3-5-7 in Figure 4.17). In analogy to Equation (4.48), the expression for the generalized potential coefficients between two quadrilateral surfaces $s_m(u_m, v_m)$ and $s_i(u_i, v_i)$ is given by

$$\underline{\Pi}_{i,m}(j\omega) = \frac{1}{\epsilon_0 \, s_{uv}^m s_{uv}^i} \int_{u_m} \int_{v_m} \int_{u_i} \int_{v_i} g(\mathbf{r}_{g_m}, \mathbf{r}_{g_i}) \mathrm{d}u_m \mathrm{d}v_m \mathrm{d}u_i \mathrm{d}v_i, \quad m, i \in M. \tag{4.89}$$

Summarizing the results of nonorthogonal geometries in this section one may state that both the structure of the PEEC models and their constitutive elements remain unchanged in comparison with the orthogonal disretization. Only the formulas for the calculation of the partial elements include the special geometrical shapes and are even more challenging for a fast and accurate evaluation than in the orthogonal case (e.g. [23]).

4.5.3 *Triangular Cells*

Triangles are well suited for geometrical meshing open and closed surfaces with arbitrary shape [24]. They accurately conform to the geometry of the surface or boundary and allow varying of the patch density according to the resolution required in the surface geometry or in the current flowing. Additionally, the patch scheme is easy to specify for the computer input.

In [25–27], triangular meshing is applied to the PEEC method. Triangular patches for modeling thin conductors and conductor surfaces as well as prisms with triangular cross section for modeling conductor and dielectric volumes are considered. In the following, the fundamentals for triangular meshing related to PEEC modeling are presented.

The current density $\underline{\mathbf{J}}(\mathbf{r})$ on a surface is developed in the form

$$\underline{\mathbf{J}}(\mathbf{r}) = \sum_{n=1}^{N} \mathbf{b}_n^{\mathrm{c}}(\mathbf{r}) \underline{I}_n. \tag{4.90}$$

N is the number of interior (nonboundary) edges of the triangulated structure. The current density is developed by vector basis functions, the so-called Rao–Wilton–Glisson (RWG) functions:

$$\mathbf{b}_n^c(\mathbf{r}) = \begin{cases} \dfrac{1}{2\Delta_n^+}\boldsymbol{\varrho}_n^+(\mathbf{r}) & \mathbf{r} \in T_n^+ \\ \dfrac{1}{2\Delta_n^-}\boldsymbol{\varrho}_n^-(\mathbf{r}) & \mathbf{r} \in T_n^- \qquad n \in N \\ 0 & \text{otherwise.} \end{cases} \tag{4.91}$$

Each basic function of Equation (4.91) is associated to a nonboundary edge and describes the current distribution in the two adjacent triangles T^+, T^-. Figure 4.18 shows such a triangle pair T_n^+, T_n^- assigned to edge n. Since a basic function is associated with each nonboundary edge, up to three basis functions may have nonzero values within each triangular patch. The normal component of $\mathbf{b}_n^c(\mathbf{r})$ at the edge n is l_n^{-1} (see Figure 4.18(b)), which is equivalent to a constant normal component of the current density $\underline{I}_n/l_n$. $\underline{I}_n$ represents the total current passing the nth edge. Because the other basis functions of the triangle are independent of each other, the current $\underline{I}_n$ is an independent value and, therefore, suitable for the development of $\underline{\mathbf{J}}$ corresponding to Equation (4.91). Each point in $T_n^\pm$ may be designated by both the global position vector $\mathbf{r}_n^\pm$ or the local position vector $\boldsymbol{\varrho}_n^\pm$ defined with respect to the free vertices of $T_n^\pm$. $\boldsymbol{\varrho}_n^+$ is directed away from the free vertex of T_n^+, $\boldsymbol{\varrho}_n^-$ is directed toward the free vertex of T_n^-. $\Delta_n^\pm$ are the areas of the triangles $T_n^\pm$, l_n is the length of the interior edge n. The positive reference direction for the current $\underline{I}_n$ is from T_n^+ to T_n^-.

The continuity of the normal component of the current density at the edge n and the fact, that normal components of the current density do not exist at the other edges of T_n^+ and T_n^- implies that all edges of T_n^+ and T_n^- are free of line charges. The surface divergence of $\mathbf{b}_n^c(\mathbf{r})$ calculated according to $\nabla_s \cdot \mathbf{b}_n^c = (\pm 1/\varrho_n^\pm)\partial(\varrho_n^\pm b_n^c)/\partial\varrho_n^\pm$ is proportional to the surface charge

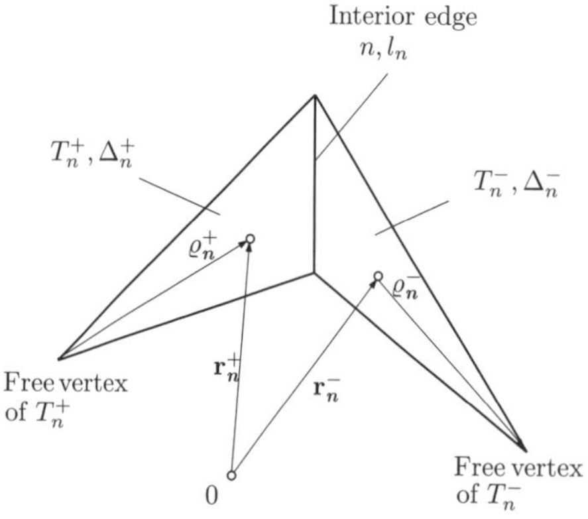

(a) Geometrical parameters of the basis function

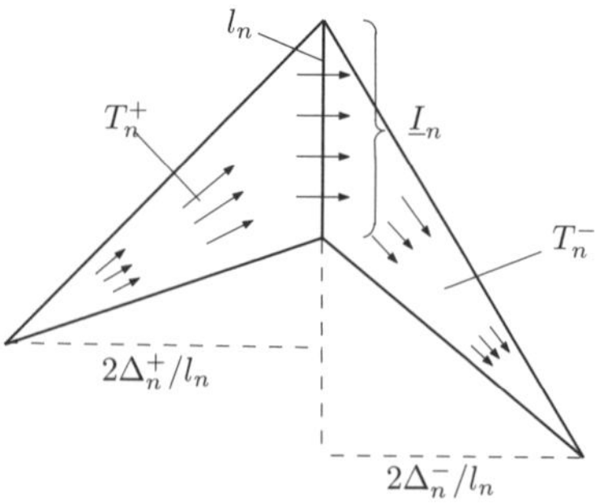

(b) Geometry of the current flow in the triangle pair

Figure 4.18 Associated triangle pair and geometrical parameters.

densities on T_n^+ and T_n^-:

$$\nabla_s \cdot \mathbf{b}_n^c(\mathbf{r}) = \begin{cases} +\dfrac{1}{\Delta_n^+} & \mathbf{r} \in T_n^+ \\ +\dfrac{1}{\Delta_n^-} & \mathbf{r} \in T_n^- \qquad n \in N \\ 0 & \text{otherwise.} \end{cases} \tag{4.92}$$

The constant charge densities in Equation (4.92) allow constant basis functions to be introduced for developing the charge density of the triangulated structure consisting of M triangles:

$$\underline{\varrho}(\mathbf{r}) = \sum_{m=1}^{M} b_m^P(\mathbf{r})\underline{q}_m \tag{4.93}$$

with

$$b_m^P(\mathbf{r}) = \begin{cases} +\frac{1}{\Delta_m} & \mathbf{r} \in T_m \qquad m \in N \\ 0 & \text{otherwise.} \end{cases} \tag{4.94}$$

The derivation process of the PEEC method for triangular meshing follows the same steps as these shown in Section 4.2 for orthogonal meshing. For applying the Galerkin method one has to build the inner product with the weighting functions $\mathbf{w}_\alpha$ corresponding to Equation (4.91):

$$\int_{T_\alpha^+ + T_\alpha^-} \mathbf{w}_\alpha \ldots \mathrm{d}s \quad , \quad \forall \alpha \in N. \tag{4.95}$$

The integral has to be taken over the surface of the triangle pair T_α^+, T_α^-. Doing this, one obtains for the generalized partial inductances

$$\underline{\Lambda}_{\alpha,n}(j\omega) = \mu_0 \int_{T_\alpha^+ + T_\alpha^-} \mathbf{b}_\alpha^c(\mathbf{r}) \int_{T_n^+ + T_n^-} g(\mathbf{r}, \mathbf{r}')\mathbf{b}_n^c(\mathbf{r}')\mathrm{d}s'\mathrm{d}s, \qquad \alpha, n \in N. \tag{4.96}$$

The derivation of the generalized potential coefficients results in

$$\underline{\Pi}_{i,m}(j\omega) = \frac{1}{\varepsilon_0} \int_{T_i} b_i^{\mathrm{P}} \int_{T_m} b_m^{\mathrm{P}} g(\mathbf{r}, \mathbf{r}')\mathrm{d}s'\mathrm{d}s, \quad i, m \in M. \tag{4.97}$$

It should be mentioned that in the derivation process of the potential coefficients the inner product is to build:

$$< \mathbf{b}_\alpha^c, \nabla \underline{\varphi} >= \int_{T_\alpha^+ + T_\alpha^-} \mathbf{b}_\alpha^c(\mathbf{r}) \times \nabla \underline{\varphi} \mathrm{d}s. \tag{4.98}$$

Using the vector identity

$$\nabla(\varphi \cdot \mathbf{f}) = \mathbf{f}\nabla\varphi + \varphi\nabla\mathbf{f}$$

and the fact that in the given case for $\mathbf{f} = \mathbf{b}_\alpha^c$ due to the properties of $\mathbf{b}_\alpha^c$ at the edges of T_α^+ and T_α^-

$$\int_{T_\alpha^+ + T_\alpha^-} \nabla(\underline{\varphi}\mathbf{b}_\alpha^c)\mathrm{d}s = 0$$

holds, Equation (4.98) can be rewritten as

$$< \mathbf{b}^{\mathrm{c}}_{\alpha}, \nabla \underline{\varphi} >= - \int_{T^+_\alpha + T^-_\alpha} \underline{\varphi} \nabla_s \cdot \mathbf{b}^{\mathrm{c}}_{\alpha} \mathrm{d}s. \tag{4.99}$$

Introducing Equation (4.92) in Equation (4.99) one obtains

$$-\frac{1}{\Delta^+_\alpha} \int_{T^+_\alpha} \underline{\varphi} \mathrm{d}s + \frac{1}{\Delta^-_\alpha} \int_{T^-_\alpha} \underline{\varphi} \mathrm{d}s = -\underline{\varphi}^+_\alpha + \underline{\varphi}^-_\alpha, \qquad \alpha^+, \alpha^- \in M. \tag{4.100}$$

The potentials $\underline{\varphi}^+_\alpha (= \underline{\varphi}_i)$ and $\underline{\varphi}^-_\alpha (= \underline{\varphi}_m)$ are mean values assigned to the centroids of the triangles T_i, $_m(i, m \in M)$. Further, for the independent voltage source $\underline{U}^{\mathrm{inc}}_\alpha$ resulting from the incident field strength $\underline{\mathbf{E}}^{\mathrm{inc}}(\mathbf{r})$ by application of the inner product of Equation (4.95) one obtains:

$$\underline{U}^{\mathrm{inc}}_\alpha = \int_{T^+_\alpha + T^-_\alpha} \mathbf{b}^{\mathrm{c}}_\alpha(\mathbf{r}) \cdot \underline{\mathbf{E}}^{\mathrm{inc}}(\mathbf{r}) \mathrm{d}s. \tag{4.101}$$

By applying the mean value theorem, Equation (4.99) can be written in the form

$$\underline{U}^{\mathrm{inc}}_\alpha = \frac{1}{2} \left\{ \underline{\mathbf{E}}^{\mathrm{inc}}(\mathbf{r}^{\mathrm{c}+}_\alpha) \cdot \boldsymbol{\rho}^{\mathrm{c}+}_\alpha + \underline{\mathbf{E}}^{\mathrm{inc}}(\mathbf{r}^{\mathrm{c}-}_\alpha) \cdot \boldsymbol{\rho}^{\mathrm{c}-}_\alpha \right\} \tag{4.102}$$

with $\mathbf{r}^{\mathrm{c}\pm}_\alpha$ and $\boldsymbol{\rho}^{\mathrm{c}\pm}_\alpha$ as the global and local vectors to the centroids of $T^\pm_\alpha$.

For imperfect conductors the finite electric field strength $\underline{\mathbf{E}}(\mathbf{r})$ causes a voltage drop $\underline{U}_\alpha$ in the triangle pair $T^\pm_\alpha$ calculated by the inner product

$$\underline{U}_\alpha = \int_{T^+_\alpha + T^-_\alpha} \mathbf{b}^{\mathrm{c}}_\alpha(\mathbf{r}) \cdot \underline{\mathbf{E}}(\mathbf{r}) \mathrm{d}s. \tag{4.103}$$

Using Equation (4.14) for the surface impedance $\underline{Z}_s$ and introducing it in Equation (4.103) yields

$$\underline{U}_\alpha = (1 + j) \sqrt{\frac{\omega \mu_0}{2\sigma}} \int_{T^+_\alpha + T^-_\alpha} \left[\mathbf{b}^{\mathrm{c}}_\alpha(\mathbf{r}) \right]^2 \mathrm{d}s \, \underline{I}_\alpha. \tag{4.104}$$

Equation (4.104) defines a local partial impedance $\underline{Z}_\alpha$ that has to be introduced as an element in the circuit model:

$$\underline{Z}_\alpha = \frac{1 + j}{\sigma \delta} \int_{T^+_\alpha + T^-_\alpha} \left[\mathbf{b}^{\mathrm{c}}_\alpha(\mathbf{r}) \right]^2 \mathrm{d}s \tag{4.105}$$

with $\delta = \sqrt{2/\omega \sigma \mu_0}$ as the so-called skin depth.

For a thin conducting plate with a thickness $d \leq \delta$, that is the skin effect cannot develop and the current distribution in the d direction is homogeneous, the impedance is real and δ has to be replaced by d:

$$R_\alpha = \Re \left\{ \underline{Z}_\alpha |_{\delta = d} \right\} = \frac{1}{\sigma d} \int_{T^+_\alpha + T^-_\alpha} \left[\mathbf{b}^{\mathrm{c}}_\alpha(\mathbf{r}) \right] \mathrm{d}s. \tag{4.106}$$

The circuit structure and the constitutive elements of the PEEC models derived for triangular meshing are the same as those derived for orthogonal cells (see Figures 4.6 and 4.7). The changed geometrical discretization and other basic functions are hidden in the formulas for

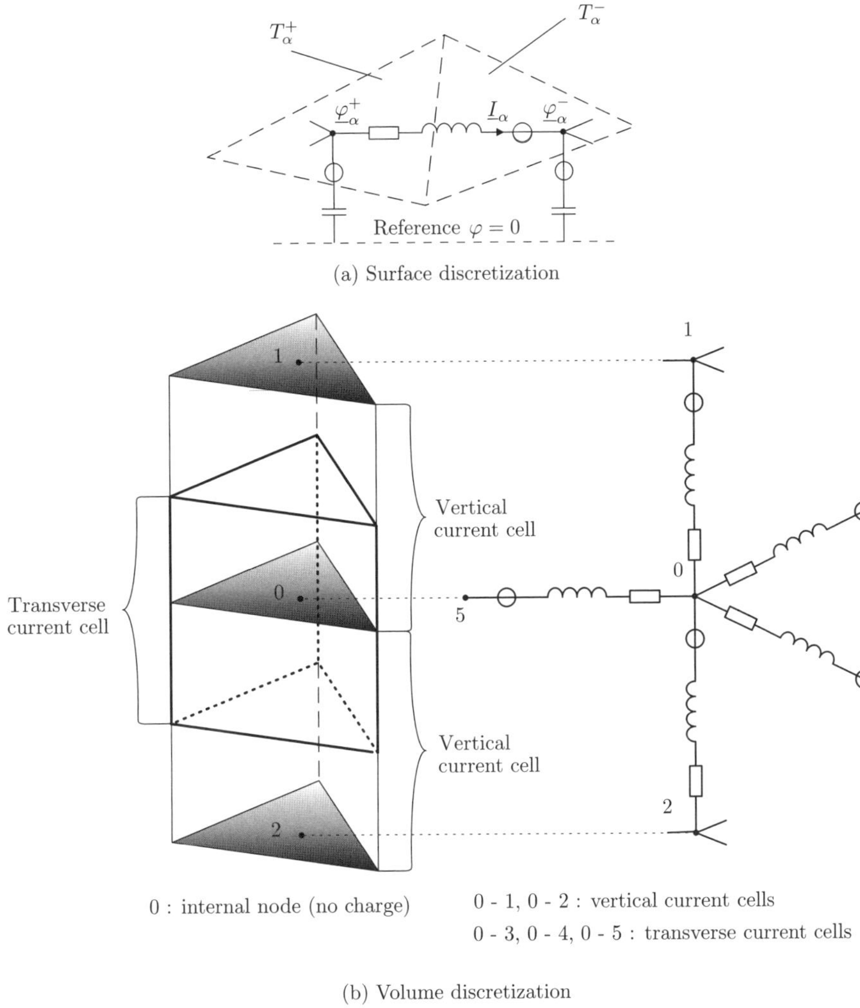

Figure 4.19 Circuit interpretation for triangular discretization.

the calculation of the partial elements. The assignment of the circuit interpretation to surface elements and to prisms with triangular cross section is shown in Figure 4.19.

4.6 PEEC Models for the Time Domain and the Stability Issue

As already mentioned in the Introduction of this book, one of the most important advantages of the PEEC method in comparison with other numerical methods is the opportunity for the creation of time domain models with circuit interpretation. This opens the way to use circuit solvers for effective simulations even in a nonlinear circuit environment. While in the previous sections the PEEC method and the PEEC models in the frequency domain have been derived, now the focus is directed at the time domain modeling.

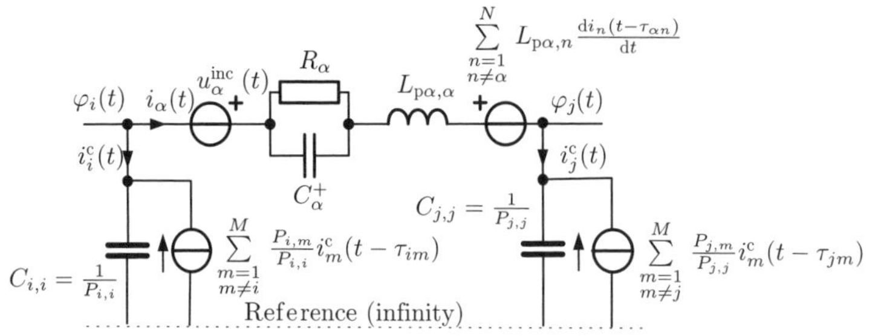

Figure 4.20 Time domain PEEC model with center-to-center retardation and constant parameters $\{R, C^+, L_p, P, \tau\}$.

4.6.1 Standard PEEC Models for the Time Domain

The starting point for the considerations in this section are the frequency domain PEEC models with center-to-center retardation according to Section 4.3.2.2 and the appropriate quasi-static PEEC models according to Section 4.3.2.3. These types of models with frequency independent partial elements $\{R, C^+, L_P, P, \tau\}$ were historically first introduced by A.E. Ruehli, (see the Introduction of this book) and are here referred to as standard PEEC models. Using the inverse Fourier transformation, the standard PEEC models can easily be transformed into the time domain. The complex variables $\underline{U}, \underline{I}, \underline{\varphi}$ are converted into time functions $u(t), i(t), \varphi(t)$. In particular it should be mentioned that complex variables multiplied by the retardation factor $\exp(-j\omega\tau)$ are transformed into time functions with a time delay τ:

$$\mathfrak{F}^{-1}\{F(j\omega)\exp(-j\omega\tau)\} = f(t-\tau). \tag{4.107}$$

The standard time domain PEEC model with center-to-center retardation and constant parameters $\{R, C^+, L_P, P, \tau\}$ is shown in Figure 4.20. Typical for the model are the controlled voltage and current sources, whose control variables are time-delayed. These controlled sources represent the magnetic and electric interaction between the cells under consideration of the finite propagation velocity of electromagnetic waves. For

$$\tau_{\alpha m} = 0 \quad \forall \alpha, m \in M \quad \text{and} \quad \tau_{im} = 0 \quad \forall i, m \in M \tag{4.108}$$

the PEEC model in Figure 4.20 represents the quasi-static time domain PEEC model. All other modifications derived before, for example the conversion of Norton into Thévenin equivalent circuits and vice versa for the capacitive branches and the introduction of reluctances and partial capacitances in the case of quasi-static models, are also valid for time domain models.

The models considered which contain constant circuit parameters as well as independent and controlled voltage and current sources with and without time delays, can be effectively implemented in circuit solvers. The use of the open circuit solver SPICE with necessary modifications was first described in [28] and [27]. One important problem, namely possible instabilities of the timedomain solutions, especially so-called late-time instabilities which are also well known from other time domain numerical methods for electromagnetic field calculations, are considered in the following section.

4.6.2 General Remarks on Stability of PEEC Model Solutions

Real interconnection structures being modeled by PEEC are physically linear passive systems and, therefore, always stable. Obviously, instabilities of PEEC models are caused by inaccuracies of the model formulation and solution process. The main reasons for inaccuracies are:

- the geometrical discretization (discretization criterion Equation (4.37), shape of cells, aspect ratio, nonuniformities in discretization, etc.),
- the calculation of the directional derivative of the scalar electric potential φ by the difference quotient (see Equation (4.46)),
- the assumption of center-to-center retardation (see Equations (4.58) and (4.59)), and
- the numerical integration technique used for the calculation of the time domain solution.

The potential instability of the discretized EFIE/MPIE formulation for the solution in the time domain is well known (e.g. [29], [32]). The appropriate PEEC models with center-to-center reatardation are described by linear, ordinary delay differential equation (LODDE) systems whose potential instability is also well known and depends on the delays (e.g. [31], [32]). From [32] it follows that by using the numerical methods backward Euler (BE) or Lobatto III-C one obtains a stable time domain solution for a stable PEEC model. This fact allows us to separate the considerations of the PEEC model stability from those of the numerical methods for obtaining stable solutions. Thus, in the following, the focus will be directed on the PEEC model stability.

Quasi-static PEEC models are linear, passive systems with lumped elements and are mathematically described by linear, ordinary differential equation (LODE) systems. Stability is given, if all complex eigenvalues of the system matrix are located inside the left complex half plane. Another criterion concerning the circuit interpretation and using the passivity is given by $\Re\{\underline{Z}_{\text{input}}\} \geq 0$ for all frequencies $\omega \geq 0$, where $\underline{Z}_{\text{input}}$ is the input impedance of the circuit. To ensure passivity of the quasi-static model the partial elements have to fulfill particular conditions. The matrix of capacitive coefficients $\mathbf{C} = \mathbf{P}^{-1}$ must be positive definite with only positive diagonal, negative off-diagonal elements and a positive sum of the elements of each row. Analogously, the same is valid for the matrix of reluctances $\mathbf{K} = \mathbf{L}_{\text{P}}^{-1}$. In [34] the impact of the partial element accuracy on the stability of quasi-static PEEC models is investigated.

Full-wave PEEC models with center-to-center retardation are described in the time domain by LODDE systems. Their stability can be analyzed by several methods known from control and system theory (e.g. [33]). The Lyapunov–Razumikhin function approach and the Lyapunov–Krasovskii functional method are applicable for the stability estimation in the time domain, whereas the matrix pencil technique is suitable for the frequency domain. The disadvantage of these methods is the difficulty to apply them to systems with a high number of delay functions such as given in PEEC models. In [31], the authors have derived an analytical approach for the transformation of a LODDE system into a LODE system and a coupled algebraic recursive system. The criterion for delay independent stability has been developed and applied to simple PEEC models. Taking into account that in PEEC models the delays are embedded in the control variables of the controlled voltage and current sources, in [7] a macromodel for the ideal retardation $\exp(-j\omega\tau)$ has been developed, whose time domain model is represented by a LODE system. Thus, a LODDE system is converted into a LODE system, whose stability can be analyzed by the well-known method of eigenvalues of the system.

Now, the difference between the stability issue for time and frequency domain PEEC models is considered. Usually, FD-PEEC models are analyzed in the frequency range up to $f_{\rm m}$ with $f_{\rm m}$ as the maximum frequency of interest that is taken as the basis for the geometrical discretization. In this context, the stability behavior is of interest up to $f_{\rm m}$ only. On the other hand, for the transient analysis of TD-PEEC models the frequency spectra of signals cannot be limited on the range up to $f_{\rm m}$. Thus, the system analysis and, implied, the stability analysis have to take place in an extended frequency range, theoretically up to infinity. Instabilities excited in the upper frequency range $f > f_{\rm m}$ are so-called 'late time instabilities'. This means that after fading of the transient response a self-excitation process takes place without external stimulation. Such instabilities are recognizable in the frequency domain by an undamped frequency response or a negative part of the input impedance for particular frequencies in the range $f > f_{\rm m}$.

In this context, it should be recalled that the real physical interconnection system is a continuous one. Therefore, from the disrete PEEC model developed for a maximum frequency of interest $f_{\rm m}$ a physically correct behavior may be expected up to this frequency only. However, for frequencies higher than $f_{\rm m}$ it is expected that at least physical principles, such as causality, passivity or stability, are not violated.

In [7], for the first time the reason for instability of the PEEC models with center-to-center retardation was discovered. As already mentioned in Section 4.3.2, the essential inaccuracy was introduced with the transition from the PEEC model with generalized, frequency dependent partial elements (see Section 4.3.2.1) to PEEC models with frequency independent partial elements multiplied by a complex factor for the center-to-center retardation. While both frequency dependencies are in very good compliance up to $\omega_{\rm m}$, a significant difference is given for frequencies $\omega > \omega_{\rm m}$ (see Figure 4.9). The loss of damping the modulus of the complex factor $\exp(-j\omega\tau)$ with increasing frequency for the frequency range $\omega > \omega_{\rm m}$ is cause for the potential instability of PEEC models with center-to-center retardation. Also in [7] it was shown that an increased refinement cannot eliminate the potential instability of these PEEC models.

Based on this insight, stable time domain PEEC models can be developed (see Section 4.6.4) and measures for improving the stability of PEEC models with center-to-center retardation, already in application, can be assessed (see Section 4.6.3).

4.6.3 Stability Improvement of PEEC Models with Center-to-Center Retardation

Resulting from the stability analysis, for reducing late-time instabilities in PEEC models with center-to-center retardation two methods are mainly proposed in the literature [34]: the damping resistor and the partitioning scheme of cells. Both methods are briefly explained in the following paragraph.

Damping Resistor

To extend the stability beyond $\omega_{\rm m}$ a damping resistor in parallel with each partial self inductance as shown in Figure 4.21(a) is included. The actual value of this resistor has to be calculated for each cell using the equation

$$R_\alpha^d = k\omega_{\rm m} L_{{\rm p}\alpha,\alpha} \tag{4.109}$$

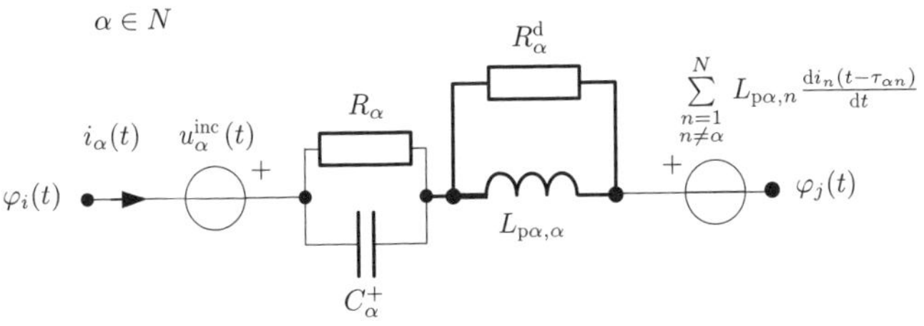

(a) Including a damping resistor R_α^d

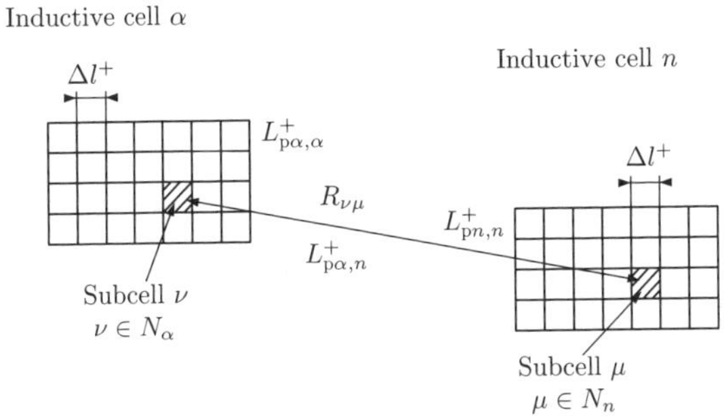

(b) Partitioning scheme

Figure 4.21 Measures for stability improvement.

with $k = 10 \ldots 100$, depending on the particular problem. In each case, R_α^d introduces additional losses in the model and, thus, increases the stability. It has to be chosen large enough to dominate the partial inductance in the extended frequency range, and small enough so as not to corrupt the solution in the active frequency range.

The advantage of introducing R_α^d is that it does not change the complexity of the model and it is applicable both for time and frequency domain. The disadvantage is that an optimal value of k can only be found heuristically. An optimal value of k means that for a stable solution the accuracy is reduced as little as possible. However, the accuracy becomes worse for each value of k.

Partitioning Scheme of Cells

By subdividing each of the cells in a finite number of partitions, as shown in Figure 4.21(b) for inductive cells, one can introduce a more accurate computing of both inductance and potential coefficient terms. The subdivision is given by $\Delta l^+ = \Delta l/n$, where Δl is computed by Equation (4.37) with $n = 1, 2, 3, \ldots$, $f_e = n \times f_m$, f_m the maximum frequency of interest, and f_e the upper end of the extended frequency range. This partitioning requires a summation over all the partitions of the cells. Taking the inductance cells α and n ($\alpha, n \in N$) as an example, the partial mutual inductance with increased accuracy, denoted by $L^+_{p\alpha,n}$, becomes

$$L^+_{p\alpha,n} = \mu_0 \frac{\cos\vartheta_{\alpha n}}{a_\alpha a_n} \sum_{\nu=1}^{N_\alpha} \sum_{\mu=1}^{N_n} e^{-jkR_{\nu\mu}} \int_{v_\nu} \int_{v_\mu} \frac{1}{R_{\nu\mu}} dv_\nu dv_\mu. \tag{4.110}$$

In Equation (4.110), N_α and N_n are the numbers of subcells of cells α and n, respectively. The phase term is not inside the integral , but is approximated by defining $R_{\nu\mu}$ as the distance between the centers of the subcells ν of cell α and μ of cell n. Although the partial inductance calculation and, in a similar manner, the calculation of potential coefficients have changed, the number of unknown currents in the PEEC model has not increased. The higher accuracy of the paramters L_P^+ and P^+ and the extended frequency range for stable solutions are balanced by a higher effort for the calculation of the partial elements. Applications of the actually complex parameters L_P^+ and P^+ to time domain models are not published in the literature.

4.6.4 *Stable Time Domain PEEC Models by Parametric Macromodeling the Generalized Partial Elements*

The basic approach for the development of stable time domain models consists of the approximation of the frequency dependency of the generalized partial elements by appropriate complex functions that permit the development of time domain macromodels for an effective implementation in circuit solvers. Two methodologies were published in the latest literature. The first one is presented in [7], [36] and [35] and named full-spectrum convolution macromodeling (FSCM). The frequency dependency of the partial elements is approximated by a special complex function, for which the inverse Fourier transformation yields an analytical expression. By applying the FSCM-technique, a time domain model can be developed. The second methodology presented in [37] seperates the delay part and approximates the remaining part of the frequency dependence of the partial elements by rational functions of Forster's canonical type as the basis for the circuit synthesis. In the following, based on the cited literature, both the methodologies are outlined with the aim of supporting the fundamental understanding.

4.6.4.1 Stable Time Domain PEEC Models by Full-Spectrum Convolution Macromodeling (FSCM)

As mentioned before, the starting point for the considerations in this section is the approximation of the frequency dependency of the generalized partial elements. The frequency dependency for both the partial inductances and potential coefficients is the same, namely given by the double integrals over the Green's functions. For simplicity, abstract elements are introduced

$$\xi\,(j\omega) = \left\{\underline{\Lambda}\,(j\omega)\,,\ \underline{\Pi}\,(j\omega)\right\}. \tag{4.111}$$

Two approximations are provided. The first one is developed for the generalized partial elements concerning remotely located cells, the second one is applied to modeling self partial elements and closely located cells.

The FSCM for Generalized Partial Elements with Large Time Delays
For the frequency approximation the following function is used:

$$\xi\,(j\omega) \approx W\,(j\omega) = \xi_0 \mathrm{e}^{-(\alpha\omega)^2} \mathrm{e}^{-j\omega\,\tilde{\tau}}\,. \tag{4.112}$$

The parameters in Equation (4.112) may be explicitely calculated using two frequency points of the original function. The parameter ξ_0 corresponds exactly to the standard partial element for $\omega = 0$. The other frequency point is

$$\omega_{\mathrm{g}} = \min\left\{\omega_{\mathrm{m}}, \frac{0.1\pi}{\tau}\right\} \tag{4.113}$$

with ω_{m} as the maximum frequency of interest and τ as the center-to-center delay time between the cells under consideration. With this, the parameters α and $\tilde{\tau}$ are calculated by

$$\alpha = \frac{1}{\omega_{\mathrm{g}}}\sqrt{\log\frac{\xi_0}{\left|\xi\left(j\omega_{\mathrm{g}}\right)\right|}}, \qquad \tilde{\tau} = -\frac{\arg\left\{\xi\left(j\omega_{\mathrm{g}}\right)\right\}}{\omega_{\mathrm{g}}}. \tag{4.114}$$

The time domain response of the controlled source $Y = W(j\omega)X$ with the transfer function $W(j\omega)$ on an arbitary excitation can be calculated via the Duhamel integral

$$y(t) = h(0)\,x(t) + \int_0^t g(t-\theta)\,x(\theta)\,\mathrm{d}\theta. \tag{4.115}$$

In Equation (4.115) , $x(t) = \mathfrak{F}^{-1}\{X\}$ is the excitation or control function, $y(t) = \mathfrak{F}^{-1}\{Y\}$ is the output function, $g(t) = \mathfrak{F}^{-1}\{W(j\omega)\}$ is the impulse response, and $s(t) = \mathfrak{F}^{-1}\{W(j\omega)/j\omega\}$ is the step response. The impulse and step responses for $W(j\omega)$ can be written in a closed form:

$$g(t) = \frac{\xi_0}{2\alpha\sqrt{\pi}}\exp[(\frac{\tilde{\tau}-t}{2\alpha})] \tag{4.116}$$

$$s(t) = \frac{1}{2}\left[1 - \operatorname{erf}\left(\frac{\tilde{\tau}-t}{2\alpha}\right)\right]. \tag{4.117}$$

$\mathfrak{F}^{-1}$ is the symbol for the inverse Fourier transform. Equation (4.116) represents the well-known Gauss function, erf(x) in Equation (4.117) is the error function.

The causality of the time response is only given for $\tilde{\tau} \geq \tau_\alpha$ with $\tau_\alpha = 2\nu\alpha$ and $\nu \leq 3$ for practical calculations. That means, the Gauss function is equal to zero out the interval $\tilde{\tau} - \tau_\alpha < t < \tilde{\tau} + \tau_\alpha$. The conservation of causality, that is $s(t) = 0$ for $t \leq 0$, requires $\tilde{\tau} \geq \tau_\alpha$. This is the reason that for small delay times $\tilde{\tau}$ one needs another function for the approximation of $\xi(j\omega)$.

Further, the time response $y(t)$ on an arbitrary input signal $x(t)$ is considered for discrete time points. For simplicity, a constant time step Δt is assumed. Thus, the notation $n\Delta t = t_n$, $x(t_n) = x_n$, and $y(t_n) = y_n$ may be introduced. The input function between the time points is approximated by a linear function

$$x(t) = a_{n+1}(t-t_n) + b_{n+1}, \quad t \in [t_n, t_{n+1}] \tag{4.118}$$

$$\text{with} \qquad a_{n+1} = \frac{x_{n+1}-x_n}{\Delta t}, \; b_{n+1} = x_n.$$

The convolution integral of Equation (4.115), simplified for $h(0) = 0$ because of the causal signal response, can be represented as a sum of integrals over the time steps Δt:

$$y_{n+1} = \sum_{k=0}^{n} \int_{0}^{\Delta t} g\left(t_{n+1} - t_k - \theta\right)\left(a_{k+1}\theta + b_{k+1}\right) \mathrm{d}\theta\,. \tag{4.119}$$

Equation (4.119) can be solved in a closed form. After several transformations a moving-average formula for the output function is obtained as follows:

$$y_{n+1} = \sum_{k=p}^{q} \Omega_k x_k. \tag{4.120}$$

The coefficients of Equation (4.120) may be computed before the numerical integration of the PEEC model by the following expressions:

$$\Omega_k = \begin{cases} \dfrac{\alpha}{\sqrt{\pi}}\dfrac{1}{\Delta t}\left[\Psi_{n+1,p+1} - \Psi_{n+1,p}\right] + \dfrac{T_{n+1,p+1}}{2\Delta t}\left[\Phi_{n+1,p+1} - \Phi_{n+1,p}\right], & k = p \\ \dfrac{\alpha}{\sqrt{\pi}}\dfrac{1}{\Delta t}\left[\Psi_{n+1,q-1} - \Psi_{n+1,q}\right] + \dfrac{T_{n+1,q-1}}{2\Delta t}\left[\Phi_{n+1,q-1} - \Phi_{n+1,q}\right], & k = q \\ \dfrac{\alpha}{\sqrt{\pi}}\dfrac{1}{\Delta t}\left[\Psi_{n+1,k+1} - 2\Psi_{n+1,k} + \Psi_{n+1,k-1}\right] + & \\ +\dfrac{T_{n+1,k-1}}{2\Delta t}\left[\Phi_{n+1,k-1} - \Phi_{n+1,k}\right] + \dfrac{T_{n+1,k+1}}{2\Delta t}\left[\Phi_{n+1,k+1} - \Phi_{n+1,k}\right], & \text{otherwise} \end{cases} \tag{4.121}$$

with

$$\Psi_{n,k} = \exp\left[-\left(\frac{T_{n,k}}{2\alpha}\right)^2\right], \quad \Phi_{n,k} = \operatorname{erf}\left(\frac{T_{n,k}}{2\alpha}\right), \quad T_{n,k} = \tau - t_n + t_k. \tag{4.122}$$

Because of the selectivity of the impulse response $g(t)$ (see Figure 4.22(a)) one has to calculate only a few summands of the moving average Equation (4.120), namely the time points between t_p and t_q. The boundaries for the subscripts p and q are chosen in the range from 0 up to n as follows (see also Figure 4.22):

$$p = \max\left\{\frac{t_{n+1} - \tilde{\tau} - \tau_\alpha}{\Delta t}, 0\right\}, \quad q = \frac{t_{n+1} - \tilde{\tau} - \tau_\alpha}{\Delta t}. \tag{4.123}$$

The FSCM for Generalized Partial Elements with Small Time Delays

For small time delays, a first-order rational function is applied to the approximation of the generalized partial elements:

$$\xi\left(j\omega\right) \approx H\left(j\omega\right) = \frac{\xi_0}{1 + j\omega\tau}\mathrm{e}^{-j\omega\tilde{\tau}}. \tag{4.124}$$

The determination of the unknown parameters requires the calculation for $\xi\left(j\omega\right)$ for two frequency points. The parameter ξ_0 is given by $\xi_0 = \xi(0)$. The second frequency point for

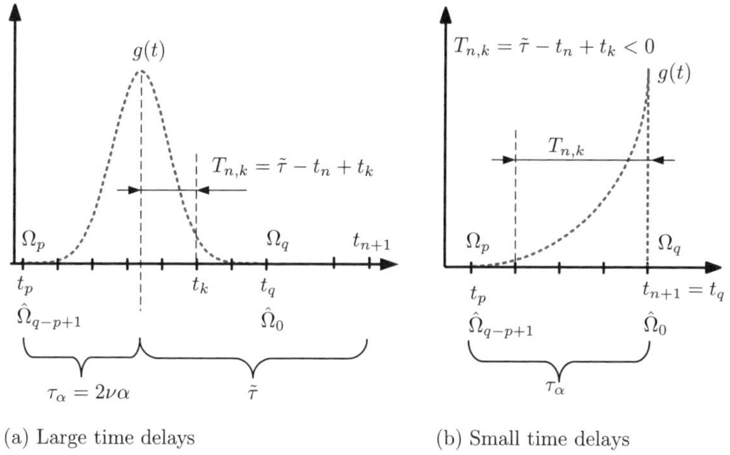

Figure 4.22 Coefficients of the moving average Equation (4.120).

calculation of ξ is ω_g according to the Equation (4.113). The equivalent time delay for the generalized partial element is

$$\tau_{eq} = -\frac{\arg\{\xi(j\omega_g)\}}{\omega_g}. \tag{4.125}$$

The equality $|\xi(j\omega_g)| = |H(j\omega_g)|$ is used for the determination of the remaining parameters:

$$\begin{aligned} \tau &= \frac{\sqrt{1-\delta^2}}{\omega_g \delta} \\ \tilde{\tau} &= \tau_{eq} - \tau \end{aligned} \tag{4.126}$$

with $\delta = |\xi(j\omega_g)|/\xi_0$.

The impulse and step responses of $H(j\omega)$ are derived in closed form:

$$g(t) = \frac{\xi_0}{\tilde{\tau}} e^{-\frac{t-\tilde{\tau}}{\tau}} h(t-\tilde{\tau}) \tag{4.127}$$

$$s(t) = \xi_0 \left(1 - e^{-\frac{t-\tilde{\tau}}{\tau}}\right) h(t-\tilde{\tau}), \tag{4.128}$$

where $h(t)$ is the Heaviside step function. By the Duhamel integral, the time response $y(t)$ on an excitation function $x(t)$ is computed analogously to Equation (4.119). In the moving average Equation (4.120), the coefficients Ω_k are calculated as follows:

$$\Omega_k = \frac{1}{\eta} \begin{cases} e^{T_{n+1,k}} - (1+\eta)\, e^{T_{n+1,k+1}} & k = p \\ e^{T_{n+1,k-1}} + (\eta - 1)\, e^{T_{n+1,k}} & k = q \\ e^{T_{n+1,k-1}} - 2e^{T_{n+1,k}} + e^{T_{n+1,k+1}} & \text{otherwise} \end{cases} \tag{4.129}$$

with

$$T_{n,k} = \tilde{\tau} - t_n + t_k, \eta = \frac{\Delta t}{\tilde{\tau}}.$$

Since $g(t)$ is a selective function shown in Figure 4.22(b), one has to calculate only a few summands in the moving average Equation (4.120), namely for the time points between t_p and t_q. The index p may be estimated by the heuristic condition

$$\Omega_p < \delta_{\text{err}} \sum_{k=p+1}^{q} \Omega_k, \tag{4.130}$$

where δ_{err} is the relative error.

A Short Discussion about both the FSC Macromodels

The majority of partial elements has to be modeled by the first model (large delay times). Because of the high selectivtiy of the Gauss function the number of summands for the calculation of the moving mean value is low (about 3...8, depending on the cell geometry). For the low number of remaining partial elements modeled by the second type (small delay times) the number of summands is about three times higher. Since the coefficients Ω_k are computed only once during the preprocessing, the run time of a step of numerical integration is approximately equal to the one of the PEEC model with center-to-center retardation multiplied by a mean number of summands of all controlled sources. This is an important advantage to other methods which introduce additional variables and, thus, increase the order of the MNA equation system.

Implementation of the FSC Macromodels into the SPICE-like Circuit Solver

The generalized frequency dependent partial elements interpreted as transfer functions of controlled sources in the frequency domain are transferred by the FSCM into algebraic form in the time domain according to Equation (4.120). For a circuit representation of this equation, the so-called associated discrete circuits (ADC) are used that are described in detail, for example in [38]. This kind of circuit is used in SPICE for algebraic modeling of the reactive elements (inductances, capacitances). ADCs are resistive models valid for each discrete time step of numerical integration. The actual ADC for a circuit element depends on the rule of numerical integration and on the time step. The principle of development of ADCs is demonstrated in Figure 4.23 for an inductance. The method for the soluton of resistive circuits is MNA. The detailed explanation of this wide spread method can be found in [38].

In the papers [36] and [35], the ADCs for partial self and mutual generalized inductances as well as self and mutual generalized potential coefficients are derived in detail. Additionally, the entries for an effective assembly of the MNA matrices for these stable PEEC models are given. The backward differentiation formula of the order two (BD2) is used for time integration. BD2 has been chosen because of the A-and L-stability [39] and, additionally, because of the simplicity and sufficient accuracy. Numerical examples [36] show that the new FSCM-PEEC is free of late-time instabilites and the solutions are of a higher accuracy at frequencies near the maximum frequency of interest.

The runtime of the FSCM-PEEC code is comparable with the one for a standard PEEC with center-to-center retardation. While the computation time for each time step is higher for

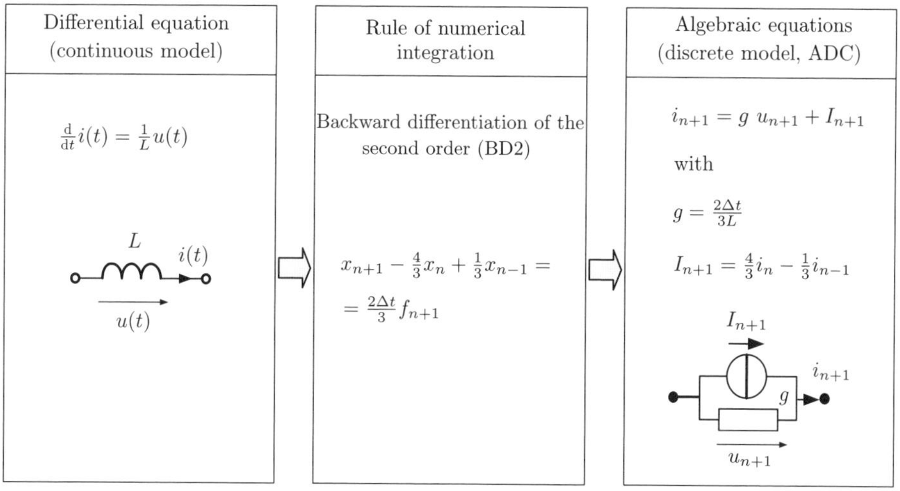

Figure 4.23 The principle of development of an ADC.

FSCM–PEEC as mentioned before, the time step can be chosen to be much bigger than the shortest time delay, which causes a compensating effect concerning the total runtime.

4.6.4.2 Stable Time Domain PEEC Models by Macromodeling Using Foster's Rational Functions and Circuit Synthesis

In this section a second methodology for obtaining stable time domain PEEC models by macromodeling the frequency dependency of the generalized partial elements presented in [37] is considered.

The first step of the procedure is the separation of the frequency dependency of the generalized partial elements into two parts, in a delay part and a remaining part without delay. To do this, one obtains

$$\Lambda(s) = \Lambda_p^{\mathrm{dl}}(s)\mathrm{e}^{-s\tau} \qquad \text{and} \qquad \Pi(s) = \Pi^{\mathrm{dl}}(s)\mathrm{e}^{-s\tau}, \tag{4.131}$$

where $s = \sigma + j\omega$ is the complex frequency, dl means delayless and $\tau = R/c$ with R as the center-to-center distance between two cells. For simplicity, the indices for denoting the cells are omitted. For the representation of the partial elemtents by current controlled voltage sources (CCVS), impedances are introduced:

$$Z_{\mathrm{L}}(s) = s\Lambda(s) = s\Lambda_p^{\mathrm{dl}}(s)\mathrm{e}^{-s\tau} = Z_{\mathrm{L}}^{\mathrm{dl}}(s)\mathrm{e}^{-s\tau} \tag{4.132}$$

and

$$Z_{\mathrm{C}}(s) = \frac{\Pi(s)}{s} = \frac{\Pi^{\mathrm{dl}}(s)}{s}\mathrm{e}^{-s\tau} = Z_{\mathrm{C}}^{\mathrm{dl}}(s)\mathrm{e}^{-s\tau}. \tag{4.133}$$

The separation of the delay term $\exp(-s\tau)$ leads to a significant reduction in the frequency dependency of the remaining delayless part and, as a consequence, to a significant reduction in the mathematical effort for the approximation. In the following, the focus is directed at the delayless part of the impedances

$$Z^{dl}(s) = \left\{Z_L^{dl}(s), Z_C^{dl}(s)\right\}, \tag{4.134}$$

whereas the subscripts L and C are omitted. The delay part $\exp(-s\tau)$ does not need a special attention because its time domain implementation is only a time shift from t to $t-\tau$. $Z^{dl}(s)$ is approximated by a rational function that may be written in the pole-residue representation in Foster's canonical form:

$$Z^{dl}(s) = \frac{N(s)}{D(s)} = d + se + \sum_{k=1}^{N_p^r} \frac{\mathrm{Res}_k^r}{s - p_k^r} + \sum_{k=1}^{N_p^C} \left(\frac{\mathrm{Res}_k^C}{s - p_k^C} + \frac{\mathrm{Res}_k^{C*}}{s - p_k^{C*}} \right), \tag{4.135}$$

where Res_k^r , p_k^r, and N_p^r refer to real poles, Res_k^C, p_k^C and N_p^C are the same quantities for the conjugate complex pairs, and $*$ denotes the conjugate complex operator. As an effective algorithm for the identification of the unknown paramteters in Equation (4.135) in [37] the orthonormal vector fitting (OVF) is proposed, which accurately approximates the impedance function $Z^{dl}(s_w)$ at multiple complex frequencies s_w for $w = 0, \ldots, W$ in a broad frequency range.

The cicuit synthesis on the basis of Equation (4.135) is a well-known process, for example it is described in [40]. The constant term d in Equation (4.135) corresponds to a resistance, the s proportional term is modeled as an inductance $L = e$. The kth term for a real pole can be synthesized easily by an *RC* parallel circuit, the kth term for a conjugate complex pair of poles can be synthesized by an *RLCG* equivalent circuit. By using a discretizing scheme, the authors apply the simple backward Euler (BE) formula, the synthesized circuit can be transformed in so-called ADCs, already introduced in Section 4.6.4.1. This discrete algebraic form is well suited for the implementation in the MNA based time domain cicuit solvers. Thus, each one-port representing a controlled source in the frequency domain is transformed in an equivalent one-port represented by ADCs in the time domain, which means that the general PEEC structure can be retained in the enhanced PEEC model.

Numerical tests carried out [37] have proved the robustness, the accuracy and improved stability of the enhanced PEEC models. Usually zero to five poles are sufficient to capture the broadband delayless impedance behavior. By using advanced techniques like multiple moments in the OVF procedure and adaptive frequency sampling the macromodeling process by rational functions can be significantly accelerated [37].

4.7 Skin Effect in PEEC Models

4.7.1 Cross-Sectional Discretization of Wires

As mentioned in Section 4.2.1, the current density has to be constant over the cross section of the cell. Further, in Section 4.3.1.1 it has been stated that the assumption of a uniform current distribution is only correct up to frequencies where the skin effect is negligible, that is roughly estimated, if the skin depth $\delta = \sqrt{2/\omega\mu_0\sigma}$ is bigger than half of the cross-sectional dimensions. Figure 4.8 shows the PEEC adequate solution for considering the skin effect: the segmentation of the generally arbitrary cross section in current filaments so that the current is uniformly distributed in each of them. In this way, both the skin effect and the proximity effect can be taken into account simultaneously in the frequency as well as in the time domain.

Numerical calculations show that for sufficiently accurate results a relatively high number of filaments for the cross-sectional discretization of wires is needed (e.g. [41], [42]). Further, it can be shown [43] that, for example for parallel wires, the additional influence of the proximity effect is low related to the skin effect for single wires. The major disadvantage of the described method for considering the skin effect in PEEC models by cross-sectional discretization of wires is the high model order for applications which are practically interesting. The use of model order reduction (MOR) methods [44], [10] may lead to a reduction in the computational effort. Because accurate skin effect models are necessary for a careful loss and dispersion calculation, especially in transient cases, there are many attempts for approximative approaches to include the skin effect into PEEC models. Two of the approaches are outlined below.

4.7.2 Skin Effect Modeling by Means of a Global Surface Impedance

In [41], the implementation of the skin effect in PEEC models by the introduction of a global surface impedance (GSI) is proposed. For an explanation of the approach, let us consider a bar conductor element analogous to Figure 4.8. The cross section is uniformly discretized in N_1 cells in the x direction and N_2 cells in the y direction, while the current flows in the z direction. The aim of introducing the GSI is to reduce the number of unknown currents for representation of one conductor element from $N = N_1 \times N_2$ in the case of volume discretization (see Section 4.7.1) to $M = 2(N_1 + N_2)$ unknown surface currents by using the GSI. The starting point for the derivation of the GSI is the two-dimensional quasi-stationary Helmholtz equation:

$$\frac{\partial^2 \underline{E}_z(x, y)}{\partial x^2} + \frac{\partial^2 \underline{E}_z(x, y)}{\partial y^2} - j\omega\mu\sigma \underline{E}_z(x, y) = 0. \tag{4.136}$$

Additionally, from the Maxwell equation $\nabla \times \underline{\mathbf{E}} = -j\omega\mu\underline{\mathbf{H}}$ results on the surface

$$\frac{\partial \underline{E}_z}{\partial n} = -j\omega\mu \underline{H}_t. \tag{4.137}$$

Further, the tangential component of the magnetic field strength can be replaced by the surface current density $\underline{\mathbf{J}}_s$:

$$\mathbf{e}_n \times \underline{\mathbf{H}}_t = \underline{J}_s \mathbf{e}_z. \tag{4.138}$$

Applying Equations (4.136) to (4.138) in a discretized form and making some modifications (see [41]) one obtains finally

$$\mathbf{E}_s = \mathbf{Z}_{\mathrm{GSI}}(j\omega)\mathbf{J}_s, \tag{4.139}$$

where $\mathbf{E}_s$ and $\mathbf{J}_s$ are vectors $(M,1)$ with z components of the surface electric field strength and the surface current density, respectively, $\mathbf{Z}_{\mathrm{GSI}}$ is the (M,M) global surface impedance matrix, $M = 2(N_1 + N_2)$) is the set of all surface cells of the conductor element considered. The diagonal elements of $\mathbf{Z}_{\mathrm{GSI}}$ have the physical meaning of 'self' surface impedances of the cells, the off-diagonal elements are 'mutual' surface impedances. The GSI matrix $\mathbf{Z}_{\mathrm{GSI}}$ describes the interior problem of current distribution over the cross section only by surface quantities $\underline{J}_s$ and $\underline{E}_s$. Because the GSI matrix is independent of the surrounding geometry, it can be calculated for a given wire cross section and range of frequency and lends itself to the development of component libraries.

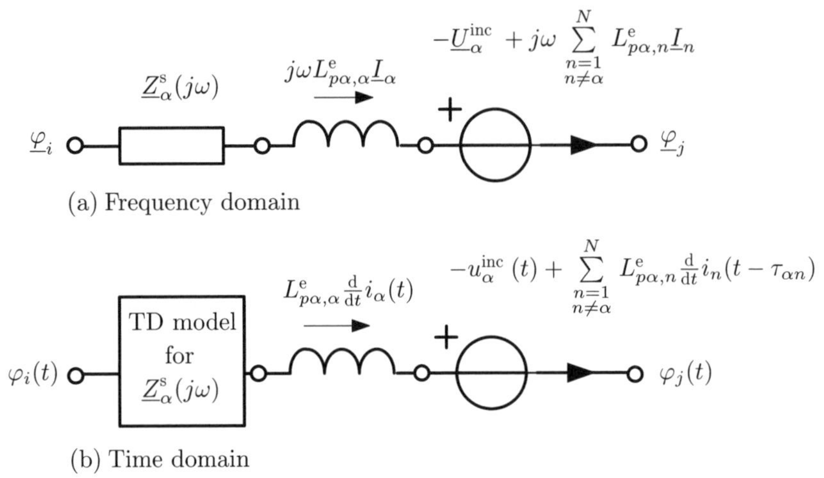

Figure 4.24 PEEC models for a current cell considering the skin effect by the MSI $\underline{Z}^{\mathrm{s}}_{\alpha}(j\omega)$ (superscript e: external).

Applying the continuity of the tangential component of the electric field strength at the boundary, the z component of the electric field strength $\left(\underline{\mathbf{E}}_{\text{interior}}\right)_z$ given by Equation (4.139) as a solution of the interior problem has to be equal to the z component of the external electric field strength $(\underline{\mathbf{E}}_{\text{exterior}} = \underline{\mathbf{E}}^{\text{inc}} + \underline{\mathbf{E}}^{\text{sc}})_z$ (see Equations (4.17) and (4.18)). Thus the GSI is included in the MPIE as a starting point of the PEEC derivation process.

The advantage of the approach according to [41] is the accurate modeling of the skin effect and the opportunity to consider the proximity effect, too. The disadvantage is the still high computational effort.

4.7.3 Skin Effect Modeling by Means of a Local Mean Surface Impedance

Unlike the numerical model presented in Section 4.7.2, the approach with the local mean surface impedance (MSI) according to [42] presented now is a behavioral macromodel.

The frequency dependent MSI is included serially in the equivalent circuit of the current cell (see Figure 4.24). Its time domain counterpart can be developed by two methods: by a staircase *LR*-circuit or by the full spectrum convolution technique. Due to the local character of the MSI the proximity effect cannot generally be considered. This is an acceptable disadvantage because of the relatively weak influence of the proximity effect (see [43]).

The derivation of the MSI is based on a surface formulation of the Galerkin approach instead of the volume formulation according to Section 4.2.1. Using the weighting functions

$$\mathbf{w}_{\alpha}(\mathbf{r}) = \mathbf{b}^{\mathrm{c}}_{\alpha} = \begin{cases} \dfrac{\mathbf{e}_{\alpha}}{d_{\alpha}} & \mathbf{r} \in s_{\alpha} \\ 0 & \text{otherwise,} \end{cases} \tag{4.140}$$

where d_α is the perimeter of the cross section and s_α is the surface of the conductor element α, one obtains

$$\int_{s_\alpha} \mathbf{w}_\alpha \frac{\underline{\mathbf{J}}_\alpha}{\sigma} \mathrm{d}s_\alpha = \frac{1}{d_\alpha} \int_{s_\alpha} \underline{E}_\alpha \mathrm{d}s_\alpha = \underline{Z}_\alpha^{\mathrm{s}}(j\omega)\underline{I}_\alpha. \tag{4.141}$$

$\underline{Z}_\alpha^{\mathrm{s}}(j\omega)$ is the local MSI and $\underline{I}_\alpha$ is the total current in the volume cell α. According to Equation (4.141), the rule for calculation of $\underline{Z}_\alpha^{\mathrm{s}}(j\omega)$ is given by

$$\underline{Z}_\alpha^{\mathrm{s}}(j\omega) = \frac{1}{d_\alpha \underline{I}_\alpha} \int_{s_\alpha} \underline{E}_\alpha \mathrm{d}s_\alpha. \tag{4.142}$$

As a consequence of the surface formulation of the Galerkin approach one has to consider that the partial inductances are calculated by double surface integrals instead of double volume integrals as shown in Section 4.2.1. As a result, both the partial inductances and potential coefficients are calculated by the same mathematical expression, exept the directional dependence of the partial inductances. From the physical point of view, the partial inductances are external ones, because the internal magnetic field is already considered in the local MSI. The structure of the equivalent circuit for the PEEC model is unchanged (see Figure 4.24). The MSI, whose value depends on the wire material, the cross-sectional geometry and the frequency, may be computed by an auxiliary problem. Let an infinitely long transmission line with a particular cross section and material be considered. The distance between the direct and the return wires is so large that the proximity effect does not influence the result. An incident electric field $\underline{\mathbf{E}}^{\mathrm{inc}}$ is impressed on the wire surface. Since the electric field intensity and the current density distributions are calculated over the cross section, the mean surface impedance can be obtained through Equation (4.142)

$$\underline{Z}_\alpha^{\mathrm{s}}(j\omega) = l_\alpha \cdot \underline{Z}_s'(j\omega) = \frac{1}{s \cdot \underline{I}} \oint_s \underline{E}(x, y) \mathrm{d}s, \tag{4.143}$$

where $\underline{Z}_s'(j\omega)$ is the per-unit-length MSI and s is the perimeter of the cross section. Solutions in closed form can be derived for circular [45] and rectangular [46] cross sections. A useful approximate closed-form solution for microstrip lines with rectangular cross section is developed in [47].

The MSI for conductors with arbitrary cross section can be calculated numerically by different methods. An original approach for numerical computation of the MSI using the PEEC method with cross-sectional discretization analogous to Section 4.7.1 is presented in [42]. As mentioned before, the first way for developing a time domain PEEC model consists of replacing the MSI $\underline{Z}_\alpha^{s}(j\omega)$ by an *L,R* staircase circuit (see Figure 4.25), where the order $q = 3 \ldots 7$ provides a fair accuracy of dependence on the bandwidth. The evaluation of the parameters of the *L,R* circuit with the given structure is based on the solution of the inverse problem

$$\min_P \left[\int_0^{\omega_m} \left(\frac{\underline{Z}_s(j\omega) - \tilde{\underline{Z}}_s(j\omega, P)}{\underline{Z}_s(j\omega)} \right)^2 \right]^{\frac{1}{2}}, \tag{4.144}$$

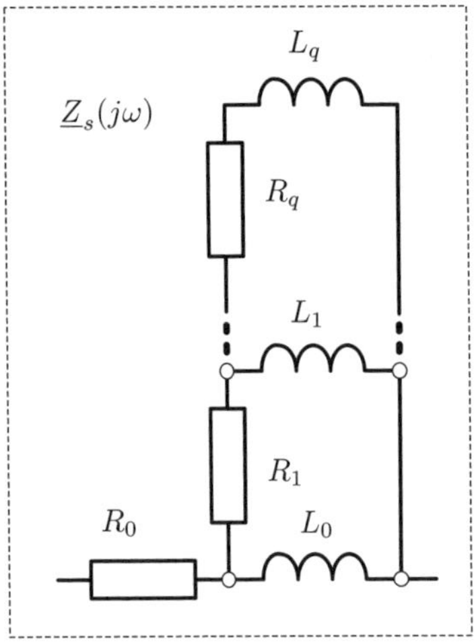

(a) Staircase circuit

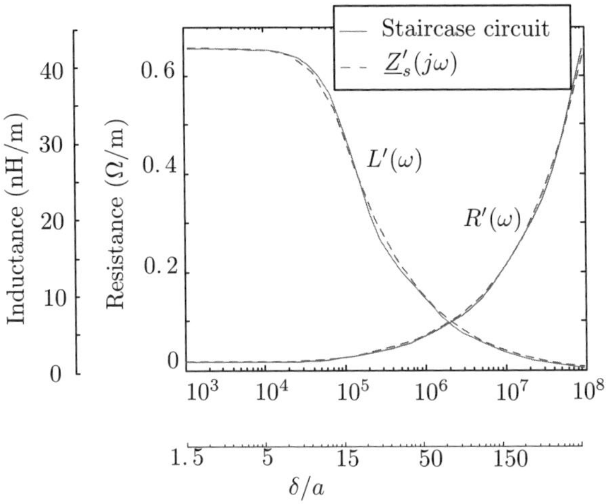

(b) Approximated parameters of the wire

Figure 4.25 Approximation of $Z_s(j\omega)$ using a L, R staircase circuit.

where $P = \left(R_0, \ldots, R_q, L_0, \ldots, L_q\right)$ is the parameter vector to be determined, $\tilde{\underline{Z}}_s(j\omega, P)$ is the impedance of the staircase circuit and ω_{m} is the maximum frequency of interest. The problem of determining the parameters can be simplified by applying the following definitions [35]:

$$R_0 = \Re\left\{\underline{Z}_s\,(\omega = 0)\right\},\, L_0 = \lim_{\omega \to 0} \Im\left\{\underline{Z}_s(j\omega)/\omega\right\}$$
$$R_k = R_0\zeta^k, \quad L_k = L_0\eta^k, \quad k \in [1, q]. \tag{4.145}$$

In this case, the vector of parameters consists only of two components $P = (\zeta, \eta)$. Figure 4.25 shows the approximation of $\underline{Z}_s(j\omega)$ by a staircase circuit with $q = 4$ for a copper wire with a cross section of 1 mm × 1 mm.

Although the computational effort of the staircase model is significantly reduced in comparison with the cross-sectional discretization of conductors, with the inclusion of the staircase circuit in the PEEC model the number of nodes in the MNA-matrix is increased and, as a consequence, the computational efficiency is reduced.

This remaining disadvantage can be avoided by using the second aforementioned approach, the macromodeling $\underline{Z}_s(j\omega)$ by the full spectrum convolution technique (FSCM) as already introduced in Section 4.6.4.1. According to the general strategy, one has to find an approximation with the same asymptotic behavior as $\underline{Z}_s(j\omega)$) and then determine the impulse response for this function. Since $\lim_{\omega\to 0} \underline{Z}_s = \infty$, $\underline{Y}_s(j\omega) = 1/\underline{Z}_s(j\omega)$ as the mean surface admittance is used. In [35], a frequency domain approximation

$$\tilde{\underline{Y}}_s(j\omega) = \frac{1}{R_0}\frac{1}{\sqrt{1 + j\omega\tau_c}}, \tag{4.146}$$

is introduced, for which a closed form inverse Fourier transform can be found. The two parameters R_0 and τ_c of $\underline{\tilde{Y}}_s(j\omega)$ have to be determined. Of course, the approximation is not unique, but it has been found as optimal concerning precision and simplicity. For comparison, the calculated frequency dependence of the real and imaginary parts of $\underline{Y}'_s(j\omega)$ is shown in Figure 4.25. Substituting $\omega = 0$ into Equation (4.146) yields

$$\underline{\tilde{Y}}_s(0) = \frac{1}{R_0}, \quad R_0 = \Re\left\{\underline{Z}_s(0)\right\}, \tag{4.147}$$

which represents the exact d.c. ohmic resistance. By the Taylor series expansion of $\underline{\tilde{Y}}_s(j\omega)$ for $\omega \to 0$ one obtains

$$\lim_{\omega\to 0} \Im\left\{1/\underline{\tilde{Y}}_s\right\} = R_0\frac{\tau_c}{2}\omega + \mathcal{O}\left(\omega^2\right) \tag{4.148}$$

and, with $L(0) = L_0$ for the staircase circuit,

$$\tau_c = 2\frac{L_0}{R_0}. \tag{4.149}$$

This is the first possible estimation of τ_c based on the accurate approximation of the static internal inductance of the wire. An alternative estimation is derived from the fact that the minimum of the imaginary part of $\underline{\tilde{Y}}_s(j\omega)$ is given for $\omega_{\min} = \sqrt{3}/\tau_c$. Therefore,

$$\tau_c = \sqrt{3}/\omega_{\min} \tag{4.150}$$

should be chosen. The second estimation provides a better accuracy in the middle of the frequency range and a certain deviation of the static inductance, while the resistance is modeled accurately at all frequencies. After determination of the parameters R_0 and τ_c one obtains the impulse and step responses for $\underline{\tilde{Y}}_s(j\omega)$ via IFT:

$$\begin{aligned} y(t) &= \frac{1}{R_0}\frac{e^{-\frac{t}{\tau_c}}}{\sqrt{\pi t \tau_c}}h(t), \\ s(t) &= \frac{1}{R_0}\operatorname{erf}\left(\frac{t}{\tau_c}\right)h(t), \end{aligned} \tag{4.151}$$

where $h(t)$ is the Heaviside step function, erf( ) is the error function, $y(t)$ is the impulse and $s(t)$ the step function response of $\underline{\tilde{Y}}_s(j\omega)$. Since $s(0) = 0$, the instantaneous value of the current $i(t)$ for an arbitrary voltage $u(t)$ is given by the convolution integral

$$i(t) = \int_0^t y(t-\tau)u(\theta)\mathrm{d}\theta. \tag{4.152}$$

Analogously to Equation (4.118), a piecewise approximation for $u(t)$ on the disrete time axis is applied and, after introducing the convolution integral of Equation (4.152), one obtains

$$i\left(t_{n+1}\right) = i_{n+1} = \sum_{k=0}^{n}\int_0^{\Delta t} y\left(t_{n+1} - t_k - \theta\right)\left(a_{k+1}\theta + b_{k+1}\right)\mathrm{d}\theta \tag{4.153}$$

for the current values in the discrete time points. The integrals over the time steps Δt are calculated in closed form [35]:

$$\int_0^{\Delta t} y(t_{n+1} - t_k - \vartheta)(a_k\vartheta + b_k)\mathrm{d}\vartheta =$$

$$+\frac{1}{R_0}\left[a_k\sqrt{\frac{\tau_c T_{n+1,k}}{\pi}}\Psi_{n+1,k} + \left(a_k T_{n+1,k} + b_k - \frac{a_k}{2}\tau_c\right)\Phi_{n+1,k}\right] \tag{4.154}$$

$$-\frac{1}{R_0}\left[a_k\sqrt{\frac{\tau_c T_{n,k}}{\pi}}\Psi_{n,k} + \left(a_k T_{n+1,k} + b_k - \frac{a_k}{2}\tau_c\right)\Phi_{n,k}\right]$$

$$\text{with} \quad \Psi_{n,k} = \mathrm{e}^{-\frac{T_{n,k}}{\tau_c}}, \ \Phi_{n,k} = \mathrm{erf}\left(\frac{\sqrt{T_{n,k}}}{\sqrt{\tau_c}}\right), \ T_{n,k} = t_n - t_k = (n-k)\Delta t.$$

Substituting Equation (4.154), one can cast Equation (4.153) in a moving average form

$$i_{n+1} = \sum_{k=p}^{n+1} \Omega_k u_k, \quad 0 \le p \le n+1. \tag{4.155}$$

The coefficients Ω_k are developed in detail in [35]. Because of the selectivity of $y(t)$, instead of all time points k with $0 \le k \le n+1$ values must be calculated only in fewer points $p \le k \le n+1$ (see Figure 4.26(a)). The width of the moving average window is $\tau_\alpha = n_\alpha \Delta t$ with $n_\alpha = n + 2 - p$. The last step in developing the time domain macromodel consists of the interpretation of the moving average Equation (4.155) as an ADC. For this, Equation (4.155) is rewritten in the form:

$$\begin{aligned} i_{n+1} &= \widehat{\Omega}_0 u_{n+1} + \widehat{\Omega}_1 u_n + \widehat{\Omega}_2 u_{n-1} + \cdots = \\ &= \widehat{\Omega}_0 u_{n+1} + \sum_{k=n-n_\alpha+2}^{n} \Omega_{n-k+1} u_k, \quad \widehat{\Omega}_k = \Omega_{n-k+1}. \end{aligned} \tag{4.156}$$

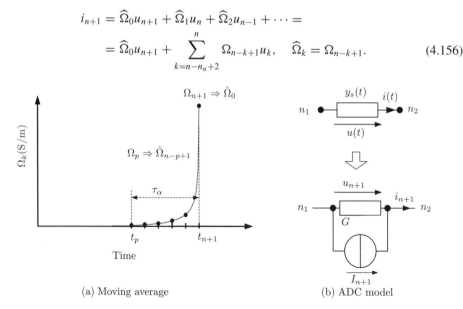

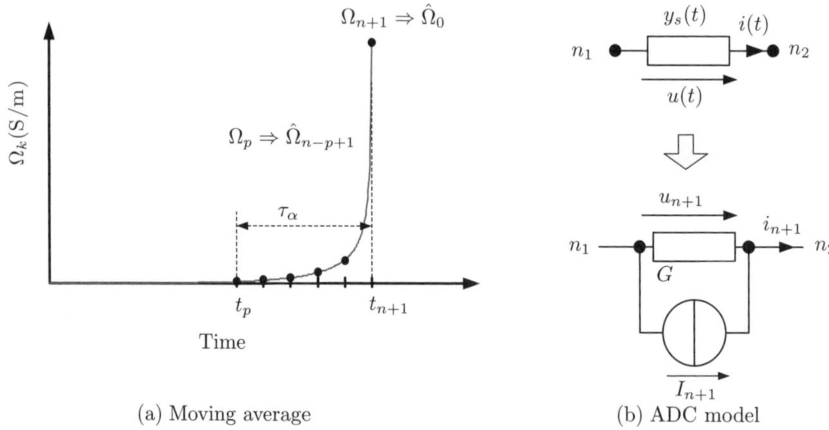

Figure 4.26 Coefficients (a) Ω_k and (b) the ADC model for the skin effect modeling.

Since the coefficients $\widehat{\Omega}_k$ do not depend on n, Equation (4.156) is applicable at all time points and can be interpreted as an ADC in the Norton equivalent form (see Figure 4.26(b))

$$i_{n+1} = Gu_{n+1} + I_{n+1} \tag{4.157}$$

with $G = \widehat{\Omega}_0$ and $I_{n+1} = \Sigma_{k=n-n_\alpha+2}^{n} \Omega_{n-k+1} u_k$. Thus, a time domain PEEC model suitable for an effective simulation on MNA circuit solvers is obtained.

4.8 PEEC Models Based on Dyadic Green's Functions for Conducting Structures in Layered Media

4.8.1 Motivation

Printed circuit boards (PCBs) are the most widespread technology for interconnections in modern electronics. They present a special class of interconnection structures consisting of a number of layers with printed traces, via holes, and ponding pads located in a stratified dielectric. To ensure functionality and reliability of electronics a thorough analysis in the frequency and time domain – including the features of the interconnections – is required. Since PCBs are used to connect the pins of integrated circuits (IC) and other active and passive devices, the direct frequency domain analysis can be impossible due to the nonlinear circuit environment. Although for linear circuits the analysis in the frequency domain is possible, in the case of broadband excitation it lasts a very long time and, as a consequence, time domain simulations are more effective. Generally it can be stated that the importance of time domain methods is growing.

In principle, the PEEC method is suitable for analyzing the electromagnetic problems of PCBs. This concerns the simulations in frequency and time domain as well as the availability of models for conductors and dielectrics. As already mentioned in Section 4.1.2, the dielectrics are represented by their unknown polarization currents and bound charges related to free space. This means in the mathematical formulations the use of the 'simple' free or half space Green's functions in a homogeneous medium with ε_0, μ_0.

A simple and rough approximation of the influence of dielectrics in PCB-like configurations is the concept of the 'effective permittivity' [48], [49]. This means, the replacement of the inhomogeneous medium of a PCB by an equivalent homogeneous one in connection with slightly changed geometrical parameters. Concerning the application to the PEEC method only conductors (not dielectrics) have to be discretized, ε_0 has to be replaced by $\varepsilon_0\varepsilon_{\mathrm{r,eff}}$, and the partial elements have to be calculated with the simple Green's functions for the free or half space as derived before, of course, under consideration of slightly changed geometrical parameters. Let this concept be explained, for example, by a microstrip line (see Figure 4.27). Since a dominant quasi-TEM mode is assumed, the maximum permissible frequency has to be lower than the boundary frequency (see [49])

$$f_{\mathrm{b}} = \frac{21.3\,\mathrm{mm}}{(W + 2h)\sqrt{\varepsilon_{\mathrm{r}} + 1}}\,\mathrm{GHz}. \tag{4.158}$$

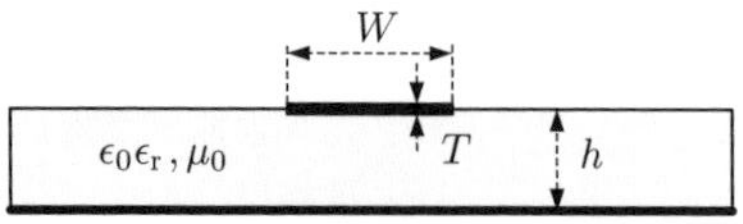

(a) Microstrip line

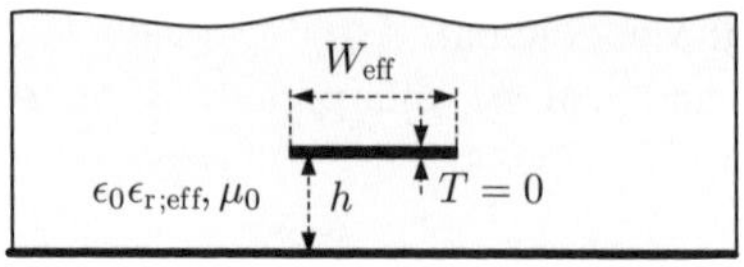

(b) Equivalent line in the half space

Figure 4.27 The concept of 'effective permittivity'.

In the case where $f < f_b$, the quasi-TEM wave propagates with a constant velocity. The permittivity corresponding to this velocity in a homogeneous medium is the so-called effective permittivity [48]

$$\varepsilon_{r,eff} = \frac{\varepsilon_r + 1}{2} + \frac{\varepsilon_r - 1}{2}\left(1 + \frac{10h}{W}\right)^{-\frac{1}{2}} - \frac{\varepsilon_r - 1}{4.6}\frac{T/h}{\sqrt{W/h}}. \tag{4.159}$$

The geometry of the equivalent line has to be changed slightly as follows [48]:

$$W_{eff} = W + 1.25\frac{T}{\pi}\begin{cases}\left(1 + \ln 4\pi\frac{W}{T}\right) & \frac{W}{h} \leq \frac{1}{2\pi} \\ \left(1 + \ln\frac{2}{T/h}\right) & \frac{W}{h} \geq \frac{1}{2\pi}\end{cases}. \tag{4.160}$$

The extension of this approach to the consideration of an external electromagnetic field proposed in connection with the method of moments [50] can be applied to PEEC as well.

The disadvantages of the described approach of 'effective permittivity', mainly consisting in the limitation to low frequencies, simple geometries and a dominant quasi-TEM mode, are the motivation for developing more sophisticated PEEC models based on the dyadic Green's functions for layered media. Roughly speaking, the stratified, inhomogeneous structure of PCBs is then hidden behind relatively complicated dyadic and scalar Green's functions for the electric and magnetic field strengths and the electromagnetic potentials, respectively. Further, only the conducting interconnection structures like traces and via holes have to be discretized. This new concept of DGFLM–PEEC (DGFLM stands for dyadic Green's functions for layered media) will be presented in the following [51], [52], [35].

4.8.2 The DGFLM–PEEC Method

Let a dyadic Green's function–MPIE (DGF–MPIE, see Equation (4.27)) be considered for an observation point $\mathbf{r}$ on a conductor with an electric conductivity σ. For simplicity, in Section 4.2 the generalized PEEC method was derived by using the scalar Green's function for free space $g\left(\mathbf{r}, \mathbf{r}'\right)$, which is the same for the calculation of both the magnetic vector potential and the electric scalar potential. For more complicated boundary conditions and for inhomogeneous media an appropriate dyadic Green's function $\overline{\overline{\mathbf{G}}}^A(\mathbf{r}, \mathbf{r}')$ for the magnetic vector potential and scalar Green's function $G^{\Phi}(\mathbf{r}, \mathbf{r}')$ for the electric scalar potential may be developed.

Applying the same derivation process with the same basis and weighting functions as in Section 4.2, expressions for the calculation of the generalized partial inductances and potential

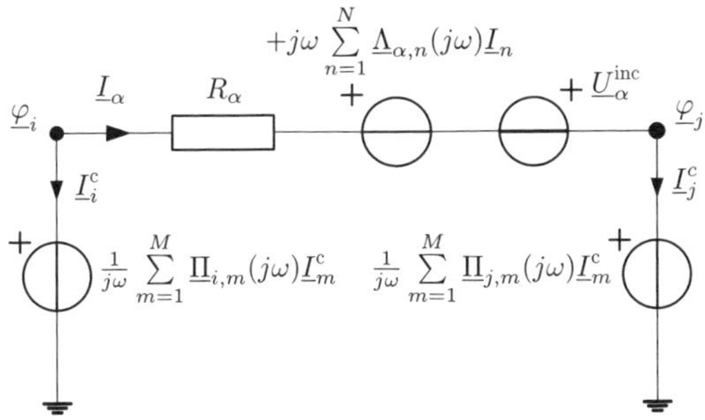

Figure 4.28 Generalized PEEC model for a conductor segment in the frequency domain.

coefficients may be obtained:

$$\underline{\Lambda}_{\alpha,n}(j\omega) = \frac{\mu_0}{a_\alpha \cdot a_n} \int_{v_\alpha} \int_{v_n} \left(\mathbf{e}_\alpha \cdot \overline{\overline{\mathbf{G}}}^A(\mathbf{r}, \mathbf{r}') \cdot \mathbf{e}_n \right) \mathrm{d}v'_n \, \mathrm{d}v_\alpha \quad \forall \alpha, n \in N \tag{4.161}$$

$$\underline{\Pi}_{i,m}(j\omega) = \frac{1}{\varepsilon s_i s_m} \int_{s_i} \int_{s_m} G^\Phi(\mathbf{r}, \mathbf{r}') \, \mathrm{d}s'_m \, \mathrm{d}s_i \quad \forall i, m \in M. \tag{4.162}$$

In Equation (4.162) ε means the permittivity of the medium where the potential cell i ($i \in M$) is located. In addition to Figures 4.6 and 4.7, in Figure 4.28 the generalized PEEC model for a conductor segment, comprising only current controlled voltage sources, is depicted.

The derivation of a DGF–PEEC model for a multilayer PCB requires the dyadic Green's functions for plane, stratified media. Since they cannot be developed for arbitrary PCB structures, several assumptions related to a stratified medium as shown in Figure 4.29 are made. The medium is composed of n dielectric layers, each of them ($i \in [1, n]$) is characterized by the permittivity ε_i, the electric conductivity $\sigma_{s,i}$ and μ_0. Each dielectric layer is homogeneous and unbounded in the x,y plane. Although the MPIE may be derived for anisotropic layers, for simplicity isotropic ones, that is the permittivity for each layer is a scalar constant, are assumed. Additionally, it is supposed that traces (x,y plane) and vias (z direction) as elements of the interconnection structure do not influence the homogeneity of the dielectric layers. The

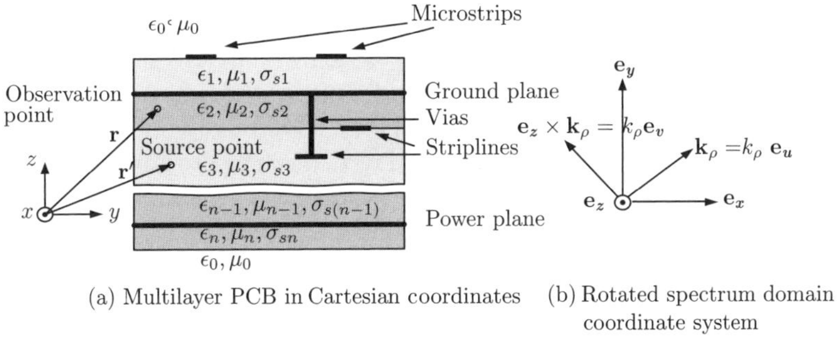

Figure 4.29 Multilayer PCB and the rotated spectrum-domain coordinate system.

space under and above the PCB is air with the parameters ε_0 , μ_0 and $\sigma_s = 0$. Some of the layers are completely metallized ground or power planes that are also infinitely expanded in the x, y plane.

Many MPIE formulations for stratified media have been developed. The formulations given in [3] and [4] are most appropriate for the development of DGFLM–PEEC models. Knowing the dyadic Greens's function $\overline{\overline{\mathbf{G}}}{}^{A}_{\mathrm{LM}}(\mathbf{r}, \mathbf{r}')$ for the magnetic vector potential and the scalar Green's function $G^{\Phi}_{\mathrm{LM}}(\mathbf{r}, \mathbf{r}')$ for the scalar electrical potential (the subscript LM stands for **l**ayered **m**edium), the partial inductances and potential coefficients for a layered medium can be calculated according to Equations (4.161) and (4.162), and a DGFLM–PEEC model in the frequency domain can be composed (see Figure 4.28). The whole derivation process of the DGFLM–MPIE was originally and outlined in detail in [35]. To establish an understanding of the results and their further application, some characteristic steps of the derivation process are pointed on.

Since the medium is homogeneous and of infinite extent in the x,y plane, an arbitrary vector $\mathbf{A}$ may be decomposed in a transversal (in the x,y plane) and a longitudinal (in the z direction) component according to $\mathbf{A} = \mathbf{A}_t + \mathbf{A}_z$ with $\mathbf{A}_z = A_z\mathbf{e}_z = \mathbf{e}_z\mathbf{e}_z \times \mathbf{A}$ and $\mathbf{A}_t = -\mathbf{e}_z \times \mathbf{e}_z \times \mathbf{A}$. In particular, the TE and TM modes of the electromagnetic waves propagate in the z direction. In such a case, the analysis is facilitated by the Fourier transformation of all fields with respect to the transverse coordinates. The two-dimensional spatial Fourier transformation is given by

$$\tilde{g}\left(z, z', \mathbf{k}_\rho\right) = \mathfrak{F}_{\mathrm{s}}\left\{g\left(\mathbf{r}, \mathbf{r}'\right)\right\} = \int_{-\infty}^{\infty}\int_{-\infty}^{\infty} g\left(\mathbf{r}, \mathbf{r}'\right) \mathrm{e}^{j\mathbf{k}_\rho\cdot\boldsymbol{\rho}}\mathrm{d}x\,\mathrm{d}y \tag{4.163}$$

with

$$\mathbf{r} - \mathbf{r}' = \left(z - z'\right)\mathbf{e}_z + \boldsymbol{\rho} \quad , \quad \boldsymbol{\rho} = \left(x - x'\right)\mathbf{e}_x + \left(y - y'\right)\mathbf{e}_y$$

$$\mathbf{k}_\rho = k_x\mathbf{e}_x + k_y\mathbf{e}_y \quad , \quad \zeta = \arctan\frac{y - y'}{x - x'}.$$

The inverse two-dimensional spatial Fourier transformation is defined as follows:

$$g\left(\mathbf{r}, \mathbf{r}'\right) = \mathfrak{F}_{\mathrm{s}}^{-1}\left\{\tilde{g}\left(z, z', \mathbf{k}_\rho\right)\right\} = \frac{1}{4\pi^2}\int_{-\infty}^{\infty}\int_{-\infty}^{\infty} \tilde{g}\left(z, z', \mathbf{k}_\rho\right) \mathrm{e}^{-j\mathbf{k}_\rho\cdot\boldsymbol{\rho}}\mathrm{d}k_x\,\mathrm{d}k_y. \tag{4.164}$$

The subsequent analysis is greatly simplified if one defines a rotated spectrum domain coordinate system based on $\left(\mathbf{k}_\rho, \mathbf{e}_z \times \mathbf{k}_\rho\right)$ (see Figure 4.29), with the unit vectors $(\mathbf{e}_u, \mathbf{e}_v)$ given by $\mathbf{e}_u = \left(k_x/k_\rho\right)\mathbf{e}_x + \left(k_y/k_\rho\right)\mathbf{e}_y$ and $\mathbf{e}_v = -\left(k_y/k_\rho\right)\mathbf{e}_x + \left(k_x/k_\rho\right)\mathbf{e}_y$.

The TM and TE modes propagating in the z direction can be described by two decoupled sets of transmission-line equations whose eight scalar Green's functions (SGFs) are

$I_i'\left(z, z'\right)$ and $I_i''\left(z, z'\right)$ for currents excited by current sources,
$I_u'\left(z, z'\right)$ and $I_u''\left(z, z'\right)$ for currents excited by voltage sources,
$V_i'\left(z, z'\right)$ and $V_i''\left(z, z'\right)$ for voltages excited by current sources,
$V_u'\left(z, z'\right)$ and $V_u''\left(z, z'\right)$ for voltages excited by voltage sources.

The SGFs with one prime belong to the TM mode, with two primes to the TE mode. After the determination of the dyadic Green's functions for the electric and magnetic field strengths, the electromagnetic potentials $\underline{\mathbf{A}}'(\mathbf{r}, \mathbf{r}')$ and $\underline{\varphi}'(\mathbf{r}, \mathbf{r}')$ are introduced applying the Lorenz gauge. In the resulting MPIE a correction term (see [3]) appears. Finally, this term is grouped with the vector potential terms (see [4], [35]). For the changed potentials $\underline{\mathbf{A}}(\mathbf{r}, \mathbf{r}')$ and $\underline{\varphi}(\mathbf{r}, \mathbf{r}')$, no longer coupled by the Lorenz gauge, one obtains the DGFLM–MPIE in the usual form

$$\underline{\mathbf{E}}^{\mathrm{inc}} = \frac{\underline{\mathbf{J}}(\mathbf{r})}{\sigma} + j\omega\mu_0 \int_{V'} \overline{\overline{\mathbf{G}}}{}^{\mathbf{A}}_{\mathrm{LM}}(\mathbf{r}, \mathbf{r}') \cdot \underline{\mathbf{J}}(\mathbf{r}')\mathrm{d}\mathbf{V}' + \frac{\nabla}{\varepsilon} \int_{\mathbf{V}'} \mathbf{G}^{\Phi}_{\mathrm{LM}} \underline{\rho}(\mathbf{r}')\mathrm{d}\mathbf{V}' \tag{4.165}$$

with $\overline{\overline{\mathbf{G}}}{}^{\mathbf{A}}_{\mathrm{LM}}(\mathbf{r}, \mathbf{r}')$ and $G^{\Phi}_{\mathrm{LM}}(\mathbf{r}, \mathbf{r}')$ as the dyadic and the scalar Green's functions, respectively, for a layered medium. The dyadic Green's function for the magnetic vector potential in a layered medium is given by [35]

$$\overline{\overline{\mathbf{G}}}{}^{\mathbf{A}}_{\mathrm{LM}}(\mathbf{r}, \mathbf{r}') = G^A_{zz}\left(\mathbf{r}, \mathbf{r}'\right) \mathbf{e}_z\mathbf{e}_z + G^A_t\left(\mathbf{r}, \mathbf{r}'\right) \overline{\overline{\mathbf{I}}}_t + G^A_{\rho z}\left(\mathbf{r}, \mathbf{r}'\right) \mathbf{e}_\rho\mathbf{e}_z + G^A_{z\rho}\left(\mathbf{r}, \mathbf{r}'\right) \mathbf{e}_z\mathbf{e}_\rho \tag{4.166}$$

with

$$G^A_{zz}\left(\mathbf{r}, \mathbf{r}'\right) = S_0\left\{\widetilde{G}^A_{zz}\left(z, z'\right)\right\}\ \widetilde{G}^A_{zz}\left(z, z'\right) = \frac{1}{j\omega\underline{\varepsilon}}\left[\frac{\underline{k}^2}{k_\rho^2} I''_u\left(z, z'\right) + \frac{k_\rho^2 - k_z^2}{k_\rho^2} I'_u\left(z, z'\right)\right]$$

$$G^A_t\left(\mathbf{r}, \mathbf{r}'\right) = S_0\left\{\widetilde{G}^A_t\left(z, z'\right)\right\}\ \widetilde{G}^A_{zz}\left(z, z'\right) = \frac{1}{j\omega\mu_o} V''_i\left(z, z'\right)$$

$$G^A_{\rho z}\left(\mathbf{r}, \mathbf{r}'\right) = S_1\left\{\frac{\widetilde{G}^A_{uz}\left(z, z'\right)}{jk_\rho}\right\}\ \widetilde{G}_{uz}\left(z, z'\right) = \frac{1}{jk_\rho}\left[V''_u\left(z, z'\right) - V'_u\left(z, z'\right)\right]$$

$$G^A_{z\rho}\left(\mathbf{r}, \mathbf{r}'\right) = S_1\left\{\frac{\widetilde{G}^A_{zu}\left(z, z'\right)}{jk_\rho}\right\}\ \widetilde{G}_{zu}\left(z, z'\right) = \frac{1}{jk_\rho}\left[I''_i\left(z, z'\right) - I'_i\left(z, z'\right)\right].$$

Further, the scalar Green's function of the electric potential for a layered medium yields

$$G^{\Phi}_{\mathrm{LM}}\left(\mathbf{r}, \mathbf{r}'\right) = S_0\left\{\widetilde{G}^{\Phi}\left(z, z'\right)\right\}, \widetilde{G}^{\Phi}\left(z, z'\right) = \frac{j\omega\underline{\varepsilon}}{k_\rho^2}\left[V'_i\left(z, z'\right) - V''_i\left(z, z'\right)\right]. \tag{4.167}$$

S_0 and S_1 in Equations (4.166) and (4.167) are Sommerfeld integrals of zeroth and first order [53]. Because of the circular symmetry of the Green's functions in the transverse plane the computation of the double integrals of the inverse spatial Fourier transformation can be simplified by using the Sommerfeld integrals:

$$\begin{aligned} \mathfrak{F}_{\mathrm{s}}^{-1}\left\{\tilde{g}\left(z, z', \mathbf{k}_\rho\right)\right\} &= S_0\left\{\tilde{g}\left(z, z', \mathbf{k}_\rho\right)\right\} \\ \mathfrak{F}_{\mathrm{s}}^{-1}\left\{jk_x\tilde{g}\left(z, z', \mathbf{k}_\rho\right)\right\} &= \cos\zeta\ S_1\left\{\tilde{g}\left(z, z', \mathbf{k}_\rho\right)\right\} \\ \mathfrak{F}_{\mathrm{s}}^{-1}\left\{jk_y\tilde{g}\left(z, z', \mathbf{k}_\rho\right)\right\} &= \sin\zeta\ S_1\left\{\tilde{g}\left(z, z', \mathbf{k}_\rho\right)\right\}. \end{aligned} \tag{4.168}$$

The Sommerfeld integral of nth order is defined as

$$S_n\left\{\tilde{g}\left(z, z', \mathbf{k}_\rho\right)\right\} = \frac{1}{4\pi}\int_{-\infty}^{\infty} \tilde{g}\left(z, z', \mathbf{k}_\rho\right) k_\rho^{n+1} H_n^{(2)}\left(k_\rho\rho\right) \mathrm{d}\rho, \tag{4.169}$$

where $H_n^{(2)}$ is a Hankel function of nth order and second kind.

The solution of Sommerfeld integrals may be obtained numerically or in a closed form (e.g. [35]). Inserting $\overline{\overline{\mathbf{G}}}^{\mathbf{A}}_{\mathrm{LM}}(\mathbf{r},\mathbf{r}')$ and $G^{\Phi}_{\mathrm{LM}}(\mathbf{r},\mathbf{r}')$ into Equations (4.161) and (4.162), the generalized partial inductances and potential coefficients for a layered medium can be calculated. Considering the particular structure of the dyadic Green's function for the magnetic vector potential, the integrand in the expression for the generalized partial inductances

$$\mathbf{e}_\alpha \cdot \overline{\overline{\mathbf{G}}}^{\mathbf{A}}_{\mathrm{LM}}(\mathbf{r},\mathbf{r}') \cdot \mathbf{e}_n$$

can be written in the form

$$\begin{aligned}\mathbf{e}_\alpha \cdot \overline{\overline{\mathbf{G}}}^{\mathbf{A}}_{\mathrm{LM}}(\mathbf{r},\mathbf{r}') \cdot \mathbf{e}_\mathbf{n} = \mathbf{G}^{\mathbf{A}}_{\mathbf{zz}}\left(\mathbf{r},\mathbf{r}'\right)\beta^{\alpha,\mathbf{n}}_{\mathbf{zz}} + \mathbf{G}^{\mathbf{A}}_{\mathbf{t}}\left(\mathbf{r},\mathbf{r}'\right)\beta^{\alpha,\mathbf{n}}_{\mathbf{t}} +\\ +G^{A}_{\rho z}\left(\mathbf{r},\mathbf{r}'\right)\beta^{\alpha,n}_{\rho z} + G^{A}_{z\rho}\left(\mathbf{r},\mathbf{r}'\right)\beta^{\alpha,n}_{z\rho}\end{aligned} \quad (4.170)$$

with

$$\begin{aligned}\beta^{\alpha,n}_{zz} &= (\mathbf{e}_\alpha\cdot\mathbf{e}_z)(\mathbf{e}_n\cdot\mathbf{e}_z) \quad , \quad \beta^{\alpha,n}_{t} = \mathbf{e}_\alpha\cdot\overline{\overline{\mathbf{I}}}_t\cdot\mathbf{e}_n\\ \beta^{\alpha,n}_{\rho z} &= \left(\mathbf{e}_\alpha\cdot\mathbf{e}_\rho\right)(\mathbf{e}_n\cdot\mathbf{e}_z) \quad , \quad \beta^{\alpha,n}_{z\rho} = (\mathbf{e}_\alpha\cdot\mathbf{e}_z)\left(\mathbf{e}_n\cdot\mathbf{e}_\rho\right).\end{aligned}$$

The substitution of Equation (4.170) into Equation (4.161) leads to the scalar formulation of the generalized partial inductances for an arbitrary structure of a layered medium:

$$\begin{aligned}\underline{\Lambda}_{\alpha,n}(j\omega) = \frac{\mu_0}{a_\alpha a_n}\int_{v_\alpha}\int_{v_n}\Big[G^{A}_{zz}\left(\mathbf{r},\mathbf{r}'\right)\beta^{\alpha,n}_{zz} + G^{A}_{t}\left(\mathbf{r},\mathbf{r}'\right)\beta^{\alpha,n}_{t}+\\ +G^{A}_{\rho z}\left(\mathbf{r},\mathbf{r}'\right)\beta^{\alpha,n}_{\rho z} + G^{A}_{z\rho}\left(\mathbf{r},\mathbf{r}'\right)\beta^{\alpha,n}_{z\rho}\Big]\,\mathrm{d}v'_n\,\mathrm{d}v_\alpha\ , \quad \forall\alpha,n\in N.\end{aligned} \quad (4.171)$$

The scalar coefficients $\beta^{\alpha,n}$ may alternatively be represented as products of trigonometric functions of the angles explained in Figure 4.30:

$$\begin{aligned}\beta^{\alpha,n}_{zz} &= \cos\theta_1\cos\theta_2\\ \beta^{\alpha,n}_{t} &= \sin\theta_1\sin\theta_2\cos\theta_5\\ \beta^{\alpha,n}_{\rho z} &= \sin\theta_1\cos\theta_2\cos\theta_3\\ \beta^{\alpha,n}_{z\rho} &= \cos\theta_1\sin\theta_2\cos\theta_4.\end{aligned}$$

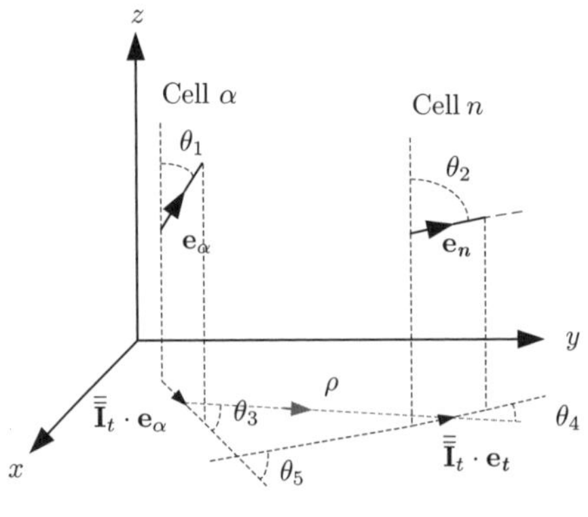

Figure 4.30 Spatial structure of the generalized partial inductance.

Thus, the frequency domain formulation of the DGFLM–PEEC method is finally derived, and this method does not differ significantly from the equivalent MoM-formulation. The circuit interpretation of the DGFLM–PEEC model does not bring many benefits in the frequency domain as compared to MoM. On the contrary, because the MoM may apply more complicated basic and weighting functions, it may yield more precise results. In this sense, the DGFLM–PEEC method in the frequency domain is interesting as a further universal computational method based on the MPIE with dyadic Green's functions.

However, the time domain DGFLM–PEEC method may be considered as preferable in comparison to the method of moments. The arbitrary interconnection structure and the features of the multilayer substrate are modeled via the frequency response of the generalized partial elements. Their time domain macromodeling results in an equivalent circuit interpretation, which may be simulated simultaneously with other active and passive devices using a circuit-simulator code.

Two methods for time domain macromodeling that can be used already have been mentioned in Section 4.6.4: the vector fitting technique and the full spectrum convolution technique. Since the Green's functions for different geometries of a layered medium have different derivations and frequency dependencies, the time domain macromodels are also different and have to be developed each for an actual configuration. Therefore, in the following the focus will be directed at the DGFLM–PEEC model for a stripline region.

4.8.3 DGFLM–PEEC Model for the Stripline Region

Generally, an arbitrary multilayer PCB may be considered as a set of subsystems separated by ground and power planes. The fields in the region between the power and ground planes may be assumed to be uncoupled. Thus, one may define two basic PCB substructures. The upper and lower parts of a PCB consisting of a dielectric substrate on the ground (or power) plane and a set of traces and vias are referred to as microstrip regions. Since a microstrip region is confined from the exterior side by air, it represents a half open problem. On the other hand, the structure composed of one or more dielectric layers confined from both sides by ground or power planes is denoted as stripline region. The stripline region corresponds to a closed problem with respect to the z-axis and to an open problem concerning the x,y-plane. Both kinds of substructures are only interconnected through vias. Thus, only two kinds of Green's functions are required for modeling the PCB using the MPIE for layered media. In the following the Green's functions for the stripline region are developed and applied in order to set up a particular realization of the DGFLM–PEEC model.

The stripline region considered is depicted in Figure 4.31. For simplification, a lossy dielectric between two ground planes is considered. The Green's functions for this structure are also applicable to the dielectric between the ground and power planes. The losses in the dielectric are included in the complex permittivity $\underline{\varepsilon}$ according to

$$\underline{\varepsilon} = \varepsilon - j\sigma_{\mathrm{s}}/\omega \tag{4.172}$$

with ε as permittivity and σ_{s} as conductivity. The thickness of the dielectric is denoted by h, and the ground planes are assumed to be perfectly conducting.

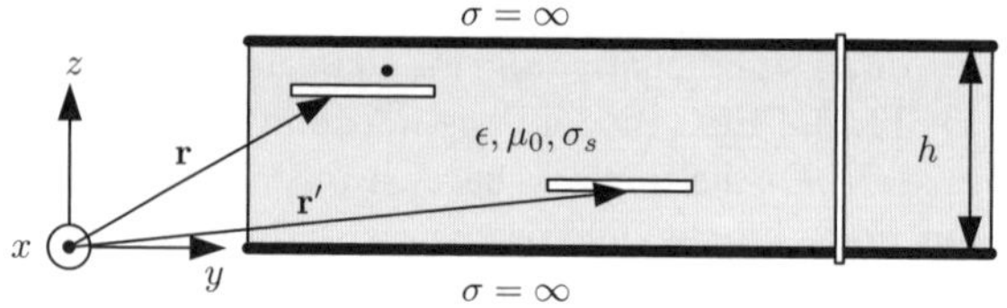

Figure 4.31 Stripline region.

4.8.3.1 Green's Functions for the Stripline Region

The scalar components of the dyadic Green's function of Equation (4.166) for the considered case of a stripline region are derived based on the general approach e.g. [35]. Here, only the final results obtained by computation of Sommerfeld integrals in closed form are presented:

$$G^{A}_{\rho z}\left(\mathbf{r},\mathbf{r}'\right) = G^{A}_{z\rho}\left(\mathbf{r},\mathbf{r}'\right) = 0 \tag{4.173}$$

$$G^{A}_{t}\left(\mathbf{r},\mathbf{r}'\right) = G^{\Phi}_{\mathrm{LM}}\left(\mathbf{r},\mathbf{r}'\right) = \frac{1}{2jh}\sum_{n=1}^{\infty}\sin k_{z,n}z'\sin k_{z,n}z H_0^{(2)}(\rho \underline{k}_{\rho,n}) \tag{4.174}$$

$$G^{A}_{zz}\left(\mathbf{r},\mathbf{r}'\right) = \frac{1}{4jh}H_0^{(2)}(\rho\underline{k}) + \frac{1}{2jh}\sum_{n=1}^{\infty}\cos k_{z,n}z'\cos k_{z,n}z H_0^{(2)}(\rho\underline{k}_{\rho,n}), \tag{4.175}$$

where h is the distance between the ground planes, $\underline{k} = \omega\sqrt{\mu_0\left(\varepsilon - j\sigma_s/\omega\right)}$ is the wave number, $k_{z,n}$ are longitudinal (with respect to $\mathbf{e}_z$) wave numbers of surface wave modes given by

$$k_{z,n} = n\pi/h. \tag{4.176}$$

They are determined by the thickness of the substrate h and do not depend on frequency. The wave numbers for transversal (with respect to $\mathbf{e}_z$) waves $\underline{k}_{\rho,n}$ are given by

$$\underline{k}_{\rho,n} = \sqrt{\underline{k}^2 - k_{z,n}^2}. \tag{4.177}$$

The investigations concerning the required number of summands N_s in Equations (4.174) and (4.175) in order to obtain a sufficient level of accuracy lead to the criterion

$$N_{\mathrm{s}} \leq \frac{h}{\pi}\sqrt{\omega^2\mu_0\varepsilon + (Q/\rho)^2}. \tag{4.178}$$

The constant Q can be evaluated by the equation $|H_0^{(2)}(-jQ)| = \delta_{\mathrm{err}}|H_0^{(2)}(\rho k_{\rho,1})|$ with δ_{err} as the required relative error. The number of summands increases if the frequency increases and by approaching to the singular point $\rho = 0$. For operating in a limited frequency range up to the maximum frequency of interest f_{m} and computing the Green's functions out of the singular point only the spatial convolution integrals of these functions have to be calculated. N_{s} is a finite value not exceeding several tens in practical cases.

4.8.3.2 Discussion of the Behavior of the Green's Functions

The Green's functions of Equations (4.174) and (4.175) have principally two kinds of frequency dependence. The first one $G_t^A(j\omega)$ represents the coupling between charges or transversally directed currents (i.e. between currents in traces). The second one $G_{zz}^A(j\omega)$ represents the coupling between vertically directed currents (i.e. between currents in vias). The behavior of both these functions is quite similar at relatively high frequencies. They have a set of resonances in frequency points corresponding to the condition $\underline{k}_{\rho,n} = 0, n \in [1, \infty]$ The resonance frequencies can be calculated in the case of negligible losses as follows:

$$\omega_{\mathrm{r}}^{(n)} = n\omega_{\mathrm{r}} \quad , \quad \omega_{\mathrm{r}} = \frac{\pi}{h\sqrt{\mu_0 \varepsilon}} \tag{4.179}$$

with ω_{r} as the frequency of the first and $\omega_{\mathrm{r}}^{(n)}$ as the frequency of the nth resonance. Resonances do not arise for high losses according to $\sigma_s > \varepsilon\omega_{\mathrm{r}}\sqrt{2}$. The low frequency behavior of $G_t^A(j\omega)$ and $G_{zz}^A(j\omega)$ is different. The function $G_t^A(j\omega)$ converges for frequencies $\omega \ll \omega_{\mathrm{r}}$ to a static value while $G_{zz}^A(j\omega)$ becomes infinite for $\omega \to 0$. The last one is caused by the pole of $H_0^{(2)}(\rho\underline{k})$ at $\underline{k} = 0$. This is a logarithmic singularity that can be physically interpreted. As the grounds are assumed to be of infinite conductivity, according to the image principle the vertical currents are reflected from the ground planes infinitely many times with equal signs. A phase shift is not obtained with d.c. current and, thus, the infinite sum of all currents diverges for a finite ρ. The relatively weak singularity of G_{zz}^A is regularized in the MPI Equation (4.165) by multiplication with $j\omega$. Because $\lim_{k\to 0} j\omega H_0^{(2)}(\rho\underline{k}) = 0$, the electric field intensity resulting from the vector potential $-j\omega\underline{\mathbf{A}}$ converges to zero at low frequencies and does not lead to a nonphysical solution.

Figure 4.32 shows an example of the frequency responses $G_{zz}^A(j\omega)$ and $G_t^A(j\omega)$ calculated with the following parameters: $h = 1$ mm, $\sigma_{\mathrm{s}} = 1$ mS/m, $\varepsilon_{\mathrm{r}} = 4.0$, and $z = z' = h/2$. The frequency of the first resonance is 75 GHz. One can notice that operational frequencies in modern digital systems are essentially below this frequency. Strong cross-talk caused by lateral wave resonances will appear if this condition is not satisfied. Therefore, the PCB-designer should keep the thickness of layers small in order to avoid this cross-talk. Thus, for

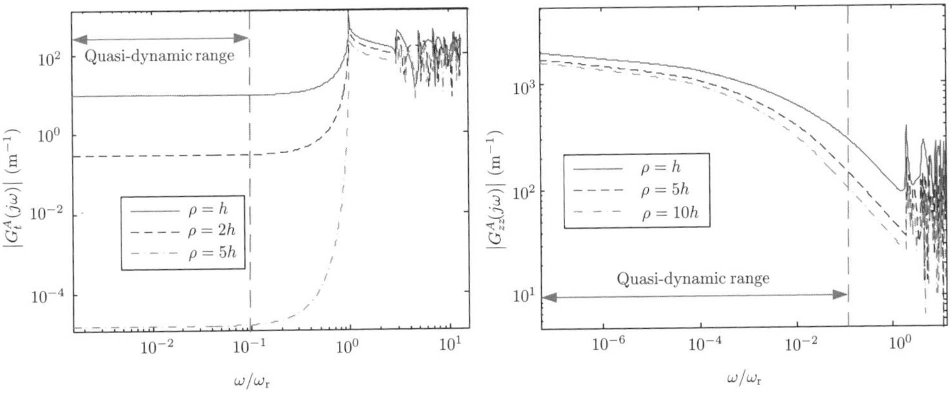

Figure 4.32 Scalar components of the dyadic Green's function for the stripline region.

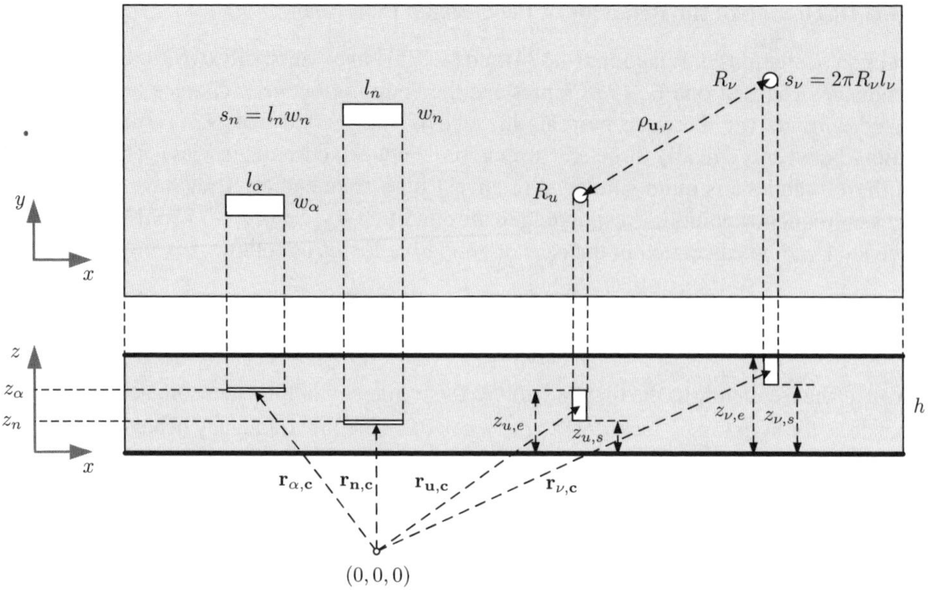

Figure 4.33 Geometry of the stripline region.

all practical problems concerning the design of PCBs this condition is implemented, while the full dynamic behavior of the Green's functions is more interesting for microwave applications. The frequency region below ω_r is referred to as the quasi-dynamic one.

4.8.3.3 Frequency Domain DGFLM-PEEC Model

The appropriate PEEC model is given by the circuit structure pictured in Figure 4.28 with controlled voltage sources whose control coefficients are represented by the frequency dependent, generalized partial elements which are derived in the following paragraphs. In Figure 4.33 all types of cells with the appropriate geometrical notations are shown. Each cell (e.g. cell n) has the length l_n, surface s_n, perimeter d_n, geometrical center $\mathbf{r}_{n,c}$. A rectangular cell located in the x,y-plane has a width w_n. A cylindrical cell (e.g. the via cell u) has a radius R_u, starts in $z = z_{u,s}$ and ends in $z = z_{u,e}$. The transversal distance between two centers of cells u and v is $\rho_{u,v}$.

Generalized Partial Inductances of Via Cells

In the following, a closed form solution for the generalized partial inductance of via cells using the standard thin wire assumption is developed (see, e.g., [54]). The dyadic Green's function of the magnetic vector potential for the stripline region has the two scalar components G^A_{zz} and G^A_t. Since the current in the vias flows in the z-direction, one obtains $\beta^{u,u}_{zz} = 1$ and $\beta^{u,u}_t = 0$ and, consequently, for the generalized partial self inductance

$$\underline{\Lambda}_{u,u}(j\omega) = \frac{\mu_0}{d_u d_u} \int\limits_{v_u} \int\limits_{v_u} G^A_{zz}\left(\mathbf{r}, \mathbf{r}'\right) \mathrm{d}v'_u \, \mathrm{d}v_u. \tag{4.180}$$

Introducing the thin wire approximation results in

$$\underline{\Lambda}_{u,u}(j\omega) = \frac{\mu_0}{d_u} \int_{s_u} \int_{l'_u} G_{zz}^A \left(\mathbf{r}, \mathbf{r}'\right) \mathrm{d}l'_u \, \mathrm{d}s_u. \tag{4.181}$$

Because the source current is in the axis of the via and the observation point on the surface s_u, the integrand depends only on z and z' and Equation (4.181) can be written in the form

$$\underline{\Lambda}_{u,u}(j\omega) = \mu_0 \int_{l_u} \int_{l_u} G_{zz}^A \left(z, z', \rho = R_u\right) \mathrm{d}l'_u \, \mathrm{d}l_u. \tag{4.182}$$

The double integral in Equation (4.182) can be calculated in closed form

$$\underline{\Lambda}_{u,u}(j\omega) = \frac{\mu_0}{2jh} \left(\frac{l_u^2}{2} H_0^{(2)} \left(R_u k\right) + \sum_{n=1}^{N} \frac{\xi_n}{k_{z,n}^2} H_0^{(2)} \left(R_u k_{\rho,n}\right) \right) \tag{4.183}$$

with

$$\xi_n = \sin\left(k_{z,n} \, z_{ue}\right) - \sin\left(k_{z,n} \, z_{us}\right).$$

The frequency response of the partial self inductance has been investigated by an example. A via with a radius $R_u = 0.1$ mm is located in a stripline region with the parameters $h = 1$ mm, $\sigma_\mathrm{s} = 1$ mS/m and $\epsilon_\mathrm{r} = 4.0$. The partial self inductance has been calculated for three values of the current cell length h, $h/2$ and $h/4$. The via starts at $z = 0$. Figure 4.34 shows the real part of the partial self inductance. The generalized partial inductance consists of both real and imaginary parts. The real part is an inductance while the imaginary part represents losses. The losses in this example are low and therefore only the inductance is considered. Obviously, all variants of the frequency response increase logarithmically towards low frequencies and show resonances at high frequencies due to the poles in Equation (4.183). For $l_u = h$, the general

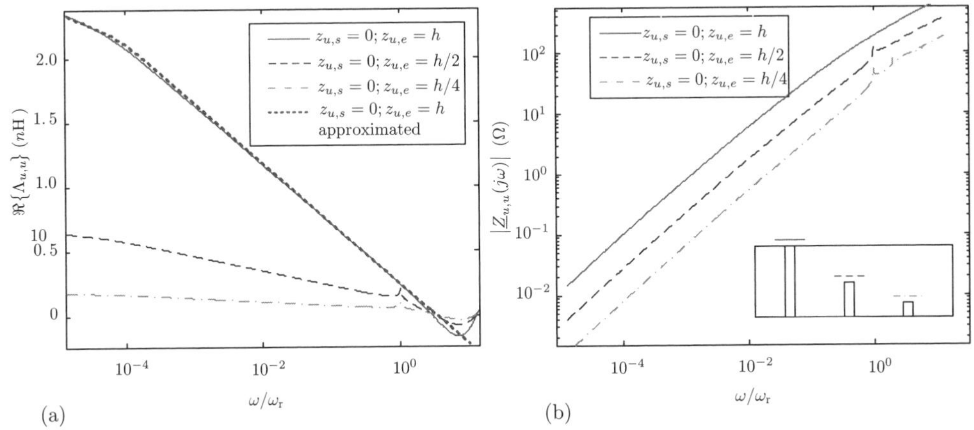

Figure 4.34 Frequency responses of a generalized partial self inductance and a transfer impedance for a via cell.

Equation (4.182) may essentially be simplified:

$$\underline{\Lambda}_{u,u}(j\omega) = \frac{h\mu_0}{2\pi}\left(\ln\frac{2}{R_u k} - \gamma\right), \tag{4.184}$$

where $\gamma \approx 0.57722$ is the Euler's constant. The frequency response obtained by this formula is shown in Figure 4.34(a) by the dotted line, which is close to the exact curve. These results may be compared with the rough approximation for the static via inductance given in [55] under the free space assumption:

$$L_{\text{via}} = 5.08 \times 10^{-9} h\,(\ln 2h/R_u + 1)\,. \tag{4.185}$$

For a via of length h, Equation (4.185) yields a value of 0.02 nH, which is much lower than the result shown in Figure 4.34(a). Moreover, this approximation does not consider the frequency dependence.

Inductive couplings in a PEEC model are represented by voltage sources controlled by the derivative of currents, in particular, the voltage drop on the current cell due to the partial self inductance is $\underline{U}^{u,u} = j\omega\underline{\Lambda}_{u,u}(j\omega)\underline{I}^u$. The factor in this controlled voltage source can be considered as an impedance $\underline{Z}_{u,u}(j\omega) = j\omega\underline{\Lambda}_{u,u}(j\omega)$. The frequency response of the magnitude of $\underline{Z}_{u,u}(j\omega)$ is shown in Figure 4.34(b). This function decreases to zero at low frequencies and increases with growing frequency.

Now, the generalized partial mutual inductances between via cells u and v are considered. Under the thin wire assumption a closed-form formula for the partial inductance may be derived:

$$\underline{\Lambda}_{u,v}(j\omega) = \frac{\mu_0}{2jh}\left(\frac{l_v l_u}{2} H_0^{(2)}(\rho_{u,v}k) + \sum_{n=1}^{N_s} \frac{\xi_n}{k_{z,n}^2} H_0^{(2)}(\rho_{u,v}k_{\rho,n})\right), \tag{4.186}$$

with

$$\xi_n = \left\{\sin(k_{z,n}z_{u,e}) - \sin(k_{z,n}z_{u,s})\right\} \times \left\{\sin(k_{z,n}z_{v,e}) - \sin(k_{z,n}z_{v,s})\right\}.$$

An important explanation should be added for this equation. It is assumed that the source current flows in the axis of the conductor. In the case of two cells located on the same line, instead of $\rho_{u,v}$ in Equation (4.186) the radius of the via has to be used. A comparison between the frequency responses of the partial self and mutual inductances is shown in Figure 4.35(a). The parameters of the stripline region are $h = 1$ mm, $\sigma_s = 1$ mS/m, $\epsilon_r = 4.0$, $z_{u,s} = 0$, $z_{v,s} = h/2$, $z_{u,e} = h/2$, $z_{v,e} = h$, $\rho_{u,v} = 0$ mm and $R_u = R_v = 0.1$ mm. The computed frequency responses have a similar behavior. The mutual inductance is less than the self inductance in the quasi-dynamic frequency range. This looks similar to the ratio between the self and mutual inductances in the free space PEEC model.

In the event of $l_v = l_u = h$, the general Equation (4.186) may be simplified using its low-frequency asymptotic expansion:

$$\underline{\Lambda}_{u,v}(j\omega) = \frac{h\mu_0}{2\pi}\left(\ln\frac{2}{\rho_{u,v}k} - \gamma\right). \tag{4.187}$$

The comparison of this approximate formula with the exact expression of Equation (4.186) is shown in Figure 4.35(b) for the parameters $h = l_v = l_u = 1$ mm, $\sigma_s = 1$ mS/m, $\epsilon_r = 4.0$, $\rho_{u,v} = 20$ mm and $R_u = R_v = 0.1$ mm. The results show good correspondence.

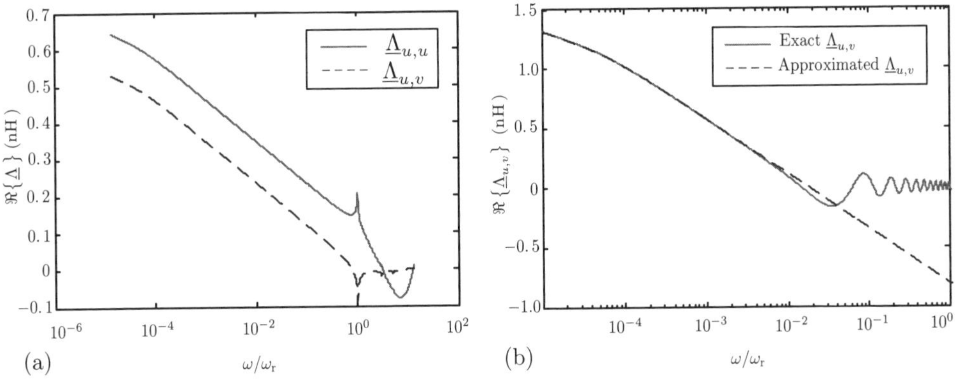

Figure 4.35 Comparison between self and mutual inductances (a) and between exact and approximated formulas for the via inductance (b).

Generalized Potential Coefficients of Via Cells

A generalized potential coefficient for a cylindrical cell directed along $\mathbf{e}_z$ is derived analogously to a partial self inductance starting from Equations (4.162) and (4.166). The final result for the self potential coefficient of the cell u according to Figure 4.33 is given by

$$\underline{\Pi}_{u,u}(j\omega) = \frac{1}{2jhl_u^2\epsilon}\left(\sum_{n=1}^{N_s}\frac{\xi_n}{k_{z,n}^2}H_0^{(2)}(R_u k_{\rho,n})\right) \tag{4.188}$$

$$\text{with} \quad \xi_n = (\cos k_{z,n}z_{u,e} - \cos k_{z,n}z_{u,s})^2.$$

A generalized mutual potential coefficient between cells u and v in Figure 4.33 is derived in a final form as:

$$\underline{\Pi}_{u,v}(j\omega) = \frac{1}{2jhl_v l_u\epsilon}\left(\sum_{n=1}^{N_s}\frac{\xi_n}{k_{z,n}^2}H_0^{(2)}(\rho_{u,v}k_{\rho,n})\right) \tag{4.189}$$

with

$$\xi_n = \left\{\cos\left(k_{z,n}z_{u,e}\right) - \cos\left(k_{z,n}z_{u,s}\right)\right\} \times \left\{\cos\left(k_{z,n}z_{v,e}\right) - \cos\left(k_{z,n}z_{v,s}\right)\right\}.$$

Generalized Partial Self Inductance of a Horizontal Cell

Rectangular horizontal current cells in the x, y plane are used for modeling traces of the stripline configuration. For simplification, the thickness of the cell is assumed to be zero. The length and width of the cell are denoted by l_α and w_α respectively. The z-coordinate of the cell is denoted by z_α (see Figure 4.33). As the current is directed perpendicular to $\mathbf{e}_z$, the general formula for partial inductances in Equation (4.171) is simplified as:

$$\underline{\Lambda}_{\alpha,\alpha}(j\omega) = \frac{\mu_0}{w_\alpha^2}\int_{s_\alpha}\int_{s_\alpha} G_t^A(\mathbf{r},\mathbf{r}')\mathrm{d}s_\alpha',\, \mathrm{d}s_\alpha. \tag{4.190}$$

The direct closed-form integration of $G_t^A(\mathbf{r}, \mathbf{r}')$ in the horizontal plane is impossible, since the Green's function contains $H_0^{(2)}(\rho k_{\rho,n})$. The direct numerical integration is computationally expensive because $G_t^A(\mathbf{r}, \mathbf{r}')$ is singular at $\rho = 0$. In order to find a solution, one considers the frequency response $G_t^A(\mathbf{r}, \mathbf{r}')$ (see Figure 4.32). Obviously, in contrast to $G_{zz}^A(\mathbf{r}, \mathbf{r}')$, this function reaches a constant value at $\omega \to 0$. Thus, one may simplify the integration in an analogous way as in the standard PEEC method. A mutual inductance in the standard retarded PEEC is calculated with the help of the following approximation:

$$\underline{\Lambda}_{\alpha,n}(j\omega) = \frac{\mu_0}{w_\alpha w_n} \int\limits_{s_\alpha} \int\limits_{s_n'} g(\mathbf{r}, \mathbf{r}') \mathrm{d}s_n', \mathrm{d}s_\alpha,$$

$$g(\mathbf{r}, \mathbf{r}') = \frac{\mathrm{e}^{-jk|\mathbf{r}-\mathbf{r}'|}}{4\pi|\mathbf{r}-\mathbf{r}'|} \approx \frac{1}{4\pi|\mathbf{r}-\mathbf{r}'|} \mathrm{e}^{-jk|\mathbf{r}_{\alpha,c}-\mathbf{r}_{n,c}|}, \tag{4.191}$$

where $\mathbf{r}_{n,c}$ and $\mathbf{r}_{\alpha,c}$ are the geometrical centers of cells. This approximation requires the integration of only the static part of the Green's function, approximating the frequency response with respect to the geometrical centers of cells. Thus, one computes exactly the usual partial inductance $L_{\alpha,n}$ and approximately its frequency dependence at high frequencies. The Green's function $G_t^A(\mathbf{r}, \mathbf{r}')$ can be approximated in the same manner:

$$\begin{aligned} G_t^A(\mathbf{r}, \mathbf{r}') &\approx G_{t,\mathrm{st}}^A(\mathbf{r}, \mathbf{r}') G_{t,\mathrm{dyn}}^A(\mathbf{r}_{\alpha,c}, \mathbf{r}_{n,c}), \\ G_{t,\mathrm{st}}^A(\mathbf{r}, \mathbf{r}') &= G_t^A(\mathbf{r}, \mathbf{r}', j\omega = 0), \\ G_{t,\mathrm{dyn}}^A(\mathbf{r}_{\alpha,c}, \mathbf{r}_{n,c}, j\omega) &= \frac{G_t^A(\mathbf{r}_{\alpha,c}, \mathbf{r}_{n,c}, j\omega)}{G_{t,st}^A(\mathbf{r}, \mathbf{r}')}. \end{aligned} \tag{4.192}$$

The Green's function is divided into two parts. The static part $G_{t,\mathrm{st}}^A(\mathbf{r}, \mathbf{r}')$ depends on the space coordinates and the dynamic part $G_{t,\mathrm{dyn}}^A(\mathbf{r}_{\alpha,c}, \mathbf{r}_{n,c}, j\omega)$ depends only on frequency. At low frequencies the dynamic part is equal to one. The substitution of Equation (4.192) into Equation (4.190) yields

$$\begin{aligned} \underline{\Lambda}_{\alpha,\alpha}(j\omega) &= \frac{\mu_0}{w_\alpha^2} G_{t,\mathrm{dyn}}^A(\mathbf{r}_{\alpha,c}, \mathbf{r}_{\alpha,c}, j\omega) \int\limits_{s_\alpha} \int\limits_{s_n} G_{t,\mathrm{st}}^A(\mathbf{r}, \mathbf{r}') \mathrm{d}s_n', \mathrm{d}s_\alpha \approx \\ &\approx \frac{\mu_0}{w_\alpha^2} G_{t,\mathrm{dyn}}^A(\mathbf{r}_{\alpha,c}, \mathbf{r}_{\alpha,c}, j\omega) s_\alpha \int\limits_{s_n} G_{t,\mathrm{st}}^A(\mathbf{r}_{\alpha,c}, \mathbf{r}') \mathrm{d}s_n'. \end{aligned} \tag{4.193}$$

Unfortunately, the integral of the static part cannot be derived in closed form. The integral of Equation (4.193) does not change the coordinates z and z', since the integral is taken over the (x, y) plane. For simplification, one approximates the dependence of $G_{t,\mathrm{st}}^A(\mathbf{r}, \mathbf{r}') = G_{t,\mathrm{st}}^A(\rho, z = z_{\alpha,c}, z' = z_{\alpha,c})$ on ρ by the functions proposed in [56] for the approximation of the Green's functions for microstrips:

$$G_{t,\mathrm{st}}^A(\rho, z, z') \approx \Phi(\rho, z, z') = \sum_{i=1}^{n} \frac{a_i}{(b_i^2 + \rho^2)^{3/2}}. \tag{4.194}$$

The integrals of each term in the sum of Equation (4.194) are computed in closed form. The unknown coefficients a_i and b_i may be estimated by the numerically efficient least square procedure [57]. The sampling points for the least square procedure have to be chosen in the range from $\rho_{\min}$ up to $\rho_{\max}$, where $\rho_{\max}$ may be determined as the largest horizontal distance on the PCB. The choice of $\rho_{\min}$ raises the problem concerning the singularity at $\rho = 0$.

To solve this, an asymptotic formula $G^A_{t,\mathrm{st}}(\rho)$ for $\rho \to 0$ has been derived:

$$\begin{aligned}
\lim\nolimits_{\rho \to 0} G^A_{t,\mathrm{st}}(\rho, z, z') &= \frac{1}{4\pi\rho} + C(z, z', h), \\
C(z, z', h) &= \frac{1}{4\pi|z+z'|} + \\
&+ \frac{1}{8\pi h}\left[\Psi\left(1 + \frac{z+z'}{2h}\right) + \Psi\left(1 - \frac{z+z'}{2h}\right)\right. \\
&\left. -\Psi\left(1 + \frac{z'-z}{2h}\right) - \Psi\left(1 - \frac{z'-z}{2h}\right)\right],
\end{aligned} \tag{4.195}$$

where $C(z, z', h)$ is a finite constant defined via the geometry and $\Psi()$ is the digamma function. Through Equation (4.195) one may see that the singularity is defined by $1/\rho$. The corollary to this fact is

$$\lim_{s \to 0} \int_s G^A_{t,\mathrm{st}}(\rho)\mathrm{d}s = 0 \quad \Rightarrow \quad \int_{S_n} G^A_{t,\mathrm{st}}(\rho)\mathrm{d}s'_n = \int\int_{S_n} \Phi(\rho)\mathrm{d}s'_n, \tag{4.196}$$

although $\Phi(0) \neq G^A_{t,\mathrm{st}}(0) = \infty$, if one follows the criterion

$$\rho_{\min} \ll \lambda_{\min}/20. \tag{4.197}$$

In the following, the sensitivity of the solution for $\underline{\Lambda}_{\alpha,\alpha}$ calculated via the approximation of the static Green's function by Equation (4.194) is investigated. For this, the static self inductance for a rectangular current cell with the parameters $l_\alpha = 0.746$ mm ($f_\mathrm{m} = 10$ GHz), $w_\alpha = 0.3$ mm, $h = 1$ mm, $\sigma_\mathrm{s} = 1$ mS/m, $\epsilon_\mathrm{r} = 4.0$, and $z_\alpha = h/2$ is computed. The approximation $\Phi(\rho)$ has been calculated for several values of $\rho_{\min}$ between 1 nm (27 terms in the sum of Equation (4.194)) and 0.1 mm (six terms in the sum of Equation (4.194)). The results are shown in Figure 4.36(a) in comparison with the reference curve $G^A_{t,\mathrm{st}}(\rho)$. Obviously, all approximations fit the reference very well at all points between $\rho_{\min}$ and $\rho_{\max}$. An error appears only if $\rho < \rho_{\min}$. Figure 4.36(b) shows the dependence of the static value of $\underline{\Lambda}_{\alpha,\alpha}$ on $\rho_{\min}$ used for the approximation. It can be observed that the values calculated with $\rho_{\min}$ equal to 1 nm and 10 nm do not differ. $\rho_{\min} = 1\,\mu$m produces an error of 0.6 %, 10 µm of 2.4 %, and 0.1 mm of 19 %. The error increases violently if $\rho_{\min}$ approaches $\lambda_\mathrm{m}/20 = 0.75$ mm. A fair accuracy of the approximation is obtained with

$$\rho_{\min} < \lambda_\mathrm{m}/10\,000. \tag{4.198}$$

The application of the approximation of Equation (4.194) leads to the following expressions for the generalized partial self inductance

$$\underline{\Lambda}_{\alpha,\alpha}(j\omega) = L_{\mathrm{p}\alpha,\alpha} G^A_{t,\mathrm{dyn}}(\mathbf{r}_{\alpha,\mathrm{c}}, \mathbf{r}_{\alpha,\mathrm{c}}, j\omega) \quad , \quad L_{\mathrm{p}\alpha,\alpha} = \frac{\mu_0}{w_\alpha^2} I_{\alpha,\alpha}, \tag{4.199}$$

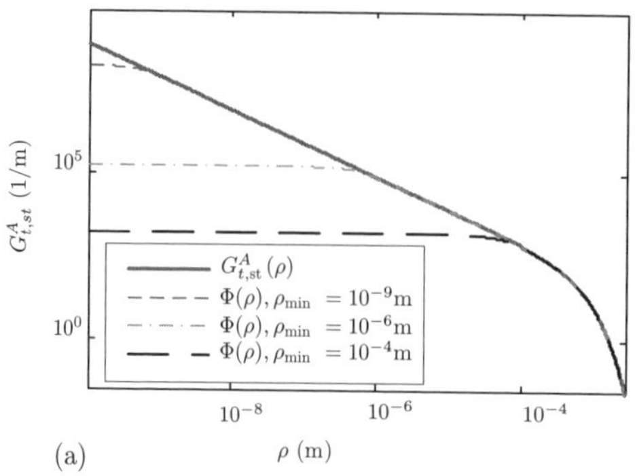

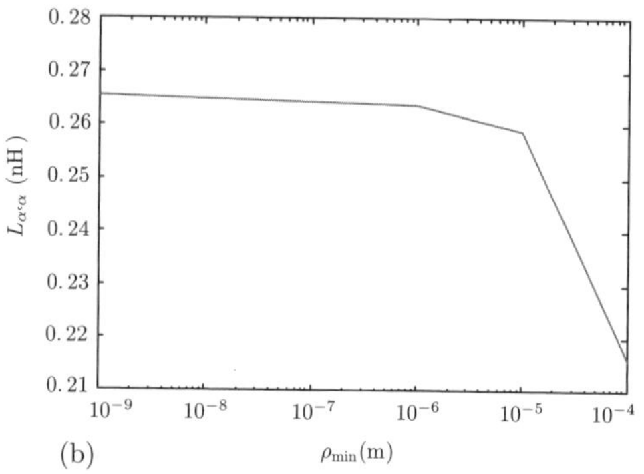

Figure 4.36 (a) Approximation of the Green's function and (b) the dependence of the approximated solution $L_{p\alpha,\alpha}$ of $\rho_{\min}$.

where $L_{p\alpha,\alpha}$ is the static partial inductance and $I_{\alpha,\alpha}$ is the space-convolution integral of the static Green's function computed through Equation (4.194) given by

$$I_{\alpha,n} = s_\alpha \int_{S_n} \Phi(\rho, z_{\alpha,c}, z_{n,c}) \mathrm{d}s'_n. \tag{4.200}$$

Remaining Generalized Partial Elements

Since $G^A_t(\mathbf{r}, \mathbf{r}') = G^\Phi_{\mathrm{LM}}(\mathbf{r}, \mathbf{r}')$, the remaining variants of partial elements are derived using the above given concepts. Here, only final formulas expressed in terms of space-convolution integrals given by Equation (4.200) are presented. The mutual inductance between two rectangular cells α and n shown in Figure 4.33 is calculated as follows:

$$\underline{\Lambda}_{\alpha,n}(j\omega) = L_{p\alpha,n} G^A_{t,\mathrm{dyn}}(\mathbf{r}_{\alpha,c}, \mathbf{r}_{n,c}, j\omega), \qquad L_{p\alpha,n} = \frac{\mu_0}{w_\alpha w_n} I_{\alpha,n}. \tag{4.201}$$

The calculation of the generalized potential coefficients is done in view of $G^A_t(\mathbf{r}, \mathbf{r}') = G^\Phi_{\mathrm{LM}}(\mathbf{r}, \mathbf{r}')$. Thus, the generalized potential coefficients of Equation (4.162) are expressed through the same approximation of integrals as given in Equation (4.200). The self potential coefficient for a flat cell i (see Figure 4.33) is

$$\underline{\Pi}_{i,i}(j\omega) = P_{i,i} G^\Phi_{\mathrm{LM,dyn}}(\mathbf{r}_{i,c}, \mathbf{r}_{i,c}, j\omega), \qquad P_{i,i} = \frac{I_{i,i}}{\epsilon s_i^2}. \tag{4.202}$$

The mutual potential coefficient for the flat cells i and m (see Figure 4.33) is given by:

$$\underline{\Pi}_{i,m}(j\omega) = P_{i,m} G^\Phi_{\mathrm{LM,dyn}}(\mathbf{r}_{i,c}, \mathbf{r}_{m,c}, j\omega), \qquad P_{i,m} = \frac{I_{i,m}}{\epsilon s_i s_m}. \tag{4.203}$$

The mutual potential coefficient for the flat cell i and the via cell u (see Figure 4.33) yields

$$\underline{\Pi}_{i,u}(j\omega) = P_{i,u} G^\Phi_{\mathrm{LM,dyn}}(\mathbf{r}_{i,c}, \mathbf{r}_{u,c}, j\omega), \qquad P_{i,u} = \frac{I_{i,u}}{\epsilon s_i s_u}. \tag{4.204}$$

4.8.3.4 DGFLM–PEEC Models in the Time Domain

The DGFLM–PEEC model in the frequency domain is a circuit consisting of linear controlled voltage sources with frequency dependent coefficients (see Figure 4.28). The direct implementation of this model in the time domain is not possible. The way for obtaining time domain models in such cases has already been discussed in Section 4.6.4. For the development of DGFLM–PEEC models in the time domain the full spectrum convolution macromodeling (FSCM) is used as well. To establish time domain PEEC models that are suitable for an effective simulation on circuit solvers the frequency dependence of the generalized partial elements is represented by macromodels in the form of associated discrete circuits (ADCs). Depending on the approximation of the frequency behavior of the partial elements one may obtain different time domain macromodels: quasi-static (qs), quasi-dynamic (qd) and full-spectrum models. In the following, quasi-static and quasi-dynamic time domain models are considered.

Quasi-Static Time Domain DGFLM–PEEC Model

The quasi-static representation of the partial elements is obtained by neglecting the dynamic part in analogy to the considerations in Section 4.3.2.3. Thus, for the partial elements derived from G_t^A and G_{LM}^{Φ} one takes the real part of the complex functions at $\omega = 0$. However, the partial inductances of vertical via cells calculated by means of G_{zz}^A tend logarithmically to infinity at $\omega = 0$. Therefore, the real part is approximately calculated at the maximum frequency of interest ω_{m}:

$$\Lambda_{u,v}^{qs} = L_{\mathrm{p}u,v}^{qs} = \Re\left\{\underline{\Lambda}_{u,v}(j\omega_{\mathrm{m}})\right\} \;\; \forall u, v \in \text{vias}. \tag{4.205}$$

The choice of ω_{m} is motivated since the inductive coupling is proportional to the frequency. Thus, the partial inductance has the maximum influence of the solution at ω_{m} which guarantees the best precision for the determination.

Since the quasi-static partial elements of the DGFLM–PEEC model are real constants, the time domain implementation is completely the same as the quasi-static PEEC model considered in Section 4.3.2.3. However, the big advantage of the DGFLM–PEEC model is that the influence of the substrate is taken into account correctly for the considered low frequency range.

Quasi-Dynamic Time Domain DGFLM–PEEC Model

The quasi-dynamic model approximates the frequency dependence of the generalized partial elements up to frequencies, where the local wave resonances of the PCB are negligible. This level of modeling takes into account dielectric losses as well as time delays of inductive and capacitive couplings. As mentioned before, the generalized partial elements developed in Section 4.8.3.3 have two types of frequency response. The first one is associated with the Green's functions $G_t^A(j\omega)$ and G_{LM}^{Φ} and tends to be constant at low frequencies. The second type of frequency response is associated with the Green's function $G_{zz}^A(j\omega)$ and concerns the partial inductances of vias. This last type of frequency response tends logarithmically to infinity for $\omega \to 0$. For both types of frequency response the generalized partial elements are developed, in particular, under the assumption of weak dielectric losses.

The concept of the quasi-dynamic modeling is formulated by the equation

$$G^{qd}(j\omega) = G(j\omega) \cdot W(j\omega). \tag{4.206}$$

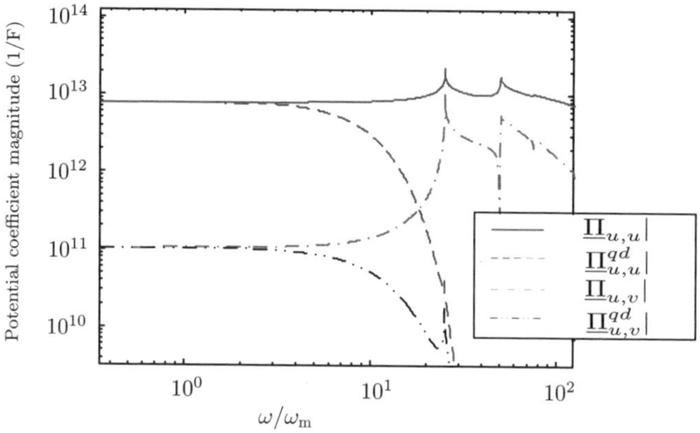

Figure 4.37 Quasi-dynamic frequency responses of the generalized partial elements of the first type.

$G(j\omega)$ is the general notation of the frequency dependencies of the Green's functions $G_t^A(j\omega)$, $G_{\mathrm{LM}}^{\Phi}(j\omega)$ and $G_{zz}^A(j\omega)$ with lateral wave resonances. $G^{qd}(j\omega)$ is the equivalent quasi-dynamic transfer function (EQDTF) corresponding to $G(j\omega)$ for $f < f_{\mathrm{m}}$ and to completely eliminated resonances for $f > f_{\mathrm{m}}$. $W(j\omega)$ is the transfer function according to Equation (4.112) with $\xi_0 = 1$, $\tilde{\tau} = 0.01$ ps, $\alpha = \sqrt{-l_n\eta}/\omega_{\mathrm{m}}$, and $\eta = 0.99$ [35]. As a consequence, the impulse response $g^{qd}(t) = \mathfrak{F}^{-1}\left\{G^{qd}(j\omega)\right\}$ does not contain high frequency oscillations as $g(t) = \mathfrak{F}^{-1}\left\{G(j\omega)\right\}$. Likewise, the calculation of the Duhamel integral for an arbitrary excitation waveform does not lead to high-frequency oscillations in the time response. This is the basis for setting up stable quasi-dynamic PEEC models. The quasi-dynamic model of the generalized partial elements of the first type (based on $G_t^A(j\omega)$ and G_{LM}^{Φ} respectively, is exemplified by the potential coefficients of vertical cells (vias):

$$\underline{\Pi}^{qd}_{u,v}(j\omega) = \underline{\Pi}_{u,v}(j\omega)\mathrm{e}^{-(\alpha\omega)^2} \quad , \quad u, v \in \text{vias}. \tag{4.207}$$

Figure 4.37 shows the comparison between the full and the quasi-dynamic frequency responses. The time domain models for this type of generalized partial element may be successfully obtained by FSCM as demonstrated in Section 4.7.4.1 under the aspect of stability.

For the second type of generalized partial element (based on $G_{zz}^A(j\omega)$), whose frequency response tends to infinity for $\omega \to 0$, according to the strategy of FSCM a suitable approximation in the frequency domain is needed and chosen as follows:

$$\underline{\Lambda}_{u,v}(j\omega) \approx \underline{\Lambda}^{qd}_{u,v}(j\omega) = \xi \ln \frac{j\omega + \beta}{j\omega w + \alpha}\mathrm{e}^{-j\omega\tilde{\tau}}. \tag{4.208}$$

$\underline{\Lambda}_{u,v}(j\omega)$ are the generalized partial inductances of vias. The parameter $\tilde{\tau}$ is a time delay between the centers of cells for partial mutual inductances ($u \neq v$)

$$\tilde{\tau} = \rho_{u,v}\sqrt{\varepsilon\mu_0} \tag{4.209}$$

and

$$\tilde{\tau} = R_u\sqrt{\varepsilon\mu_0} \tag{4.210}$$

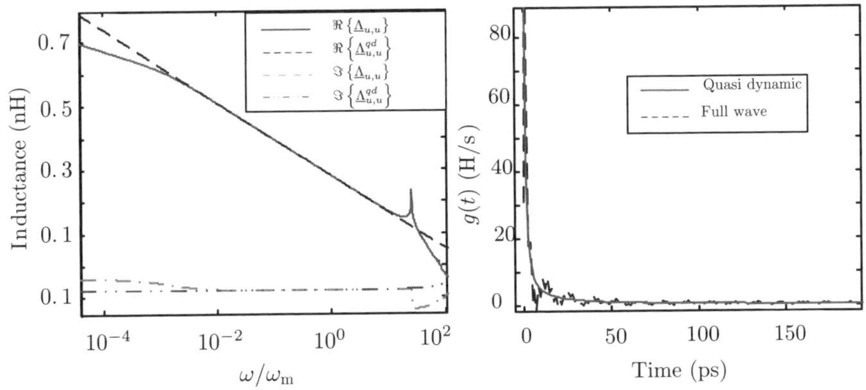

(a) Transfer function (b) Impulse response

Figure 4.38 Quasi-dynamic model for the generalized partial self inductance of a via cell.

for the partial self inductances ($u = v$). The remaining parameters of $\underline{\Lambda}_{u,v}^{qd}(j\omega)$ are calculated using the original frequency response $\underline{\Lambda}_{u,v}(j\omega)$ after the delay extraction (multiplying by $\exp(j\omega\tau)$). The parameter α has to be much less than ω_m and is chosen as $\alpha = 10^{-6}\omega_m$. The remaining parameters ξ and β are determined by comparison of the real parts of the original and the approximated functions without delays in two points ω_s:

$$\Re\left\{\underline{\Lambda}_{u,v}(j\omega_s)e^{j\omega_s\tilde{\tau}}\right\} = \frac{1}{2}\xi \ln \frac{\omega_s^2 + \beta^2}{\omega_s^2}. \tag{4.211}$$

The two chosen points are $\omega_{s1} = \omega_m$ and $\omega_{s2} = \omega_m/100$. As a result, Equation (4.211) with the parameters determined as described, is a perfect approximation in the frequency range $0.001\,\omega_m < \omega < 10\,\omega_m$ without high frequency resonances.

Figure 4.38 shows the results for the partial self inductance of a via in a stripline region with the following parameters : $h = 1$ mm, $\sigma_s = 1$ mS/m, $\varepsilon_r = 4.0$, $z_{u,s} = 0$, $z_{u,e} = h/2$, $R_u = 0.1$ mm. The little deviations between $\underline{\Lambda}_{u,u}(j\omega)$ and $\underline{\Lambda}_{u,u}^{qd}(j\omega)$ at low frequencies do not influence the transients see Figure 4.38(b). While the impulse response of $\underline{\Lambda}_{u,u}(j\omega)$ is calculated numerically using the IFFT, the impulse and step responses of $\underline{\Lambda}_{u,u}^{qd}(j\omega)$ are derived in closed form (without time delay, which may be added later in the calculation of the Duhamel integral (see Equation (4.214)):

$$g(t) = \xi \frac{e^{-\alpha t} - e^{-\beta t}}{t} h(t), \tag{4.212}$$

$$s(t) = \xi \left(\ln \frac{\beta}{\alpha} + \mathrm{Ei}_1(\beta t) - \mathrm{Ei}_1(\alpha t) \right) h(t). \tag{4.213}$$

$\mathrm{Ei}_1(t)$ are exponential integrals (see [68]). Obviously , the only difference between the numerically and the closed form calculated impulse responses in Figure 4.38(b) is the absence of oscillations in the quasi-dynamic one. Thus, Equation (4.208) may be accepted as a fair quasi-dynamic approximation. Considering $s(0) = 0$, the time response of the generalized partial inductance $y(t)$ on an arbitrary excitation $x(t)$ may be calculated by the Duhamel integral

$$y(t) = \int_0^t f(t - \theta)\, x(\theta - \tilde{\tau})\, \mathrm{d}\theta, \qquad (4.214)$$

where $\tilde{\tau}$ is the time delay of the approximation function of Equation (4.208). Using the approach already explained in Section 4.7.4.1, the moving average discrete formulation of Equation (4.214) can be derived for the time point $t = t_{n+1}$:

$$y_{n+1} = \sum_{k=p}^{q} \Omega_k x_k. \qquad (4.215)$$

The coefficients of Equation (4.215) may be calculated before the numerical integration of the PEEC model:

$$\Omega_k = \frac{\xi}{2} \begin{cases} \mathrm{Ei}_1(\beta T_{n+1,k}) - \mathrm{Ei}_1(\alpha T_{n+1,k}) - \mathrm{Ei}_1(\beta T_{n,k}) + \mathrm{Ei}_1(\alpha T_{n,k}) & k = p \\ \mathrm{Ei}_1(\beta T_{n,k}) - \mathrm{Ei}_1(\alpha T_{n,k}) - \mathrm{Ei}_1(\beta T_{n-1,k}) + \mathrm{Ei}_1(\alpha T_{n-1,k}) & k = q \\ \mathrm{Ei}_1(\beta T_{n+1,k}) - \mathrm{Ei}_1(\alpha T_{n+1,k}) - \mathrm{Ei}_1(\beta T_{n-1,k}) + \mathrm{Ei}_1(\alpha T_{n-1,k}) & \text{else} \end{cases} \qquad (4.216)$$

with $T_{n,k} = \tilde{\tau} - t_n + t_k$. Because of the selectivity of $g(t)$ only a few summands in the moving average of Equation (4.215) have to be calculated (see Figure 4.39), namely, for the time points from t_p until t_q. The subscript q of the time point is defined under consideration of $\tilde{\tau}$ as

$$q = n - \left\lfloor \frac{\tilde{\tau}}{\Delta t} \right\rfloor. \qquad (4.217)$$

The subscript p depends on the width of the moving average denoted by τ_α (see Figure 4.39). Since the coefficients Ω_k may be calculated subsequently from q until p, the subscript p may be obtained from the heuristic condition

$$\Omega_p < \delta_{\mathrm{err}} \sum_{k=p+1}^{q} \Omega_k \qquad (4.218)$$

with δ_{err} as the relative error.

The moving average formulation of Equation (4.215) has the same form as Equation (4.120) with only quantitative differences. That means, the ADC models developed in Section 4.7.4.1

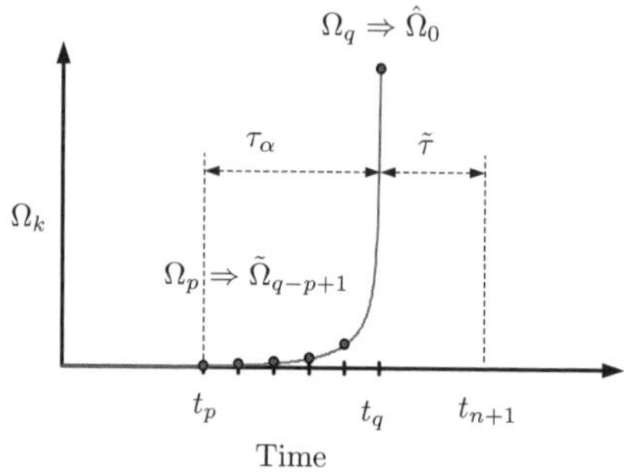

Figure 4.39 Moving average for the FSCM of the quasi-dynamic model (see Equation (4.208)).

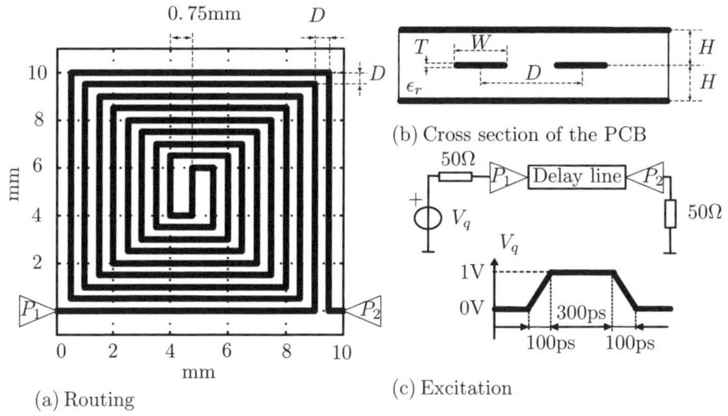

Figure 4.40 Spiral delay line.

may be applied for time domain modeling the generalized partial inductances according to Equation (4.208). For validation of the theoretical considerations to the quasi-dynamic DGFLM–PEEC modeling for a stripline configuration an example taken from the benchmark catalog for numerical field calculations [59] is investigated. The results in the time domain are compared with the ones obtained by an MoM code using dyadic Green's functions in the frequency domain and subsequently applying IFFT for transformation in the time domain. In Figure 4.40 the geometrical configuration and the electric excitation are shown. The following parameters are used for calculation: $H = 0.4\,\text{mm}$, $\varepsilon_r = 6$, $W = 0.1\,\text{mm}$, $T = 10\,\mu\text{m}$, $\sigma = 5.8 \times 10^7\,\text{S/m}$, $D = 0.5\,\text{mm}$. The delay line has been modeled by PEEC in two variants: with and without vias. The results of the simulation are presented in Figure 4.41 The MoM/IFFT and the PEEC solutions correspond very well. A minor influence of the vias can be made out.

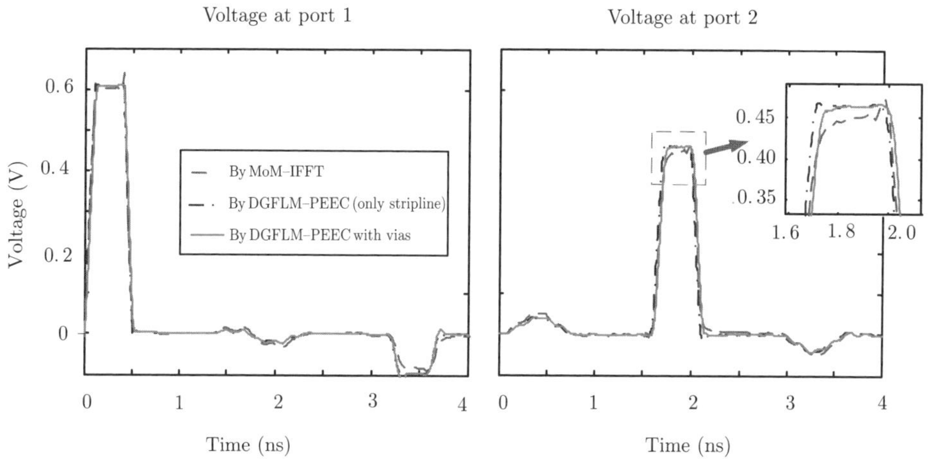

Figure 4.41 Voltages at ports 1 and 2 calculated by DGFLM–PEEC and MoM/IFFT.

4.9 PEEC Models and Uniform Transmission Lines

The PEEC method provides full wave models for geometrically arbitrary wiring structures, of course, with an appropriate expense for modeling and simulation. In contrast, the Transmission-line (TL) classical theory is simple and computationally very effective; however, it requires stringent restrictions concerning the structure to be modeled. Generally, real interconnection structures are complex and are composed of both simple and complicated structures. Therefore, in this section the attention is focused on applying PEEC models to uniform transmission lines with the aim of simplifying the PEEC models, of discovering similarities between PEEC and TL models and using them for reducing the computational expense.

Let an infinitely long, perfectly conducting, straight wire with circular cross section (radius a) running with the constant distance h above an infinite, perfectly conducting ground plane (see Figure 4.42) be considered. Additionally it is required that

$$a \ll 2h \quad \text{and} \quad 2h \ll \lambda_{\text{min}} \tag{4.219}$$

(thin wire approximation) with λ_{min} as the wave length of the maximum frequency of interest. $\underline{\mathbf{E}}^i$ and $\mathbf{k}$ characterize an incoming electromagnetic plane wave exciting the wire. Applying the DGF–MPI Equation(4.32) for the plane half space to the geometry of the configuration in Figure 4.42 and considering the boundary condition $\underline{E}_z\,(h, 0, z) = 0$ for $-\infty < z < \infty$, one obtains

$$\underline{E}_z^{\text{inc}} = j\omega \underline{A}_z(z) + \nabla_z \underline{\varphi}(z), \tag{4.220}$$

where $\underline{E}_z^{\text{inc}}(z) = \underline{E}_z^{\text{i}}(z) + \underline{E}_z^{\text{r}}(z)$ is the incident electric field strength on the wire, which is composed of the incoming and reflected waves (in absence of the wire).

The electromagnetic potentials on the wire $(h, 0, z)$ are given by

$$\underline{A}_z(z) = \mu_0 \int_{-\infty}^{\infty} g_{\text{H}}\left(z, z'\right) \underline{I}(z')\text{d}z', \tag{4.221}$$

$$\underline{\varphi}(z) = \frac{1}{\varepsilon_0} \int_{-\infty}^{\infty} g_{\text{H}}\left(z, z'\right) \underline{q}(z')\text{d}z', \tag{4.222}$$

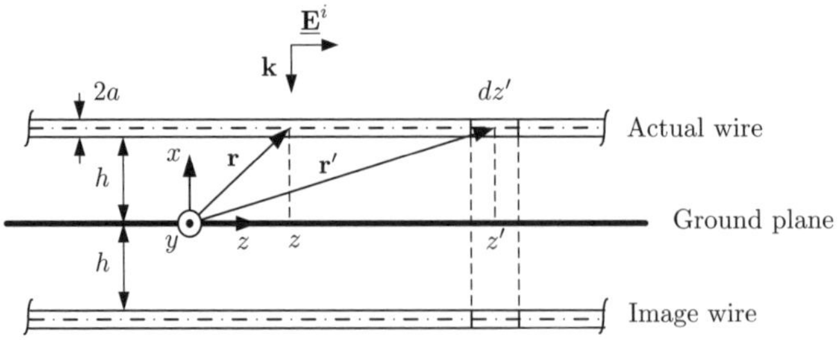

Figure 4.42 Infinitely long, straight wire above ground plane.

with

$$g_{\mathrm{H}}\left(z, z'\right) = \frac{1}{4\pi}\left[\frac{\exp(-jkR_1)}{R_1} - \frac{\exp(-jkR_2)}{R_2}\right] \tag{4.223}$$

and

$$R_1 = \sqrt{(z - z')^2 + a^2} \quad , \quad R_2 = \sqrt{(z - z')^2 + 4h^2}.$$

The analysis of the Green's function Equation (4.223) under the condition of Equation (4.219) results in $\Im\left\{g_{\mathrm{H}}\left(z, z'\right)\right\} \approx 0$ and, consequently, the complex Green's function $g_{\mathrm{H}}\left(z, z'\right)$ may be replaced by the static one

$$g_{\mathrm{H}}^{\mathrm{s}}\left(z, z'\right) = \frac{1}{4\pi}\left[(1/R_1) - (1/R_2)\right], \tag{4.224}$$

that is the retardation is negligible. Further, the analysis shows a high selectivity of Equation (4.224) (see Figure 4.43).

In the small region $-2h \leq z - z' \leq +2h$ the current and charge distributions on the wire can be assumed to be constant and pulled out of the integrals in Equations (4.221) and (4.222):

$$\underline{A}_z(z) = \mu_0 \int_{-\infty}^{+\infty} g_{\mathrm{H}}^{\mathrm{s}}\left(z, z'\right) \mathrm{d}z' \, \underline{I}(z'), \tag{4.225}$$

$$\underline{\varphi}(z) = \frac{1}{\varepsilon_0} \int_{-\infty}^{+\infty} g_{\mathrm{H}}^{\mathrm{s}}\left(z, z'\right) \mathrm{d}z' \, \underline{q}(z'). \tag{4.226}$$

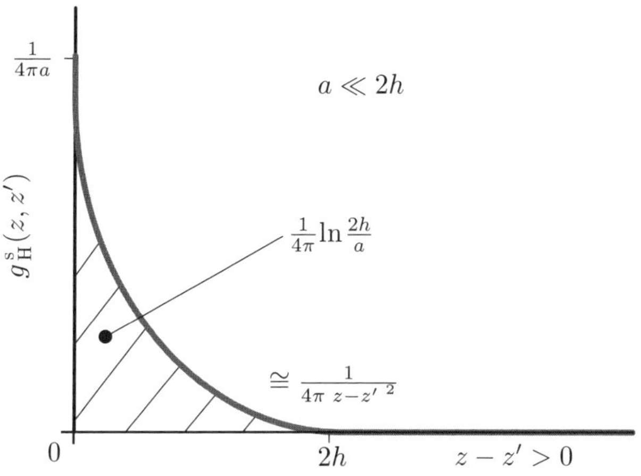

Figure 4.43 Approximate representation of $g_{\mathrm{H}}^{\mathrm{s}}(z, z')$ for $z - z' > 0$.

With

$$\int_{-\infty}^{+\infty} g_{\mathrm{H}}^{\mathrm{s}}\left(z, z'\right) \mathrm{d}z' = \frac{1}{2\pi} \ln \frac{2h}{a} \tag{4.227}$$

one may introduce L' as inductance and C' as capacitance per unit length:

$$\begin{aligned} L' &= \mu_0 \int_{-\infty}^{+\infty} g_{\mathrm{H}}^{\mathrm{s}}\left(z, z'\right) \mathrm{d}z' = \frac{\mu_0}{2\pi} \ln \frac{2h}{a}, \\ C' &= \left[\frac{1}{\varepsilon_0} \int_{-\infty}^{+\infty} g_{\mathrm{H}}^{\mathrm{s}}\left(z, z'\right) \mathrm{d}z'\right]^{-1} = \frac{2\pi\varepsilon_0}{\ln \frac{2h}{a}}. \end{aligned} \tag{4.228}$$

Introducing Equations (4.225) to (4.228) into Equation (4.220) yields

$$j\omega L' \underline{I}(z) + \frac{\mathrm{d}}{\mathrm{d}z} \underline{U}(z) = \underline{E}_z^{\mathrm{inc}}(z). \tag{4.229}$$

Since the magnetic vector potential has only a z component, the electric potential $\underline{\varphi}(z)$ can be replaced by the voltage $\underline{U}(z)$ between wire and ground. The continuity Equation (4.16), applied to the wire element, results in

$$j\omega C' \underline{U}(z) + \frac{\mathrm{d}}{\mathrm{d}z} \underline{I}(z) = 0. \tag{4.230}$$

Equations (4.229) and (4.230) are the classical TL equations for electric field excitation according to Agrawal *et al.*[60]. Because of the infinite length of the transmission line considered, L' and C' do not depend on z. However, for a semi-infinite line, for example for $0 \leq z' < +\infty$, in a small region of $z \geq 0$ a nonuniformity arises:

$$L'(z) = \mu_0 \int_0^{\infty} g_{\mathrm{H}}^{\mathrm{s}}\left(z, z'\right) \mathrm{d}z' = \frac{\mu_0}{4\pi} \ln \frac{\sqrt{z^2 + 4h^2} - z}{\sqrt{z^2 + a^2} - z}, \tag{4.231}$$

$$P'(z) = \frac{1}{C'(z)} = \frac{1}{\varepsilon_0} \int_0^{\infty} g_{\mathrm{H}}^{\mathrm{s}}\left(z, z'\right) \mathrm{d}z' = \frac{1}{4\pi\varepsilon_0} \ln \frac{\sqrt{z^2 + 4h^2} - z}{\sqrt{z^2 + a^2} - z} \tag{4.232}$$

$$L'(0) = \frac{L'(\infty)}{2} \quad \text{and} \quad P'(0) = \frac{1}{C'(0)} = \frac{P'(\infty)}{2}$$

(see Figure 4.44). For a line of finite length such a nonuniformity occurs at both ends of the line. The classical TL Equations (4.229) and (4.230) do not take into account such nonuniformities. There have been many attempts to extend and generalize the TL equations for application to nonuniform configurations. Chapter 2 is dedicated to this issue systematically and thoroughly.

Now, attention is focused on PEEC models derived under the conditions of Equation (4.219) for uniform transmission lines. Because of the high selectivity of the Green's function $g_H\left(z, z'\right)$ and its replacing by the static one $g_{\mathrm{H}}^{\mathrm{s}}\left(z, z'\right)$ (Equation (4.224)) one obtains quasi-static PEEC

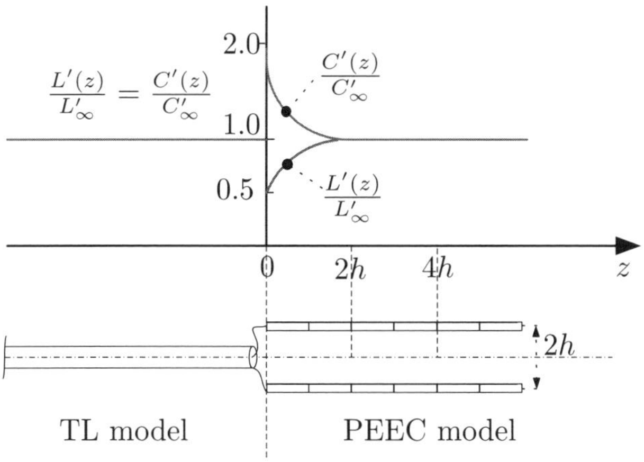

Figure 4.44 Discontinuity at the interface between PEEC and TL models.

models with partial elements:

$$L_{\mathrm{p}\alpha,n} = \int_{l_\alpha} \left[\frac{\mu_0}{4\pi} \int_{l_n} g_{\mathrm{H}}^{\mathrm{s}}\left(z, z'\right) \mathrm{d}z' \right] \mathrm{d}z, \quad \alpha, n \in N \tag{4.233}$$

$$P_{i,m} = \int_{l_i} \left[\frac{1}{\varepsilon_0} \int_{l_m} g_{\mathrm{H}}^{\mathrm{s}}\left(z, z'\right) \mathrm{d}z' \right] \mathrm{d}z, \quad i, n \in N. \tag{4.234}$$

Since Equations (4.233) and (4.234) have the same structure, the interpretations are focussed only on one, for example the inductances. The inner integral of Equation (4.233) represents the contribution of cell n to the inductance per unit length at the point z belonging to the cell α. The outer integral of Equation (4.233) summarizes all these contributions to $L_{\mathrm{p}\alpha,n}$.

If the selectivity of the Green's function is much smaller than the discretization length l, that is

$$2h \ll l_\alpha, \tag{4.235}$$

the contributions of the neighbor elements of cell α are negligible because of

$$L_{\mathrm{p}_{\alpha,\alpha}} \gg L_{\mathrm{p}\alpha,\alpha\pm1}, \tag{4.236}$$

and the actual inductance of cell α is given by

$$L_{\mathrm{p}\alpha} \approx L_{\mathrm{p}\alpha,\alpha} \approx L' l_\alpha. \tag{4.237}$$

For this extreme case, the quasi-static PEEC circuit degrades into an LC circuit with L and P matrices having only diagonal entries. For example in the case of $2h = l_\alpha$, the neighbor element on both sides contribute to the inductance:

$$L_{\mathrm{p}\alpha} \approx L_{\mathrm{p}\alpha,\alpha} + 2L_{\mathrm{p}\alpha,\alpha\pm1} \approx L' \cdot l_\alpha. \tag{4.238}$$

Of course, the values of L' in Equations (4.237) and (4.238) are different. Increasing the frequency for a given geometrical wire configuration leads to a weakening of the selectivity of the Green's function in relation to the discretization length. In consequence, more and more couplings have to be considered, which means that more and more controlled sources have to be introduced in the PEEC models and the appropriate $\mathbf{L}_\mathrm{p}-$ and $\mathbf{P}-$ matrices are filled up with more and more off-diagonal elements. For very high frequencies, if the condition of Equation (4.219) is no longer fulfilled, the quasi-static model even commutes into a full wave model with retardations.

Based on this inside, in [61] and [62] a method is developed starting from a PEEC model of a whole interconnection structure and introducing transmission-line features to appropriate substructures to thin out the coupling matrices and, thus, essentially to reduce the computation time without worsening the accurancy. A further interesting problem for reducing the modeling and simulation effort for wiring structures is the hybrid use of PEEC and TL models. For this, one has to overcome the discontinuity resulting from a direct coupling of a PEEC with a TL model even in a uniform region of an actual line (see Figure 4.44). While the parameters of the TL model are constant over the length, the parameters $L'(z)$ and $P'(z) = 1/C'(z)$ on the PEEC side change and fall to half their values at $z = 0$ as shown in Figure 4.44. The reason for this are the missing electromagnetic couplings from the TL side to the PEEC model in a small region on both sides of $z = 0$ according to the selectivity of the Green's function. This discontinuity is the cause of artificial reflections. To smooth this discontinuity two approaches are developed and applied in [63] and [62].

The first one consists in the introduction of so-called 'hidden cells' which represent a virtual extension of the PEEC model for generating the necessary couplings. The second one can be used, if the TL is implemented as an *LC* model, at least near the interface to the PEEC model. In this case, the electric and magnetic couplings from the *LC* elements into the PEEC model near the interface occur undirectionally in the PEEC-like manner. The mutual partial inductances and potential coefficients are involved in the PEEC model and the appropriate sources are controlled by the currents and voltages of the *LC* elements of the TL side. This solution is more flexible than the first one, particularly, if the selectivity of the Green's function becomes weaker and more elements have to be included in the coupling region.

4.10 Power Considerations in PEEC Models

4.10.1 General Remarks

Power calculations constitute a fundamental part of the analysis of electromagnetic systems like antenna systems and interconnection structures. The power distribution within the system, the incident power, the radiated power, but also other parameters like input impedance or radiation efficiency are of great interest for both the design and the EMC analysis.

On the electromagnetic field level the power relations are described by Poynting's law in differential and integral forms well known from textbooks about electromagnetics (e. g. [15]). In particular, for the calculation of the radiated power distribution of antenna structures there are two approaches in the literature: the method of induced EMF (IEMF [15]) and the method of far-field analysis of radiation sources (FARS [58], [64]). The IEMF method is based on the analysis of the near-field close to the wires; the Poynting vector flow through the surface around the wire is calculated. In contrast, with the FARS method the radiated power as the integral over the Poynting vector flow through a spherical surface with a big distance to the radiating system is calculated. Both methods deliver the same results for the radiated power of the entire system. However, they show a different distribution of the power inside the structure, because the IEMF method computes the radiated power and the power interaction between the elements of the structure in the near-field. In each case, the power analysis on the field level is computationally very expensive.

Based on the circuit interpretation of a PEEC model, in the following paragraphs a new approach for the power analysis of electromagnetic structures without field calculations is presented. As Poynting's theorem describes the power balance in the electromagnetic field, Tellegen's theorem (e.g. [65]) represents the conservation of power for electric circuits. In the frequency domain Tellegen's theorem can be written in the form

$$\underline{S} = P + jQ = \mathbf{I}^{*T}\mathbf{U} = 0. \tag{4.239}$$

In Equation (4.239) are: $\underline{S}$ complex power, P active power, Q reactive power $\mathbf{I}^*$ vector of the branch currents (conjugate complex, effective values) and $\mathbf{U}$ vector of the branch voltages (complex effective values). The PEEC model to be analyzed consists of N inductive branches, M capacitive branches and, maybe, additional branches with lumped sources for excitation connected to the model externally. The excitation by an incident electric field $\underline{\mathbf{E}}^{\text{inc}}$ is already considered as lumped voltage sources $\underline{U}_\alpha^{\text{inc}}$ ($\alpha \in N$) in the inductive branches of the PEEC model.

The new approach to be presented provides not only the opportunity for fast computation of the different kinds of power (e. g. incident power, radiation power, transfer power and power losses) and their distribution over the structure, but also allows the mechanism of the generation of radiation power and its representation on circuit level to be discovered.

4.10.2 Power Analysis of Magnetic and Electric Couplings

For analysis, a standard PEEC model for conductors with center-to-center retardation according to Section 4.3.2.2 is considered. The characteristic of PEEC models are the inductive and capacitive couplings with retardation represented by appropriate controlled sources. At first, let the focus directed at inductive couplings, in particular, on an elementary inductive coupled pair (α, n) with $\alpha, n \in N$ (see Figure 4.10). The voltage–current equations may be written in two-port form:

$$\begin{bmatrix} \underline{U}_{\alpha(n)} \\ \underline{U}_{n(\alpha)} \end{bmatrix} = \begin{bmatrix} 0 & \underline{Z}_{\alpha n} \\ \underline{Z}_{n\alpha} & 0 \end{bmatrix} \begin{bmatrix} \underline{I}_\alpha \\ \underline{I}_n \end{bmatrix} =$$
$$= \omega L_{\mathrm{p}\alpha,n} \begin{bmatrix} 0 & \sin\omega\tau_{\alpha n} + j\cos\omega\tau_{\alpha n} \\ \sin\omega\tau_{n\alpha} + j\cos\omega\tau_{n\alpha} & 0 \end{bmatrix} \begin{bmatrix} \underline{I}_\alpha \\ \underline{I}_n \end{bmatrix}. \tag{4.240}$$

Since $L_{\mathrm{p}\alpha,n} = L_{\mathrm{p}n,\alpha}$ and $\tau_{\alpha n} = \tau_{n\alpha}$ it follows that

$$\underline{Z}_{\alpha n} = \underline{Z}_{n\alpha}, \tag{4.241}$$

that is the matrix $\mathbf{Z}$ of the mutual couplings is reciprocal. For a quasi-static PEEC, when $\tau_{\alpha n} = 0$, the mutual impedance $\underline{Z}_{\alpha n}$ is purely imaginary and Equation (4.240) describes a transformer coupling as is usually applied in quasi-static circuit theory and technique. Moving the inductive cells α and n (or appropriate transformer coils) away from each other and considering the retardation $\tau_{\alpha n}$ leads to an increase in the real part of the transfer impedance $\Re\left\{\underline{Z}_{\alpha n}\right\} = \omega L_{\mathrm{p}\alpha,n} \sin \omega\tau_{\alpha n}$, while the imaginary part $\Im\left\{\underline{Z}_{\alpha n}\right\} = \omega L_{\mathrm{p}\alpha,n} \cos \omega\tau_{\alpha n}$ changes with $\cos \omega\tau_{\alpha n}$.

Now, the complex powers of both controlled sources of the the pair (α, n) are calculated. For branch α coupled with branch n (superscript L means 'inductive')

$$\underline{S}^{L}_{\alpha(n)} = P^{L}_{\alpha(n)} + j\,Q^{L}_{\alpha(n)} \tag{4.242}$$

is obtained with

$$P^{L}_{\alpha(n)} = \omega L_{\mathrm{p}\alpha,n} I_\alpha I_n \left[\sin(\varphi_\alpha - \varphi_n)\cos\omega\tau_{\alpha n} + \cos(\varphi_\alpha - \varphi_n)\sin\omega\tau_{\alpha n}\right], \tag{4.243}$$

$$Q^{L}_{\alpha(n)} = \omega L_{\mathrm{p}\alpha,n} I_\alpha I_n \left[\cos(\varphi_\alpha - \varphi_n)\cos\omega\tau_{\alpha n} - \sin(\varphi_\alpha - \varphi_n)\sin\omega\tau_{\alpha n}\right]. \tag{4.244}$$

Analogously, the complex coupling power of branch n coupled with branch α yields

$$\underline{S}^{L}_{n(\alpha)} = P^{L}_{n(\alpha)} + j\,Q^{L}_{n(\alpha)} \tag{4.245}$$

with

$$P^{L}_{n(\alpha)} = \omega L_{\mathrm{p}\alpha,n} I_\alpha I_n \left[-\sin(\varphi_\alpha - \varphi_n)\cos\omega\tau_{\alpha n} + \cos(\varphi_\alpha - \varphi_n)\sin\omega\tau_{\alpha n}\right], \tag{4.246}$$

$$Q^{L}_{n(\alpha)} = \omega L_{\mathrm{p}\alpha,n} I_\alpha I_n \left[\cos(\varphi_\alpha - \varphi_n)\cos\omega\tau_{\alpha n} + \sin(\varphi_\alpha - \varphi_n)\sin\omega\tau_{\alpha n}\right]. \tag{4.247}$$

Let the focus be directed at the active coupling powers $P^{L}_{\alpha(n)}$ (Equation (4.243)) and $P^{L}_{n(\alpha)}$ (Equation (4.246)). A comparison shows that they each consist of two parts:

$$P^{L}_{\alpha(n)} = \mathrm{P}^{L,t}_{\alpha(n)} + P^{L,\mathrm{r}}_{\alpha(n)} \quad , \qquad P^{L}_{n(\alpha)} = \mathrm{P}^{L,\mathrm{t}}_{n(\alpha)} + P^{L,\mathrm{r}}_{n(\alpha)} \tag{4.248}$$

with

$$P^{L,\mathrm{t}}_{\alpha(n)} = -P^{L,\mathrm{t}}_{n(\alpha)} = \omega L_{\mathrm{p}\alpha,n} I_\alpha I_n \sin(\varphi_\alpha - \varphi_n)\cos\omega\tau_{\alpha n}, \tag{4.249}$$

$$P^{L,\mathrm{r}}_{\alpha(n)} = P^{L,\mathrm{r}}_{n(\alpha)} = \omega L_{\mathrm{p}\alpha,n} I_\alpha I_n \cos(\varphi_\alpha - \varphi_n)\sin\omega\tau_{\alpha n}. \tag{4.250}$$

The superscripts t and r denote 'transfer' and 'radiation' respectively. In Equation (4.249) $P^{L,\mathrm{t}}_{\alpha(n)} > 0$ means an active power absorbed by the controlled source in branch α. The same amount of active power, however, with the opposite sign $P^{L,\mathrm{t}}_{n(\alpha)} < 0$, is generated from the controlled source in branch n. These relations in the circuit model represent an active power transport between the two branches α and n, which is physically supported by the magnetic field and is well known from transformers in the quasi-static case ($\tau_{\alpha n} = 0$). Because of Equation (4.249), the net contribution of the coupled pair (α, n) to the active transfer power of the whole PEEC model is zero.

While the active transfer powers described above are to a certain extent the differential mode parts of the coupling powers $P^{L}_{\alpha(n)}$ and $P^{L}_{n(\alpha)}$, the active powers according to Equation (4.250) represent the common mode parts of $P^{L}_{\alpha(n)}$ and $P^{L}_{n(\alpha)}$ always with the same signs. These parts $P^{L,\mathrm{r}}_{\alpha(n)} = P^{L,\mathrm{r}}_{n(\alpha)}$ may be interpreted as the contribution of each partner of the coupled pair (α, n) to the radiation power of the whole PEEC model. A necessary condition for arising radiation parts of active powers $P^{L,\mathrm{r}}_{\alpha(n)}$ and $P^{L,\mathrm{r}}_{n(\alpha)}$ is the retardation $\tau_{\alpha n} > 0$.

The analysis of an electric or capacitive coupled pair (i, m) with $i, m \in M$ may be carried through analogously. Thus, for the interesting active powers of the current controlled voltage sources one obtains

$$P^C_{i(m)} = P^{C,\mathrm{t}}_{i(m)} + P^{C,\mathrm{r}}_{i(m)} \quad , \quad P^C_{m(i)} = P^{C,\mathrm{t}}_{m(i)} + P^{C,\mathrm{r}}_{m(i)} \tag{4.251}$$

with

$$P^{C,\mathrm{t}}_{i(m)} = -P^{C,\mathrm{t}}_{m(i)} = -\frac{1}{\omega} P_{i,m} I_i I_m \sin(\varphi_i - \varphi_m) \cos \omega\tau_{im}, \tag{4.252}$$

$$P^{C,\mathrm{r}}_{i(m)} = P^{C,\mathrm{r}}_{m(i)} = -\frac{1}{\omega} P_{i,m} I_i I_m \cos(\varphi_i - \varphi_m) \sin \omega\tau_{im}. \tag{4.253}$$

$P_{i,m}$ on the right-hand side of Equations (4.252) and (4.253) are the potential coefficients, the superscript C denotes the 'capacitive (electric)' coupling. Equation (4.252) describes the active power transfer between the partners of the coupled pair (i, m), whose net contribution to the whole PEEC model is zero. The common mode parts $P^{C,\mathrm{r}}_{i(m)}$ and $P^{C,\mathrm{r}}_{m(i)}$ arising only for a retardation $\tau_{\mathrm{im}} > 0$ represent contributions to the radiation power.

Negative signs of radiation power contributions at the circuit level (e.g. Equation (4.250) and (4.253)) may be interpreted as a result of interferences at the field or wave level. Further, regarding physical interpretations one has to take into account that the partial inductances and potential coefficients are gauge-dependent. PEEC uses the Lorenz gauge (see Section 4.3.2).

4.10.3 Power Analysis of PEEC Models

After considering elementary magnetic and electric couplings and discovering the controlled sources with retardation as origin and representation of radiation power at the circuit level, let us now calculate the power distributions and total powers for a whole PEEC model. For this, one can use the well-known relations for power calculation in electric circuits with lumped elements in connection with Tellegen's theorem (Equation (4.239)).

The total active coupling power of an inductive cell α ($\alpha \in N$) is given by

$$P^L_\alpha = \sum_{\substack{n=1 \\ n\neq\alpha}}^{N} \left(P^{L,\mathrm{t}}_{\alpha(n)} + P^{L,\mathrm{r}}_{\alpha(n)} \right), \tag{4.254}$$

where $P^{L,t}_{\alpha(n)}$ and $P^{L,r}_{\alpha(n)}$) are calculated from Equations (4.249) and (4.250) respectively.

Analogously, the total active coupling power of a capacitive cell i ($i \in M$) yields

$$P^C_i = \sum_{\substack{m=1 \\ m\neq i}}^{M} \left(P^{C,\mathrm{t}}_{i(m)} + P^{C,\mathrm{r}}_{i(m)} \right) \tag{4.255}$$

with $P^{C,\mathrm{t}}_{i(m)}$ and $P^{C,\mathrm{r}}_{i(m)}$ according to Equations (4.252) and (4.253) respectively.

The total active incident power absorbed from the incident electric field by the whole stucture results in

$$P^{\mathrm{inc}}_{\mathrm{total}} = \sum_{\alpha=1}^{N} \Re\left\{ \underline{I}^*_\alpha \underline{U}^{\mathrm{inc}}_\alpha \right\} \tag{4.256}$$

and is composed of the contributions of all N inductive cells.

The total ohmic losses of the PEEC model are given by the contributions of all N inductive cells as follows (lossless dielectrics are assumed):

$$P_{\text{total}}^{\Omega} = \sum_{\alpha=1}^{N} R_{\alpha} I_{\alpha}^{2}. \tag{4.257}$$

Since for the total active transfer power of the whole PEEC model yields

$$P_{\text{total}}^{L,\text{t}} = \sum_{\alpha=1}^{N} \sum_{\substack{n=1 \\ n\neq\alpha}}^{N} P_{\alpha(n)}^{L,\text{t}} = 0, \tag{4.258}$$

$$P_{\text{total}}^{C,\text{t}} = \sum_{i=1}^{M} \sum_{\substack{m=1 \\ m\neq i}}^{M} P_{i(m)}^{C,\text{t}} = 0, \tag{4.259}$$

the total active coupling power of the whole structure $P_{\text{total}}^{\text{coup}}$ is only composed by the radiation parts $P_{\alpha(n)}^{L,\text{r}}$ and $P_{i(m)}^{C,\text{r}}$ of all couplings and represents $P_{\text{total}}^{\text{r}}$, the total power radiated into the far-field:

$$\begin{aligned} P_{\text{total}}^{\text{r}} = P_{\text{total}}^{\text{coup}} &= \sum_{\alpha=1}^{N} P_{\alpha}^{L} + \sum_{i=1}^{M} P_{i}^{C} \\ &= \omega \sum_{\alpha=1}^{N} \sum_{\substack{n=1 \\ n\neq\alpha}}^{N} L_{\text{p}\alpha,n} I_{\alpha} I_{n} \cos\left(\varphi_{\alpha} - \varphi_{n}\right) \sin \omega\tau_{\alpha n} \\ &\quad - \frac{1}{\omega} \sum_{i=1}^{M} \sum_{\substack{m=1 \\ m\neq i}}^{M} P_{i,m} I_{i} I_{m} \cos\left(\varphi_{i} - \varphi_{m}\right) \sin \omega\tau_{im}. \end{aligned} \tag{4.260}$$

Applying Tellegen's theorem (Equation (4.239)), the balance of active powers of the whole PEEC model can be formulated:

$$P_{\text{total}}^{\text{inc}} + P_{\text{total}}^{\Omega} + P_{\text{total}}^{\text{r}} = 0. \tag{4.261}$$

If the PEEC model is embedded in a circuit environment, Equation (4.261) is to be extended by $P_{\text{total}}^{\text{src}}$ as the active power of all lumped sources and by $P_{\text{total}}^{\text{load}}$ as the active power of all load impedances.

In addition to the fast computation of power based on the known branch currents the PEEC method also allows the determination of the spatial power distribution within and the total power of the structure. A further advantage is to distinguish between ohmic and radiation losses and to differentiate the coupling power into transfer and radiation powers.

The method presented is especially advantageously applicable to the analysis of electromagnetic coupled structues, for example coupled antennas [66]. The total radiated far-field power calculated by the PEEC method agrees fully with the ones obtained by the FARS and IEMF methods.

It is worth remembering that the input impedance relating to any port of a circuit can also be calculated by means of power according to

$$\underline{Z}_{\text{in}} = \frac{\underline{U}_{\text{in}}}{\underline{I}_{\text{in}}} = R_{\text{in}} + jX_{\text{in}} = \frac{1}{I_{\text{in}}^2}(P_{\text{total}} + jQ_{\text{total}}) \tag{4.262}$$

with P_{total} and Q_{total} as the powers of all passive elements and controlled sources of the PEEC model. The real part of $\underline{Z}_{\text{in}}$ of a structure (e. g. of an antenna) can be expressed by the powers calculated as before:

$$R_{\text{in}} = \Re\left\{\underline{Z}_{\text{in}}\right\} = \frac{1}{I_{\text{in}}^2}(P_{\text{total}}^{\Omega} + P_{\text{total}}^{L,r} + P_{\text{total}}^{C,r}). \tag{4.263}$$

Equation (4.263) shows that the input resistance of a radiating structure is composed of an ohmic part characterizing the conductor losses as well as an inductive and a capacitive part representing the radiation losses (radiated far-field power). Such relations, however, in closed form for the special case of a simple thin circular conductor loop, were already formulated by Wessel in 1936 [67].

Also, the radiation efficiency η_{rad} of an antenna-like structure defined as

$$\eta_{\text{rad}} = \frac{P_{\text{total}}^{\text{r}}}{P_{\text{total}}^{\Omega} + P_{\text{total}}^{\text{r}}} \tag{4.264}$$

may easily be calculated.

References

[1] Felsen, L. B. and Marcowitz, N. *Radiation and Scattering of Waves*, IEEE Press, (1994).

[2] Tai, C. T. *Dyadic Green's Functions in Electromagnetic Theory*. USA, IEEE Press, Piscataway, NJ, 1997.

[3] Michalski, K. A. 'Mixed Potential Electric Field Integral Equation for Objects in Layered Media'. *Archiv Elektr. Übertragung*, **39**, 1985, 317–322.

[4] Michalski, K. A. and Mosig, J. R. 'Multilayered Media Green's Functions in Integral Equation Formulation'. *IEEE Transactions on Antennas and Propagation*, **45**, 1997, 508–519.

[5] Kochetov, S. V., Leone, M. and Wollenberg, G. 'PEEC Formulation Based on Dyadic Green's Functions for Layered Media in Time and Frequency Domain'. *IEEE Trans. on EMC*, **50**, 2008, 953–965.

[6] Ruehli, A. E. and Heeb, H. 'Circuit Models for Three-Dimensional Geometries Including Dielectrics'. *IEEE Transactions on Microwave Theory and Techniques*, **40**, 1992, 1507–1516.

[7] Kochetov, S. V. and Wollenberg, G. 'Stability of Full-Wave PEEC Models: Reason for Instabilities and Approach for Correction'. *IEEE Transactions on EMC*, **47**, 2005, 738–748.

[8] Kochetov, S. V. and Wollenberg, G. 'PEEC Models with Multipoint Approximations of Derivatives'. 28th General Assembly of the International Union of Radio Science, September 23–29, 2005, New-Dehli, India.

[9] Paul, C. R. *Analysis of Multiconductor Transmission Lines*, John Wiley and Sons, Inc., New York, USA, 1994.

[10] Chiprout, E. and Nakhla, M. S. *Asymptotic Waveform Evaluation*. Kluwer, The Netherlands, 1994.

[11] Mei, K. K. 'Theory of Maxwellian Circuits'. *Radio Science Bulletin*, **305**, 2003, 6–13.

[12] Thamm, S., Kochetov, S. V., Wollenberg, G. and Leone, M. 'Alternative PEEC Modeling with Partial Reluctances and Capacitances for Power Electronics Applications'. Proceedings of the 7th International Symposium on EMC, Saint- Petersburg, Russia, 2007, 56–59.

[13] Scott, K. J. *Practical Simulation of Printed Circuit Boards and Related Structures*, John Wiley and Sons, Inc., New York, USA, 1994.

[14] Krauder, B. and Pileggi, L. T. 'Generating Sparse Partial Inductance Matrices with Guaranteed Stability'. Proceedings of the International Conference on Computer Aided Design, November 1995.

[15] Balanis, C. A. *Advanced Engineering Electromagnetics*, John Wiley and Sons, Inc., New York, USA, 1989.

[16] Hoer, C. and Love, C. 'Exact Inductance Equations for Rectangular Conductors with Applications to More Complicated Geometries'. J. Res. Natl. Bur. Stand., **69C**, 1965, 127–137.

[17] Ruehli, A. E. 'Inductance Calculations in a Complex Integrated Circuit Environment'. *IBM J. Res. Dev.*, **16** 1972, 470–481.

[18] Ruehli, A. E. and Brennan, P. A. 'Efficient Capacitance Calculations for Three-Dimensional Multiconductor Systems'. *IEEE Trans. MTT*, **21**, 1973, 76–82.

[19] Rosa, E. B. and Grover, F. W. 'Formulas and Tables for the Calculation of Mutual and Self-Induction'. *Bull. Bur. Stand.,* **8**, No. 1, 1911.

[20] Antonini, G., Ruehli, A. E. and Esch, J. 'Nonorthogonal PEEC Formulation for Time and Frequency Domain Modeling'. Proceedings of IEEE International Symposium on Electromagnetic Compatibility, Minneapolis, MN, USA, August 2002, pp. 452–456.

[21] Ruehli, A. E., Antonini, G. and Orlandi, A. 'Extension of the Partial Element Equivalent Circuit Method to Non Rectangular Geometries'. Proceedings of IEEE International Symposium on Electromagnetic Compatibility, Seattle, WA, USA, August 1999, pp. 728–732.

[22] Ruehli, A. E., Antonini, G., Esch, J. *et al.*, 'Nonorthogonal PEEC Formulation for Time- and Frequency-Domain EM and Circuit Modeling'. *IEEE Trans. EMC*, **45**, 2003, 167–176.

[23] Antonini, G., Orlandi, A. and Ruehli, A. E. 'Analytical Integration of Quasi-Static Potential Integrals on Nonorthogonal Coplanar Quadrilaterals for the PEEC Method'. *IEEE Trans. EMC*, **44**, 2002, 399–403.

[24] Rao, S. M., Wilton, D. R. and Glisson, A. W. 'Electromagnetic Scattering by Surfaces of Arbitrary Shape'. *IEEE Trans. Antennas and Propagation*, **30**, 1982, 409–418.

[25] Rong, A. and Cangellaris, A. C. 'Generalized PEEC Models for Three-Dimensional Interconnect Stuctures and Integrated Passives of Arbitrary Shapes'. Procedings of the Digital Electrical Performance Electronic Packaging Conference, Vol. 10, Boston, MA, October 2001, pp. 225–228.

[26] Jandhyala, V., Wang, Y., Gope, D. and Shi, R. 'Coupled Electromagnetic-Circuit Simulation of Arbitrarily-Shaped Conducting Strucures using Triangular Meshes'. Proceedings of the IEEE International Symposium on Quality Electronic Design, San Jose, California, USA, 2002, 38–42.

[27] Wang, Y., Jandhyala, V. and Shi, R. 'Coupled Electromagnetic-Circuit Simulation of Arbitrarily-Shaped Conducting Structures'. Proceedings of the Electrical Performance of Electronic Packaging, Cambridge, MA, October 2001, pp. 233–236.

[28] Wollenberg, G. and Goerisch, A. 'Analysis of 3-D Interconnect Structures with PEEC Using SPICE'. *IEEE Trans. EMC*, **41**, 1999, 412–417.

[29] Ruehli, A. E., Miekkala, U. and Heeb, H. 'Stability of Discretized Partial Element Equivalent EFIE Circuit Models'. *IEEE Trans. Antennas and Propagation*, **43**, 1995, 553–559.

[30] Tai, C. T. *Generalized Vector and Dyadic Analysis: Applied Mathematics in Field Theory*, USA, IEEE Press, Piscataway, NJ, 1997.

[31] Bellen, A., Guglielmi, N. and Ruehli, A. E. 'Methods for Linear Systems of Circuit Delay Differential Equations of Neutral Type'. *IEEE Trans. Circuits and Systems I, Fun. Theory Appl.*, **46**, 1999, 212–215.

[32] Ruehli, A. E., Miekkala, U., Bellen, A. and Heeb, H. 'Stable Time Domain Solutions for EMC Problems using PEEC Circuit Models'. Proceedings of the IEEE International Symposium on EMC, Chicago, Illinois, USA, 1994, pp. 371–376.

[33] Dugard, L. and Verriest, E. J. 'Stability and Control of Time-Delay Systems'. *Lecture Notes in Control and Information Sciences,* Vol. 228, Springer, New York, USA, 1998.

[34] Garrett, J., Ruehli, A. E. and Paul, C. 'Accuracy and Stability Advancements of the Partial Element Equivalent Circuit Model'. Proceedings of the International Symposium on EMC, Zurich, Switzerland, March 1997, pp. 529–534.

[35] Kochetov, S. V. 'Time- and Frequency-Domain Modeling of Passive Interconnection Structures in Field and Circuit Analysis. Habilitation Dissertation, Otto-von-Guericke-University Magdeburg, Germany, 2008.

[36] Kochetov, S. V. and Wollenberg, G. 'Stable and Effective Full-Wave PEEC Models by Full-Spectrum Convolution Macromodeling'. *IEEE Trans*. EMC, **49**, 2007, 25–34.

[37] Antonini, G., Deschrijver, D. and Dhaene, T. 'Broadband Macromodels for Retarded Partial Element Equivalent Circuit (rPEEC) Method'. *IEEE Trans. EMC*, **49**, 2007, 35–48

[38] Chua, L. O. *Computer-Aided Analysis of Electronic Circuits*, Prentice-Hall, Inc., N.J, USA, 1975.

[39] Lambert, J. D. *Numerical Methods for Ordinary Differential Equation Systems: The Initial Value Problem.* John Wiley and Sons, Ltd., Chichester, UK, 1997.

[40] Antonini, G. 'SPICE Compatible Equivalent Circuits of Rational Approximation of Frequency Domain Responses'. *IEEE Trans. EMC*, **45**, 2003, 502–512.
[41] Coperich, K. M., Ruehli, A. E. and Cangellaris, A. 'Enhanced Skin Effect for Partial Element Equivalent Circuit (PEEC) Models'. *IEEE Trans. Microwave Theory and Technique*, **48**, 2000, 1435–1442.
[42] Wollenberg, G. and Kochetov, S. V. 'Modeling the Skin Effect in Wire-Like 3D Interconnection Structures with Arbitrary Cross Section by a New Modification of the PEEC Method'. Proceedings of the 15th International Symposium on EMC, Zurich, Switzerland, February 2003, pp. 609–614.
[43] Ruehli, A., Paul, C. and Garrett, J. 'Inductance Calculations using Partial Inductances and Macromodels'. Proceedings of the IEEE International Symposium on EMC, Atlanta, GA, USA, August 1995, pp. 23–28.
[44] Kamon, M., Marques, L., Silveira, L. M. and White, J. 'Generating Reduced Order Models via PEEC for Capturing Skin and Proximity Effects'. Proceedings of the Electrical Performance of Electronic Packaging Conference, West Point, NY, USA October 1998, pp. 259–262.
[45] Tesche, F. M., Ianoz, M. and Karlsson, T. *EMC Analysis Methods and Computational Models*, John Wiley and Sons, Inc., Hoboken, USA, 1997.
[46] Giacoletto, L. J. 'Frequency and Time-Domain Analysis of Skin Effect'. *IEEE Trans. Mag.* **32**, 1996, 220–229.
[47] Djordhevic, A. R. and Sarkar, T. K. 'Closed-Form Formulas for Frequency-Dependent Resistance and Inductance per Unit Length of Microstrip and Strip Transmission Lines'. *IEEE Trans. Microwave Theory Tech.*, **42**, 1994, 241–248.
[48] Gupta, K. C., Gard, R. and Chada, R. *Computer-Aided Design of Microwave Circuits*, Artech House Inc., Norwood, USA, 1981.
[49] Hoffmannn, R. K. *Handbook of Microwave Integrated Circuits*, Artech House Inc., Norwood, USA, 1987.
[50] Leone, M. and Singer, H. L. 'On the Coupling of an External Electromagnetic Field to a Printed Circuit Board Trace'. *IEEE Trans. EMC*, **41**, 1999, 418–424.
[51] Kochetov, S. V., Wollenberg, G. and Leone, M. 'PEEC Models Based on Dyadic Green's Functions for Structures in Layered Media'. Proceedings of the 7th International Symposium on EMC, Saint-Petersburg, Russia, June 2000, pp.179–182.
[52] Kochetov, S. V., Leone, M. and Wollenberg, G. 'PEEC Formulation Based on Dyadic Green's Functions for Layered Media in Time and Frequency Domain'. *IEEE Trans. EMC*, **50**, 2008, 953–965.
[53] Sommerfeld, A. *Partial Differential Equations*, Academic Press, New York, USA, 1949.
[54] Sadiku, M. N. *Numerical Methods in Electromagnetics*, 2nd edn., CRC Press, Boca Raton, USA, 2000.
[55] Hall, S. H., Hall, G. and McCall, J. *High-Speed Digital System Design*, John Wiley and Sons, Inc., New York, USA, 2000.
[56] Ge, Y. and Essele, K. 'New Closed-Form Green's Functions for Microstrip Structures - Theory and Results'. *IEEE Trans. Microw. Theory and Techn.*, **50**, 2002, 1556–1560.
[57] Hildebrand, F. B. *Introduction to Numerical Analysis*. Unabridged Dover, New York, USA, 1987.
[58] Miller, E. K. 'PCs for AP and other EM Reflection'. *IEEE Antennas and Propagation Mag.*, **41** 1999, 82–86.
[59] Benchmark Catalog for Numerical Field Calculations in the Area of EMC. Benchmark Catalog of the German IEEE/EMC Chapter, http://www.ewh.ieee.org./r8/germany/emc/ag-num/benchmark.html
[60] Agrawal, A. K., Price, H. J. and Gurbaxani, S. H. ' Transient Response of Multiconductor Transmission Lines Excited by a Nonuniform Electromagnetic Field'. *IEEE Trans. EMC*, **22**, 1980, 119–129.
[61] Goerisch, A. and Wollenberg, G. 'PEEC Models for Interconnection Structures Considering Features of Transmission Lines'. Proceedings of the International Symposium on EMC, Zurich, Switzerland, February 2001, pp. 531–536.
[62] Goerisch, A. 'Netzwerkorientierte Modellierung und Simulation elektrischer Verbindungsstrukturen mit der Methode der partiellen Elemente'. Dissertation, Otto-von-Guericke University Magdeburg, Germany, 2002.
[63] Wollenberg, G. and Goerisch, A. 'Coupling of PEEC with Transmission Line Models of Wiring Structures'. Proceedings of the IEEE International Symposium on EMC, Seattle, WA, USA, 1999, pp. 497–502.
[64] Miller, E. K. 'Further Investigation using Far-Field Analysis of Radiation Sources (FARS)'. Proceedings of the IEEE International Symposium on Antennas and Propagation, Vol. 38, July 2000, pp. 1546–1549.
[65] Simonyi, K. *Theoretische Elektrotechnik*, Deutscher Verlag der Wissenschaften, Berlin, Germany, 1977.

[66] Wollenberg, G. and Kochetov, S. V. 'Fast Computation of Radiated Power Distribution in Coupled Wire Systems by the PEEC Method'. Proceedings of the IEEE International Symposium on EMC, Istanbul, Turkey, May 2003, pp. 1152–1155.

[67] Wessel, W. 'Über den Einfluss des Verschiebungsstromes auf den Wechselstromwiderstand'. *Zeitschrift für Technische Physik*, 1936, pp. 472–475.

[68] Abramovitz, M. and Stegun, I.A, *Handbook of Mathematical Functions*, Dover, New York, USA, 1965.

Appendix A: Tensor Analysis, Integration and Lie Derivative

Within a theoretical formulation physical quantities are modeled as mathematical objects. The understanding and application of appropriate mathematics yields, in turn, the properties of physical quantities. In the development of the axiomatic approach, repeated use has been made of integration, of the Poincaré lemma, and of the Stokes' theorem. With the use of these mathematical concepts it is straightforward to derive the basics of electromagnetism from a small number of axioms.

Integration is an operation that yields coordinate independent values. It requires an integration measure, the dimension of which depends on the type of region that is being integrated over. Integration is needed over one-dimensional curves, two-dimensional surfaces or three-dimensional volumes that are embedded in three-dimensional space. Therefore, line, surface, and volume elements have to be defined as integration measures. Then one can think of suitable objects as integrands that can be integrated to yield coordinate independent physical quantities.

A.1 Integration Over a Curve and Covariant Vectors as Line Integrands

To begin with, a one-dimensional curve $\boldsymbol{c} = \boldsymbol{c}(t)$ in three-dimensional space is considered. In a specific coordinate system x^i, with indices $i = 1, 2, 3$, a parameterization of $\boldsymbol{c}$ is given by the vector

$$\boldsymbol{c}(t) = (c^1(t), c^2(t), c^3(t))\,. \tag{A.1}$$

The functions $c^i(t)$ define the shape of the curve. For small changes of the parameter t, with $t \to t + \Delta t$, the difference vector between $\boldsymbol{c}(t + \Delta t)$ and $\boldsymbol{c}(t)$ is given by

$$\Delta \boldsymbol{c}(t) = \left(\frac{\Delta c^1}{\Delta t}, \frac{\Delta c^2}{\Delta t}, \frac{\Delta c^3}{\Delta t} \right) \Delta t\,, \tag{A.2}$$

Radiating Nonuniform Transmission-Line Systems and the Partial Element Equivalent Circuit Method Jürgen Nitsch, Frank Gronwald, and Günter Wollenberg

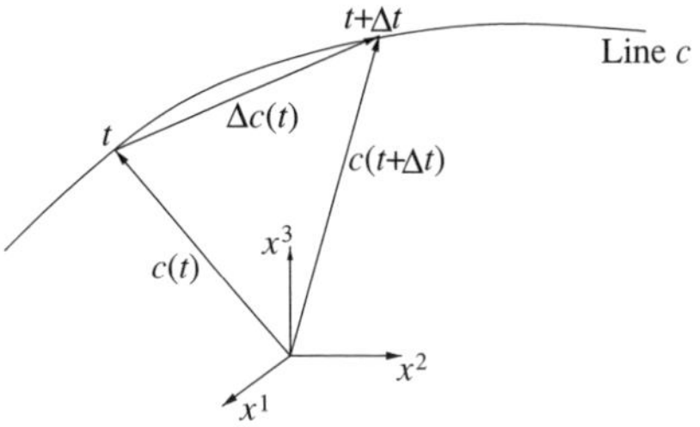

Figure A.1 Parameterization of a curve $\boldsymbol{c}(t)$. The difference vector $\Delta\boldsymbol{c}(t)$ between $\boldsymbol{c}(t+\Delta t)$ and $\boldsymbol{c}(t)$ yields, in the limit $\Delta t \to 0$, the line element $\mathrm{d}\boldsymbol{c}(t)$.

(see Figure A.1). In the limit where Δt becomes infinitesimal one obtains the line element

$$\begin{aligned}\mathrm{d}\boldsymbol{c}(t) &= (\mathrm{d}c^1(t), \mathrm{d}c^2(t), \mathrm{d}c^3(t)) \\ &:= \left(\frac{\partial c^1(t)}{\partial t}, \frac{\partial c^2(t)}{\partial t}, \frac{\partial c^3(t)}{\partial t}\right) \mathrm{d}t \end{aligned} \tag{A.3}$$

which is characterized by an infinitesimal length and an orientation.

The next step is the construction of objects that can be integrated along the curve $\boldsymbol{c}$ in order to obtain a coordinate invariant scalar. The line element $\mathrm{d}\boldsymbol{c}$ contains three independent components $\mathrm{d}c^i$. If one shifts from old coordinates x^i to new coordinates $y^{j'} = y^{j'}(x^i)$ these components transform according to

$$\mathrm{d}c^{j'} = \frac{\partial y^{j'}}{\partial x^i}\mathrm{d}c^i \,. \tag{A.4}$$

Therefore, invariant expression can be formed if one introduces objects $\boldsymbol{\alpha} = \boldsymbol{\alpha}(x^i)$, with three independent components α_i, that transform in the opposite way,

$$\alpha_{j'} = \frac{\partial x^i}{\partial y^{j'}}\alpha_i \,. \tag{A.5}$$

This transformation behavior characterizes a vector or, more precisely, a covariant vector (a 1-form). It follows that the expression

$$\alpha_i \, \mathrm{d}c^i = \alpha_{j'} \, \mathrm{d}c^{j'} \tag{A.6}$$

yields the same value in each coordinate system.

Thus, it is immediately possible to define integration over a curve by the expression

$$\begin{aligned}\int \alpha_i \, \mathrm{d}c^i &= \int \alpha_1 \, \mathrm{d}c^1 + \alpha_2 \, \mathrm{d}c^2 + \alpha_3 \, dc^3 \\ &= \int \left(\alpha_1 \frac{\partial c^1}{\partial t} + \alpha_2 \frac{\partial c^2}{\partial t} + \alpha_3 \frac{\partial c^3}{\partial t}\right) \mathrm{d}t \,. \end{aligned} \tag{A.7}$$

The last line explicitly shows how to carry out the integration since α_i and c^i are functions of the parameter t.

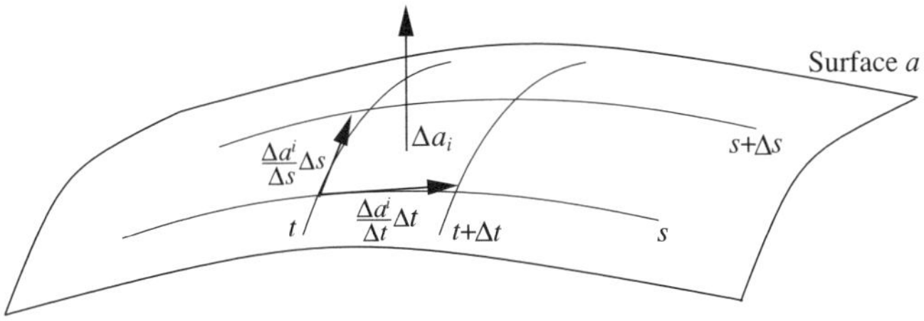

Figure A.2 Parameterization of a surface $\boldsymbol{a}(t, s)$. The lines t =const., $t + \Delta t$ =const., s =const. and $s + \Delta s$ =const. circumscribe a surface Δa_i which is spanned by the edges $\frac{\Delta a^i}{\Delta t}\,\mathrm{d}t$ and $\frac{\Delta a^i}{\Delta s}\,\mathrm{d}s$. In the limit $\Delta t \to \mathrm{d}t$, $\Delta s \to \mathrm{d}s$, it becomes an elementary surface element $\mathrm{d}a_i$.

A.2 Integration Over a Surface and Contravariant Vector Densities as Surface Integrands

Now a two-dimensional surface $\boldsymbol{a} = \boldsymbol{a}(t, s)$ is considered. Within a specific coordinate system x^i, a parameterization of $\boldsymbol{a}$ is of the form

$$\boldsymbol{a}(t, s) = (a^1(t, s), a^2(t, s), a^3(t, s)) \tag{A.8}$$

with parameters t, s and functions $a^i(t, s)$ which define the shape of the surface.

An elementary surface element is bound by lines t =const., $t + \mathrm{d}t$ =const., s =const., and $s + \mathrm{d}s$ =const. (see Figure A.2). It is characterized by the two edges $\frac{\partial a^i}{\partial t}\,\mathrm{d}t$ and $\frac{\partial a^i}{\partial s}\,\mathrm{d}s$. These edges span an infinitesimal surface, the area and orientation of which is characterized by a covariant vector $\mathrm{d}a_i$ that is normal to the infinitesimal surface. The vector $\mathrm{d}a_i$ is given by the vector product of $\frac{\partial a^i}{\partial t}\,\mathrm{d}t$ and $\frac{\partial a^i}{\partial s}\,\mathrm{d}s$,

$$\mathrm{d}a_i = \epsilon_{ijk}\frac{\partial a^j}{\partial t}\frac{\partial a^k}{\partial s}\,\mathrm{d}t\,\mathrm{d}s\,. \tag{A.9}$$

In order to know how the components $\mathrm{d}a_i$ transform under coordinate transformations $y^{j'} = y^{j'}(x^i)$, it is necessary to know the transformation behavior of the symbol ϵ_{ijk}. Since in any coordinate system ϵ_{ijk} assumes the values 0, 1 or -1 by definition it is obvious that in general

$$\epsilon_{i'j'k'} \neq \frac{\partial x^i}{\partial y^{i'}}\frac{\partial x^j}{\partial y^{j'}}\frac{\partial x^k}{\partial y^{k'}}\epsilon_{ijk}\,. \tag{A.10}$$

This is because the determinant of the transformation matrix, that is

$$\det(\partial x/\partial y) = \epsilon_{ijk}\frac{\partial x^i}{\partial y^{i'}}\frac{\partial x^j}{\partial y^{j'}}\frac{\partial x^k}{\partial y^{k'}}\,, \tag{A.11}$$

is, in general, not equal to one. However, it follows from Equation (A.11) that the correct transformation rule for ϵ_{ijk} is given by

$$\epsilon_{i'j'k'} = \frac{1}{\det(\partial x/\partial y)}\frac{\partial x^i}{\partial y^{i'}}\frac{\partial x^j}{\partial y^{j'}}\frac{\partial x^k}{\partial y^{k'}}\epsilon_{ijk}$$

$$= \det(\partial y/\partial x) \frac{\partial x^i}{\partial y^{i'}} \frac{\partial x^j}{\partial y^{j'}} \frac{\partial x^k}{\partial y^{k'}} \epsilon_{ijk} \,. \tag{A.12}$$

With Equation (A.9) this yields the transformation rule for the components $\mathrm{d}a_i$,

$$\mathrm{d}a_{j'} = \det(\partial y/\partial x) \frac{\partial x^i}{\partial y^{j'}} \mathrm{d}a_i \,. \tag{A.13}$$

Now quantities are constructed that can be integrated over a surface. Since a surface element is determined from three independent components $\mathrm{d}a_i$ an integrand with three independent components β^i is introduced that transforms according to

$$\beta^{j'} = \frac{1}{\det(\partial y/\partial x)} \frac{\partial y^{j'}}{\partial x^i} \beta^i \,. \tag{A.14}$$

Transformation rules that involve the determinant of the transformation matrix characterize so-called *densities*. Densities are sensitive towards changes of the scale of elementary volumes. In physics they represent additive quantities, also called extensities, that describe how much of a quantity is distributed within a volume or over the surface of a volume. This is in contrast to intensities. The covariant vectors that are introduced as natural line integrals are intensive quantities that represent the strength of a physical field.

The transformation behavior of Equation (A.14) of the components β^i characterizes a contravariant vector density. With this transformation behavior the surface integral

$$\int \beta^i \mathrm{d}a_i = \int \beta^i \epsilon_{ijk} \frac{\partial a^j}{\partial t} \frac{\partial a^k}{\partial s} \, \mathrm{d}t \, \mathrm{d}s \tag{A.15}$$

yields a scalar value that is coordinate independent.

A.3 Integration Over a Volume and Scalar Densities as Volume Integrands

Finally, integration over a three-dimensional volume $\boldsymbol{v}$ in three-dimensional space is considered. Again a specific coordinate system x^i is chosen and a parameterization of $\boldsymbol{v}$ is specified by

$$\boldsymbol{v}(t, s, r) = \left(v^1(t, s, r), v^2(t, s, r), v^3(t, s, r)\right), \tag{A.16}$$

with three parameters t, s and r.

An elementary volume element $\mathrm{d}v$ is characterized by three edges $\frac{\partial v^i}{\partial t}\mathrm{d}t$, $\frac{\partial v^i}{\partial s}\mathrm{d}s$, and $\frac{\partial v^i}{\partial r}\mathrm{d}r$. The volume, which is spanned by these edges, is given by the determinant

$$\begin{aligned} \mathrm{d}v &= \det\left(\frac{\partial v^i}{\partial t}\mathrm{d}t, \frac{\partial v^i}{\partial s}\mathrm{d}s, \frac{\partial v^i}{\partial r}\mathrm{d}r\right) \\ &= \epsilon_{ijk} \frac{\partial v^i}{\partial t} \frac{\partial v^j}{\partial s} \frac{\partial v^k}{\partial r} \, \mathrm{d}t \, \mathrm{d}s \, \mathrm{d}r \,. \end{aligned} \tag{A.17}$$

It is not coordinate invariant but transforms under coordinate transformations $y^{j'} = y^{j'}(x^i)$ according to

$$\mathrm{d}v' = \det(\partial y/\partial x)\,\mathrm{d}v\,. \tag{A.18}$$

Since the volume element $\mathrm{d}v$ constitutes one independent component, a natural object to integrate over a volume has one independent component as well. Such an integrand is denoted by γ. It transforms according to

$$\gamma' = \frac{1}{\det(\partial y/\partial x)}\,\gamma\,. \tag{A.19}$$

This transformation rule characterizes a scalar density and yields

$$\int \gamma\,\mathrm{d}v = \int \gamma\,\epsilon_{ijk}\frac{\partial v^i}{\partial t}\frac{\partial v^j}{\partial s}\frac{\partial v^k}{\partial r}\,\mathrm{d}t\,\mathrm{d}s\,\mathrm{d}r \tag{A.20}$$

as a coordinate independent value.

A.4 Poincaré Lemma

The axiomatic approach takes advantage of the Poincaré lemma. The Poincaré lemma states under which conditions a mathematical object can be expressed in terms of a derivative, that is in terms of a potential.s

Integrands α_i, β^i and γ of line, surface, and volume integrals, respectively, are considered and it is assumed that they are defined in an open and simply connected region of three-dimensional space. Then the Poincaré lemma yields the following conclusions:

(1) If α_i is curl free, it can be written as the gradient of a scalar function f,

$$\epsilon^{ijk}\partial_j\alpha_k = 0 \qquad \Longrightarrow \qquad \alpha_i = \partial_i f\,. \tag{A.21}$$

(2) If β^i is divergence free, it can be written as the curl of the integrand α_i of a line integral,

$$\partial_i\beta^i = 0 \qquad \Longrightarrow \qquad \beta^i = \epsilon^{ijk}\partial_j\alpha_k\,. \tag{A.22}$$

(3) The integrand γ of a volume integral can be written as the divergence of an integrand β^i of a surface integral,

$$\gamma \text{ is a volume integrand} \qquad \Longrightarrow \qquad \gamma = \partial_i\beta^i\,. \tag{A.23}$$

While conclusions given by Equations (A.21) and (A.22) are familiar from elementary vector calculus, this might not be the case for the conclusion given by Equation (A.23). However, Equation (A.23) is rather trivial since, in Cartesian coordinates x, y, z, for a given volume integrand $\gamma = \gamma(x, y, z)$ the vector β^i with components $\beta^x = \int_0^x \gamma(t, y, z)/3\,\mathrm{d}t$, $\beta^y = \int_0^y \gamma(x, t, z)/3\,\mathrm{d}t$ and $\beta^z = \int_0^z \gamma(x, y, t)/3\,\mathrm{d}t$ fulfills Equation (A.23). Of course, the vector β^i is not uniquely determined from γ since any divergence-free vector field can be added to β^i without changing γ. It is further noted that γ, as a volume integrand, constitutes a scalar density. It can be integrated as above to yield the components of β^i as components of a contravariant vector density. Therefore, the integration does not yield a coordinate invariant scalar such that γ cannot be considered as a natural integrand of a line integral.

A.5 Stokes' Theorem

In the current notation, Stokes' theorem, if applied to line integrands α_i or surface integrands β^i, yields the identities:

$$\int_V \partial_i \beta^i \, \mathrm{d}v = \int_{\partial V} \beta^i \, \mathrm{d}a_i \,, \tag{A.24}$$

$$\int_S \epsilon^{ijk} \partial_j \alpha_k \, \mathrm{d}a_i = \int_{\partial S} \alpha_i \, \mathrm{d}c^i \,, \tag{A.25}$$

where ∂V denotes the two-dimensional boundary of a simply connected volume V and ∂S denotes the one-dimensional boundary of a simply connected surface S.

A.6 Lie Derivative

The Lie derivative l_v describes the change of an object T between two infinitesimally neighboring points p (with coordinates x^i) and $\tilde{p}$ (with coordinates $x^i + \varepsilon v^i(p)$), as noticed by an observer who applies a coordinate system x' that is dragged along the vector field v^i [1, 2]:

$$l_v T := \lim_{\tilde{p}\to p} \frac{T'(\tilde{p}) - T(p)}{||\tilde{p} - p||} \tag{A.26}$$

$$= \lim_{\varepsilon\to 0} \frac{T'(x^i + \varepsilon v^i) - T(x^i)}{\varepsilon} \,. \tag{A.27}$$

At the point $\tilde{p}$ the relation between the coordinate system x and the dragged coordinate system x' is given by the coordinate transformation

$$x^n = x'^n + \varepsilon v^n(x^i) \tag{A.28}$$

with the transformation matrix

$$\frac{\partial x^n}{\partial x'^i} = \delta_i^n + \varepsilon \partial_i v^n \,. \tag{A.29}$$

Example A.1: The definition given in Equation (A.26) is applied to a contravariant vector density β^n with transformation behavior

$$\beta'^n = \det\left(\frac{\partial x^n}{\partial x'^i}\right) \frac{\partial x'^n}{\partial x^i} \beta^i \,. \tag{A.30}$$

To evaluate Equation (A.26) only terms linear in ε are kept and higher-order terms are neglected. It is noted that

$$\det\left(\frac{\partial x^n}{\partial x'^i}\right) = \det\left(\delta_i^n + \varepsilon \partial_i v^n\right) \tag{A.31}$$

$$= 1 + \varepsilon \underbrace{\mathrm{Trace}(\partial_i v^n)}_{=\partial_j v^j} + \mathcal{O}(\epsilon^2) . \tag{A.32}$$

It follows that

$$\beta'^n(\tilde{p}) = \det\left(\frac{\partial x^n}{\partial x'^i}\right) \frac{\partial x'^n}{\partial x^i} \beta^i(\tilde{p}) \tag{A.33}$$

$$= (1 + \varepsilon\, \partial_j v^j)(\delta_i^n - \varepsilon\, \partial_i v^n)(\beta^i(p) + \varepsilon\, v^j \partial_j \beta^n(p)) \tag{A.34}$$

$$= \beta^n(p) + \varepsilon\big(v^j \partial_j \beta^n(p) - \partial_i v^n \beta^i(p) + \partial_j v^j \beta^n(p)\big) , \tag{A.35}$$

and one obtains from Equation (A.27) for the Lie derivative of β^n the expression

$$l_v \beta^n = v^j \partial_j \beta^n - \beta^i \partial_i v^n + \beta^n \partial_j v^j . \tag{A.36}$$

In a similar way, the Lie derivative of a covariant vector α_n is obtained as

$$l_v \alpha_n = v^j \partial_j \alpha_n + \alpha_j \partial_n v^j \tag{A.37}$$

and the Lie derivative of a scalar density γ turns out to be

$$l_v \gamma = v^j \partial_j \gamma + \gamma \partial_j v^j \tag{A.38}$$

$$= \partial_j(\gamma v^j) . \tag{A.39}$$

References

[1] Schouten, J.A. *Tensor Analysis for Physicists*, 2nd edn, reprinted Dover, New York, USA, 1989.

[2] Truesdell, C. and Toupin, R.A. 'The classical field theories', in *Handbuch der Physik*, vol. III/1, (ed. S. Flügge Springer, Berlin, Germany, 1960, 226–793.

Appendix B: Elements of Functional Analysis

This appendix brings together some material on functional analysis. First function spaces are defined which are appropriate to accommodate functions that represent solutions of (electromagnetic) boundary value problems. These spaces are characterized by a number of algebraic, geometric and analytic properties. In most cases the function spaces of physical interest are given by so-called *Hilbert spaces* [1].

The next step will be to define the notion of a *linear operator* that acts on the elements of a Hilbert space. This serves to reformulate a linear boundary value problem in terms of a linear operator equation. A linear operator equation will be viewed as a linear mapping between two Hilbert spaces. Then the solution of a linear boundary value problem reduces to the construction of the corresponding inverse mapping. This formulation is necessary for a proper understanding of the Green's function method and numerical solution procedures.

There is a wealth of literature on the subject of functional analysis. Many treatments extend on a mathematically solid and abstract level. Thorough introductions include [2, 3]. Historically, functional analytic methods have been developed for the solution of partial differential equations that are important in mathematical physics and many books focus on these applications, see for instance [1, 4–6]. Functional analysis has also been of fundamental importance for the formulation of quantum mechanics [7–9]. The mathematical framework of quantum mechanics is necessarily a functional analytic one and any serious book on quantum mechanics needs to introduce the corresponding mathematical concepts to some degree. Electromagnetic theory can be formulated without prior knowledge of functional analysis. However, functional analysis unifies the various solution methods for electromagnetic boundary value problems into a single framework and makes it possible to find solutions in a systematic way. This is the main motivation to study functional analysis which, at first sight, seems to be a rather abstract topic and unrelated to electrical engineering problems. Monographs that explain the advantages of functional analysis in an electrical engineering context are given by [10–13], among others.

B.1 Function Spaces

In this section, *sets* are considered. These are equivalently termed *spaces*, with elements f, g and h. For the following mathematical definitions these elements are not required to represent physical quantities (even though one should have in mind that they do).

Example B.1: A function space that is important in both mathematics and physics is denoted by $\boldsymbol{L}^p(\Omega)^m$ [12, 14]. This is the set of all functions $\boldsymbol{f} = \boldsymbol{f}(\boldsymbol{r}) = (f_1(\boldsymbol{r}), f_2(\boldsymbol{r}), \ldots, f_m(\boldsymbol{r})) \in \mathbb{C}^m$ that are Lebesgue integrable[1] to the pth power on $\boldsymbol{r} = (x_1, x_2, \ldots x_n) \in \Omega \subseteq \mathbb{R}^n$ such that

$$\int_\Omega |\boldsymbol{f}(\boldsymbol{r})|^p \, \mathrm{d}\Omega < \infty\,, \tag{B.1}$$

with $1 \leq p \leq \infty$. Here the absolute value $|\boldsymbol{f}(\boldsymbol{r})| \in \mathbb{R}$ is defined by

$$|\boldsymbol{f}(\boldsymbol{r})| := \left(\boldsymbol{f}(\boldsymbol{r}) \cdot \boldsymbol{f}^*(\boldsymbol{r})\right)^{1/2} = \left(\sum_{i=1}^{m} f_i(\boldsymbol{r}) f_i{}^*(\boldsymbol{r})\right)^{1/2}. \tag{B.2}$$

Again, the asterisk * denotes complex conjugation. Of particular interest in electromagnetic applications is the function space $\boldsymbol{L}^2(\Omega)^3$. It accommodates electromagnetic field configurations that are represented by both a three-component electric field and a three component magnetic field and, additionally, are square integrable. Square integrability implies that within the integration domain the electromagnetic energy is finite (see Equation (1.234)). This is a necessary physical requirement [16].

Next, structures are added to general spaces in order to be able to define Hilbert spaces. This, in turn, requires the definitions of:

(a) metric spaces,
(b) linear spaces or, equivalently, vector spaces,
(c) normed spaces, and
(d) inner product spaces.

These different categories of spaces are, in fact, interrelated. This is illustrated in Figure B.1. In the following, metric spaces are defined first and then it is proceeded towards the definition of Hilbert spaces.

B.1.1 Metric Spaces

A space S is called a *metric space* if there exists a mapping $d : S \times S \to \mathbb{R}$, $d(f, g) \in \mathbb{R}$ with properties

$$d(f, g) \geq 0\,, \tag{B.3}$$

$$d(f, g) = 0 \quad \text{if and only if} \quad f = g\,, \tag{B.4}$$

[1] The theory of Lebesgue integration is necessary if one wants to integrate functions that are *discontinuous* or *unbounded* [15]. It extends the theory of Riemann integration which is limited to *continuous* functions. In practice, if one wishes explicitly to calculate the value of an integral, the differences between Lebesgue integration and Riemann integration play no major role and it is enough to keep in mind the few facts on Lebesgue integration that are listed in [5].

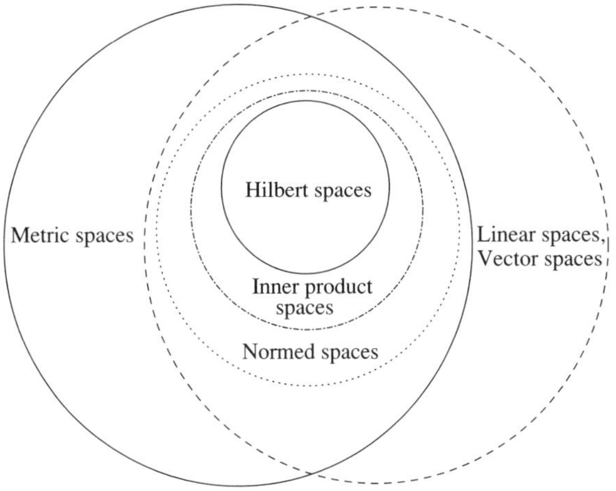

Figure B.1 Function spaces that are important in functional analysis: a Hilbert space turns out to be a special case of an inner product space which, in turn, is a special case of a normed space; a normed space is both a linear space and a metric space, but a linear space is not necessarily a metric space and vice versa

$$d(f, g) = d(g, f), \tag{B.5}$$

$$d(f, g) \leq d(f, h) + d(g, h) \qquad \text{(triangle inequality)}. \tag{B.6}$$

A metric space introduces the concept of *distance* (see Figure B.2). The mapping d is called a *metric*.

Example B.2: The function space $\boldsymbol{L}^p(\Omega)^m$ is a metric space. A class of metrics for $\boldsymbol{L}^p(\Omega)^m$ is given by the p-metric d_p,

$$d_p(\boldsymbol{f}, \boldsymbol{g}) := \left(\int_\Omega |\boldsymbol{f}(\boldsymbol{r}) - \boldsymbol{g}(\boldsymbol{r})|^p \, \mathrm{d}\Omega \right)^{1/p} \tag{B.7}$$

with $1 \leq p < \infty$. In this definition the absolute value of Equation (B.2) has been used. The p-metric is also defined for $p = \infty$ via

$$d_\infty(\boldsymbol{f}, \boldsymbol{g}) := \sup_{\boldsymbol{r} \in \Omega} |\boldsymbol{f}(\boldsymbol{r}) - \boldsymbol{g}(\boldsymbol{r})| \,. \tag{B.8}$$

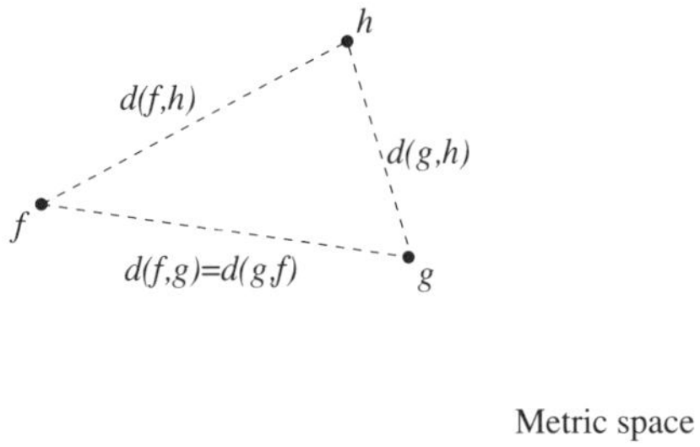

Figure B.2 In a metric space a distance is associated to any two elements. In particular, the metric d respects the triangle inequality of Equation (B.6)

In a metric space the topological notions of (1) continuity, (2) convergence, (3) Cauchy convergence, (4) completeness, (5) denseness, (6) closure, (7) boundedness and (8) compactness can be defined:

(1) If $L : S_1 \to S_2$ is a mapping between two metric spaces S_1, S_2 with metrics d_{S_1}, d_{S_2}, respectively, then L is *continuous* at $f \in S_1$ if for every number $\varepsilon > 0$ there exists a number $\delta > 0$ such that $d_{S_2}(L(f), L(g)) < \varepsilon$ whenever $d_{S_1}(f, g) < \delta$ with $g \in S_1$.
(2) A sequence of elements $\{f_n\} = f_1, f_2, \ldots$ in a metric space S with metric d is *convergent* if there is an element f such that for every number $\varepsilon > 0$ there is an integer N with $d(f_n, f) < \varepsilon$ whenever $n > N$.
(3) A sequence of elements $\{f_n\} = f_1, f_2, \ldots$ in a metric space S with metric d is *Cauchy convergent* or a *Cauchy series* if for every number $\varepsilon > 0$ there is an integer N with $d(f_n, f_m) < \varepsilon$ whenever $n, m > N$.
(4) A metric space S with metric d is *complete* if each Cauchy series in S is a convergent sequence in S.
(5) If one considers two subspaces S_1 and S_2 of a metric space S with $S_1 \subset S_2$ then the set S_1 is said to be *dense* in S_2 if for each $g \in S_2$ and each $\epsilon > 0$ there exists an element $f \in S_1$ such that $d(f, g) < \epsilon$. This states that every element of S_2 can be approximated arbitrarily close by elements of the set S_1.
(6) The *closure* $\overline{S}$ of a set S consists of the limits of all sequences that can be constructed from S. Then a set S is called *closed* if $\tilde{S} = S$.
(7) A set S is *bounded* if there is a real number k such that $d(f, g) < k$ for all $f, g \in S$.
(8) A set S is called *compact* if each sequence of elements in S (which is not necessarily convergent) has a subsequence that converges to an element of S. Compact sets are closed and bounded.

For simple illustrations and examples of these notions refer to [9].

B.1.2 Linear Spaces, Vector Spaces

In a *linear space* S, also called a *vector space*, the operations *addition* and *scalar multiplication* are defined (see Figure B.3). Moreover, with real or complex scalars α, β the following

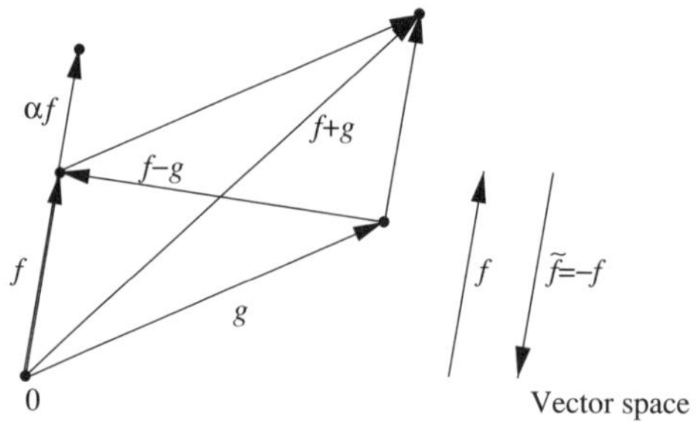

Figure B.3 In a linear space the operations *addition* and *scalar multiplication* can be performed. Also the notions of a *zero element* and the *negative of an element* are defined

properties are valid:

$$f + g = g + f. \tag{B.9}$$

$$(f + g) + h = f + (g + h). \tag{B.10}$$

$$\text{There is a zero element 0 such that} \quad f + 0 = f. \tag{B.11}$$

$$\text{For every } f \text{ there is an element } \tilde{f} = -f \text{ such that} f + \tilde{f} = 0. \tag{B.12}$$

$$\alpha(f + g) = \alpha f + \alpha g, \tag{B.13}$$

$$(\alpha + \beta)f = \alpha f + \beta f, \tag{B.14}$$

$$1 \times f = f, \tag{B.15}$$

$$\alpha(\beta f) = (\alpha\beta)f. \tag{B.16}$$

Example B.3: The function space $\boldsymbol{L}^p(\Omega)^m$ is a linear space by the common method of adding functions and multiplying functions with scalars. These additions and scalar multiplications are induced by the usual rules for addition and scalar multiplication in $\mathbb{C}^n$. Therefore, the functions of $\boldsymbol{L}^p(\Omega)^m$, or the functions of some other linear function space, can be referred to as vectors.

Also the spaces $\mathbb{R}^n$ and $\mathbb{C}^n$ are linear spaces. In particular, from the common use of vector calculus one is very much used to consider $\mathbb{R}^3$ as a vector space and its elements as vectors.

B.1.3 Normed Spaces

A linear space S is *normed* if there is a real valued function $\|f\|$, the norm of f, with properties

$$\|f\| \geq 0 \quad \text{and} \quad \|f\| = 0 \text{ if and only if } f = 0\,, \tag{B.17}$$

$$\|\alpha f\| = |\alpha|\ \|f\|\,, \tag{B.18}$$

$$\|f + g\| \leq \|f\| + \|g\| \quad \text{for all} \quad f, g \in S\,. \tag{B.19}$$

A normed space introduces the concept of *length*, as illustrated in Figure B.4. It is noted that any normed space is a metric space with $d(f, g) := \|f - g\|$.

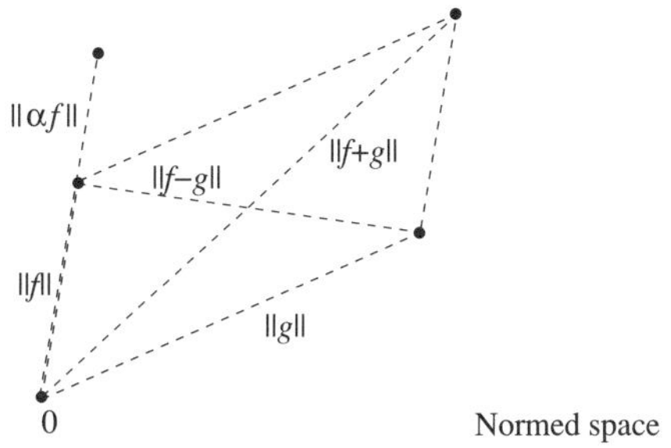

Figure B.4 In a normed space a *length* $\|f\|$ is associated with each vector f. A norm induces via $d(f, g) := \|f - g\|$ a metric in a natural way. Therefore, any normed space is also a metric space

Example B.4: The function space $\boldsymbol{L}^p(\Omega)^m$ is a normed space. With $\boldsymbol{f} \in \boldsymbol{L}^p(\Omega)^m$ a class of norms for $\boldsymbol{L}^p(\Omega)^m$ is defined by the p-norm according to

$$\|\boldsymbol{f}\|_p := d_p(\boldsymbol{f}, \boldsymbol{0}) = \left(\int_\Omega |\boldsymbol{f}(\boldsymbol{r})|^p \, \mathrm{d}\Omega \right)^{1/p} \tag{B.20}$$

with $1 \leq p < \infty$. For $p = \infty$ one defines

$$\|\boldsymbol{f}\|_\infty := d_\infty(\boldsymbol{f}, \boldsymbol{0}) = \sup_{\boldsymbol{r} \in \Omega} |\boldsymbol{f}(\boldsymbol{r})| \,. \tag{B.21}$$

B.1.4 Inner Product Spaces and Pseudo Inner Product Spaces

An *inner product space* is a linear space S with an inner product. An inner product is a mapping $S \times S \to \mathbb{C}$ which associates to each ordered pair $f, g \in S$ a complex scalar $\langle f, g \rangle$ with

$$\langle f, g \rangle = \langle g, f \rangle^* \,, \tag{B.22}$$

$$\langle f, f \rangle \geq 0 \quad \text{and} \quad \langle f, f \rangle = 0 \text{ if and only if } f = 0 \,, \tag{B.23}$$

$$\langle \alpha f, g \rangle = \alpha \langle f, g \rangle \,, \tag{B.24}$$

$$\langle f + g, h \rangle = \langle f, h \rangle + \langle g, h \rangle \,. \tag{B.25}$$

Within an inner product space the notion of *orthogonality* is defined. Two elements f, g are defined to be orthogonal if $\langle f, g \rangle = 0$. An inner product space is also a normed space since an inner product induces a norm by means of

$$\|f\| := \langle f, f \rangle^{1/2} \,. \tag{B.26}$$

Example B.5: The function space $\boldsymbol{L}^2(\Omega)^m$ is an inner product space. For $\boldsymbol{f}, \boldsymbol{g} \in \boldsymbol{L}^2(\Omega)^m$ an inner product is given by

$$\langle \boldsymbol{f}, \boldsymbol{g} \rangle := \int_\Omega \boldsymbol{f}(\boldsymbol{r}) \cdot \boldsymbol{g}^*(\boldsymbol{r}) \, \mathrm{d}\Omega \,. \tag{B.27}$$

The norm induced by this inner product is the two norm $\|\boldsymbol{f}\|_2$,

$$\langle \boldsymbol{f}, \boldsymbol{f} \rangle^{1/2} = \left(\int_\Omega |\boldsymbol{f}(\boldsymbol{r})|^2 \, \mathrm{d}\Omega \right)^{1/2} = \|\boldsymbol{f}\|_2 \,. \tag{B.28}$$

One should note that the definition of Equation (B.27) presupposes square integrability and, therefore, it is required $\boldsymbol{f}, \boldsymbol{g} \in \boldsymbol{L}^2(\Omega)^m$ rather than $\boldsymbol{f}, \boldsymbol{g} \in \boldsymbol{L}^p(\Omega)^m$ for arbitrary p. However, if $\boldsymbol{f}, \boldsymbol{g} \in \boldsymbol{L}^p(\Omega)^m$ for $p \neq 2$ one can still have $\boldsymbol{f}, \boldsymbol{g} \in \boldsymbol{L}^2(\Omega)^m$, but this is not true in general.

Further geometric interpretation can be attributed to the inner product if one considers the expressions

$$f^{\|_g} := \frac{\langle f, g \rangle g}{\|g\|^2} \,, \tag{B.29}$$

$$f^{\perp_g} := f - \frac{\langle f, g \rangle g}{\|g\|^2} = f - f^{\|_g} \,. \tag{B.30}$$

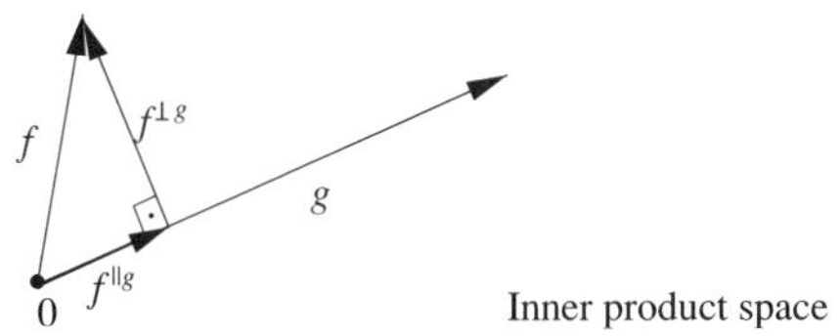

Figure B.5 Within an inner product space the inner product $\langle\, , \,\rangle$ allows the *projection* $f^{\|_g}$ of one element f onto another element g to be defined. The complement $f^{\perp_g}$ is orthogonal to $f^{\|_g}$ and one has $f = f^{\|_g} + f^{\perp_g}$

Here it is assumed that $\|g\| \neq 0$ and that the norm $\| \; \|$ is defined according to Equation (B.26). From the properties of the inner product one then obtains

$$\langle f^{\|_g}, f^{\|_g} \rangle + \langle f^{\perp_g}, f^{\perp_g} \rangle = \langle f, f \rangle \,, \tag{B.31}$$

or

$$\| f^{\|_g} \|^2 + \| f^{\perp_g} \|^2 = \| f \|^2 \,. \tag{B.32}$$

It is also easy to check that

$$\langle f^{\perp_g}, g \rangle = 0 \,. \tag{B.33}$$

These relations are explained in terms of the Pythagoras theorem in Figure B.5 where the expression $f^{\|_g}$ is explained geometrically as the *projection* of f onto g. In the special case of $\|g\| = 1$ the absolute value $|\langle f, g \rangle|$ of the inner product $\langle f, g \rangle$ coincides with the norm of this projection:

$$\| f^{\|_g} \| = \left\| \frac{\langle f, g \rangle g}{\|g\|^2} \right\| \tag{B.34}$$

$$= |\langle f, g \rangle| \frac{\|g\|}{\|g\|^2} \tag{B.35}$$

$$= |\langle f, g \rangle| \,, \tag{B.36}$$

where in the last line $\|g\| = 1$ was used.

In electromagnetic theory the notion of a *pseudo inner product space* is also of importance. A pseudo inner product space is a linear space S which is equipped with a *pseudo inner product*. A pseudo inner product, in turn, is a mapping $S \times S \to \mathbb{C}$ which associates to each pair $f, g \in S$ a scalar $\langle f, g \rangle_p$ with

$$\langle f, g \rangle_p = \langle g, f \rangle_p \,, \tag{B.37}$$

$$\langle \alpha f, g \rangle_p = \alpha \langle f, g \rangle_p \,, \tag{B.38}$$

$$\langle f + g, h \rangle_p = \langle f, h \rangle_p + \langle g, h \rangle_p \,. \tag{B.39}$$

These properties imply that $\langle f, f \rangle_p$ is not necessarily positive or real-valued. Therefore, a pseudo inner product does not always generate a norm.

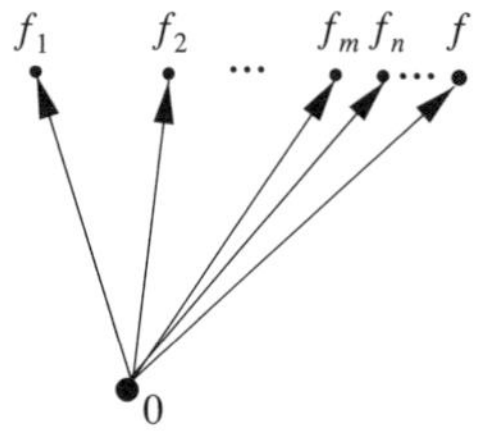

Figure B.6 An inner product space is a Hilbert space H if any Cauchy sequence $\{f_n\} \subset H$ converges to an element $f \in H$

Example B.6: The function space $\boldsymbol{L}^2(\Omega)^m$ is a pseudo inner product space. For $\boldsymbol{f}, \boldsymbol{g} \in \boldsymbol{L}^2(\Omega)^m$ a pseudo inner product is defined via

$$\langle \boldsymbol{f}, \boldsymbol{g} \rangle_p := \int_\Omega \boldsymbol{f}(\boldsymbol{r}) \cdot \boldsymbol{g}(\boldsymbol{r}) \, \mathrm{d}\Omega \,. \tag{B.40}$$

B.1.5 Hilbert Spaces

It is often necessary to expand a physical quantity in terms of an infinite series of known basis functions. From a physical point of view it is important to choose a basic function with physically meaningful properties. A physical meaningful property can be 'continuity', 'differentiability' or 'square-integrability', for example. Once appropriate basic functions are chosen one wants to know that an infinite linear combination of these functions shares the same properties. This requires that the space of the basic functions is complete.[2]

As a simple illustration the set $\mathbb{Q}$ of rational numbers is considered. This set is not complete. To show this one considers the sequence $\{\sum_{m=1}^{n} \frac{1}{m!}\}_{n=1}^{\infty}$. With the metric $d(f, g) = |f - g|$ this sequence is easily recognized as a Cauchy series. However, due to the relation

$$\sum_{n=1}^{\infty} \underbrace{\frac{1}{n!}}_{\text{rational}} = \underbrace{e}_{\text{irrational}} = 2.71828\ldots\,, \tag{B.41}$$

it does not converge in $\mathbb{Q}$ and it follows that $\mathbb{Q}$ is not complete. Therefore, the result of an infinite linear combination of rational numbers is not necessarily a rational number.

Now a Hilbert space is defined: an inner product space is called a *Hilbert space* H if it is complete in the induced norm $\|f\| = \langle f, f \rangle^{1/2}$. This means that every Cauchy sequence in H converges to an element of H, that is, for every sequence $\{f_n\} \subset H$ with $\|f_n - f_m\| \to 0$ there exists an $f \in H$ such that $\|f_n - f| \to 0$. This is illustrated in Figure B.6.

Example B.7: The function space $\boldsymbol{L}^2(\Omega)^m$ with the inner product of Equation (B.27) is a Hilbert space. This implies that $\boldsymbol{L}^2(\Omega)^m$ needs to be complete. The completeness of $\boldsymbol{L}^2(\Omega)^m$ follows from the completeness of $\boldsymbol{L}^p(\Omega)^m$ as a metric space. The proof that $\boldsymbol{L}^p(\Omega)^m$ is complete as a

[2] Completeness has been defined in Section B.1.1.

metric space is known as the Riesz–Fischer theorem and can be found in [14, pp. 99–100], for example.

B.1.6 Finite Expansions and Best Approximation

Next the approximation of a given element f of a Hilbert space H by a *finite* set of mutually orthonormal elements $g_m \in H, m = 1, \ldots, N$ is considered. To this end one forms the linear combination $\tilde{f} := \sum_{m=1}^{N} \alpha_m g_m$ with a sequence of coefficients α_m. The difference $e := f - \tilde{f}$ is defined as an *error* with norm $\|e\| = \|f - \tilde{f}\|$.

Obviously, it is of interest to know which choice of coefficients α_m minimizes the error. One has

$$\|f - \tilde{f}\| = \langle f - \tilde{f}, f - \tilde{f}\rangle \tag{B.42}$$

$$= \langle f, f\rangle + \langle \tilde{f}, \tilde{f}\rangle - \langle \tilde{f}, f\rangle - \langle f, \tilde{f}\rangle \tag{B.43}$$

$$= \|f\|^2 + \sum_{m=1}^{N} |\alpha_m|^2 - \sum_{m=1}^{N} \alpha_m \langle f, g_m\rangle^* - \sum_{m=1}^{N} \alpha_m{}^* \langle f, g_m\rangle \tag{B.44}$$

$$= \|f\|^2 + \sum_{m=1}^{N} |\alpha_m - \langle f, g_m\rangle|^2 - \sum_{m=1}^{N} |\langle f, g_m\rangle|^2 , \tag{B.45}$$

and this leads to the conclusion that the choice

$$\alpha_m = \langle f, g_m\rangle \tag{B.46}$$

minimizes the norm of the error. These coefficients are known as *generalized Fourier coefficients* and the linear combination

$$\tilde{f} = \sum_{m=1}^{N} \langle f, g_m\rangle g_m \tag{B.47}$$

is the *expansion* of f with respect to the elements g_m. It is the best possible expansion of f in terms of the set g_m. A finite-dimensional example of this circumstance is provided by Figure B.7.

The error e is orthogonal to the finite expansion, $\langle e, \tilde{f}\rangle = 0$. Since the set of all elements that can be obtained from linear combinations of the set g_m forms a closed linear subspace M of H one arrives at an illustration of the projection theorem.

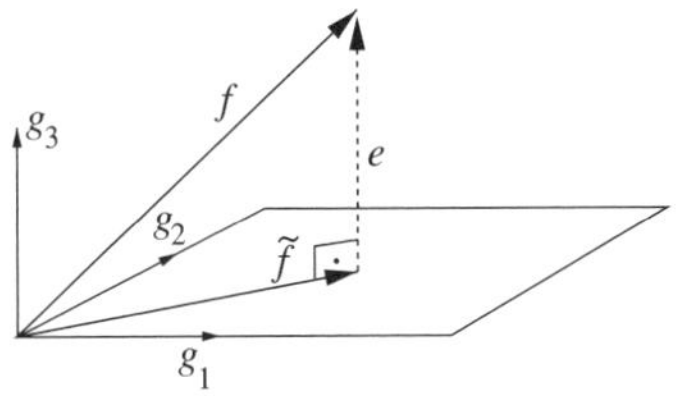

Figure B.7 Illustration of the "best approximation" of an element $f = \sum_{i=1}^{3} f^i g_i$ by an element $\tilde{f} = \sum_{i=1}^{2} \langle f, g_i\rangle g_i$. The error $e = f - \tilde{f}$ is orthogonal to the approximation $\tilde{f}$

B.1.7 The Projection Theorem

The projection theorem can be stated as:

- If M is a closed linear subspace of a Hilbert space H any element $f \in H$ can be written uniquely as the sum $f = \tilde{f} + e$ of an element $\tilde{f} \in M$ and an element $e \in M^{\perp}$.

Here the *orthogonal complement* $M^{\perp}$ is the set of all $f \in H$ such that $\langle f, g \rangle = 0$ for all $g \in M$. A proof of the projection theorem can be found in [9, p. 123].

B.1.8 Basis of a Hilbert Space

The basis of a Hilbert space H is defined as a set of orthonormal elements f_n such that any $f \in H$ can be written uniquely as

$$f = \sum_{n=1}^{\infty} \langle f, f_n \rangle f_n \,. \tag{B.48}$$

Here it is assumed that the Hilbert space H is infinite-dimensional and, thus, requires an infinite number of basic elements. However, there also exist finite dimensional Hilbert spaces. Examples of finite dimensional Hilbert spaces are $\mathbb{R}^n$ and $\mathbb{C}^n$.

B.2 Linear Operators

With the definition of a Hilbert space a function space has been provided that is suitable to accommodate the solutions of (electromagnetic) boundary value problems. However, the structures of a Hilbert space are not sufficient to model the equations of electromagnetic field theory. Therefore, it is necessary to introduce operators that relate elements of a Hilbert space in a general way.

In what follows linear operators are considered. This is sufficient as long as the equations which need to be modeled and, eventually, to be solved are linear. Electromagnetic field theory is a linear theory as long as the constitutive relations are linear. In this case one may apply the methods that are provided by linear operator theory.

B.2.1 Definition of a Linear Operator, Domain and Range of an Operator

A *linear operator* $\mathcal{L}$ is defined as a linear mapping $\mathcal{L} : S_1 \to S_2$ between linear spaces S_1, S_2. With $f, g \in S_1$, $\mathcal{L}f, \mathcal{L}g \in S_2$ and $\alpha, \beta \in \mathbb{C}$ linearity implies

$$\mathcal{L}(\alpha f + \beta g) = \alpha\, \mathcal{L}f + \beta\, \mathcal{L}g \,. \tag{B.49}$$

The *domain* $D_{\mathcal{L}}$ of an operator $\mathcal{L}$ is the set of all elements $f \in S_1$ for which the operator is defined, while the *range* $R_{\mathcal{L}}$ of the operator $\mathcal{L}$ is the set of elements of S_2 that result from the mapping of the domain.

In the following, linear operators are considered that act between Hilbert spaces H_1 and H_2.

B.2.2 Bounded Operators and the Norm of an Operator

A linear operator $\mathcal{L} : H_1 \to H_2$ is defined as being *bounded* if for all elements $f \in H_1$ there is a real number k such that

$$\|\mathcal{L}f\| \leq k \, \|f\| \, . \tag{B.50}$$

Operators that are not bounded are called *unbounded*. Closely related to the definition of bounded operators is the definition of the *norm* of an operator: the norm $\|\mathcal{L}\|$ of a linear operator $\mathcal{L} : H_1 \to H_2$ is the smallest number k that satisfies $\|\mathcal{L}f\| \leq k \, \|f\|$ for all $f \in H_1$,

$$\|\mathcal{L}\| := \sup_{\|f\| \neq 0} \frac{\|\mathcal{L}f\|}{\|f\|} \, . \tag{B.51}$$

From this definition the relation $\|\mathcal{L}f\| \leq \|\mathcal{L}\| \, \|f\|$ follows immediately.

B.2.3 Continuous Operators

The *continuity* of an operator that acts between two Hilbert spaces is defined in analogy to the continuity of a mapping between two metric spaces: A linear operator $\mathcal{L} : H_1 \to H_2$ is continuous at an element $f_0 \in H_1$ if for every ε there is a δ such that $\|\mathcal{L}f - \mathcal{L}f_0\| < \varepsilon$ if $\|f - f_0\| < \delta$.

It can be shown that a linear operator $\mathcal{L} : H_1 \to H_2$ is continuous if and only if it is bounded [5, p. 318]. Moreover, if a linear operator $\mathcal{L} : H_1 \to H_2$ is defined on a finite dimensional Hilbert space it is continuous and, thus, also bounded.

B.2.4 Linear Functionals

Linear functionals are special cases of linear operators. They map elements of a linear space into the set $\mathbb{C}$ of complex numbers. If one focusses on mappings between Hilbert spaces one may define a functional $\mathcal{I}$ as a mapping $\mathcal{I} : H \to \mathbb{C}$ with the property

$$\mathcal{I}(\alpha f + \beta g) = \alpha \mathcal{I}(f) + \beta \mathcal{I}(g) \tag{B.52}$$

for $f, g \in H$ and $\alpha, \beta \in \mathbb{C}$. Clearly, the notions of boundedness, norm and continuity are defined for linear functionals in the same way as for linear operators. For each element g of a Hilbert space there is a natural bounded linear functional which is defined via the inner product and given by

$$\mathcal{I}_g(f) = \langle f, g \rangle \, . \tag{B.53}$$

That the converse is also true is the content of the Riesz representation theorem.

B.2.5 The Riesz Representation Theorem

The Riesz representation theorem can be stated as:

- For a bounded linear functional $\mathcal{I}$ on a Hilbert space H there is a unique element $g \in H$ such that $\mathcal{I}_g(f) = \langle f, g \rangle$ for all $f \in H$. In this case it also follows that $\|\mathcal{I}\| = \|g\|$.

The Riesz representation theorem is proven in [9, p. 126], for example.

B.2.6 Adjoint and Pseudo Adjoint Operators

To approach the definition of an adjoint operator one first considers a bounded linear operator $\mathcal{L} : H_1 \to H_2$. Then $\mathcal{I}_g(f) = \langle \mathcal{L}f, g \rangle$ is a bounded linear functional $\mathcal{I}_g : H_1 \to \mathbb{C}$ for all $f \in H_1$ and it follows from the Riesz representation theorem that there is a unique $g^* \in H_1$ such that for all $f \in H_1$

$$\underbrace{\langle \mathcal{L}f, g \rangle}_{\in H_2} = \underbrace{\langle f, g^* \rangle}_{\in H_1} . \tag{B.54}$$

Since g^* depends on g one introduces a new operator $\mathcal{L}^* : H_2 \to H_1$, the so-called *adjoint operator* of $\mathcal{L}$, which is defined by

$$\mathcal{L}^* g := g^* . \tag{B.55}$$

This implies

$$\langle \mathcal{L}f, g \rangle = \langle f, \mathcal{L}^* g \rangle \tag{B.56}$$

and it follows that for any bounded linear operator $\mathcal{L}$ there is a unique adjoint $\mathcal{L}^*$.

For an unbounded linear operator the Riesz representation theorem does not necessarily hold. However, even if $\mathcal{L}$ is unbounded we may still relate an element $g \in H_2$ to an element $g^* \in H_1$ such that for $f \in H_1$ the property of Equation (B.56) is valid.

Pseudo adjoint operators $\mathcal{L}^{*\mathrm{p}} : S_1 \to S_2$ that act between pseudo inner product spaces S_1 and S_2 can also be considered. By means of a pseudo inner product they are introduced by the relation

$$\langle \mathcal{L}f, g \rangle_{\mathrm{p}} = \langle f, \mathcal{L}^{*\mathrm{p}} g \rangle_{\mathrm{p}} . \tag{B.57}$$

B.2.7 Compact Operators

A bounded linear operator $\mathcal{L} : H_1 \to H_2$ is *compact* if it maps any bounded set of H_1 into a compact set of H_2. From the definition of a compact set, as given in Section B.1.1, it follows that for a compact operator for each bounded sequence $\{f_n\} \subset H_1$ there is a subsequence $\{f_{n_i}\} \subset H_1$ such that $\mathcal{L}\{f_{n_i}\}$ converges in H_2. Compact operators that act on infinite dimensional spaces are comparatively well understood and have advantageous properties if compared to other operators that act on infinite dimensional spaces and are not compact.

B.2.8 Invertible Operators, Resolvent Operator

One often wants to solve operator equations of the form $\mathcal{L}f = g$. If $\mathcal{L}$ possesses a continuous inverse $\mathcal{L}^{-1}$ one finds the unique solution $f = \mathcal{L}^{-1}g$. This formal solution procedure naturally leads to the notion of invertible operators: an operator $\mathcal{L} : H_1 \to H_2$ is *invertible* if there exists an operator $\mathcal{L}^{-1} : H_2 \to H_1$ such that

$$\mathcal{L}^{-1}\mathcal{L}f = f \tag{B.58}$$

for all $f \in H_1$ and

$$\mathcal{L}\mathcal{L}^{-1}g = g \tag{B.59}$$

for all $g \in H_2$. The operator $\mathcal{L}^{-1}$ is the *inverse* of $\mathcal{L}$. It is easy to see that if $\mathcal{L} : H_1 \to H_2$ is an invertible linear operator then $\mathcal{L}^{-1}$ is a linear operator.

A class of operators that often occurs in the formulation of boundary value problems in terms of integral equations has the form $\mathcal{L} - \lambda I$, where $\mathcal{L}$ is a compact operator, I is the identity operator and $\lambda \in \mathbb{C}$, $\lambda \neq 0$. Operators of this form are invertible under certain conditions. In particular, there is the result [5, p. 401] that an operator $\mathcal{L} - \lambda I$, with $\mathcal{L} : H \to H$ bounded and $|\lambda| > \|\mathcal{L}\|$, is invertible with a bounded inverse

$$(\mathcal{L} - \lambda I)^{-1} = -\sum_{n=0}^{\infty} \frac{1}{\lambda^{n+1}} A^n \tag{B.60}$$

and, furthermore,

$$\|(\mathcal{L} - \lambda I)^{-1}\| \leq (|\lambda| - \|\mathcal{L}\|)^{-1} . \tag{B.61}$$

The series expansion of Equation (B.60) is often called the *Neumann series* and the operator $(\mathcal{L} - \lambda I)^{-1}$ is known as the *resolvent operator.*

B.2.9 Self-Adjoint, Normal and Unitary Operators

A linear operator $\mathcal{L} : H \to H$ is called *self-adjoint* or *Hermitian* if

$$\mathcal{L} = \mathcal{L}^* . \tag{B.62}$$

A linear operator $\mathcal{L} : H \to H$ is called *normal* if it is bounded and

$$\mathcal{L}\mathcal{L}^* = \mathcal{L}^*\mathcal{L} . \tag{B.63}$$

Moreover, a linear operator $\mathcal{L} : H \to H$ is *unitary* if and only if the adjoint operator $\mathcal{L}^*$ is equal to the inverse $\mathcal{L}^{-1}$,

$$\mathcal{L}\mathcal{L}^* = \mathcal{L}^*\mathcal{L} = I . \tag{B.64}$$

Unitary operators preserve sizes, distances and angles since

$$\langle \mathcal{L}f, \mathcal{L}g \rangle = \langle f, \mathcal{L}^*\mathcal{L}g \rangle = \langle f, g \rangle . \tag{B.65}$$

B.3 Spectrum of a Linear Operator

B.3.1 Standard Eigenvalue Problem, Spectrum and Resolvent Set

Consider a linear operator $\mathcal{L} : H \to H$ that maps a Hilbert space onto itself. The *standard eigenvalue problem* involves finding the nontrivial solutions of the equation

$$\mathcal{L}f = \lambda f \tag{B.66}$$

with $\lambda \in \mathbb{C}$ an *eigenvalue*, $f \in D_{\mathcal{L}}$ and $f \neq 0$ an *eigenfunction* or *eigenvector*. Trivially, the standard eigenvalue problem can also be written as

$$\mathcal{L}_\lambda f := (\mathcal{L} - \lambda I) f = 0 \tag{B.67}$$

with I the identity operator on H.

For finite dimensional Hilbert spaces H the properties of eigenvalues and eigenvectors are well known from elementary linear algebra. Hence, for the definition of the spectrum of a linear operator it is instructive to first consider the finite dimensional case and then move on to the infinite dimensional case.

(i) Finite Dimensional Case

Suppose the resolvent operator $\mathcal{L}_\lambda^{-1}$ exists for a particular $\lambda \in \mathbb{C}$. Then λ cannot be an eigenvalue since

$$\mathcal{L}_\lambda^{-1} \mathcal{L}_\lambda f = 0 \tag{B.68}$$

implies $f = 0$, that is Equations (B.66) or (B.67) would only have trivial solutions. Conversely, if $\mathcal{L}_\lambda^{-1}$ does not exist one has a nontrivial solution for $\mathcal{L}_\lambda f = 0$ and λ is an eigenvalue. The set of all eigenvalues of $\mathcal{L}$ makes up the *spectrum* of $\mathcal{L}$ and, in the finite dimensional case considered here, is denoted by $\sigma_{\text{finite}}(\mathcal{L})$,

$$\sigma_{\text{finite}}(\mathcal{L}) := \left\{ \lambda \in \mathbb{C} : \mathcal{L}_\lambda^{-1} = (\mathcal{L} - \lambda I)^{-1} \text{ does not exist} \right\}. \tag{B.69}$$

For all other values of λ the operator $\mathcal{L}_\lambda$ is invertible and Equations (B.66) or (B.67) only admit trivial solutions. This complement of the spectrum is called the *resolvent set* and denoted by $\rho(\mathcal{L})$,

$$\rho(\mathcal{L}) := \mathbb{C} - \sigma_{\text{finite}}(\mathcal{L}). \tag{B.70}$$

It follows that for finite dimensional H the resolvent set is the set of all $\lambda \in \mathbb{C}$ that make $\mathcal{L}_\lambda^{-1}$ exist. In this case $\mathcal{L}_\lambda^{-1}$ is bounded since all linear operators that act between finite dimensional Hilbert spaces are bounded. Additionally, the range of $\mathcal{L}_\lambda$ constitutes the complete space H, that is the dimension of $R_{\mathcal{L}_\lambda}$ is equal to the dimension of H since the kernel of the linear mapping $\mathcal{L}_\lambda$ only contains the zero element $\{0\}$. In mathematical terms the latter statement is written as $\overline{R_{\mathcal{L}_\lambda}} = H$ where in this context the bar indicates the closure of a set.[3] As a result, the inverse mapping $\mathcal{L}_\lambda^{-1}$ is defined on the complete space H and not just on a subset of H.

(ii) Infinite Dimensional Case

It has just been observed that in the finite dimensional case the resolvent set is the set of $\lambda \in \mathbb{C}$ for which $\mathcal{L}_\lambda^{-1}$ exists and it followed that $\mathcal{L}_\lambda^{-1}$ is bounded and is defined on the whole space H, that is $\overline{R_{\mathcal{L}_\lambda}} = H$. In the infinite dimensional case it will no longer be true that the existence of $\mathcal{L}_\lambda^{-1}$ implies boundedness and $\overline{R_{\mathcal{L}_\lambda}} = H$. If the existence of $\mathcal{L}_\lambda^{-1}$ implies boundedness and $\overline{R_{\mathcal{L}_\lambda}} = H$ one calls λ a *regular value* of $\mathcal{L}$. Then the resolvent set is defined as the set of all regular values of $\mathcal{L}_\lambda$,

$$\rho(\mathcal{L}) := \left\{ \lambda \in \mathbb{C} : \lambda \text{ a regular value of } \mathcal{L}_\lambda \right\}. \tag{B.71}$$

[3] The closure of a set was defined in Section B.1.1.

The set of all $\lambda \in \mathbb{C}$ which are not regular values of $\mathcal{L}$ is defined as the spectrum $\sigma(\mathcal{L})$ of $\mathcal{L}$,

$$\sigma(\mathcal{L}) := \mathbb{C} - \rho(\mathcal{L})\,. \tag{B.72}$$

These definitions are generalizations of the definitions of a resolvent set and a spectrum in the finite dimensional case.

B.3.2 Classification of Spectra by Operator Properties

In the following a specific classification of operators and their corresponding spectra is outlined which has been introduced and discussed by a number of authors, see, for example, [17, p. 125], [14, p. 194], [18, p. 412] and [12, p. 223]. Other types of classification do exist [5, p. 347]. An advantage of the classification which is described below is the division of the corresponding spectra into disjoint sets. It is based on the consideration of $\mathcal{L}_\lambda$, that is, one first fixes both $\mathcal{L}$ and λ, put $\mathcal{L}_\lambda$ into one of the following categories and then identify λ as part of a spectrum or part of the resolvent set.

(1) *The operator $\mathcal{L}_\lambda$ is not invertible.* Then the standard eigenvalue problem of Equation (B.66) admits a nontrivial solution f. For a fixed $\mathcal{L}$ the set of all eigenvalues λ which makes $\mathcal{L}_\lambda$ not invertible form the *point spectrum* of $\mathcal{L}$ which is denoted by $\sigma_\mathrm{p}(\mathcal{L})$. That is, the point spectrum is exactly the set of all eigenvalues. For a given eigenvalue λ the corresponding nontrivial solution f is an eigenvector corresponding to that eigenvalue. Eigenvalues λ of an operator $\mathcal{L}$ are also often called *poles* of the resolvent operator $\mathcal{L}_\lambda^{-1}$.
(2) *The operator $\mathcal{L}_\lambda$ is invertible.* Then the standard eigenvalue problem of Equation (B.66) admits only the trivial solution $f = 0$ and one distinguishes the following subcases:
   (a) *The range of $\mathcal{L}_\lambda$ is dense in H, $\overline{R_{\mathcal{L}_\lambda}} = H$.* In this subcase one further distinguishes between the possibilities that the resolvent $\mathcal{L}_\lambda^{-1}$ is bounded or unbounded.
      (i) *The resolvent $\mathcal{L}_\lambda^{-1}$ is bounded.* Then the value λ is called a *regular value* of $\mathcal{L}$. The set of all regular values forms the resolvent set $\rho(\mathcal{L})$.
      (ii) *The resolvent $\mathcal{L}_\lambda^{-1}$ is unbounded.* Then the value λ is part of the so-called *continuous spectrum* which is denoted by $\sigma_\mathrm{c}(\mathcal{L})$.
   (b) *The range of $\mathcal{L}_\lambda$ is not dense in H, $\overline{R_{\mathcal{L}_\lambda}} \neq H$.* Then the range of $\mathcal{L}_\lambda$ is a proper subset of H, $R_{\mathcal{L}_\lambda} \subset H$, $R_{\mathcal{L}_\lambda} \neq H$, and the corresponding set of values of λ forms the *residual spectrum* of $\mathcal{L}$ which is denoted by $\sigma_\mathrm{r}(\mathcal{L})$.

The diagram of Figure B.8 illustrates these definitions. The *total spectrum* is the union of the disjoint sets that form the point spectrum, continuous spectrum and residual spectrum,

$$\sigma(\mathcal{L}) = \sigma_\mathrm{p}(\mathcal{L}) \cup \sigma_\mathrm{c}(\mathcal{L}) \cup \sigma_\mathrm{r}(\mathcal{L})\,, \tag{B.73}$$

$$= \mathbb{C} - \rho(\mathcal{L})\,. \tag{B.74}$$

This mathematical classification is important because it defines and categorizes the spectrum of a linear operator. However, it does not actually tell us how to calculate the spectrum.

For illustrative examples of the point, continuous and residual spectrum one is referred to [18]. In view of the applications to electromagnetics one is interested in the calculation of the spectrum of linear operators that occur in the formulation of electromagnetic boundary value problems. An important class of such operators is given by second-order differential

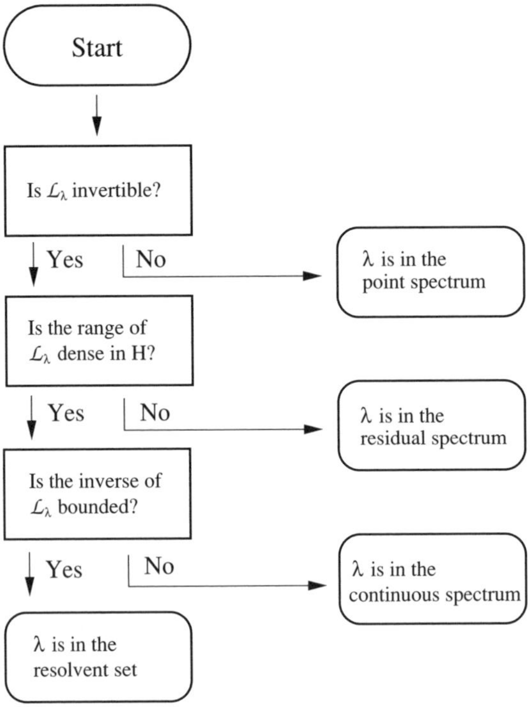

Figure B.8 Illustration of the definition of the point spectrum, residual spectrum, continuous spectrum and the resolvent set, as adapted from [18]

operators that are known as *Sturm–Liouville operators*. The properties of these operators have been investigated in the context of electromagnetic applications and this includes the calculation of their spectra for specified boundary value problems. It then turns out that the residual spectrum usually does not occur, but the point and continuous spectrum are of great importance in electromagnetic field analysis. Relevant examples and calculations are not detailed here but can be found in the literature, see [10] and, in particular, [12]. The main motivation is to introduce the spectrum of a linear operator and to arrive at the method of eigenfunction expansion.

B.4 Spectral Expansions and Representations

B.4.1 Linear Independence of Eigenfunctions

The objective is to construct expansions of elements of a Hilbert space in terms of eigenfunctions of a specific operator. In this context the following statements are important:

- For a linear operator $\mathcal{L} : H \rightarrow H$ the eigenvectors $f_1, f_2, \ldots, f_n$ that correspond to distinct eigenvalues $\lambda_1, \lambda_2, \ldots, \lambda_n$ form a linearly independent set in H. A simple proof of this statement is given in [12, p. 228].

- If the operator $\mathcal{L} : H \to H$ is not only linear but also self-adjoint and has eigenvalues then these eigenvalues are real and the eigenvectors corresponding to distinct eigenvalues are orthogonal. This is proven in [9, p. 182], for example.
- Moreover, if the linear operator $\mathcal{L} : H \to H$, with H infinite-dimensional, is both compact and self-adjoint the *Hilbert–Schmidt theorem* states that there exists an orthonormal system of eigenvectors $\{e_n\}$ corresponding to nonzero eigenvalues $\{\lambda_n\}$ such that any $f \in H$ can uniquely be represented in the form

$$f = f_0 + \sum_{n=1}^{\infty} \langle f, e_n \rangle e_n, \tag{B.75}$$

where the element f_0 satisfies $\mathcal{L} f_0 = 0$.

A proof of the Hilbert–Schmidt theorem can be found in [9, p. 188]. It should be noted that the eigenvectors e_n do not necessarily form a basis of H. In this case the element f_0 is the projection of f on the space which is orthogonal to the closed linear subspace of H which is spanned by the eigenvectors $\{e_n\}$.

The Hilbert–Schmidt theorem can be generalized if the eigenvectors corresponding to zero eigenvalues are included. This leads to the spectral theorem for compact and self-adjoint operators.

B.4.2 Spectral Theorem for Compact and Self-Adjoint Operators

The spectral theorem for compact and self-adjoint operators can be stated as follows.

- If $\mathcal{L} : H \to H$ is a compact and self-adjoint linear operator acting on an infinite-dimensional Hilbert space H then there exists an orthonormal basis H of eigenvectors $\{f_n\}$ with corresponding eigenvalues $\{\lambda_n\}$. It follows for every $f \in H$

$$f = \sum_{n=1}^{\infty} \langle f, f_n \rangle f_n \tag{B.76}$$

and

$$\mathcal{L} f = \sum_{n=1}^{\infty} \lambda_n \langle f, f_n \rangle f_n \, . \tag{B.77}$$

This is the spectral theorem for compact and self-adjoint operators which is a standard piece of linear operator theory. A proof is given in [9, p. 190] and [19, p. 243], for example. The *eigenfunction expansions* of Equations (B.76) and (B.77) are particularly useful because they allow the action of a linear operator to be reduced to an algebraic mapping. The corresponding matrix with respect to the eigenvectors $\{f_n\}$ is given by Equation (B.77). It is diagonal and, in the usual cases, infinite dimensional.

B.4.3 Remarks on the Relation Between Differential and Integral Operators

The differential operators that are involved in electromagnetic boundary value problems can be self-adjoint (see Section 1.4.6.2 below) but they are usually not compact. This is unfortunate since in these situations one cannot apply the spectral theorem for compact and self-adjoint operators. Consequently, one does not know if the eigenfunctions of a differential operator, if they exist, form a basis within the Hilbert space considered.

At this point, a loophole is found if one considers the following statement:

- If $\mathcal{L} : H \to H$ is an invertible linear operator with associated eigenvalues λ_n and eigenfunctions f_n then $\mathcal{L}^{-1} : H \to H$ has eigenvalues $1/\lambda_n$ corresponding to the eigenfunctions f_n.

A proof of this statement can be found in [12, p. 233]. Therefore, it can be shown that the eigenfunctions of a differential operator form a basis if it is possible to construct its inverse operator and to demonstrate that this inverse operator is compact and self-adjoint.

It might be clear from intuition that the inverse operators of many differential operators are given by integral operators.

- If the inverse integral operator of a self-adjoint differential operator can be constructed it has a symmetric kernel and it follows as a consequence that it is self-adjoint and compact.

This fundamental result is obtained in [1]. The mathematical details of this circumstance, like the precise assumptions that are needed to derive this results, are elaborated in the relevant mathematical literature [3, 4, 17, 19] among others.

It is concluded that the standard way to justify the eigenfunction expansion method, if applied to a linear differential operator, is to construct the corresponding inverse integral operator and to show that it is self-adjoint and compact. It is also noted that this general strategy to convert a differential boundary value problem to an integral boundary value problem is contained in the Green's function method.

In electromagnetics, the negative Laplace operator $\mathcal{L}_{\mathrm{D}} = -\Delta$ and the double-curl operator $\mathcal{L}_{\mathrm{D}} = \nabla \times \nabla \times$ are of particular interest, as is evident from Section 1.4.1.3. Solutions of the homogeneous Helmholtz equations

$$\Delta \boldsymbol{F} + k^2 \boldsymbol{F} = 0\,, \tag{B.78}$$

$$\nabla \times \nabla \times \boldsymbol{F} - k^2 \boldsymbol{F} = 0\,, \tag{B.79}$$

are determined from solutions of the spectral problems

$$-\Delta \boldsymbol{F} = k^2 \boldsymbol{F}\,, \tag{B.80}$$

$$\nabla \times \nabla \times \boldsymbol{F} = k^2 \boldsymbol{F}\,, \tag{B.81}$$

respectively. Therefore, it is important to study the spectral properties of these operators. There are no general or simple results that characterize the spectra of the negative Laplace operator or the double-curl operator because the definition of an operator includes the definition of its domain which, in turn, is determined from the specific boundary value problem. Self-adjointness of the negative Laplace operator and the double-curl operator will depend on their domains and one first has to specify these in order to be able to arrive at explicit results.

B.4.4 A Comment on Sobolev Spaces

It seems to be inconvenient that the spectral theorem for compact and self-adjoint operators cannot be applied directly to differential operators which, usually, are not compact. This reflects the fundamental difficulty that differential operators often produce singular effects if acting on functions that are not differentiable in their whole domain. As a rule, integral operators are more well-behaved, they tend to 'smooth out' singular behavior and therefore it is plausible that integral operators are required in order to show that the eigenfunctions of a differential operator are complete and form a basis.

However, it is possible to avoid the reduction from differential to integral equations if one passes to *distributional Hilbert spaces* which contain generalized functions that are defined in a distributional sense. Such spaces are called *Sobolev spaces*. The Sobolev-space approach can be applied to the direct solution of partial differential equations [4, 19]. This approach is not pursued here since in this book focus is put on the conventional Green's function approach which is based on integral equations.

References

[1] Courant, R. and Hilbert, D. *Methoden der mathematischen Physik*, 4th edn, Springer, Berlin, Germany, 1993.

[2] Dunford, N. and Schwartz, J.T. *Linear Operators – Part I: General Theory, Part II: Spectral Theory, Part III: Spectral Operators*, John Wiley & Sons, Inc., New York, USA, 1988.

[3] Achieser, N.I. and Glasmann, I.M. *Theorie der linearen Operatoren im Hilbert-Raum*, 8th edn, Akademie-Verlag, Berlin, Germany, 1981.

[4] Michlin, S.G. *Lehrgang der mathematischen Physik*, 2nd edn, Akademie-Verlag, Berlin, Germany, 1975.

[5] Stakgold, I. *Green's Functions and Boundary Value Problems*, 2nd edn, John Wiley & Sons, Inc. New York, USA, 1998.

[6] Byron, F.W. and Fuller, R.W. *Mathematics of Classical and Quantum Physics*, Dover, New York, USA, 1992.

[7] Mackey, G.W. *The Mathematical Foundations of Quantum Mechanics*, Benjamin, New York, USA, 1963.

[8] Merzbacher, E. *Quantum Mechanics*, 2nd edn, John Wiley & Sons, Inc., New York, USA, 1961.

[9] Debnath, L. and Mikusiński, P. *Introduction to Hilbert Spaces with Applications*, Academic Press, San Diego, USA, 1999.

[10] Dudley, D.G. *Mathematical Foundations for Electromagnetic Theory*, IEEE Press, New York, USA, 1994.

[11] Jones, D.S. *Methods in Electromagnetic Wave Propagation*, IEEE Press, New York, USA, 1995.

[12] Hanson, G.W. and Yakovlev, A.B. *Operator Theory for Electromagnetics*, Springer, New York, USA, 2002.

[13] Zhou, P. *Numerical Analysis of Electromagnetic Fields*, Springer, Berlin, Germany, 1993.

[14] Friedman, A. *Foundations of Modern Analysis*, Dover, New York, USA, 1982.

[15] Sobolev, S.L. *Partial Differential Equations of Mathematical Physics*, Dover, New York, USA, 1989.

[16] Van Bladel, J. *Singular Electromagnetic Fields and Sources*, Clarendon Press, Oxford, UK, 1991.

[17] Friedman, B. *Principles and Techniques of Applied Mathematics*, John Wiley & Sons, Inc., New York, USA, 1957.

[18] Naylor, A.W. and Sell, G.R. *Linear Operator Theory in Engineering and Science*, 2nd edn, Springer, New York, USA, 1982.

[19] Griffel, D.H. *Applied Functional Analysis*, John Wiley & Sons, Inc., New York, USA, 1981.

Appendix C: Some Formulas of Vector and Dyadic Calculus

This appendix collects together a few identities that are useful for working out expressions that involve vector and dyadic quantities. A number of these identities has been used in the previous chapters. A much more exhaustive collection of vector and dyadic identities is provided by the appendices of the classic book of Van Bladel [1].

C.1 Vector Identities

The following vector identities involve three-component vector functions $\boldsymbol{a}, \boldsymbol{b}, \boldsymbol{c}, \boldsymbol{d}$, a scalar function ψ and the differential operator ∇.

$$\boldsymbol{a} \cdot (\boldsymbol{b} \times \boldsymbol{c}) = \boldsymbol{b} \cdot (\boldsymbol{c} \times \boldsymbol{a}) = \boldsymbol{c} \cdot (\boldsymbol{a} \times \boldsymbol{b}) \tag{C.1}$$

$$\boldsymbol{a} \times (\boldsymbol{b} \times \boldsymbol{c}) = (\boldsymbol{a} \cdot \boldsymbol{c})\boldsymbol{b} - (\boldsymbol{a} \cdot \boldsymbol{b})\boldsymbol{c} \tag{C.2}$$

$$(\boldsymbol{a} \times \boldsymbol{b}) \cdot (\boldsymbol{c} \times \boldsymbol{d}) = (\boldsymbol{a} \cdot \boldsymbol{c})(\boldsymbol{b} \cdot \boldsymbol{d}) - (\boldsymbol{a} \cdot \boldsymbol{d})(\boldsymbol{b} \cdot \boldsymbol{c}) \tag{C.3}$$

$$\nabla \times \nabla \psi = 0 \tag{C.4}$$

$$\nabla \cdot (\nabla \times \boldsymbol{a}) = 0 \tag{C.5}$$

$$\nabla \times (\nabla \times \boldsymbol{a}) = \nabla(\nabla \cdot \boldsymbol{a}) - \Delta \boldsymbol{a} \tag{C.6}$$

$$\nabla \cdot (\psi \boldsymbol{a}) = \boldsymbol{a} \cdot \nabla \psi + \psi \nabla \cdot \boldsymbol{a} \tag{C.7}$$

$$\nabla \times (\psi \boldsymbol{a}) = \nabla \psi \times \boldsymbol{a} + \psi \nabla \times \boldsymbol{a} \tag{C.8}$$

$$\nabla(\boldsymbol{a} \cdot \boldsymbol{b}) = (\boldsymbol{a} \cdot \nabla)\boldsymbol{b} + (\boldsymbol{b} \cdot \nabla)\boldsymbol{a} + \boldsymbol{a} \times (\nabla \times \boldsymbol{b}) + \boldsymbol{b} \times (\nabla \times \boldsymbol{a}) \tag{C.9}$$

$$\nabla \cdot (\boldsymbol{a} \times \boldsymbol{b}) = \boldsymbol{b} \cdot (\nabla \times \boldsymbol{a}) - \boldsymbol{a} \cdot (\nabla \times \boldsymbol{b}) \tag{C.10}$$

$$\nabla \times (\boldsymbol{a} \times \boldsymbol{b}) = \boldsymbol{a}(\nabla \cdot \boldsymbol{b}) - \boldsymbol{b}(\nabla \cdot \boldsymbol{a}) + (\boldsymbol{b} \cdot \nabla)\boldsymbol{a} - (\boldsymbol{a} \cdot \nabla)\boldsymbol{b}. \tag{C.11}$$

Radiating Nonuniform Transmission-Line Systems and the Partial Element Equivalent Circuit Method Jürgen Nitsch, Frank Gronwald, and Günter Wollenberg

C.2 Dyadic Identities

In the following $\overline{G}$ denotes a dyadic and $\overline{G}^{\mathrm{T}}$ its transpose.

$$a \cdot (\overline{G} \cdot b) = (a \cdot \overline{G}) \cdot b = a \cdot \overline{G} \cdot c \tag{C.12}$$

$$(a \cdot \overline{G}) \times b = a \cdot (\overline{G} \times b) = a \cdot \overline{G} \times b \tag{C.13}$$

$$(a \times \overline{G}) \cdot b = a \times (\overline{G} \cdot b) \tag{C.14}$$

$$(a \times b) \cdot \overline{G} = a \cdot (b \times \overline{G}) = -b \cdot (a \times \overline{G}) \tag{C.15}$$

$$(\overline{G} \times a) \cdot b = \overline{G} \cdot a \times b = -(\overline{G} \times b) \cdot a \tag{C.16}$$

$$a \times (b \times \overline{G}) = b(a \cdot \overline{G}) - (a \times b)\overline{G} \tag{C.17}$$

$$a \cdot \overline{G} = \overline{G}^{\mathrm{T}} \cdot a = (a \cdot \overline{G})^{\mathrm{T}}. \tag{C.18}$$

C.3 Integral Identities

The following integral identities involve volume integrals that extend over a regular volume $\Omega \subset \mathbb{R}^3$ which is bounded by a closed surface $\Gamma = \partial\Omega$. The unit normal vector e_n is defined to point inward from the surface Γ into the volume Ω.

C.3.1 Vector-Dyadic Green's First Theorem

$$\int_\Omega \left[(\nabla \times a) \cdot (\nabla \times \overline{G}) - a \cdot (\nabla \times \nabla \times \overline{G})\right] \mathrm{d}^3 r = \oint e_n \times (a \times \nabla \times \overline{G}) \, \mathrm{d}^2 r. \tag{C.19}$$

C.3.2 Vector-Dyadic Green's Second Theorem

$$\int_\Omega \left[(\nabla \times \nabla \times a) \cdot \overline{G} - a \cdot (\nabla \times \nabla \times \overline{G})\right] \mathrm{d}^3 r$$
$$= -\oint_\Gamma \left[(n \times a) \cdot (\nabla \times \overline{G}) + (e_n \times \nabla \times a) \cdot \overline{G}\right] \mathrm{d}^2 r. \tag{C.20}$$

By means of vector and dyadic identities the vector-dyadic Green's second theorem can be put into the equivalent form

$$\int_\Omega \left[(\Delta a) \cdot \overline{G} - a \cdot \Delta \overline{G}\right] \mathrm{d}^3 r = \oint_\Gamma [(e_n \times a) \cdot (\nabla \times \overline{G}) - (\nabla \times a) \cdot (e_n \times \overline{G})$$
$$+ e_n \cdot a(\nabla \cdot \overline{G}) - e_n \cdot \overline{G}(\nabla \cdot a)] \, \mathrm{d}^2 r. \tag{C.21}$$

Reference

[1] Van Bladel, J. *Electromagnetic Fields*, revised printing, Hemisphere Publishing, New York, USA, 1985.

Appendix D: Adaption of the Integral Equations to the Conductor Geometry

Here the manipulations of the integral Equation (2.31), that are necessary to adapt this equation to the geometry of a nonuniform multiconductor line, are discussed. The original equation, see also Equation (2.31), reads:

$$\nabla \frac{1}{\epsilon} \int_V G\left(\mathbf{x}, \mathbf{x}'\right) \rho\left(\mathbf{x}'\right) \mathrm{d}v' + j\omega\mu \int_V G\left(\mathbf{x}, \mathbf{x}'\right) \mathbf{J}\left(\mathbf{x}'\right) \mathrm{d}v' + \frac{\mathbf{J}(\mathbf{x})}{\sigma} = \mathbf{E}^{(i)}(\mathbf{x}). \tag{D.1}$$

First the volume integrals are split into a sum of integrals over the individual conductor volumes. Moreover, the observation position $\mathbf{x}$ is placed at the surface of one conductor (e.g. j) at ζ, that is $\mathbf{x} = \hat{\mathbf{x}}_j(\zeta, \alpha)$:

$$\begin{aligned} &\nabla \frac{1}{\epsilon} \sum_{i=1}^{N} \int_{V_i} G\left(\hat{\mathbf{x}}_j, \mathbf{x}_i'\right) \rho\left(\mathbf{x}_i'\right) \mathrm{d}v' \\ &\quad + j\omega\mu \sum_{i=1}^{N} \int_{V_i} G\left(\hat{\mathbf{x}}_j, \mathbf{x}_i'\right) \mathbf{J}\left(\mathbf{x}_i'\right) \mathrm{d}v' + \frac{\mathbf{J}\left(\hat{\mathbf{x}}_j\right)}{\sigma} = \mathbf{E}^{(i)}\left(\hat{\mathbf{x}}_j\right). \end{aligned} \tag{D.2}$$

Now this expression is dot multiplied with the tangential unit vector $\mathbf{T}_j^{\mathrm{u}}$ at the observation point. This extracts the tangential component along the conductors, all other components are small and, therefore, neglected.

$$\begin{aligned} &\mathbf{T}_j^{\mathrm{u}}(\zeta) \cdot \nabla \frac{1}{\epsilon} \sum_{i=1}^{N} \int_{V_i} G\left(\hat{\mathbf{x}}_j, \mathbf{x}_i'\right) \rho\left(\mathbf{x}_i'\right) \mathrm{d}v' \\ &+ \mathbf{T}_j^{\mathrm{u}}(\zeta) \cdot j\omega\mu \sum_{i=1}^{N} \int_{V_i} G\left(\hat{\mathbf{x}}_j, \mathbf{x}_i'\right) \mathbf{J}\left(\mathbf{x}_i'\right) \mathrm{d}v' + \mathbf{T}_j^{\mathrm{u}}(\zeta) \cdot \frac{\mathbf{J}\left(\hat{\mathbf{x}}_j\right)}{\sigma} = \mathbf{T}_j^{\mathrm{u}}(\zeta) \cdot \mathbf{E}^{(i)}\left(\hat{\mathbf{x}}_j\right). \end{aligned} \tag{D.3}$$

Radiating Nonuniform Transmission-Line Systems and the Partial Element Equivalent Circuit Method Jürgen Nitsch, Frank Gronwald, and Günter Wollenberg

Eventually, this expression is averaged over the surface by integration over the angle α:

$$\frac{1}{2\pi}\int_0^{2\pi} \mathbf{T}_j^{\mathrm{u}}(\zeta)\cdot\nabla\frac{1}{\epsilon}\sum_{i=1}^{N}\int_{V_i} G\left(\hat{\mathbf{x}}_j,\mathbf{x}_i'\right)\rho\left(\mathbf{x}_i'\right)\mathrm{d}v'\,\mathrm{d}\alpha$$
$$+\frac{1}{2\pi}\int_0^{2\pi} \mathbf{T}_j^{\mathrm{u}}(\zeta)\cdot j\omega\mu\sum_{i=1}^{N}\int_{V_i} G\left(\hat{\mathbf{x}}_j,\mathbf{x}_i'\right)\mathbf{J}\left(\mathbf{x}_i'\right)\mathrm{d}v'\,\mathrm{d}\alpha$$
$$+\frac{1}{2\pi}\int_0^{2\pi} \mathbf{T}_j^{\mathrm{u}}(\zeta)\cdot\frac{\mathbf{J}\left(\hat{\mathbf{x}}_j\right)}{\sigma}\mathrm{d}\alpha = \frac{1}{2\pi}\int_0^{2\pi} \mathbf{T}_j^{\mathrm{u}}(\zeta)\cdot\mathbf{E}^{(i)}\left(\hat{\mathbf{x}}_j\right)\mathrm{d}\alpha. \tag{D.4}$$

It is now possible to replace the current and charge density with Equations (2.27) and (2.28) and to investigate the individual terms of the expression. After some reordering and the application of

$$\mathbf{T}_j^{\mathrm{u}}\cdot\nabla = \frac{1}{u_{jj}}\frac{\partial}{\partial\zeta} \tag{D.5}$$

the first term of the left-hand side can be written as

$$\frac{1}{2\pi}\int_0^{2\pi} \mathbf{T}_j^{\mathrm{u}}(\zeta)\cdot\nabla\frac{1}{\epsilon}\sum_{i=1}^{N}\int_{V_i} G\left(\hat{\mathbf{x}}_j,\mathbf{x}_i'\right)\rho\left(\mathbf{x}_i'\right)\mathrm{d}v'\,\mathrm{d}\alpha$$
$$= \frac{1}{u_{jj}}\frac{\partial}{\partial\zeta}\sum_{i=1}^{N}\int_{\zeta_0}^{\zeta_l} k_{c_{ji}}\left(\zeta,\zeta'\right)q_i\left(\zeta'\right)\mathrm{d}\zeta', \tag{D.6}$$

where

$$k_{c_{ji}}\left(\zeta,\zeta'\right) = \frac{1}{2\pi\epsilon}\int_0^{2\pi}\int_{S_i(\zeta')} G\left(\hat{\mathbf{x}}_j,\mathbf{x}_i'\right)d_{\rho_i}\left(\mathbf{x}_i'\right)f_{c_i}\mathrm{d}a'\,\mathrm{d}\alpha\,. \tag{D.7}$$

The second term can be treated in a very similar way resulting in

$$\frac{1}{2\pi}\int_0^{2\pi} \mathbf{T}_j^{\mathrm{u}}(\zeta)\cdot j\omega\mu\sum_{i=1}^{N}\int_{V_i} G\left(\hat{\mathbf{x}}_j,\mathbf{x}_i'\right)\mathbf{J}\left(\mathbf{x}_i'\right)\mathrm{d}v'\,\mathrm{d}\alpha$$
$$= \frac{1}{u_{jj}}\sum_{i=1}^{N}\int_{\zeta_0}^{\zeta_l} k_{l_{ji}}\left(\zeta,\zeta'\right)i_i\left(\zeta'\right)\mathrm{d}\zeta', \tag{D.8}$$

with

$$k_{l_{ji}}\left(\zeta,\zeta'\right) = \frac{\mu}{2\pi}u_{jj}(\zeta)\,u_{ii}\left(\zeta'\right)\mathbf{T}_j^{\mathrm{u}}(\zeta)\cdot\mathbf{T}_i^{\mathrm{u}}\left(\zeta'\right)\int_0^{2\pi}\int_{S_i(\zeta')} G\left(\hat{\mathbf{x}}_j,\mathbf{x}_i'\right)d_{\mathbf{J}_i}\left(\mathbf{x}_i'\right)f_{c_i}\mathrm{d}a'\,\mathrm{d}\alpha\,. \tag{D.9}$$

Evaluating the third term is rather easy and yields

$$\frac{1}{2\pi}\int_0^{2\pi} \mathbf{T}_j^{\mathrm{u}}(\zeta)\cdot\frac{\mathbf{J}\left(\hat{\mathbf{x}}_j\right)}{\sigma}\mathrm{d}\alpha = \frac{1}{u_{jj}}z_{jj}i_j, \tag{D.10}$$

where z_{jj} is the surface impedance defined as

$$z_{jj}(\zeta) = u_{jj}(\zeta)\frac{1}{2\pi}\int_0^{2\pi}\frac{d_{\mathbf{J}_i}(\hat{\mathbf{x}}_i)}{\sigma}\,\mathrm{d}\alpha. \tag{D.11}$$

Finally, the term on the right-hand side, multiplied by u_{jj}, will be assigned to a new variable, that is

$$v_j^{(i)'}(\zeta) = \frac{1}{2\pi}u_{jj}(\zeta)\int_0^{2\pi}\mathbf{T}_j^{\mathrm{u}}(\zeta)\cdot\mathbf{E}^{(i)}\left(\hat{\mathbf{x}}_j\right)\mathrm{d}\alpha. \tag{D.12}$$

These results can now be combined to give the new adapted integral equation

$$\frac{\partial}{\partial\zeta}\sum_{i=1}^{N}\int_{\zeta_0}^{\zeta_l}k_{c_{ji}}\left(\zeta,\zeta'\right)q_i\left(\zeta'\right)\mathrm{d}\zeta' + j\omega\sum_{i=1}^{N}\int_{\zeta_0}^{\zeta_l}k_{l_{ji}}\left(\zeta,\zeta'\right)i_i\left(\zeta'\right)\mathrm{d}\zeta' + z_{jj}i_j = v_j^{(i)'}(\zeta). \tag{D.13}$$

Appendix E: The Product Integral/Matrizant

E.1 The Differential Equation and Its Solution

Many processes in nature can be described with the aid of a system of first-order ordinary differential equations of the form

$$\frac{\partial}{\partial \zeta}\mathbf{X}(\zeta) - \mathbf{C}(\zeta)\mathbf{X}(\zeta) = \mathbf{X}_s'(\zeta) \tag{E.1}$$

where $\mathbf{X}$ is an N-element column vector containing the unknown physical quantities, $\mathbf{C}$ is the known $N \times N$ coefficient matrix characterizing the physical system and $\mathbf{X}_s'(\zeta)$ is the known excitation. The solution is constructed with the product integral $\mathcal{M}_{\zeta_0}^{\zeta}\{\mathbf{C}\}$ [1, 2],

$$\mathbf{X}(\zeta) = \mathcal{M}_{\zeta_0}^{\zeta}\{\mathbf{C}\}\mathbf{X}(\zeta_0) + \int_{\zeta_0}^{\zeta} \mathcal{M}_{\xi}^{\zeta}\{\mathbf{C}\}\mathbf{X}_s'(\xi)\,\mathrm{d}\xi. \tag{E.2}$$

The column vector $\mathbf{X}(\zeta_0)$ is determined by the boundary conditions. The product integral itself fulfills the equation

$$\frac{\partial}{\partial \eta}\mathcal{M}_{\zeta_0}^{\eta}\{\mathbf{C}\} = \mathbf{C}(\eta)\mathcal{M}_{\zeta_0}^{\eta}\{\mathbf{C}\}, \tag{E.3}$$

and the boundary condition

$$\mathcal{M}_{\zeta_0}^{\zeta_0}\{\mathbf{C}\} = \mathbf{1}. \tag{E.4}$$

E.2 The Determination of the Product Integral

One way to calculate the product integral is to integrate Equation (E.3)

$$\mathcal{M}_{\zeta_0}^{\zeta}\{\mathbf{C}\} = \mathbf{1} + \int_{\zeta_0}^{\zeta} \mathbf{C}(\eta)\mathcal{M}_{\zeta_0}^{\eta}\{\mathbf{C}\}\,\mathrm{d}\eta \tag{E.5}$$

and successively replace $\mathcal{M}_{\zeta_0}^{\eta}\{\mathbf{C}\}$ on the right-hand side with Equation (E.5). This yields Picard's series

$$\mathcal{M}_{\zeta_0}^{\zeta}\{\mathbf{C}\} = \mathbf{1} + \int_{\zeta_0}^{\zeta} \mathbf{C}(\eta)\,\mathrm{d}\eta + \int_{\zeta_0}^{\zeta} \mathbf{C}(\eta) \int_{\zeta_0}^{\eta} \mathbf{C}(\xi)\,\mathrm{d}\xi\,\mathrm{d}\eta + \ldots. \tag{E.6}$$

Volterra developed a different formula for the product integral. He divided the interval $[\zeta_0, \zeta]$ into k subintervals, with $\zeta_i, i = 0 \ldots k$ being the interval boundaries. Then by letting the number of subintervals go to infinity and the subinterval size to zero, the product integral is given by

$$\mathcal{M}_{\zeta_0}^{\zeta}\{\mathbf{C}\} = \lim_{k\to\infty} \prod_{i=0}^{k-1} \mathrm{e}^{(\zeta_{i+1}-\zeta_i)\mathbf{C}(\zeta_i)} \tag{E.7}$$

$$= \prod_{\zeta_0}^{\zeta} \mathrm{e}^{\mathbf{C}(\eta)\mathrm{d}\eta}. \tag{E.8}$$

It was this operation that inspired the name 'product integral', because it is similar to the ordinary integral, but instead of the summation there is a multiplication. This formula is also suitable for a numerical evaluation of the product integral, in this case k is kept finite.

A third way to determine the product integral is to develop its Taylor series at ζ_0:

$$\mathcal{M}_{\zeta_0}^{\zeta}\{\mathbf{C}\} = \mathbf{M}_0 + \mathbf{M}_1(\zeta - \zeta_0) + \frac{1}{2}\mathbf{M}_2(\zeta - \zeta_0)^2 + \cdots + \frac{1}{n!}\mathbf{M}_n(\zeta - \zeta_0)^n + \ldots. \tag{E.9}$$

With the aid of Equation (E.3) and corresponding derivatives, it is easy to find a recursive expression for the coefficient matrices:

$$\mathbf{M}_0 = \mathbf{1} \tag{E.10}$$

$$\mathbf{M}_n = \sum_{i=0}^{n-1} \binom{n-1}{i} \left(\frac{\partial^{n-i-1}}{\partial \zeta^{n-i-1}} \mathbf{C}(\zeta) \right)\Bigg|_{\zeta=\zeta_0} \mathbf{M}_i\,. \tag{E.11}$$

For a commuting parameter matrix function, that is $\mathbf{C}(\eta_1)\,\mathbf{C}(\eta_2) = \mathbf{C}(\eta_2)\,\mathbf{C}(\eta_1)$, the product integral simplifies to the matrix exponential and one obtains

$$\mathcal{M}_{\zeta_0}^{\zeta}\{\mathbf{C}\} = \mathrm{e}^{\int_{\zeta_0}^{\zeta} \mathbf{C}(\eta)\,\mathrm{d}\eta}. \tag{E.12}$$

E.3 Inverse Operation

The inverse operation of product integration is product differentiation. The product derivative is defined by

$$\mathcal{D}_{\zeta}\{\mathbf{C}\} := \frac{\partial \mathbf{C}(\zeta)}{\partial \zeta} \mathbf{C}(\zeta)^{-1}. \tag{E.13}$$

Thus one may write

$$\mathcal{D}_{\zeta}\left\{\mathcal{M}_{\zeta_0}^{\zeta}\{\mathbf{C}\}\right\} = \mathbf{C}. \tag{E.14}$$

E.4 Calculation Rules for the Product Integral

There are several rules that can be applied to the product integral, the most important are:

$$\mathcal{M}_{\zeta_0}^{\zeta}\{\mathbf{C}\} = \mathcal{M}_{\zeta_1}^{\zeta}\{\mathbf{C}\}\,\mathcal{M}_{\zeta_0}^{\zeta_1}\{\mathbf{C}\} \tag{E.15}$$

$$\mathcal{M}_{\zeta_0}^{\zeta}\{\mathbf{C}\} = \left(\mathcal{M}_{\zeta}^{\zeta_0}\{\mathbf{C}\}\right)^{-1} \tag{E.16}$$

$$\mathcal{M}_{\zeta_0}^{\zeta}\{\mathbf{A}+\mathbf{B}\} = \mathcal{M}_{\zeta_0}^{\zeta}\{\mathbf{A}\}\,\mathcal{M}_{\zeta_0}^{\zeta}\{\mathbf{C}\} \tag{E.17}$$

$$\text{where} \quad \mathbf{C}(\xi) = \left(\mathcal{M}_{\zeta_0}^{\xi}\{\mathbf{A}\}\right)^{-1}\mathbf{B}\mathcal{M}_{\zeta_0}^{\xi}\{\mathbf{A}\}$$

$$\mathcal{M}_{\zeta_0}^{\zeta}\left\{\mathbf{X}\mathbf{C}\mathbf{X}^{-1}\right\} = \mathbf{X}\mathcal{M}_{\zeta}^{\zeta_0}\{\mathbf{C}\}\,\mathbf{X}^{-1} \quad (\mathbf{X} = \text{const.}) \tag{E.18}$$

$$\mathcal{M}_{\zeta_0}^{\zeta}\left\{\mathbf{C}+\mathcal{D}_{\eta}\mathbf{X}\right\} = \mathbf{X}(\zeta)\,\mathcal{M}_{\zeta_0}^{\zeta}\left\{\mathbf{X}^{-1}\mathbf{C}\mathbf{X}\right\}\mathbf{X}(\zeta_0)^{-1} \quad (\mathbf{X} = \mathbf{X}(\zeta)). \tag{E.19}$$

References

[1] Gantmacher, F.R. *The Theory of Matrices*, Vol. 2, Chelsea Publishing Company, New York, USA, 1984.

[2] Dollard, J.D. and Friedman, C.N. *Product Integration with Application to Differential Equations,* Addison-Wesley, Reading, Massachusetts, USA, 1979.

Appendix F: Solutions for Some Important Integrals

F.1 Integrals Involving Powers of $\sqrt{x^2+b^2}$

For the numerical evaluation of the parameters of the generalized Telegrapher equations the following integral must be solved:

$$I_{np} = \int \frac{\left(\sqrt{x^2+b^2} - \sqrt{x_c^2+b^2}\right)^p}{\sqrt{x^2+b^2}} (x - x_c)^n \, dx. \tag{F.1}$$

It is possible to derive a general solution involving the hypergeometric function [1], which is, however, rather complicated. Here, simpler solutions are given for individual numbers n and p. One can distinguish between two cases here, namely $b > 0$ and $b = 0$. Table F.1 below shows the results for $n = 0, 1, 2$ and $p = 0, 1, 2$ for the first case. These and higher-order terms can easily be obtained with a computer algebra system or with the aid of integral tables and partial integration.

If $b = 0$ the integral becomes

$$I_{np} = \int \frac{(x - x_c)^{p+n}}{x} dx \tag{F.2}$$

and the solution is much simpler. The results are given in Table F.2.

F.2 Integrals Involving Exponential and Power Functions

The solution of the integral

$$\int \frac{e^{-jk\left(\sqrt{(\zeta'-\zeta)^2+b^2}+\zeta'-\zeta\right)}}{\sqrt{(\zeta'-\zeta)^2+b^2}} (\zeta'-\zeta)^p \, d\zeta', \tag{F.3}$$

can be found by substituting

$$t = -jk\left(\sqrt{(\zeta'-\zeta)^2+b^2} + \zeta' - \zeta\right) \tag{F.4}$$

Table F.1

n	p	I_{np}
0	0	$\operatorname{arcsinh}\dfrac{x}{b}$
1	0	$\sqrt{x^2+b^2} - x_c \operatorname{arcsinh}\dfrac{x}{b}$
2	0	$\left(\dfrac{1}{2}x - 2x_c\right)\sqrt{x^2+b^2} + \left(x_c^2 - \dfrac{1}{2}b^2\right)\operatorname{arcsinh}\dfrac{x}{b}$
n	1	$\dfrac{(x-x_c)^{n+1}}{n+1} - \sqrt{x_c^2+b^2}\, I_{n0}$

n	p	I_{np}
0	2	$\dfrac{1}{2}x\sqrt{x^2+b^2} + \left(\dfrac{3}{2}b^2 + x_c^2\right)\operatorname{arcsinh}\dfrac{x}{b} - 2(x-x_c)\sqrt{x_c^2+b^2}$
1	2	$\left(\dfrac{1}{3}x - \dfrac{1}{2}xx_c + x_c^2 + \dfrac{4}{3}b^2\right)\sqrt{x^2+b^2} - x_c\left(\dfrac{3}{2}b^2 + x_c^2\right)\operatorname{arcsinh}\dfrac{x}{b} - (x-x_c)^2\sqrt{x_c^2+b^2}$
2	2	$\left(\dfrac{1}{4}x^3 - \dfrac{2}{3}x^2x_c + xx_c^2 - 2x_c^3 - b^2\left(\dfrac{5}{8}x - \dfrac{8}{3}x_c\right)\right)\sqrt{x^2+b^2}+$ $\left(x_c^4 - \dfrac{5}{8}b^4 + x_c^2b^2\right)\operatorname{arcsinh}\dfrac{x}{b} - \dfrac{2}{3}(x-x_c)^3\sqrt{x_c^2+b^2}$

Table F.2

$n+p$	I_{np}
0	$\ln x$
1	$x - x_c \ln x$
2	$\dfrac{1}{2}x^2 - 2xx_c + x_c^2 \ln x$
3	$\dfrac{1}{3}x^3 - \dfrac{3}{2}x^2x_c + 3xx_c^2 + x_c^3 \ln x$
4	$\dfrac{1}{4}x^4 - \dfrac{4}{3}x^3x_c + 3x^2x_c^2 - 4xx_c^3 + x_c^4 \ln x$

giving:

$$\int \frac{e^t}{t}\left(\frac{(jkb)^2 + t^2}{2jkt}\right)^p \mathrm{d}t. \tag{F.5}$$

For the first three values of p the results are summarized in Table F.3.

Table F.3

p	$\int \frac{e^t}{t}\left(\frac{(jkb)^2+t^2}{2jkt}\right)^p \mathrm{d}t$
0	$-E_1(-t)$
1	$\frac{e^t}{2jk} - (jkb)^2 E_1(-t)$
2	$\frac{e^t(t-1)}{(2jk)^2} + jkb^2 e^t - (jkb)^4 E_1(-t)$
3	$\frac{e^t(t^2-2t+2)}{(2jk)^3} + \frac{3}{2}(jk)^3 b^4 e^t + \frac{3}{4} b^2 e^t (t-1) - (jkb)^6 E_1(-t)$

F.3 Integrals Involving Trigonometric and Exponential Functions

$$\int_{-\infty}^{\infty} \frac{\cos\left(k\sqrt{\zeta^2+b^2}\right)}{\sqrt{\zeta^2+b^2}} \cos(k_1\zeta)\,\mathrm{d}\zeta = \begin{cases} -\pi Y_0\left(b\sqrt{k^2-k_1^2}\right) & k > k_1 > 0 \\ 2K_0\left(b\sqrt{k_1^2-k^2}\right) & k_1 > k > 0 \end{cases} \tag{F.6}$$

$$\int_{-\infty}^{\infty} \frac{\sin\left(k\sqrt{\zeta^2+b^2}\right)}{\sqrt{\zeta^2+b^2}} \cos(k_1\zeta)\,\mathrm{d}\zeta = \begin{cases} -\pi J_0\left(b\sqrt{k^2-k_1^2}\right) & k > k_1 > 0 \\ 0 & k_1 > k > 0. \end{cases} \tag{F.7}$$

Then:

$$\int_{-\infty}^{\infty} \frac{e^{-jk\sqrt{(\zeta'-\zeta)^2+b^2}}}{\sqrt{(\zeta'-\zeta)^2+b^2}} e^{-jk_1(\zeta'-\zeta)}\mathrm{d}z'$$

$$= \begin{cases} -\pi Y_0\left(a\sqrt{k^2-k_1^2}\right) - j\pi J_0\left(a\sqrt{k^2-k_1^2}\right) & \\ +\pi Y_0\left(2h\sqrt{k^2-k_1^2}\right) + j\pi J_0\left(2h\sqrt{k^2-k_1^2}\right) & k > k_1 > 0 \\ 2K_0\left(a\sqrt{k_1^2-k^2}\right) - 2K_0\left(2h\sqrt{k_1^2-k^2}\right) & k_1 > k > 0. \end{cases} \tag{F.8}$$

With $H_0^{(2)}(z) = J_0(z) - jY_0(z)$

$$\int_{-\infty}^{\infty} \frac{e^{-jk\sqrt{(\zeta'-\zeta)^2+b^2}}}{\sqrt{(\zeta'-\zeta)^2+b^2}} e^{-jk_1(\zeta'-\zeta)}\mathrm{d}\zeta'$$

$$= \begin{cases} -j\pi\left[H_0^{(2)}\left(a\sqrt{k^2-k_1^2}\right) - H_0^{(2)}\left(2h\sqrt{k^2-k_1^2}\right)\right] & k > k_1 > 0 \\ 2\left[K_0\left(a\sqrt{k_1^2-k^2}\right) - K_0\left(2h\sqrt{k_1^2-k^2}\right)\right] & k_1 > k > 0. \end{cases} \tag{F.9}$$

The function J_0 is the Bessel function of the first kind, Y_0 the Bessel function of the second kind, K_0 the modified Bessel function of the second kind and $H_0^{(2)}$ the Hankel function of the second kind.

In free space one has $k_1 = k$, thus the limits of the above formulas must be taken. For both of the above cases the result is:

$$\lim_{k_1 \to k} \int_{-\infty}^{\infty} \frac{\mathrm{e}^{-jk\sqrt{(\zeta'-\zeta)^2+b^2}}}{\sqrt{(\zeta'-\zeta)^2+b^2}} \mathrm{e}^{-jk_1(\zeta'-\zeta)} \mathrm{d}\zeta' = 2\ln\frac{2h}{a}. \tag{F.10}$$

Reference

[1] Abramowitz, M. and Stegun, I. A. *Handbook of Mathematical Functions with Formulas, Graphs, and Mathematical Tables*, 9th printing, Dover, New York, USA, 1972.

Index

Radiating Nonuniform Transmission-Line Systems and the Partial Element Equivalent Circuit Method Jürgen Nitsch, Frank Gronwald, and Günter Wollenberg